PHYSICAL CHEMISTRY OF MACROMOLECULES

Macro to Nanoscales

PHYSICAL CHEMISTRY OF MACROMOLECULES

Macro to Nanoscales

Edited by

Chin Han Chan, PhD, Chin Hua Chia, PhD, and Sabu Thomas, PhD

Apple Academic Press

TORONTO NEW JERSEY

Apple Academic Press Inc.
3333 Mistwell Crescent
Oakville, ON L6L 0A2
Canada

Apple Academic Press Inc.
9 Spinnaker Way
Waretown, NJ 08758
USA

First issued in paperback 2021

Exclusive worldwide distribution by CRC Press, a member of Taylor & Francis Group

ISBN 13: 978-1-77463-318-2 (pbk)
ISBN 13: 978-1-926895-64-2 (hbk)

Library of Congress Control Number: 2013955818

Library and Archives Canada Cataloguing in Publication

Physical chemistry of macromolecules: macro to nanoscales/edited by Chin Han Chan, Chin Hua Chia, and Sabu Thomas.

Includes bibliographical references and index.
ISBN 978-1-926895-64-2 (bound)
1. Macromolecules. I. Chan, Chin Han, editor of compilation II. Chia, Chin Hua, editor of compilation III. Thomas, Sabu, editor of compilation

QD381.8.P49 2014 547'.7 C2013-907892-4

Apple Academic Press also publishes its books in a variety of electronic formats. Some content that appears in print may not be available in electronic format. For information about Apple Academic Press products, visit our website at **www.appleacademicpress.com** and the CRC Press website at **www.crcpress.com**

ABOUT THE EDITORS

Chin Han Chan, PhD

Chin Han Chan is a registered chemist with research interests in physical properties of polymer blends. She has been elected as council member of the Malaysian Institute of Chemistry and she has been appointed as the Chair of the Polymer Committee of the Institute of Materials, Malaysia. After earning her doctorate from Universiti Sains Malaysia (University of Science, Malaysia) in the field of semicrystalline polymer blends, she spent one year for her postdoctorate on reactive blends of themoplastic elastomers. Currently, she is an associate professor at the Faculty of Applied Sciences of Universiti Teknolgi MARA (MARA University of Technology), Malaysia. She has been teaching elementary physical chemistry, advanced physical chemistry, physical chemistry of macromolecular systems, and general chemistry at undergraduate and graduate levels.

She has published more than 45 papers in international and national refereed journals, more than 60 publications in conference proceedings, and more than 20 invited lectures for international conferences. She has been one of the editors of *Malaysian Journal of Chemistry*, *Berita IKM – Chemistry in Malaysia*, and books published by Royal Society of Chemistry entitled *Natural Rubber Materials, Volume 1: Blends and IPNs* and *Volume 2: Composites and Nanocomposites*. She peer-reviews a few international journals on polymer science. Her research interest is devoted to modified natural rubber-based thermoplastic elastomers, biodegradable polyester/polyether blends, and solid polymer electrolytes

Chin Hua Chia, PhD

Chin Hua Chia is currently an Associate Professor in the Materials Science Programme, School of Applied Physics, Universiti Kebangsaan Malaysia (UKM) (also known as National University of Malaysia). He obtained his PhD in 2007 in Materials Science (UKM, Malaysia). His core research interests include developing polymer nanocomposites, bio-polymers, magnetic nanomaterials, bio-adsorbents for wastewater treatment, etc. He has published more than 50 research articles and more than 60 publications in conference proceeding. He has recently received the Best Young Scientist Award (2012) and the Excellent Service Award (2013) from UKM.

Sabu Thomas, PhD

Sabu Thomas is the Director of the School of Chemical Sciences, Mahatma Gandhi University, Kottayam, India. He is also a full professor of polymer science and engineering and the Honorary Director of the Centre for Nanoscience and Nanotechnology of the same university. He is a fellow of many professional bodies. He has authored or co-authored many papers in international peer-reviewed journals in the area of polymer processing. He has organized several international conferences and has more than 420 publications, 11 books and two patents to his credit. He has been involved in a number of books both as author and editor. He is a reviewer to many international journals and has received many awards for his excellent work in polymer processing. His h-index is 42. He is listed as the 5th position in the list of Most Productive Researchers in India, in 2008.

CONTENTS

LIST OF CONTRIBUTORS

Abdul Kariem Arof
Centre for Ionics University of Malaya, Physics Department, Faculty of Science, University of Malaya, 50603 Kuala Lumpur, Malaysia

Dušan Berek
Polymer Institute, Slovak Academy of Sciences, 84541 Bratislava, Slovakia

Khairiah Haji Badri
School of Chemical Sciences and Food Technology, Faculty of Science and Technology, Universiti Kebangsaan Malaysia, 43600 UKM Bangi, Selangor, Malaysia

Chin Han Chan
Faculty of Applied Sciences, Universiti Teknologi MARA, 40450 Shah Alam, Malaysia

Seng Neon Gan
Department of Chemistry, Universiti Malaya, 50603 Kuala Lumpur, Malaysia

Animesh Ghosh
Department of Pharmaceutical Sciences, Birla Institute of Technology, Mesra Ranchi-835215, India

Jun-Ichi Kadokawa
Graduate School of Science and Engineering, Kagoshima University, 1-21-40 Korimoto, Kagoshima 890-0065, Japan

Hans-Werner Kammer
Faculty of Applied Sciences, Universiti Teknologi MARA, 40450 Shah Alam, Malaysia

SitiRozana Abdul Karim
Faculty of Applied Sciences, Universiti Teknologi MARA, 40450 Shah Alam, Malaysia

K. Sudesh Kumar
School of Biological Sciences, Universiti Sains Malaysia, 11700 Penang, Malaysia

Siang Yin Lee
Pharmaceutical Chemistry Department, International Medical University, Bukit Jalil, 57000 Kuala Lumpur, Malaysia

Amamer Musbah Omran Redwan
School of Chemical Sciences and Food Technology, Faculty of Science and Technology, Universiti Kebangsaan Malaysia, 43600 UKM Bangi, Selangor, Malaysia

Yoga Sugama Salim
Department of Chemistry, Universiti Malaya, 50603 Kuala Lumpur, Malaysia

C. Sarathchandran
Centre for Nanoscience and Nanotechnology, Mahatma Gandhi University, Kottayam, Kerala, India

Lai Har Sim
Centre of Foundation Studies, Universiti Teknologi MARA, 42300, PuncakAlam, Malaysia

Sabu Thomas
Centre for Nanoscience and Nanotechnology, Mahatma Gandhi University, Kottayam, Kerala, India

Tan Winie
Faculty of Applied Sciences, Universiti Teknologi MARA, 40450 Shah Alam, Malaysia

Tin Wui Wong
Non-Destructive Biomedical and Pharmaceutical Research Centre, Universiti Teknologi MARA, 42300, Puncak Alam, Selangor, Malaysia

LIST OF ABBREVIATIONS

AA	Acrylic acid
ABS	Acrylonitrile-butadiene-styrene
AC	Alternating current
AFM	Atomic force microscopy
AMIMBr	1-Allyl-3-methylimidazolium bromide
ATHAS	Advanced thermal analysis system
BA	Butyl acrylate
BCF	Bulk continuous fibers
BN	Boron nitride
BR	Butadiene rubber
CAP	Critical adsorption point
CCD	Charge coupled device
CFC	Chlorofluorocarbon
CGF	Chopped glass fiber
CNT	Carbon nanotube
CPE	Constant phase element
CPP	Critical partition point
CR	Chloroprene rubber
CT	Cream time
CTCGs	Content of terminal carboxyl groups
DBTO	Dibutyl tin oxide
DC	Direct current
DE	Degree of esterification
DMA	Dynamic mechanical analysis
DMP	Dimethyl phthalate
DMT	Dimethyl terephthalate
DNA	Deoxyribonucleic acid
DRS	Dielectric relaxation spectroscopy
DSC	Differential Scanning Calorimetry
EBBA	p-ethoxy benzylidene-bis-4-n-butylaniline
EC	Ethylene carbonate
EDM	Electric discharge machining
ENR	Epoxidized natural rubber
EO	Ethylene oxide

EPDM Ethylene propylene diene monomer
EVAc Poly(ethylene-co-vinyl acetate)
EVOH Poly(ethylene-co-vinyl alcohol)
FFB Fresh fruit bunches
FR Fire retardant
FTIR Fourier transform infrared
FWHM Full width at half maximum
GPC Gel permeation chromatography
GPE Gelled polymer electrolytes
HDT Heat distortion temperature
2HEA 2-Hydroxy ethylacrylate
HFFR Halogen-free flame retardants
HIP Hot Isostatic Press
HPLC High-performance liquid chromatography
IBMA Isobutyl methacrylate
IDT Initial decomposition temperature
IR Isoprene rubber
IS Impedance Spectroscopy
it-PMMA isotactic poly(methyl methacrylate)
KWW Kohlrausch–Williams–Watts
LC Liquid crystalline
LCP Liquid crystal polymer
LCST Lower critical solution temperature
MBBA 4-Methyloxylbenzylidene – 4'-butylaniline
MMA Methyl methacrylate
MWCNT Multi-wall carbon nanotube
NBR Acrylonitrile butadiene rubber
NMR Nuclear magnetic resonance
NR Natural rubber
OMMT Organically modified montmorillonite
OPCs Organophosphorus compounds
PAN Poly(acrylonitrile)
PArM Poly(aryl methacrylate)
PBA Poly(butyl acrylate)
PBE Poly(bisphenol A-co-epichlorohydrin)
PBMA Poly(butyl methacrylate)
PBS Poly(butadiene-co-styrene)
PBT Poly(butylene terephthalate)
PBzMA Poly(benzyl methacrylate)
PCL Poly(ε-caprolactone)
PCMA Poly(cyclohexyl methacrylate)
PEA Poly(ethyl acrylate)

PEAT	Point of exclusion – adsorption transition
PEG	Poly(ethylene glycol)
PEGMA	Poly(ethylene glycol) methacrylate
PEGMe	Poly(ethylene glycol) methyl ether
PEHA	Poly-2-ethylhexyl acrylate
PEO	Poly(ethylene oxide)
PER	Polyester resins
PET	Poly(ethylene terephthalate)
PHAs	Polyhydroxyalkanoates
PHB	Poly(hydroxy butyrate)
PHBV	Poly(hydroxyl butyrate –co– hydroxyl valerate)
PHV	Poly(hydroxyvalerate)
PiBMA	Poly(iso-butyl methacrylate)
PKO	Palm kernel oil
PLA	Polylactide
PMA	Poly(methy acrylate)
PMMA	Poly(methyl methacrylate)
PMVE-Mac	Poly(methyl vinyl ether-maleic acid)
PnBMA	Poly(n-butyl methacrylate)
PNIPAM	Poly(N-isopropylacrylamide)
PO	Propylene oxide
POE	Poly(oxyethylene)
POM	Polarized optical microscopy
PPG	Poly(propylene glycol)
PPhMA	Poly(phenyl methacrylate)
PPMA	Poly(propyl methacrylate)
PPO	Poly(propylene oxide)
PPO-PU	Polypropylene oxide-based polyurethane
PtBMA	Poly(tert-butyl methacrylate)
PTMS	Poly(tetramethylene succinate)
PTT	Poly(trimethyleneterephthlate)
PUNL	Polyurethane-nitrolignin
PVA	Poly(vinyl alcohol)
PVDF	Poly(vinylidene fluoride)
PVME	Poly(vinyl methyl ether)
PVPh	Poly(vinyl phenol)
RC	Resistor-capacitor
RP	Red phosphorus
RPA	Random phase approximation
RT	Rise time
SAN	Poly(styrene-co-acrylonitrile)
SANS	Small angle neutron scattering

SAXS	Small angle X-ray scattering
SBM	Styrene-butadiene-maleic
SBR	Styrene-butadiene rubber
SEC	Size exclusion chromatograms
SEM	Scanning electron microscope
SFG	Short glass fiber
SPE	Solid polymer electrolyte
sPS	Syndiotactic poly(styrene)
TEM	Transmission electron microscope
TEMPO	2,2,6,6-tetramethylpiperidine-1-oxyl radical
TFT	Tack-free-time
TGA	Thermal gravimetrical analysis
TGIC	Temperature gradient interaction chromatography()
TMDSC	Temperature-modulated differential scanning calorimetry
TMPSF	Tetra methyl poly(sulfone)
TR-SAXS	Temperature-resolved small-angle X-ray scattering
UCST	Upper critical solution temperature
UM	University of Malaya
VTF	Vogel-Tamman-Fulcher
WAXS	Wide-angle X-ray diffraction
XRD	X-ray diffraction

PREFACE

The honor of this book shall be credited to Prof. Dr. Hans-Werner Kammer, who served as the Senior Visiting Professor at Universiti Teknologi MARA, Shah Alam, Malaysia (UiTM), from 2008 to 2012. Prof. Dr. Kammer was one of prime driving forces in the initiation of compiling the lectures that are aimed at young reseachers and practitioners. The first part of the book is an elaboration of keynote lectures presented by him and the other authors during the Workshops on Macromolelcules I, II and III (2009, 2010 and 2011). These workshops were organized by UiTM and co-organized by the Malaysian Institute of Chemistry. In this book, Chapters 1 to 12 present a coherent view of a broad number of topics pertaining to basic concepts of polymer science. These chapters comprise polymer characterization, polymer thermodynamics, and the behavior of polymers (melts, solutions, and solids). They emphasize basic science and terms and concepts that are critical to polymer science and technology. These chapters provide a secure ladder for young reseachers and practitioners to progress from the primary level to an advanced level without much difficulty. We note here, physical chemistry of polymer science does require a familiarity with mathematics. However, many of the basic concepts are understandable to researchers who have experienced elementary courses of physical chemistry for tertiary education. The mathematics in these chapters is minimized, and hence, undergraduates and graduates should be able to master the discussion in the chapters.

Nowadays, there is a growing tendency for researchers to attempt to analyze selected phenomena to the greatest depth with increased specialization. The participants of the Workshops on Macromolelcules I, II and III were inspired and have benefited from the keynote lectures, which provided broader perspective at a given domain. The understanding of the basic principles on polymer science resulted in thought-provoking impulses on the experimental design coupled with the results and discussion of research. Some of the participants of the workshops have subsequently presented their valuable research findings at the International Symposium on Advanced Polymeric Materials 2012 (ISAPM 2012). ISAPM 2012 was a joint international symposium on polymeric materials between the

Institute of Materials, Malaysia (IMM), Malaysia, and Mahatma Gandhi University (MGU), Kottayam, Kerala, India, under the auspices of the 8th International Materials Technology Conference and Exhibition (IMTCE 2012) in Kuala Lumpur, Malaysia. The second part of the chapters are the collections of lectures from the ISAPM 2012. Chapters 13 to 19 focus on application areas emphasizing emerging trends and applications of polymeric materials, which cover the advances in the fields of polymer blends, micro- to nanocomposites, and biopolymers.

Finally, we wish to express our sincere gratitude and appreciation to the contributors of the chapters. All criticism, comments, and additional information from reviewers are gratefully appreciated. Special thanks are due to Prof. Dr. Hans-Werner Kammer, the main contributor of the book, who made valuable suggestions for the content of this book. This book is an outcome of the initiative taken by Prof. Dr. Hans-Werner Kammer. We also would like to extend our thanks to Siti Rozana Abdul Karim and Fatin Harun in formatting some of the chapters.

— Chin Han Chan, PhD, Chin Hua Chia, PhD, and Sabu Thomas, PhD

Part 1: Physical Chemistry of Macromolecules

CHAPTER 1

INTRODUCTION

HANS-WERNER KAMMER

CONTENTS

The high molecular mass compounds or polymers consist of large molecules having molecular masses in the order of 10^4 to 10^6 g/mol. The molecules of these compounds are formed by low-molecular units of identical chemical structure, called monomers. Monomers are covalently linked to build up a polymer molecule or a macromolecule, frequently like a chain. Therefore, macromolecules are also termed chain molecules. The combination of a large number of monomers to a polymer molecule generates completely new properties, such as elasticity or the ability to form fibers or films. The large molecules also display flexibility.

In the beginning of 20th Century, it was believed that molecular masses of many thousand dalton are impossible and macromolecules were seen as physically bounded associates or colloids. Indeed for a stable macromolecule the bonding energy RT must exceed approximately 2.48 kJ/mol at room temperature. The measurements of vapor or osmotic pressure provide strong arguments for existence of chemically bounded large molecules. If macromolecules would exist just as colloids or associates one could find conditions in increasingly diluted solutions where they decay into their constituents, that is one would find lower molecular masses. However, this was not observed. Historically, Staudinger (around 1930) proved by so-called polymer analogue reactions, where only side groups change not the backbone that macromolecules really exist. Hydrogenation of natural rubber removed double bonds that were seen as source of intermolecular attraction leading to associates. Hence, again lowering of molecular mass should result. This effect was not observed giving rise to Staudinger's famous conclusion about the structure of chain molecules.

We may distinguish natural and synthetic or man-made polymers. Examples for natural systems are proteins, polysaccharides, and natural rubber whereas polyethylene, polystyrene, and polyamide are examples for synthetic polymers.

A polymer molecule consists of monomer units A of its low-molecular analogues linked covalently N times. It might be symbolized by $-(A)_N-$, where N is called degree of polymerization. It is for high molar mass polymers in the order of 1000, $N \approx 1000$.

1.1 GLOBAL STRUCTURE OF MACROMOLECULES

The macromolecules may form linear, branched, cross-linked, and network-like structures (see Figure 1). Usually, a chain is seen as linear, if it comprises per 1000 C atoms in the backbone less than 10 branches whereas a branched macromolecule contains more than 40 branches. For three-dimensional network polymers, the concept of molecules loses its meaning.

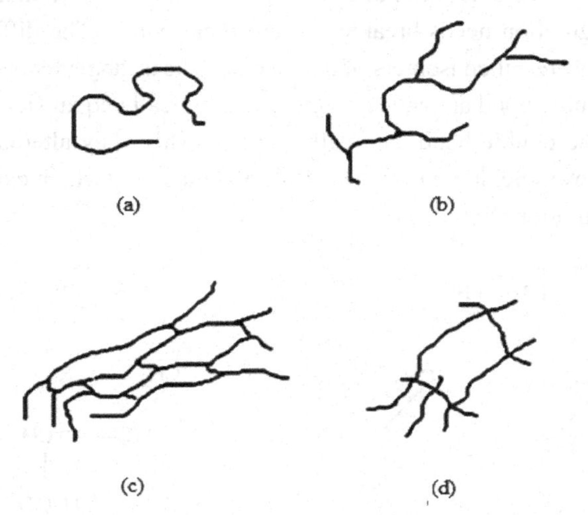

FIGURE 1 Linear, branched, cross-linked, and network-like structures of macromolecules.

1.2 CHEMICAL STRUCTURE OF MACROMOLECULES

The polymers are classified with respect to chemical structure of their main chain (or backbone). Organic polymers form the most important class, their chain consists of carbon atoms. In inorganic polymers the chain does not contain carbon. The organoelement macromolecules comprise silicon or phosphorus in the backbone.

When the backbone consists of chemically identical units, the compound is called homopolymer or polymer, in short. The copolymers comprise two or more chemically different monomers in the main chain. They are symbolized by $-(A_xB_{1-x})_N-$, where x gives the content of monomers A in the chain. Also sequence distribution of A and B along the chain determines properties of copolymers. Hence, two more degrees of freedom exist, composition x and sequence distribution. For the latter, we may distinguish four arrangements:

- Random *-AABBBBABBABAA-*
- Alternating *-ABABAB-*
- Block *-AAAAAABBBB-*
- Graft main chain formed by monomers A, side chains formed by monomer B

The structure of a macromolecule as a whole is characterized by its configuration and conformation. Configuration is the definite spatial arrangement of the atoms in the molecule. It does not change in the course of thermal motion. Alteration of configuration needs breaking of chemical bonds. The different kinds of configurations are called isomers of a molecule. The arrangements of substituents relative to double bond are called *cis*-, when chemically equal side groups are on one side to the double bond, and called *trans*-, when they alternate at different sides. As an example, it mention here 1,4-polybutadiene, which exists in *cis*- and *trans*- configuration (Figure 2).

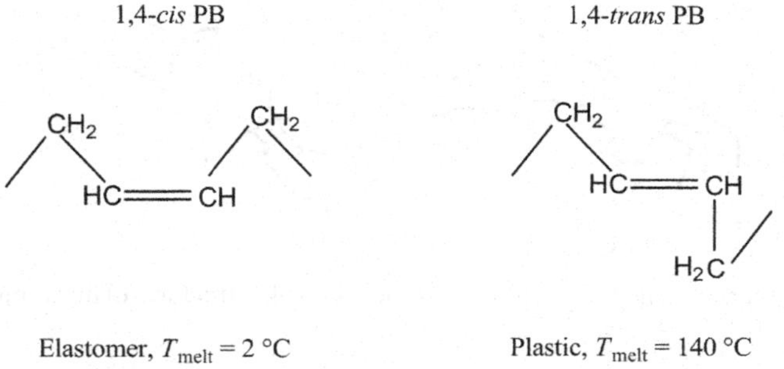

1,4-*cis* PB

CH_2 CH_2

HC═CH

Elastomer, $T_{melt} = 2$ °C

1,4-*trans* PB

CH_2

HC═CH

H_2C

Plastic, $T_{melt} = 140$ °C

FIGURE 2 *Cis*- and *trans*- configuration of 1,4-polybutadiene.

The stereoregular polymers occur due to asymmetric carbon atoms in the main chain. The isotactic structures have side groups on one side of the plane through chain axis, in syndiotactic molecules attached substituents alternate regularly at different sides and in atactic molecules there is an irregular arrangement of side groups (Figure 3 and Figure 4).

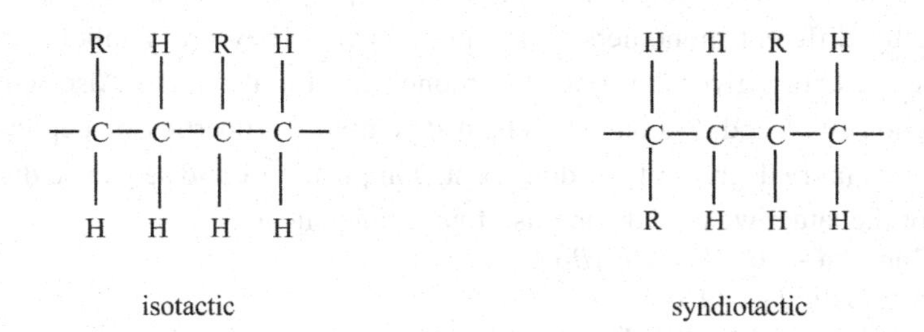

FIGURE 3 Isotactic and syndiotactic molecular structures.

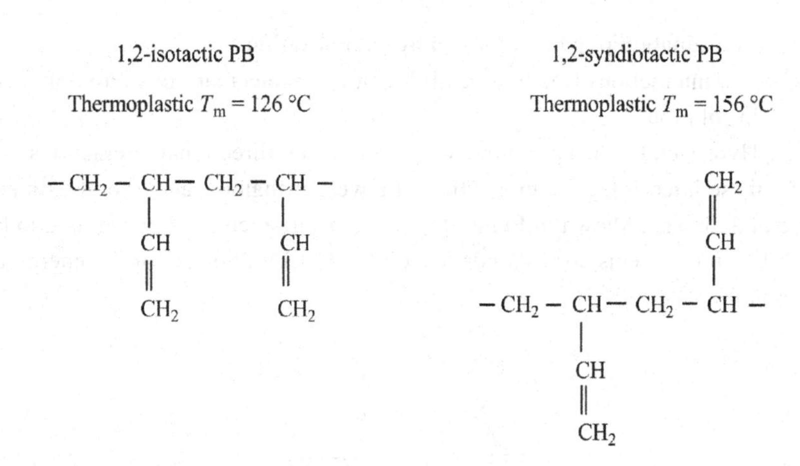

FIGURE 4 1,2-isotactic and syndiotactic polybutadiene.

The mutual repulsion between substituents may cause some displacement. As a result, the plane of symmetry is bent in the form of a helix. This occurs also in biopolymers (double-helix of deoxyribonucleic acid (DNA)). Different stereoisomers have different mechanical and thermal properties. For example, atactic polystyrene is an amorphous polymer whereas syndiotactic polystyrene is a crystalline substance. The chemical design of macromolecules determines their properties as extent of crystallization, melting point, softening (glass transition temperature), and chain flexibility which in turn strongly influence mechanical properties of the materials.

The conformation of a macromolecule is its shape in space that alters as a result of thermal motion without breaking of bonds.

The complementary to Staudinger's covalently linked chain molecules we mention here briefly supramolecular polymers (Lehn, 1995). These polymers are designed from low or high molar mass molecules that are capable of strong unidirectional association. As a result, chains develop by reversibly self-assembled molecules through unidirectional noncovalent interactions.

In covalently bound macromolecules units are irreversibly linked with high bonding energy, C–C bond amounts to 360 kJ/mol. The elements in supramolecular polymers are reversibly linked with bonding energy of more than one order of magnitude lower as in covalently linked chains, H bonds vary in the range from 10 to 20 kJ/mol. Owing to reversibility of intermolecular interactions, these systems are always in equilibrium. They mention here three classes of supramolecular polymers:

1. Coordination polymers formed by complexation
2. π–π interactions lead to assembling of low molecular units into polymers in solution
3. Hydrogen bonded polymers with multiple unidirectional interactions

All these intermolecular interactions are weaker than covalent bonds. An example of a dimer is shown in Figure 8. The association constant was found to be $K_a = 6 \cdot 10^7$ $(mol)^{-1}$. This corresponds to $\Delta G^o = -45$ kJ/mol or 1/8 of the energy of C–C bond.

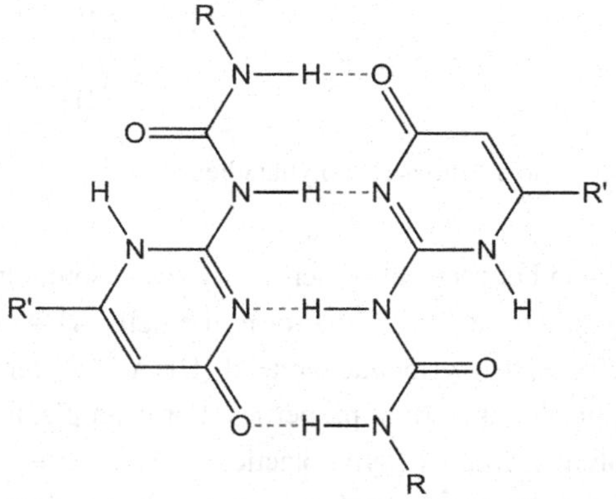

FIGURE 5 Quadrupole hydrogen bonding dimer of 2-ureido-4-pyrimidone groups.

KEYWORDS

- **Global structure**
- **Chemical structure**
- **Configuration**
- **Conformation**
- **Stereoregularity**

REFERENCES

1. Lehn, J. M. *Supramolecular Chemistry: Concepts and Perspectives*, VCH, Weinheim (1995).

CHAPTER 2

MOLECULAR MASS OF MACROMOLECULES AND ITS DISTRIBUTION

HANS-WERNER KAMMER

CONTENTS

2.1 AVERAGES OF MOLECULAR MASS

In naive approximation, one could say the degree of polymerization, N, determines the molecular mass of a polymer sample.

$$M_{\text{polymer}} = NM_{\text{monomer}} \tag{1}$$

This relationship simplifies largely the real situation. In contrast to low molecular substances that have well-defined molecular masses, macromolecular substances do not have. Due to the statistical nature of polymerization processes, it is impossible to obtain polymer samples comprising macromolecules of identical size or chain length. Samples are produced that are non-uniform with respect to molecular mass. Therefore, polymers have molecular mass distributions and one characterizes the molecular mass of a polymer by suitable averages.

Let us consider an example. A polymer sample was separated by fractionation into 12 fractions having the amounts and molecular masses as given in Table 1.

Using the data of Table 1, we may formulate two different averages of molecular mass:

$$\overline{M_{\text{w}}} = \sum_{\text{i}=1}^{12} W_{\text{i}} M_{\text{i}} \qquad \overline{M_{\text{n}}} = \sum_{\text{i}=1}^{12} X_{\text{i}} M_{\text{i}} \tag{2}$$

TABLE 1 Molecular mass distribution of a polymer sample

Relative amount in wt%	5.3	12.3	15.3	15.4	13.7	10.7	8.5	6.3	4.8	3.5	2.4	1.8
Molecular mass in 10^4 g/mol	1	2	3	4	5	6	7	8	9	10	11	12

The quantities W_{i} and X_{i} symbolize mass fraction and mole fraction of fraction *i*. The molecular masses $\overline{M_{\text{w}}}$ and $\overline{M_{\text{n}}}$ are called weight average and number average molecular mass, respectively. If the sample has a sharp molecular mass, we get $M_{\text{w}} = M_{\text{n}}$. The second equation of Equation (2) can be easily recast in mass fraction. It is:

$$\overline{M_n} = \left(\sum_{i=1}^{12} W_i / M_i \right)^{-1} \tag{3}$$

With data of Table 1, we get according to Equation (2), $\overline{M_w} = 50620$ g/mol and $\overline{M_n} \cong 35490$ g/mol. The first equation for Equation (2) can be recast in averages of M_n. It follows:

$$\overline{M_w} = \sum_i W_i M_i = \frac{1}{\overline{M_n}} \sum_i X_i M_i^2 = \frac{\overline{M_n^2}}{\overline{M_n}} > \overline{M_n} \tag{4}$$

For the average of the scatter around the average, we have with Equation (4):

$$\overline{\left(M_n - \overline{M_n}\right)^2} = \overline{M_n^2} - \overline{M_n}^2, \tag{5}$$

$$\frac{\overline{M_n^2} - \overline{M_n}^2}{\overline{M_n}^2} = \frac{\overline{M_w}}{\overline{M_n}} - 1 \tag{6}$$

The ratio $\overline{M_w}/\overline{M_n}$ serves usually as measure of fluctuation of molecular mass around the average or as measure of polydispersity of the polymer sample. In example, the polydispersity amounts to 1.4.

2.2 MOLECULAR MASS DISTRIBUTIONS

We consider an ensemble of similar molecules, having certain properties, and we ask for the average of a property over the ensemble. In generalization of Equation (2), we have for property p the average.

$$\overline{p} = \sum_{i=1}^{n} p_i \frac{m_i}{m} = \sum_{i=1}^{n} p_i w_i \qquad \text{with} \sum_i w_i = 1 \tag{7}$$

where w_i is the probability for occurrence of property p_i in the ensemble under discussion. The Equation (7) is called normalization condition of probability since we find any property with certainty. Equation (7) holds true for a discrete spectrum of properties p_i. If the property p is continuously distributed over the ensemble, we have to replace the summation in Equation (7) by integration. It is:

$$\overline{p} = \int \mathrm{d}wp = \int_0^\infty \mathrm{d}p\rho(p)\,p \quad \text{with} \quad \int_0^\infty \mathrm{d}p\rho(p) = 1 \tag{8}$$

The function $\rho(p)$ is called distribution function of property p. It tells us the probability for occurrence of a certain value of property p. One may also formulate an integral distribution of property p.

$$\Phi_o(p) = \int_0^p \mathrm{d}\pi\rho(\pi) \tag{9}$$

If $\pi = M$, the molecular mass, quantity $\Phi_o(M)$ represents the integral molecular mass distribution. It is the fraction of molecules having a molecular mass between 0 and M in the sample.

The average or expectation value of a quantity is the value of that property we expect after carrying out a series of trials or a series of measurements on similar objects. In addition to the average, calculated after Equations (7) or (8), we are interested in the quality of that average or to what extent the individual values deviate from average. Hence, a measure of the quality of average or of the variability of the individual events or their scatter around the average is called the variance or fluctuation $\langle\Delta p^2\rangle$ of property p. It is defined by the average of (positive) deviation of individual events from average

$$\langle \Delta p^2 \rangle \equiv \overline{\left(p - \overline{p}\right)^2} = \overline{p^2} - \overline{p}^2 \geq 0 \tag{10}$$

Probability distributions are characterized by two quantities, average and variance. It is precisely what we did as we calculated average $\overline{M}_n$ and variance $\langle\Delta M_n^2\rangle = \overline{M_n^2} - \overline{M}_n^2$ (compare Equation (6)) for the molecular mass distribution given in Table 1.

Let us start with a very important distribution function in analysis of random events, the Gaussian or normal distribution. They mention this distribution here just for illustration. It is not suitable for comprehension of molecular mass distributions in synthetic polymers. The Gaussian distribution function reads:

$$\rho(x) = n\exp\left[-\frac{\left(x-\bar{x}\right)^2}{2\left\langle \Delta x^2 \right\rangle}\right] \quad \text{with} \quad n = \left[2\pi\left\langle \Delta x^2 \right\rangle\right]^{-1/2} \tag{11}$$

For normalization of the distribution, see Appendix A. The Figure 1 shows the normal distribution for $\bar{x} = 2$ and $\langle \Delta x^2 \rangle = 3$. One easily finds that the maximum of $\rho(x)$ is given by $x_{max} = \bar{x}$ and the width of the curve, the distance between the inflection points, amounts to $2\langle \Delta x^2 \rangle^{1/2}$.

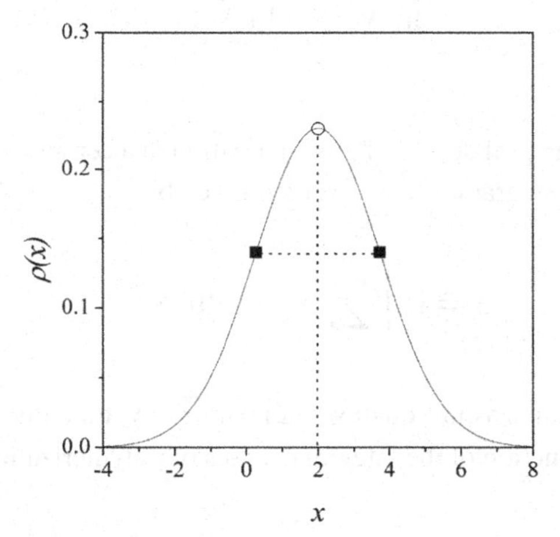

FIGURE 1 Gaussian distribution Equation (11) with $\bar{x} = 2$ and $\langle \Delta x^2 \rangle = 3$.

If we choose for x in Equation (11) the degree of polymerization belonging to $\overline{M_n}$, N_n, then $\rho(x)$ gives the mole fraction X of molecules with N_n in the sample, analogously, for $x = N_w$, it is $\rho(x) = W$, the mass fraction of molecules having N_w.

In slight generalization of Gaussian distribution, logarithmic normal distributions were introduced by replacing x in Equation (11) by ln x. In simplest version, it reads:

$$\rho(x) = \left(2\pi\sigma^2\right)^{-1/2} \frac{1}{x} \exp\left[-\frac{\left(\ln x - \ln \bar{x}\right)^2}{2\sigma^2}\right] \quad \text{with} \quad \sigma^2 = \ln \overline{x^2} - \ln \bar{x}^2 \tag{12}$$

(For normalization of logarithmic normal distribution, see Appendix A) Starting from Equation (12), we may formulate an integral mass distribution according to Equation (8) as follows:

$$\Phi_0\left(\frac{x}{\sigma}\right)=\frac{1}{\sqrt{2\pi}}\int_0^{x/\sigma}\mathrm{d}\tau\exp\left(-\frac{\tau^2}{2}\right) \tag{13}$$

After replacing x in Equation (12) by the degree of polymerization N, the symbols in Equation (13) have the following meaning. The integration variable τ is

$$\tau=\frac{\ln N_{\mathrm{cum}}-\ln\overline{N}}{\sigma}$$

where N_{cum} means the integral degree of polymerization in a certain range between $N = 0$ and $N = N_i$. The integration limit x is then given by:

$$x=\ln\left(\sum_{k=0}^{i}N_k\right)-\ln\overline{N} \tag{14}$$

The quantity σ remains as in Equation (12) with $x = N$. Function Φ_o turns out to be the distribution function of the integral degrees of polymerization. For small x/σ, we have:

$$\Phi_o=\Phi_o'|_{x=0}\frac{x}{\sigma}=\left(2\pi\sigma^2\right)^{-1/2}x \qquad \text{for } x<<1 \tag{15}$$

since the first derivative of $\Phi_o(x)$ at position $x = 0$ results to $(2\pi)^{-1/2}$. Equation (15) demonstrates that function Φ_o varies linearly for $x << 1$ with a slope depending solely on σ. Hence, a plot of $\Phi_o(x/\sigma)$ *versus* ($\ln N_{\mathrm{cum}}$) crosses the abscissa at $\left(\ln\overline{N}\right)$ and the slope of the tangent at that point yields variance σ. The Figure 2 presents the integral distribution Equation (13) (solid curve) with average and variance as calculated for the distribution given in Table 1, $\overline{N_n}=3.55$ and $\sigma = 0.596$ where degrees of polymerization have been reduced by $N = M / (10^4 \text{ g mol}^{-1})$. The dashed curve shows the tangent on the distribution at $x \to 0$ with the slope $\left(2\pi\sigma^2\right)^{-1/2}$. A linear fit through the initial data points belonging to the distribution of Table 1 yields $\overline{N_n}$ = 3.53,

$\sigma = 0.631$, and $\overline{N_w} = 5.26$ showing that the molecular masses are approximately distributed according to a logarithmic normal distribution.

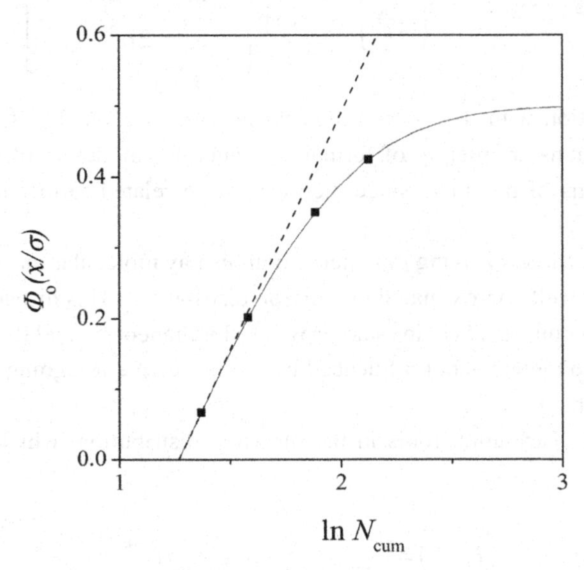

FIGURE 2 Integral distribution of degrees of polymerization, solid squares (■) illustrate data points calculated from molecular mass distribution of Table 1.

For modeling of molecular distributions of polymers, the logarithmic normal distribution was slightly generalized by changing the Equation (12) as follows. With $x = N_n$, we have:

$$\sigma^2 = \frac{\ln \overline{N_n^2} - \ln \overline{N_n}^2}{A+2} \tag{16}$$

where A is a number acting as fitting parameter for adjusting the width of the distribution. Combination of Equations (16) and (6) immediately yields:

$$\overline{N_w} = \overline{N_n} e^{(A+2)\sigma^2} \tag{17}$$

Formulated for the mole fraction, the normalized logarithmic normal distribution reads with relation Equation (16):

$$X\left(N_{\mathrm{n}}\right)=\left(2\pi\sigma^{2}\right)^{-1/2} e^{-(\sigma \mathrm{A})^{2}/2}\left(\frac{N_{\mathrm{n}}}{\overline{N_{\mathrm{n}}}}\right)^{\mathrm{A}}\frac{1}{N_{\mathrm{n}}}\exp\left[-\frac{\left(\ln N_{\mathrm{n}}-\ln \overline{N_{\mathrm{n}}}\right)^{2}}{2\sigma^{2}}\right] \tag{18}$$

The distribution with $A = -1$ is called Weslau distribution. The logarithmic normal distributions are merely of formal relevance for modeling of molecular mass distributions of polymers since they cannot be related to mechanisms of polymerization.

Polymers prepared by living polymerization display molecular mass distributions that can be well approximated by Poisson distribution. This distribution occurs if a constant number of chains start growing simultaneously and if successive attachment of monomers is not influenced by the respective foregoing monomer in the growing chain.

The Poisson distribution roots in the binomial distribution, which reads as follows:

$$P_{\mathrm{N}}(k)=\frac{N!}{(N-k)!k!}p^{\mathrm{k}}(1-p)^{\mathrm{N-k}} \tag{19}$$

Probability Equation (19) refers to the following situation. Suppose we have an urn comprising a sufficiently large number of black and white spheres. The probability of drawing a black sphere from the urn might be p and that one for a white sphere $(1-p)$. Thus, the probability for getting after N turns k black and $N-k$ white spheres in any (random) succession is given by Equation (19). The relation between binomial and normal distribution is discussed in Appendix B.

The Poisson distribution follows from Equation (19) in the limit $N \rightarrow \infty$ and $p \rightarrow 0$ under condition $Np \equiv \alpha = const$. One may say we are looking for the probability for getting k white spheres after α trials. After Equation (19), it is:

$$\lim_{\substack{N\to\infty \\ p\to 0 \\ Np=const}} P_{\mathrm{N}}(k)=\frac{\alpha^{\mathrm{k}}}{k!}\lim_{N\to\infty}\frac{N!}{(N-k)!N^{\mathrm{k}}}\left(1-\frac{Np}{N}\right)^{\mathrm{N-k}}$$

Applying

$$\lim_{N\to\infty}\left(1-\frac{x}{N}\right)^{\mathrm{N}} = e^{-\mathrm{x}}$$

We find

$$P_{\mathrm{k}}(\alpha) = \frac{\alpha^{\mathrm{k}}}{k!}e^{-\alpha} \tag{20}$$

We note the Poisson distribution holds true for discrete values of α. Speaking in terms of macromolecules, Equation (20) yields the probability for attaching $N - 1$ monomers to a given monomer. Hence, the number fraction of chains having a degree of polymerization N reads:

$$X_{\mathrm{N}} = \frac{\alpha^{\mathrm{N-1}}}{(N-1)!}e^{-\alpha} \quad \text{with } \alpha = \overline{N_{\mathrm{n}}} - 1 \tag{21}$$

Parameter α is called kinetic chain length. One easily proofs the second relationship of Equation (21).

The average degree of polymerization formulated in terms of Poisson distribution is given by:

$$\overline{N_{\mathrm{n}}} = \sum_{\mathrm{N}} N X_{\mathrm{N}}(\alpha) = e^{-\alpha}\frac{\partial}{\partial\alpha}\sum_{\mathrm{N}}\frac{\alpha^{\mathrm{N}}}{(N-1)!} = e^{-\alpha}\frac{\partial}{\partial\alpha}\left(\alpha\sum_{\mathrm{N}}\frac{\alpha^{\mathrm{N-1}}}{(N-1)!}\right) = e^{-\alpha}\frac{\partial}{\partial\alpha}\left(\alpha e^{\alpha}\right) = 1+\alpha \tag{21'}$$

In an analogous way one can calculate $\overline{N_{\mathrm{n}}^2}$:

$$\overline{N_{\mathrm{n}}^2} = \sum_{\mathrm{N}} N^2 \frac{\alpha^{\mathrm{N-1}}e^{-\alpha}}{(N-1)!} = \alpha^2 + 3\alpha + 1 \tag{22}$$

Equations (21) and (22) yield

$$\overline{N_n^2} - \overline{N_n}^2 = \alpha \tag{23}$$

Combination of Equations (6) and (23) results in:

$$\frac{\overline{N_w}}{\overline{N_n}} = 1 + \frac{1}{\overline{N_n}} - \frac{1}{\overline{N_n}^2} \tag{24}$$

For molecular masses distributed according to Poisson distribution, the ratio of the degrees of polymerization depends only on the number average degree of polymerization and the ratio tends to unity for $\overline{N_n} \to \infty$. They recognize that Poisson distribution is a narrow distribution. It is plotted for kinetic chain length = 50 in Figure 3. When one calculates the sum of the mole fraction X_N in steps of one unit between 30 and 70, it results

$$\sum_{N=30}^{70} X_N = 0.9952$$

whereas for the situation in Figure 3 the sum amounts only to approximately 0.2.

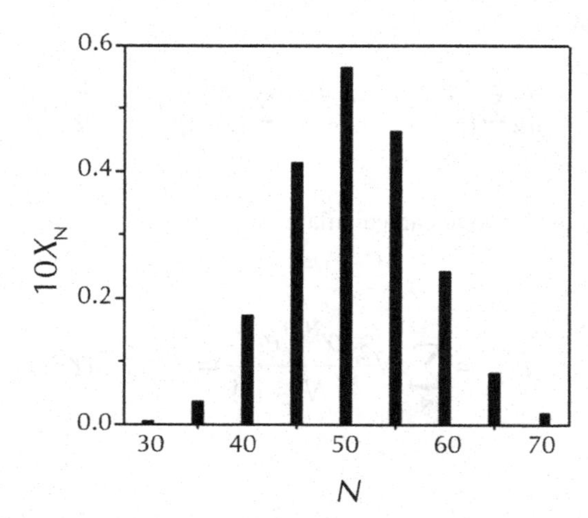

FIGURE 3 Poisson distribution for $\alpha = 50$ calculated for degrees of polymerization from $N = 30$ up to $N = 70$ in steps of 5.

Now, we discuss the Schulz-Flory distribution. It represents frequently the molecular mass distribution of polymer samples. In a steady state of nuclei concentration, a constant number of growing chains add monomers until individual growing chains cease in doing so by termination. In contrast to mechanisms leading to Poisson distributions, the initially existing nuclei are not preserved individually. Moreover, they do not start simultaneously chain growth. The sketched mechanism is observed for free-radical polymerization and polycondensation as well. Qualitatively, one sees that the resulting molecular mass distribution must be governed by two parameters. Firstly, the number of growing chains that combine to a "dead" chain rules it. Secondly, the number of attached monomers affects it that is the number of monomers added until termination occurs. The latter effect we may call simply the "waiting time" until recombination. The resulting mass distribution obeys adequately the γ-distribution. Formulated in mole fraction, it reads for integral numbers k.

$$X(N) = \frac{\beta^{k}}{(k-1)!} N^{k-1} e^{-\beta N} \tag{25}$$

They recognize that the Poisson distribution (Equation 21) turns formally in the γ-distribution for kinetic chain length $\alpha = \beta N$. The symbols in Equation (25) have the following meaning.[1] The parameter k is the coupling constant. It gives the number of growing chains that combine to a single non-reactive chain. An example for $k = 2$, may be formulated as follows:

$$P_{i}^{\bullet} + P_{N-i}^{\bullet} \rightarrow P_{N}$$

The parameter β gives the rate of recombination to dead chains or $1/\beta$ is the "waiting time" until termination occurs. It becomes obvious, the greater the "waiting time" β^{-1} the broader the molecular mass distribution. The γ-distribution was first time adjusted to molecular mass distributions of polymers by Schulz (1935) and Flory (1936). Therefore, it is called Schulz-Flory distribution in macromolecular science.

[1] In general mathematical formulation, the γ-distribution comprises the gamma function instead of the factorial. For integers k, it reads $\Gamma(k) = (k-1)!$. Since in the context of discussion only integers of quantity k make sense, we stay with the factorial.

The integral over $N^k e^{-\beta N}$ can be calculated by integration by parts. It results:

$$\int_0^\infty \mathrm{d}N N^k e^{-\beta N} = \frac{k!}{\beta^{k+1}} \equiv I_k \tag{26}$$

With Equation (26), one easily shows that $\int_0^\infty \mathrm{d}N X(N) = 1$. The averages $\overline{N_n}$ and $\overline{N_n^2}$ become

$$\overline{N_n} = \int_0^\infty \mathrm{d}N N X(N) = \frac{\beta^k}{(k-1)!} I_k = \frac{k}{\beta} \tag{27}$$

$$\overline{N_n^2} = \frac{(k+1)k}{\beta^2} \tag{28}$$

$$\overline{N_n^2} - \overline{N}^2 = \frac{k}{\beta^2} \tag{29}$$

It is consistent that with ascending "waiting time" both average and variance increases. This is depicted in Figure 4. With Equation (6), we get for the weight average degree of polymerization

$$\overline{N_w} = \overline{N_n} \frac{k+1}{k} \tag{30}$$

The γ-distribution formulated for mass fraction is given by:

$$W(N) = \frac{N}{\overline{N_n}} X(N) = \frac{\beta^{k+1}}{k!} N^k e^{-\beta N} \tag{31}$$

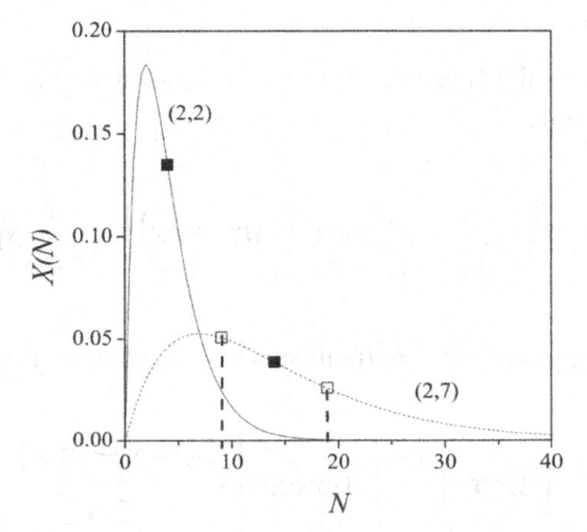

FIGURE 4 Schulz-Flory distributions for the indicated parameter values (k, β^{-1}), the solid squares mark the average degrees of polymerization and the open squares refer to variance according to Equation (27) and (29), respectively.

We note that in (2.31) $N_{max} = \overline{N_n}$.

Figure 5 presents the distributions listed in Table 1 and after Equation (31). We observe good agreement between data points of the Table and γ-distribution with $k = 2$ and $\beta^{-1} = 1.75$. The averages of degree of polymerization amount to $\overline{N_n} = 3.5$ and $\overline{N_w} = 5.25$ in fair agreement with the averages calculated. As a result, the distribution given in Table 1 can be well approximated by both logarithmic normal distribution and γ-distribution.

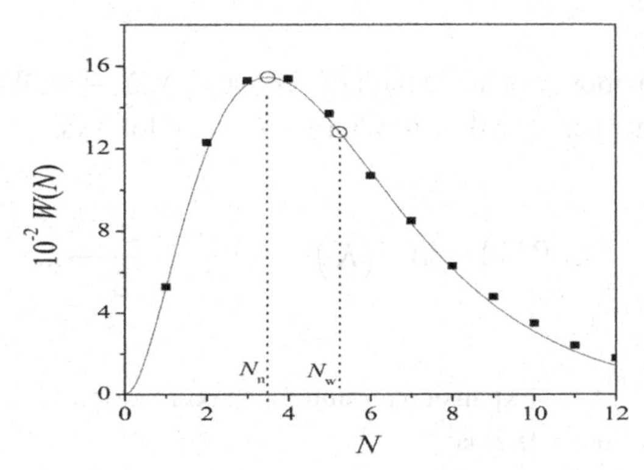

FIGURE 5 The γ-distribution, Equation (31), with $k = 2$ and $\beta^{-1} = 1.75$, solid curve, data points (■) correspond to the distribution of Table 1.

APPENDIX A

The normalization of the Gaussian distribution follows simply by application of the following integral.

$$\int_{-\infty}^{+\infty} \mathrm{d}x \exp\left[-ax^2 + bx\right] = \sqrt{\frac{\pi}{a}} \exp\left[\max_x \left(-ax^2 + bx\right)\right] = \sqrt{\frac{\pi}{a}} \exp\left[\frac{b^2}{4a}\right] \quad \text{(A1)}$$

This integral is also used for normalization of distribution Equation (12). It is:

$$\overline{x^2} = \left(2\pi\sigma^2\right)^{-1/2} \int_0^{\infty} \mathrm{d}x x \exp\left[-\frac{\left(\ln x - \ln \overline{x}\right)^2}{2\sigma^2}\right]$$

The integral can be easily transformed into an integral of type Equation (A1) by the substitutions $\ln x - \ln \overline{x} = t - \alpha = \tau$. It follows:

$$\overline{x^2} = e^{2\alpha} \left(2\pi\sigma^2\right)^{-1/2} \int_{-\infty}^{+\infty} \mathrm{d}\tau \exp\left[-\frac{\tau^2}{2\sigma^2} + 2\tau\right] = e^{2\alpha} e^{2\sigma^2} = \overline{x}^2 e^{\sigma^2} \quad \text{(A2)}$$

For getting the last expression, $\overline{x}^2\ e^{2\alpha}\ e^{\sigma^2}$ has been used. The result (A2) agrees precisely with the second relation of Equation (12).

APPENDIX B

Let us discuss Equation (19) in the limit of very large N, $N \to \infty$. We develop $P(k)$ near its maximum where, $\mathrm{d}P/\mathrm{d}k = 0$. With $k = K + \varepsilon$, it follows:

$$\ln P(k) = \ln P(K) + A_1 \varepsilon + A_2 \frac{\varepsilon^2}{2} + \ldots \quad \text{(B1)}$$

where $A_i = \frac{\mathrm{d}^i \ln P(k)}{\mathrm{d}k^i}\Big|_{k=K}$. Expansion around the maximum implies $A_1 = 0$ and also $A_2 < 0$. From Equation (19), it is:

$$\ln P(k) = \ln N! - \ln k! - \ln(N-k)! + k \ln p + (N-k)\ln(1-p)$$

For large N and k we apply Stirling's formula:

$$\ln N! = N \ln N - N \tag{B2}$$

Hence, it is:

$$\frac{\mathrm{d}\ln k!}{\mathrm{d}k}|_K = \ln K \quad \text{and} \quad \frac{\mathrm{d}\ln(N-k)!}{\mathrm{d}k}|_K = -\ln(N-K)$$

and for A_1, we have:

$$A_1 = -\ln K + \ln(N-K) + \ln p - \ln(1-p) = 0 \tag{B3}$$

One sees that it follows:

$$\ln\left(\frac{N-K}{K}\frac{p}{1-p}\right) = 0 \Rightarrow \frac{N-K}{K}\frac{p}{1-p} = 1 \text{ or } Np = K \tag{B4}$$

From Equation (B2), we easily calculate $A_{2:}$

$$A_2 = -\frac{1}{K} - \frac{1}{N-K} = -\frac{1}{Np(1-p)}$$

where the last expression of Equation (B4) has been used. Next, we look for the sum over the probabilities $P(k)$.

$$\lim_{N\to\infty}\sum_{k=0}^{N} P(k) = \int \mathrm{d}kP(k) = \int_{-\infty}^{+\infty} \mathrm{d}\varepsilon P(K+\varepsilon) = 1$$

Since each term of (B1) is of order $1/N$ smaller than the previous one, we ignore terms higher than of second order:

$$\ln P(k) = \ln P(K) - |A_2| \frac{\varepsilon^2}{2}$$

It follows:

$$P(k) = P(K) \exp\left(-\frac{|A_2| \varepsilon^2}{2}\right)$$

Normalization of the probability yields according to Equation (A1):

$$\int_{-\infty}^{\infty} \mathrm{d}\varepsilon P(k) = P(K) \sqrt{\frac{2\pi}{|A_2|}} = 1$$

With $|A_2|^{-1} = \sigma^2$ we get eventually:

$$P(k) = \left(2\pi\sigma^2\right)^{-1/2} \exp\left[-\frac{(k-K)^2}{2\sigma^2}\right]$$

This is in complete agreement with Gaussian distribution, Equation (11). It means, binomial distribution turns into normal distribution in the limit $N \to \infty$ and any fixed p.

KEYWORDS

- **Average**
- **Normal distribution**
- **Poisson distribution**
- **Schulz-Flory distribution**
- **Variance**

REFERENCES

1. Flory, P. J. *J. Amer. Chem. Soc.*, **58**, 1877–1885 (1936).
2. Schulz, G. V. Z. *Phys. Chem.*, **B30**, 379–398 (1935).

CHAPTER 3

MACROMOLECULES IN SOLUTION

HANS-WERNER KAMMER

CONTENTS

3.1 MACROMOLECULES IN SOLUTION

Macromolecules may exist in the solid state, the molten state and in solutions, but not in the gaseous state. This is so because the boiling temperature of a polymer melt is proportional to the degree of polymerization N.

$$T_{\text{boiling}} \propto \frac{\varepsilon}{k_{\text{B}}} N$$

with interaction energy ε between segments and Boltzmann's constant k_{B}. This temperature is for $N >> 1$ much higher than the temperature of thermal decomposition of the chains. For studying properties of single macromolecules, solutions are suitable. In highly diluted solutions, interactions between solvent molecules and chain molecules are dominant whereas interactions between different chains can be neglected. Hence, dilute solutions reflect properties of single chains to a very good approximation. Statistics of chains can be discussed in terms of classic theory since we have always $h\nu/k_{\text{B}}T << 1$. All what we discuss next refers to the state of high dilution.

3.2 SIZE OF A PERFECT CHAIN

We discuss a chain molecule consisting of N segments each of length a. Its contour length is $L_{\text{c}} = Na$. In a freely rotating chain, segments are uncorrelated that is the orientation of a segment is independent of any other segment. Even intersections of segments are allowed. This kind of chain, we call perfect or ideal chain. One may model it as a random walk with steps of length a as illustrated in Figure 1.

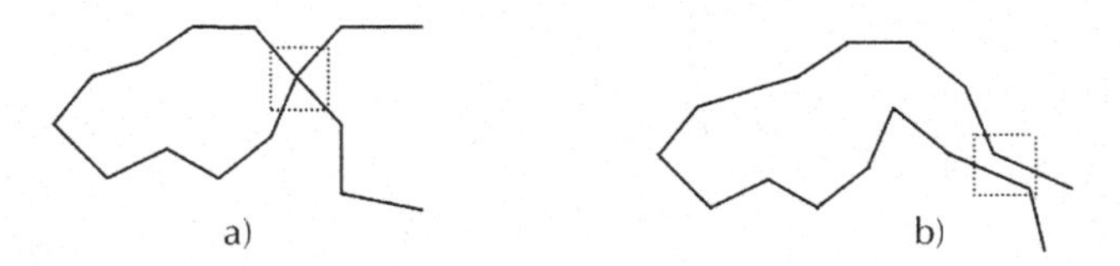

FIGURE 1 Chain conformations (a) ideal chain and (b) real chain.

The freely rotating chain is formed by randomly distributed segments around some reference origin $\vec{r} = 0$. The segment distribution density is given by the

Gaussian or normal distribution, Equation (11) (see chapter 2, Molecular Mass of Macromolecules and its Distribution). We formulate it here in three dimensions:

$$\rho(r)=\left(\frac{2\pi}{3}\left\langle\Delta\vec{r}^{2}\right\rangle\right)^{-3/2}\exp\left[-\frac{3r^{2}}{2\left\langle\Delta\vec{r}^{2}\right\rangle}\right] \tag{1a}$$

Using Equation (1), one obtains easily the probability for finding the end of a chain, which starts at origin, in a spherical volume element of radius r:

$$\mathrm{d}w=\rho(r)4\pi r^{2}\mathrm{d}r \tag{1b}$$

The variance is given by $\left\langle\Delta\vec{r}^{2}\right\rangle=\left\langle R^{2}\right\rangle$ since $\vec{r}_{o}=0$. Quantity R characterizes the end-to-end distance. We note that the distribution (1b) is isotropic. It depends only on r^2. The isotropy of the distribution causes that we have for the average distance of segments in x-direction from origin $\bar{x}=0$ since there are as many segments in distance $-x$ from locus $r = 0$ as in distance $+x$. The same is true for directions y and z. Hence, we have for the average distance from origin.

$$\left\langle\vec{r}\right\rangle=0 \tag{2}$$

For calculating the variance $\left\langle\Delta\vec{r}^{2}\right\rangle$, we suppose that we have the same number of particles in $+x$ and in $-x$ direction. The projection of segment length a on x-direction is $a/\sqrt{3}$ (Compare Appendix A). It follows for the variance in x-direction:

$$\left\langle\Delta x^{2}\right\rangle=\left\langle x^{2}\right\rangle=\frac{a^{2}}{3}N$$

We get the same for the other directions. It results:

$$\left\langle\Delta\vec{r}^{2}\right\rangle=\left\langle R^{2}\right\rangle_{o}=a^{2}N \tag{3}$$

This is the end-to-end distance of a perfect chain. As can be seen not R varies with N but R^2. Later, we will provide experimental evidence that perfect behavior, $R_o \propto N^{1/2}$, can be indeed observed under certain conditions, it is not only the approximate result of a simplified model discussion.

Let us repeat derivation of Equation (3) by projection of the random walk on a periodic lattice (see Figure 2). As before, we suppose the chain molecule consists of N chemically identical units each of length a. Its contour length is Na. Owing to its flexibility the chain forms a coil. What is the average end-to-end distance of the coil $\langle R^2 \rangle_o$?

FIGURE 2 The coil modeled as a random walk across a periodic lattice.

The walk is a succession of N steps, starting from one end (A) and reaching in an arbitrary way the end point (Z). The length of one step is a. Each step is a vector, characterized by length and direction, and it is assumed in the model under discussion that different vectors $\boldsymbol{a}$ have completely independent orientations that is orientation correlations between them are neglected. A random walker achieves after N steps each of length a in one dimension the average displacement $\langle \Delta x \rangle = 0$ analogous to Equation (2) because $-|\vec{a}|$ and $+|\vec{a}|$ occur with the same probability. Then, the end-to-end vector $\vec{\boldsymbol{R}}$ is simply:

$$\vec{\boldsymbol{R}} = \vec{\boldsymbol{a}}_1 + \vec{\boldsymbol{a}}_2 + ... + \vec{\boldsymbol{a}}_\mathrm{N} = \sum_{\mathrm{n}=1}^{\mathrm{N}} \vec{\boldsymbol{a}}_\mathrm{n} \quad (4)$$

Due to independence of orientations, the average square of the end-to-end distance on all possible pathways from A to Z is[1]:

[1] Note

$$\langle \vec{\boldsymbol{a}}_\mathrm{n} \vec{\boldsymbol{a}}_\mathrm{m} \rangle = \begin{cases} 0 \text{ for } n \neq m \\ a^2 \text{ for } n = m \end{cases}$$

$$\left\langle R^2 \right\rangle_o = \sum_{n,m=1}^{N} \left\langle \vec{a}_n \vec{a}_m \right\rangle = \sum_{n=1}^{N} a_n^2 + 2\sum_{n<m}^{1...N} \left\langle \vec{a}_n \vec{a}_m \right\rangle = Na^2 \quad (5)$$

Frequently, a shorthand notation is used for $\left\langle R^2 \right\rangle_o$ it reads simply R_o^2. Result (5) is identical with Equation (3). This is not surprising since both derivations relay on random arrangement of segments.

The experimentally accessible quantity is not the end-to-end distance, but the radius of gyration, which is closely related to R_o^2. The radius of gyration represents the mean square distance of segments from the centre of mass of the chain. For a perfect coil it is given by:

$$\left\langle S^2 \right\rangle_o = \frac{1}{6} R_o^2 \quad (6)$$

(see Appendix A). We may say N segments are enclosed in a sphere of diameter R_o or of radius S_o:

$$R_o \approx 2\, S_o \quad (7)$$

For a rigid-rod of length L, the radius of gyration is simply:

$$S^2 = \frac{1}{L} \int_{-L/2}^{L/2} \mathrm{d}l l^2 = \frac{L^2}{12} \quad (8)$$

3.2 SIZE OF CHAINS WITH SHORT-RANGE INTERACTIONS

Relationship Equation (5) gives the end-to-end distance of a chain consisting of freely rotating segments or of completely uncorrelated neighboring segments. This result remains invariant even when short-range interactions between segments exist and correlations decay exponentially along the chain (see Equation 24). Suppose, it is:

$$\left\langle \vec{a}_n \vec{a}_m \right\rangle = \beta_{nm} \neq 0 \text{ for } n \neq m \quad (9)$$

We assume that β_{nm} is an exponentially decreasing function of distance $|n-m|$. Correlations decay within a finite range to a good approximation. Therefore, we combine k consecutive units a to a new segment c:

$$a_n + a_{n+1} + \ldots + a_{n+k-1} = c_n$$

and assume that the range of correlations is smaller than the space the k a-units occupy. This means the new segments c are uncorrelated and we have analogously to Equation (5).

$$\langle R^2 \rangle_o = \frac{N}{k}\langle c^2 \rangle = Na^2 \tag{10}$$

where $\langle c^2 \rangle / k \equiv a^2$. Equation (10) is a good approximation provided $\frac{N}{k} >> 1$. In conclusion, if correlations between segments slow down exponentially along the chain, one gets, independent of the chemical structure of the chain, ideal chain behavior if N is sufficiently large.

Let us discuss the conformation of a freely rotating chain a little bit more in detail. A part of the chain is depicted in Figure 3, quantity a symbolizes the length of the C-C bond. For the end-to-end distance, we get Equation (5). Without fixed orientation between two successive bonds, it is $\langle \vec{a}_n \vec{a}_m \rangle = 0$ for $n \neq m$. However with $x \equiv \cos \theta'$ and bond angle $\theta = \pi - \theta'$, Equation (5) becomes:

$$\langle R^2 \rangle = \sum_{n=1}^{N} a_n^2 + 2a^2 \sum_{n=1}^{N-1} \sum_{m=n+1}^{N} x^{(m-n)} = a^2 N \left[\frac{1+x}{1-x} - 2\frac{x}{N}\frac{1-x^N}{(1-x)^2} \right] \tag{11}$$

Appendix B gives the derivation of Equation (11). For large N, we obtain for the end-to-end distance (5).

$$\langle R^2 \rangle = Na^2 \frac{1+\cos\theta'}{1-\cos\theta'} \qquad \text{for } N >> 1 \tag{12}$$

We find precisely, the same result as given with Equation (10). Equation (12) agrees with Equation (5), when we choose for the segment length.

$$a'^2 = a^2\left(\frac{1+\cos\,\theta'}{1-\cos\,\theta'}\right) \tag{13}$$

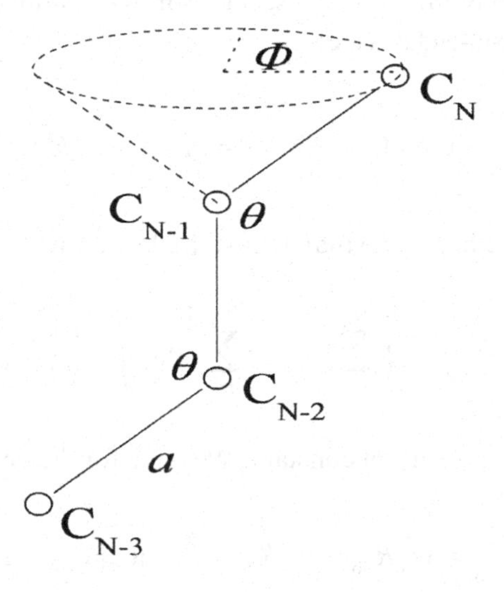

FIGURE 3 Freely rotating chain with fixed bond length *a* and bond angle *θ*, free rotation around single bond is characterized by angle *Φ*.

For polyethylene, we have $a = 0.154$ nm and $\theta' = 180° - 109.5°$. It follows

$$a'^2 = 2a^2$$

The discussion reveals that fixing of the bond angle or short-range interactions that die away rapidly along the chain do not destroy perfect behavior of the chain. However, we have to underline, exponential decay of correlations is indeed a critical approximation since interactions between segments with differences $|n - m|$ large are not necessarily and automatically negligible. Intersections of the chain with itself are even more likely with increasing difference $|n - m|$. Strict inclusion of those large loop interactions leads to non-Gaussian behavior of the chain.

3.3 SIZE OF A PERFECT BLOCK COPOLYMER

Let us calculate, in the same approximation as in Equation (5), the end-to-end distance of a perfect block copolymer chain, poly(A)-*block*-poly(B), with the degrees of polymerization N_A and N_B of the blocks and segment lengths a and b, respectively. The probability for meeting a segment of block A might be $w_A = N_A/N$. It follows for the end-to-end distance

$$\left\langle R^2 \right\rangle_o = \sum_{i=1}^{N_A} \sum_{j=1}^{N_B} \left\langle \left(w_A a_i + w_B b_j \right) \cdot \left(w_A a_i + w_B b_j \right) \right\rangle = w_A^2 a^2 N_A + w_B^2 b^2 N_B + 2 w_A w_B \sum_{i,j=1}^{N_A, N_B} \left\langle a_i b_j \right\rangle$$

Adopting Figure 2 here, one may recast the last term

$$\sum_{i,j=1}^{N_A, N_B} \left\langle a_i b_j \right\rangle = \frac{c}{2} \left(\sum_{i=1}^{N_A} \left\langle a_i^2 \right\rangle + \sum_{i=1}^{N_A} \left\langle b_j^2 \right\rangle \right) = \frac{c}{2} \left(R_{Ao}^2 + R_{Bo}^2 \right)$$

with c being a normalization constant. We have for the end-to-end distance

$$\left\langle R^2 \right\rangle_o = w_A^2 R_{Ao}^2 + w_B^2 R_{Bo}^2 + c w_A w_B \left(R_{Ao}^2 + R_{Bo}^2 \right)$$

When $w_A = ½$ and $a = b$, it is $R_{Ao}^2 = R_{Bo}^2$ and $\left\langle R^2 \right\rangle_o = 2R_{Ao}^2$. Hence, the normalization constant is $c = 3$. This leads to

$$\left\langle R^2 \right\rangle = w_A R_{Ao}^2 + w_B R_{Bo}^2 + 2 w_A w_B \left(R_{Ao}^2 + R_{Bo}^2 \right) \tag{14}$$

3.4 SEMIRIGID CHAINS—THE PERSISTENCE LENGTH

We discussed freely rotating chains with fixed bond angle. These chains assume in dilute solutions coil shape. There are also linear chains with lower flexibility of the backbone. Therefore, they are called semirigid or wormlike polymers. Rigidity of the backbone might be caused by delocalized π electrons along the main chain or hydrogen bonds acting between monomers. These effects lead to more straight conformations. Chains with completely straight conformation are called rodlike polymers.

One easily recognizes that the freely rotating chain with fixed bond angle $\pi - \theta'$ and bond length a turns into wormlike or more straight conformation for $\theta' \ll 1$. We have two lengths characterizing the semirigid chain, contour length $L_c = Na$ and persistence length Λ. Persistence length is the average projection of the end-to-end distance to the first segment. We assume, it points in z direction, then we have in the limit $N \to \infty$ (see Appendix B).

$$\Lambda \equiv \left\langle \vec{R} \cdot \vec{e}_z \right\rangle = \frac{a}{1 - \cos \theta'} \tag{15a}$$

For small θ', it follows:

$$\Lambda = \frac{2a}{\theta'^2} \tag{15b}$$

We evaluate Equation (11) in the limit:

$$\theta' \ll 1, \text{ bond length } a \to 0 \text{ and } \frac{2a}{\theta'^2} = const \tag{16}$$

It follows ($x \equiv \cos \theta'$):

$$\left\langle R^2 \right\rangle = L_c \Lambda \left[2 - \frac{a}{\Lambda} - 2\frac{\Lambda}{L_c}\left(1 - \frac{a}{\Lambda}\right)\left\{1 - \left(1 - \frac{L_c / \Lambda}{N}\right)^N\right\}\right]$$

We use here:

$$\lim_{N \to \infty} \left(1 - \frac{x}{N}\right)^N = e^{-x} \tag{17}$$

Eventually, one finds for $\left\langle R^2 \right\rangle$ in the limit $N \to \infty$ and $a \to 0$:

$$\langle R^2 \rangle = 2L_c\Lambda\left[1-\frac{\Lambda}{L_c}\left\{1-\exp\left(-\frac{L_c}{\Lambda}\right)\right\}\right] \tag{18}$$

Two limits of the ratio formed by the characteristic lengths, L_c/Λ, are important. For $L_c/\Lambda \ll 1$, it follows:

$$\lim_{L_c/\Lambda\to 0}\langle R^2 \rangle = 2L_c\Lambda\left[\frac{L_c}{2\Lambda}+\ldots\right] = L_c^2 \tag{19}$$

(cf. Appendix B). In the opposite case, $L_c/\Lambda \gg 1$, we find:

$$\lim_{L_c/\Lambda\to \infty}\langle R^2 \rangle = 2\Lambda L_c\left[1-\frac{\Lambda}{L_c}\right] = 2\Lambda L_c \propto a^2 N \tag{20}$$

We recognize that the effective segment length in the limit L_c/Λ is equal to 2Λ. The two limits reveal: Only short chains exist in a rigid-rod conformation. Sufficiently long chains always coil according to Equation (20). As a consequence, all chain molecules coil for sufficiently high molecular masses. Quantitatively, the degree of coiling might be defined as follows:

$$D_{\text{coiling}} = 1-\frac{\langle R^2 \rangle}{L_c^2} \cong 1-\frac{1}{N} \xrightarrow{N\to\infty} 1 \tag{21}$$

The coils are very loose structures. The volume of a chain is $V_{\text{chain}} = a^3 N$ and that of a coil under perfect conditions follows to $V_{\text{coil}} = a^3 N^{3/2}$ after Equation (20). The ratio of the volumes amounts to:

$$\frac{V_{\text{chain}}}{V_{\text{coil}}} = \frac{1}{\sqrt{N}} \tag{22}$$

For $N = 1000$, segments occupy only around 3 % of the coil volume.

It is also possible to model the semirigid chain in a continuous way. This wormlike model is shown in Figure 4. The overall length or its contour length is again L_c. The unit tangential vector $\vec{t}(s)$ at location $\vec{r}(s)$ of the chain is defined by:

$$\vec{t}(s) = \frac{\partial \vec{r}}{\partial s} \quad \text{with } t^2 = 1 \tag{23}$$

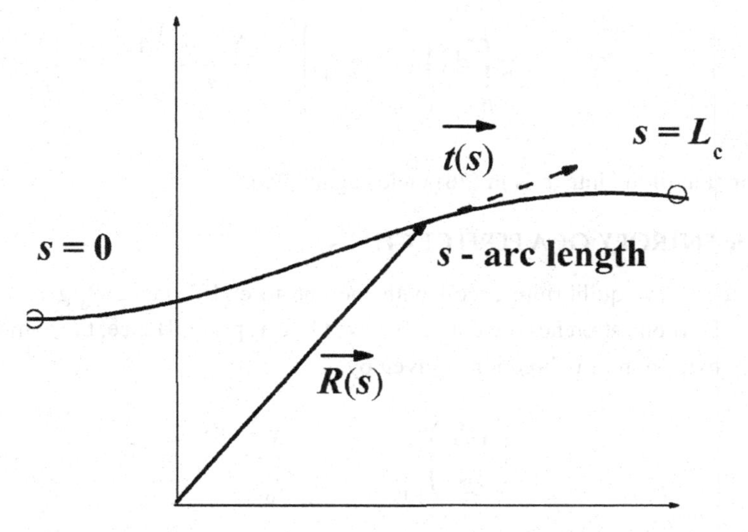

FIGURE 4 Semirigid chain with contour length L_c as continuous curve $\vec{R}(s)$.

The average orientation correlation of two tangential vectors is assumed to decay exponentially along the chain with the distance $|s - s'|$ between the vectors

$$\langle \vec{t}(s)\vec{t}(s') \rangle = \exp\left(-\frac{|s - s'|}{\Lambda} \right) \tag{24}$$

where Λ denotes persistence length as before. The average end-to-end distance of the chain reads with Equation (23) and Equation (24) (see Appendix B)

$$\left\langle R^2 \right\rangle = \int_0^{L_c} \mathrm{d}s \int_0^{L_c} \mathrm{d}s' \left\langle \vec{t}(s)\vec{t}(s') \right\rangle \tag{25}$$

Relation (Equation (25)) will be discussed in the limit Equation (16) that is for small angles θ' and segment length a under $\Lambda = 2a^2/\theta' = const.$ (Compare also Appendix A of chapter 5) Expression (25) for the mean end-to-end distance reads.

$$\left\langle R^2 \right\rangle = 2\int_0^{L_c} \mathrm{d}s \int_0^{s} \mathrm{d}s' \exp\left[-\frac{(s-s')}{\Lambda} \right] \tag{26}$$

Evaluation of the integrals in (26) yields again (18).

3.5 THE ENTROPY OF A PERFECT CHAIN

A chain forms in equilibrium a coil with the end-to-end distance R_o given by Equation (5). If one stretches the coil, it behaves like a spring. Hence, the entropy change by extension in x-direction is given by:

$$\left(\frac{\partial S}{\partial x} \right)_{\mathrm{T,P}} = -k_{\mathrm{B}} \frac{x}{x_o^2} \tag{27}$$

The minus sign is natural since entropy is maximal for the undeformed chain. Integration of Equation (27) yields for the entropy change in one direction $-k_B/2$ $(x^2 - x_o^2)/x_o^2$. In three dimensions, we get for the entropy change of the coil.

$$\Delta S = -\frac{3}{2} k_{\mathrm{B}} \frac{R^2 - R_o^2}{R_o^2} \tag{28}$$

We may also say $-T\Delta S$ represents elastic energy of the coil. The change in free energy $\Delta F = \Delta U - T\Delta S$ with $\Delta U = 0$, since the energy of the chain is independent of its conformation, follows by:

$$\frac{\Delta F}{k_B T} = \frac{3}{2} \frac{(R^2 - R_o^2)}{R_o^2} \tag{29}$$

As can be seen, the chain behaves like a spring with spring constant

$$\kappa = \frac{3}{2} \frac{k_B T}{R_o^2} \tag{30}$$

3.6 THE REAL CHAIN IN A GOOD SOLVENT – FLORY'S APPROXIMATION

Two far-reaching approximations characterize the discussion of chain extension in chapter 3.1:

- All interactions between segments have been neglected, that is the chain can intersect itself.
- Random walk implies neglecting of correlations between segments.

For a random walker, it is no problem to cross its pathway, for a randomly arranged chain molecule, however, it is impossible. Therefore, especially the first assumption is unrealistic since chain segments cannot be in close proximity due to repulsions between them. Experimental studies on dilute solutions by light scattering or viscosity measurements suggested that the size of macromolecules is:

$$R \propto N^{\nu} \text{ with } \nu > 1/2 \text{ for good solvents}$$

That is, in good solvents the segment-segment interaction is repulsive and tends to swell the coil. These facts led Flory to lift the first approximation by taking into account segment-segment interaction. The second approximation remains untouched in his approach to keep the problem tractable. It means the chain is still represented by a random walk.

The coiled chain, consisting of N units, occupies a volume R^3. Two forces in balance rule its size. The repulsive energy owing to monomer-monomer repulsion must be proportional to the number of pairs per coil volume or to square of segment concentration. Neglecting correlations between segments and ignoring numerical factors, one may formulate:

$$\frac{F_{\text{repulsion}}}{k_{\text{B}}T} = v(T)c^2 \tag{31}$$

Adopting the so-called mean-field approximation means in our context ignoring of correlations or $\langle c^2 \rangle = \langle c \rangle^2 = c^2$. The segment concentration c within the coil is:

$$c = \frac{N}{R^3}$$

The proportionality factor $v > 0$ in Equation (31) represents the volume where repulsions between segments occur, it is called excluded volume. We recognize quantity $F_{\text{repulsion}}$ has actually the unit energy per volume. Integration of Equation (31) over the coil volume yields for the energy:

$$\frac{f_{\text{repulsion}}}{k_{\text{B}}T} = v(T)c^2R^3 = v(T)\frac{N^2}{R^3} \tag{32}$$

If the coil size would be governed only by repulsion between the segments the coil would tend to large R for minimizing the energy. As we know from Equation (28) large values of R would dramatically reduce the entropy of the chain. This calls the elastic energy after Equation (29) on the stage. It balances the repulsive energy. The second energy contribution tends to coil the chain, it is:

$$\frac{f_{\text{elastic}}}{k_{\text{B}}T} = \frac{R^2}{Na^2} \tag{33}$$

Combination of Equations (32) and (33) gives the total energy as a function of coil size R.

$$\frac{f(R)}{k_{\text{B}}T} = v(T)\frac{N^2}{R^3} + \frac{R^2}{a^2N} \tag{34}$$

Minimization of the energy with respect to coil size yields the equilibrium size of the coil, $R_{equilibrium} = R_F$; index F is in honor of Flory. This approach can only provide with accuracy the molecular mass dependence of coil size.[1] Therefore, we ignore again numerical factors and find

$$R_F^5 \cong v a^2 N^3 \tag{35}$$

For good solvents, one expects the excluded volume v to agree with the segment volume a^3 in order of magnitude. Hence, we have:

$$R_F \cong a N^{3/5} \tag{36}$$

Comparison of Equations (5) and (36) shows that the size of chains in a good solvent is greater than that of perfect chains, $R_F > R_o$. It means chains are swollen in a good solvent.

When we insert in Equation (35) $v \cong a^3/N^{1/2}$, that is the excluded volume is very small as compared to segment volume, we find the size of a perfect chain:

$$R_o \cong a N^{1/2} \tag{37}$$

It means the chain behaves perfectly if attractions and repulsions cancel each other. Solvents where the size of chains is reduced to ideal dimension, Equation (37), are called Θ-solvents.

In contrast to good solvents, where chains swell, there are poor solvents that force chains to collapse. The average monomer concentration inside the coil and its size is controlled entirely by attraction and steric repulsion between segments. As a result, the concentration becomes independent of N. For the size we have:

$$R^3 \cong N a^3 \text{ or } R \cong N^{1/3} a \tag{38}$$

In summary, coils swell in good solvents and collapse in poor solvents. Somewhere in between these limits, under Θ-conditions, coils behave ideally. These important results are summarized in Table 1.

[1] Since numerical factors are not in focus of this theory, we use in relations for R the symbol $\cong$ instead of the equilibrium sign.

TABLE 1 Coil sizes in solvents of different quality

Quality of the solvent	Size of the coil	Interactions
Good	$R_F \cong aN^{3/5}$	Attractions segments-solvent ⇒ swelling of the coils
Θ-conditions	$R_o \cong aN^{1/2}$	Attractions and repulsions cancel each other ⇒ ideal behavior
Poor	$R \cong aN^{1/3}$	Repulsions segments-solvent ⇒ collapse of the coils

3.7 THE EXPANSION PARAMETER

Flory approached the real chain problem also in a slightly different version by introducing an expansion factor α:

$$\left\langle R^2 \right\rangle = \alpha^2 \left\langle R^2 \right\rangle_o \tag{39}$$

After Equation (39), the expansion factor is given by the ratio of the respective end-to-end distances. Since it is assumed that the segments are randomly arranged in both the real and the perfect chain, the problem might be formulated in terms of corresponding Gaussian distributions. Application of Equation (1b) yields the elastic contribution.

$$\ln\left(\frac{\mathrm{d}w_{\mathrm{real}}}{\mathrm{d}\tau}\right) = -\frac{3}{2} C \frac{\left\langle R^2 \right\rangle}{\left\langle R^2 \right\rangle_o}$$

where $\mathrm{d}\tau$ represents the volume element and C is a constant. We obtain for the probability ratio by adding the contribution of repulsion energy

$$\frac{\mathrm{d}w_{\mathrm{real}}}{\mathrm{d}w_{\mathrm{perfect}}}=\left(\frac{\mathrm{d}\tau}{\mathrm{d}\tau_{\mathrm{o}}}\right)\exp\left(-\frac{3}{2}\frac{\langle R^2\rangle-\langle R^2\rangle_{\mathrm{o}}}{\langle R^2\rangle_{\mathrm{o}}}-\frac{f_{\mathrm{repulsion}}}{k_{\mathrm{B}}T}\right) \quad (40)$$

One easily recognizes that the first term of the exponent corresponds to elastic energy of the chain (29) whereas the second term is given by Equation (32). The ratio of the volume elements ($\mathrm{d}\tau/\mathrm{d}\tau_{\mathrm{o}}$) equals α^3. Equation (40) can be easily recast by use of Equation (39). It follows

$$\frac{\mathrm{d}w_{\mathrm{real}}}{\mathrm{d}w_{\mathrm{perfect}}}=\exp\left(\ln\alpha^3-\frac{3}{2}\left(\alpha^2-1\right)-\frac{vN^{1/2}}{a^3}\frac{1}{\alpha^3}\right) \quad (41)$$

Ratio $v\sqrt{N}/a^3$ may be replaced by $vN^2/\langle R^2\rangle_{\mathrm{o}}^{3/2}$. The equilibrium value of α maximises (41). The expansion factor becomes:

$$\alpha^5-\alpha^3\cong\frac{vN^2}{\langle R^2\rangle_{\mathrm{o}}^{3/2}} \quad (42)$$

This result reveals that two parameters govern the expansion factor both of them comprise the number of segments forming the chain: the excluded volume, vN^2, and the end-to-end distance of a perfect chain, $\langle R^2\rangle_{\mathrm{o}}^{1/2}=a\sqrt{N}$.

Moreover, perfect behavior, that is $\alpha = 1$, requires disappearance of excluded volume, $v = 0$. Figure 5 presents the expansion parameter α *versus* total excluded volume for very large values of $v\sqrt{N}$. One observes a power-law dependence of α on $v\sqrt{N}$ that leads to

$$\alpha\propto N^{1/10} \quad (43)$$

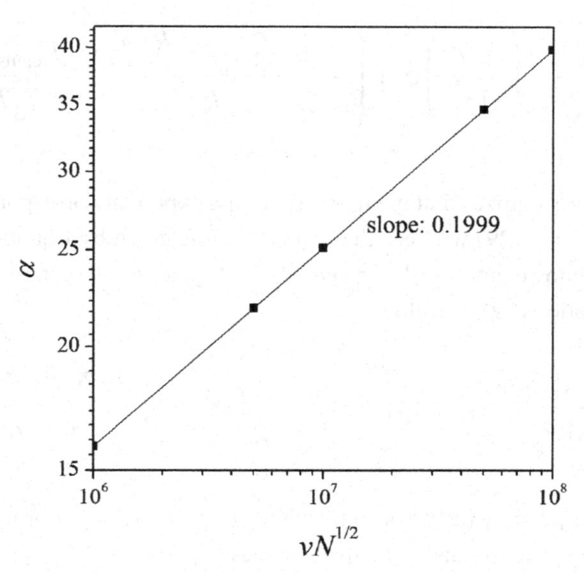

FIGURE 5 Expansion factor α as a function of $vN^{1/2}$ for large values of excluded volume.

Increasing parameter α, with N for large $vN^{1/2}$, points towards dominance of large loop interactions in governing the size of the chain. Flory introduced for the expression on the right-hand side of Equation (42) the dimensionless excluded volume parameter z.

$$z = \left(\frac{3}{2\pi}\right)^{3/2} \frac{vN^2}{\left\langle R^2 \right\rangle_o^{3/2}} \tag{44}$$

It results for Equation (42):

$$\alpha^5 - \alpha^3 \cong \left(\frac{2\pi}{3}\right)^{3/2} z \tag{45}$$

More sophisticated approaches show that approximation (42) is the first term in a power series in z.

APPENDIX A

To Equation (3) and derivation of Equation (6)

To Equation (3): Projection of segment length a in x direction is:

$$a_x = a \cos\theta$$

Hence, we get for the average of orientation under isotropic conditions:

$$\left\langle \cos^2 \theta \right\rangle = \int_0^{\pi/2} d\theta \sin\theta \cos^2\theta \overset{\sin\theta=t}{=} \int_0^1 dt t\sqrt{1-t^2} = -\frac{1}{3}\left(1-t^2\right)\Big|_{t=0}^{t=1} = \frac{1}{3}$$

It follows

$$\left\langle a_x^2 \right\rangle^{1/2} = \frac{a}{\sqrt{3}}$$

To equation (6): The centre of gravity of a chain molecule is defined by:

$$\sum_{i=1}^{N} \boldsymbol{r}_i = 0 \tag{A1}$$

Since all segment masses are equal. Quantity $\boldsymbol{r}_i$ is the vector from the centre to segment i (compare Figure A1). Radius of gyration is the second moment of $\boldsymbol{r}_i$. For its average, it follows:

$$\left\langle S^2 \right\rangle = \frac{1}{N}\sum_{i=1}^{N} \overline{\boldsymbol{r}_i^2} = \frac{1}{N}\sum_i \overline{\left(\boldsymbol{r}_1 + \boldsymbol{R}_i\right)^2} \tag{A2}$$

where $\boldsymbol{R}_i$ is the end-to-end distance of a chain with i segments. Figure A1 shows that the last expression is correct. From Equation (A2) we get:

$$\left\langle S^2 \right\rangle = \frac{1}{N}\sum_i \left(\overline{\boldsymbol{r}_1^2} + \overline{\boldsymbol{R}_i^2} + 2\overline{\boldsymbol{r}_1 \boldsymbol{R}_i}\right) = \overline{\boldsymbol{r}_1^2} + \frac{1}{N}\sum_i \overline{\boldsymbol{R}_i^2} + \frac{2}{N}\sum_i \overline{\boldsymbol{r}_1 \boldsymbol{R}_i} \tag{A3}$$

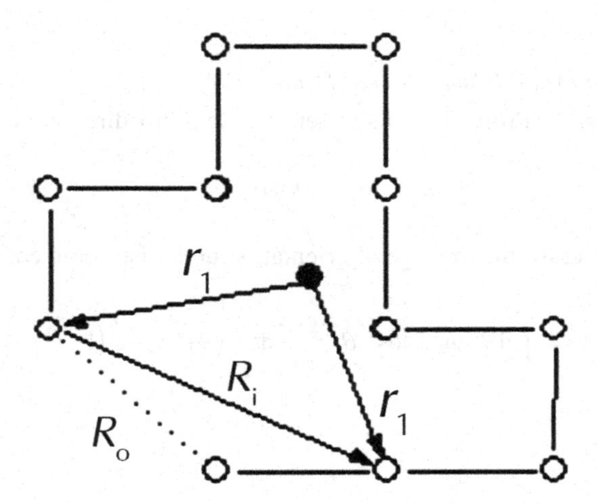

FIGURE A1 Chain molecule modeled as a random walk in a lattice. The random walk consists of N steps each of length a. The dotted line $\boldsymbol{R}_o$ symbolizes the end-to-end distance. The solid circle indicates the centre of gravity. Vectors $\boldsymbol{r}_i$ and $\boldsymbol{R}_i$ give the distance of segment i from the centre of gravity and the end-to-end distance of the chain with $N = i$, respectively.

One may replace $\boldsymbol{r}_i$ in (A1) by $\boldsymbol{r}_1 + \boldsymbol{R}_i$ (compare Figure A1). It results:

$$\boldsymbol{r}_1 = -\frac{1}{N}\sum_i \boldsymbol{R}_i$$

Inserting this in (A3) yields

$$\langle S^2 \rangle = \frac{1}{N}\sum_i \overline{\boldsymbol{R}_i^2} - \frac{1}{N^2}\sum_{i,j} \overline{\boldsymbol{R}_i \boldsymbol{R}_j}$$

The scalar product of the last term reads $\boldsymbol{R}_{ij}^2 = \boldsymbol{R}_i^2 + \boldsymbol{R}_j^2 - 2\boldsymbol{R}_i\boldsymbol{R}_j$ and we have

$$\langle S^2 \rangle = \frac{1}{2N^2}\sum_{i,j} \overline{\boldsymbol{R}_{ij}^2} = \frac{a^2}{2N^2}\sum_i \sum_j |i - j| \tag{A4}$$

The summations over i and j can be replaced by corresponding integrations. It follows:

$$\sum_i \sum_j |i-j| = 2\int_{s=0}^{N} ds \int_{s'=0}^{s} ds' |s-s'| = \frac{N^3}{3}$$

This inserted in (A4) leads to agreement with Equation (6).

APPENDIX B

Derivation of Equations (11) and (15):

Equation (11): The end-to-end distance reads:

$$\langle R^2 \rangle = a^2 N + 2a^2 \sum_{n=1}^{N-1} \sum_{m=n+1}^{N} x^{(m-n)} \quad \text{(B1)}$$

The double sum amounts to :

$$\sum_{n,m} x^{m-n} = (N-1)x + (N-2)x^2 + \ldots + (N-(N-1))x^{N-1} \quad \text{(B2)}$$

We apply $(1-x)^{-1} = 1 + x + x^2 + \ldots$ and the analogous series for $(1-x)^{-2} = 1 + 2x + 3x^2 + \ldots$ The respective first terms yield:

$$Nx(1 + x + \ldots + x^{N-2} + \ldots) = \frac{Nx}{1-x} - Nx^N(1 + x + \ldots) = Nx(1-x)^{-1} - Nx^N(1-x)^{-1} \quad \text{(B3)}$$

and for the respective second terms it follows:

$$-x(1 + 2x + \ldots + (N-1)x^{N-2} + \ldots) = -\frac{x}{(1-x)^2} + [Nx^N + (N+1)x^{N+1} + \ldots] =$$

$$-\frac{x}{(1-x)^2} + \frac{Nx^N}{1-x} + \frac{x^{N+1}}{(1-x)^2} \quad \text{(B4)}$$

Equation (11) results when we insert (B3) and (B4) in (B1).

Equation (15a):

$$\left\langle \vec{\boldsymbol{R}} \cdot \vec{e}_z \right\rangle = \frac{1}{a} \sum_{i=1}^{N} \left\langle \vec{\boldsymbol{r}}_1 \cdot \vec{\boldsymbol{r}}_i \right\rangle = a \sum_{i=1}^{N} x^{i-1} = \frac{a}{1-x}\left(1 - x^N\right) \overset{N \to \infty}{=} \Lambda$$

KEYWORDS

- **Flory's expansion parameter**
- **Quality of solvent**
- **Semirigid chains**
- **Size of block copolymer**
- **Size of perfect chain**

CHAPTER 4

CHARACTERIZATION OF POLYMERS BY FLOWING BEHAVIOR

HANS-WERNER KAMMER

CONTENTS

4.1 VISCOSITY OF POLYMER SOLUTIONS

4.1.1 DEFINITIONS AND INSTRUMENTATION

The studies of viscosity of polymer solutions are a powerful tool for getting access to microscopic quantities that characterize polymers. With solvent and solution viscosities (at $T = constant$), η_o and η, respectively, several quantities can be defined characterizing the flowing behavior of polymer solutions. The most important quantities are listed in Table 1.

TABLE 1 The quantities characterizing flowing behavior of polymer solutions

Name	Symbol and definition	Unit
Viscosity of solvent	η_o	Pa s
Viscosity of solution	η	Pa s
Relative viscosity	$\eta_{rel} = \dfrac{\eta}{\eta_o}$	1
Specific viscosity	$\eta_{spec} = \dfrac{\eta - \eta_o}{\eta_o} = \eta_{rel} - 1$	1
Inherent viscosity	$\eta_{inherent} = \dfrac{\ln \eta_{rel}}{c}$	$cm^3\ g^{-1}$
Reduced viscosity	$\eta_{red} = \dfrac{\eta_{spec}}{c}$	$cm^3\ g^{-1}$
Intrinsic viscosity	$[\eta] = \lim_{c \to 0} \left(\dfrac{\eta_{spec}}{c} \right)$	$cm^3\ g^{-1}$

Many experimental studies have been carried out with capillary viscometers, especially with a viscometer after Ubbelohde. It is operated by filling with a certain amount of liquid and measuring the time required for the liquid meniscus to fall from an upper to a lower mark. The effluence time t is related to viscosity η of the fluid according to the Hagen-Poiseuille equation. It reads for sufficiently long effluence times

$$\eta = \alpha\rho \cdot t \tag{1}$$

where α is an apparatus constant for the particular viscometer used and ρ symbolizes the density of the liquid. The relative viscosity reads with Equation (1)

$$\eta_{\mathrm{rel}} = \frac{\eta}{\eta_{\mathrm{o}}} = \frac{\rho \cdot t}{\rho_{\mathrm{o}} t_{\mathrm{o}}} \tag{2}$$

The relative viscosity simplifies in the case of diluted solutions with $\rho \cong \rho_{\mathrm{o}}$ (ρ_{o}-density of solvent) just to the ratio of the corresponding effluence times in a capillary viscometer

$$\eta_{\mathrm{rel}} = \frac{t}{t_{\mathrm{o}}} \tag{3}$$

Effluence time should be $t > 200$ sec. Accordingly, specific viscosity results from:

$$\eta_{\mathrm{spec}} = \eta_{\mathrm{rel}} - 1 = \frac{t}{t_{\mathrm{o}}} - 1 \tag{4}$$

For application of Equation (4), the ratio of effluent times should be in the range 1.2–2 in order to neglect shear corrections.

Taking into account concentration of the polymer, c, we get two more quantities, inherent and reduced viscosity. These quantities have the unit of specific volume. In the limit of infinite dilution, reduced viscosity represents intrinsic viscosity $[\eta]$.

4.1.2 INTRINSIC VISCOSITY

The intrinsic viscosity, $[\eta]$, represents a fundamental quantity. It is related to molecular mass, the structure and shape of dissolved macromolecules and reflects also the power of the solvent. Einstein (1905) found a relationship for specific viscosity of a highly diluted dispersion of spherical particles dissolved in a solvent.

$$\eta_{\text{spec}} = 2.5\Phi \qquad \text{for} \quad \Phi \to 0 \tag{5}$$

where Φ is the volume fraction of particles in the dispersion. Equation (5) has been applied in a slightly modified version to polymer solutions.

In the limit of infinite dilution, reduced viscosity, η_{spec}/c, represents the intrinsic viscosity $[\eta]$. This quantity reveals an important aspect. We consider, for the moment, c as the number of monomers per volume or c/N the number of chains per volume with N being the degree of polymerization. Then, we get for the volume fraction of chains in the solution

$$\Phi = \frac{v_{\text{coil}}c}{N} \tag{6}$$

where v_{coil} denotes the coil volume. It follows after Equation (5)

$$\frac{\eta - \eta_o}{\eta_o} \cong c\frac{R^3}{N} \tag{7}$$

The volume of a chain molecule is given by $v_{\text{coil}} = R^3$. In the limit of infinite dilution, Equation (7) reads:

$$[\eta] \cong \frac{R^3}{N} \tag{8}$$

Hence, measurements of intrinsic viscosity yield the hydrodynamic radius (or volume) of the chain molecule. Under Θ conditions and in a good solvent, respectively, we expect for solutions of flexible chain molecules after relationships Equations (36) and (37) of chapter 3

$$[\eta] \propto N^{1/2} \quad \text{and} \quad [\eta] \propto N^{4/5} \tag{9}$$

For dilute solutions of rod-like molecules, Equation (19) of chapter 3 suggests:

$$[\eta] \propto N^2 \tag{10}$$

From a multitude of experiments, it is known that the first relationship of Equation (9) is in good agreement with experimental findings. For flexible molecules in good solvents, it turns out that the exponent is frequently less than 0.8 and for rod-like molecules it is less than 2. Several scientists proposed an Equation (9). We mention Kuhn, Mark, Staudinger, Houwink, and Sakurada. It reads in the usually applied version.

$$[\eta] = KM^{\alpha} \tag{11}$$

and is called Mark-Houwink-Sakurada equation. In Equation (11), molecular mass is seen as dimensionless quantity and parameter K carries the unit of intrinsic viscosity. A few values for parameters α and K are listed in Table 2.

TABLE 2 Parameters of the Mark-Houwink Equation (11) for polymers in solvents

Polymer	Solvent	α (T / K)	$K \cdot 10^3$ / cm^{3}g^{-1}
Polystyrene	Toluene	0.73 (298)	10
Polystyrene	Cyclohexane	0.5 (308)	78
Poly(methyl methacrylate)	Toluene	0.73 (298)	7.1
Poly(ethylene)	Tetralin	0.83 (378)	16.2
Poly(propylene)	Decalin	0.8 (408)	10

For a polydisperse sample, we generalize Equation (11) as follows:

$$\overline{[\eta]} = \frac{\sum_i c_i [\eta]_i}{\sum_i c_i} = \sum_i W_i [\eta]_i = K \sum_i W_i M_i^{\alpha} \tag{12}$$

Hence, the viscosity average of molecular mass, M_η, is given by:

$$M_{\eta} = \left(\sum_{i} W_{i} M_{i}^{\alpha} \right)^{1/\alpha} \tag{13}$$

It follows, $M_{\eta} = M_{w}$ for $\alpha = 1$ and $M_{\eta} < M_{w}$ for $\alpha < 1$. Inserting $W_{i} = X_{i} M_{i} / \overline{M_{n}}$ in Equation (13), we obtain:

$$\overline{M_{\eta}}^{\alpha} - \overline{M_{\eta}}^{\alpha} = \frac{\overline{M_{n}^{\alpha+1}}}{\overline{M_{\eta}}} - \overline{M_{n}}^{\alpha} > 0 \tag{14}$$

Comparison of Equations (13) and (14) yields:

$$M_{n} < M_{\eta} < M_{w} \qquad \text{for } \alpha < 1 \tag{15}$$

In conclusion, chain molecules are isolated in the limit of high dilution. They behave as hard spheres. Hence, intrinsic viscosity reflects in that limit the effective hydrodynamic volume of isolated chains. In other words, intrinsic viscosity characterizes the increase in viscosity compared to solvent due to isolated chain molecules of solute.

4.1.3 HUGGINS EQUATION

What happens with ascending concentration of polymer? Chain molecules are in average closer to each other. As a result, reduced viscosity η_{red} increases due to mutual interference of solvent's flow patterns around the solute molecules. The simplest generalization of Equation (7) taking into account interactions between solvent molecules and segments of the chain, is given by:

$$\eta_{spec} = \Phi + K_{H} \Phi^{2} \tag{16}$$

where K_{H} is a constant, usually termed as Huggins constant. Replacing volume fraction by expressions Equations (6) and (8), we get for the reduced viscosity.

$$\eta_{red} = [\eta] + K_{H} [\eta]^{2} c \tag{17}$$

This equation was first time formulated by Huggins. The Huggins equation is a linear relationship between η_{red} and concentration c with intercept $[\eta]$, and slope related to dimensionless Huggins constant K_H. These parameters $[\eta]$ and K_H characterize the ability of the solvent to solvate a polymer. More precisely, intrinsic viscosity is a quantity expressing the hydrodynamic interference between polymer and solvent, thus, reflecting the ability of the solvent to swell the polymer, and intrinsic viscosity depends on molecular mass of the solute. On the other hand, constant K_H describes chiefly interactions originating from differences in chemical structure of polymer and solvent. It is to a good approximation independent of molecular mass and rigidity of the polymer chain since these properties are absorbed by $[\eta]$. Therefore, quantity K_H may serve as a criterion for selection of a good solvent for a particular polymer. The low values of K_H mean weak interaction among the dissolved chain molecules or high solvent power for the particular polymer. Experimental experience shows that for good solvents K_H lies in the range 0.3–0.4.

4.2 TERNARY SOLUTIONS

Dissolving of a polymer blend in a low molecular solvent forms a ternary solution. Accordingly, we formulate the Huggins Equation (17)

$$\eta_{\text{spec,b}} = [\eta]_{\text{b}} c_{\text{b}} + K_{\text{b}} [\eta]_{\text{b}}^2 c_{\text{b}}^2 \tag{18}$$

where c is the total mass concentration of macromolecules in the solvent and subscript b refers to polymer blend. The intrinsic viscosity $[\eta]_b$ and Huggins constant K_b refer to blend solution. The total polymer concentration c_b is simply.

$$c_{\text{b}} = c_1 + c_2 \tag{19}$$

The blend composition expressed in mass fraction reads:

$$W_1 = \frac{m_1}{m_1 + m_2} = \frac{c_1}{c_{\text{b}}} \qquad 1 - W_1 = \frac{m_2}{m_1 + m_2} = \frac{c_2}{c_{\text{b}}} \tag{20}$$

According to Equation (12), the intrinsic viscosity of a blend solution is given by:

$$[\eta]_b c_b = [\eta]_1 c_1 + [\eta]_2 c_2 \tag{21}$$

Experimental results confirm this linear relationship. An example is depicted in Figure 1. The second order term in Equation (18), we approximate by a second order polynomial in blend composition W. It follows with respect to Equation (21) and formulated in concentrations c_i.

$$K_b[\eta]_b^2 c_b^2 = K_1[\eta]_1^2 c_1^2 + K_2[\eta]_2^2 c_2^2 + 2\, K_{12}[\eta]_1[\eta]_2 c_1 c_2 \tag{22}$$

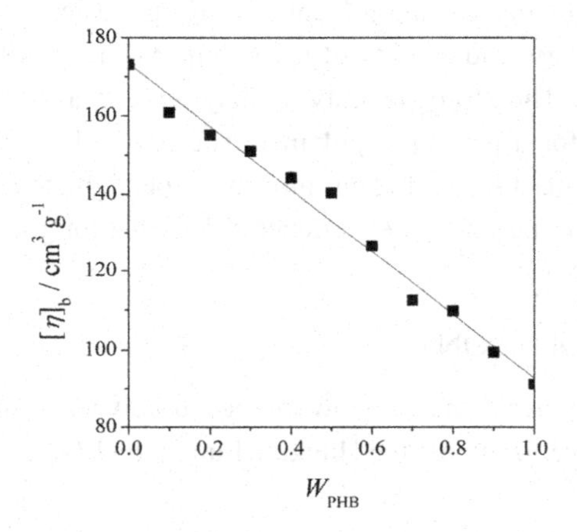

FIGURE 1 The intrinsic viscosity of polymer blend solutions *versus* composition of blend in mass fraction W, solvent-chloroform, polymer blend of poly(hydroxy butyrate) (PHB), and poly(ethylene oxide), the solid curve represents the linear regression curve.

Hence, a plot of $K_b[\eta]_b^2$ over mass fraction W_1 yields Huggins constants K_i (i = 1,2) and K_{12}. One may recast Equation (22) in the following way.

$$K_b = \left(\sqrt{K_1}Z_1 + \sqrt{K_2}Z_2\right)^2 + 2\left(K_{12} - \sqrt{K_1 K_2}\right)Z_1 Z_2 \text{ with } z_i \equiv \frac{[\eta]_i}{[\eta]_b} w_i \ (i = 1,2) \tag{23}$$

Equation (21) demonstrates that intrinsic viscosity of the blend is composed additively of the constituents' intrinsic viscosities. This is not the case for Huggins constant K_b. In blends, an excess contribution to K_b may occur, as Equation (23) manifests. We may define a blend solution as ideal when the excess term disappears:

$$K_{12} = (K_1 K_2)^{1/2} \tag{24}$$

Under this condition, Huggins coefficients of the neat constituents govern this coefficient of the blend. Pursuant to Equation (23), experimental Huggins coefficient, K_b, consists of two contributions, the Huggins coefficient of the perfect blend solution, K_{id}, and the excess quantity ΔK_b

$$\Delta K_b = K_{12} - \sqrt{K_1 K_2} \tag{25}$$

A perfect blend is characterized by $\Delta K_b = 0$. It means the chains cannot recognize differences between blend constituents. One may state, deviations from $\Delta K_b = 0$ are to a good approximation caused by specific interactions between the different macromolecular species. If $\Delta K_b < 0$, chains of different chemical structure avoid close proximity to each other which is equivalent to immiscibility of the constituents in the amorphous state. The opposite is true for $\Delta K_b > 0$, it reveals intermingling of the chains or miscibility. Figure 2 demonstrates this for two polymer blend solutions. One easily recognizes that $\Delta K_b > 0$ for the blend of PHB and PEO and $\Delta K_b < 0$ for the blend with PCL which means miscibility in the former case and immiscibility in the latter. This agrees quite well with different studies on the phase behavior of these polymers. Explicitly, the parameters ΔK_b read as follows for the blends of Figure 2: $\Delta K_{b\,PHB/PEO} = 0.037$ and $\Delta K_{b\,PHB/PCL} = -0.033$.

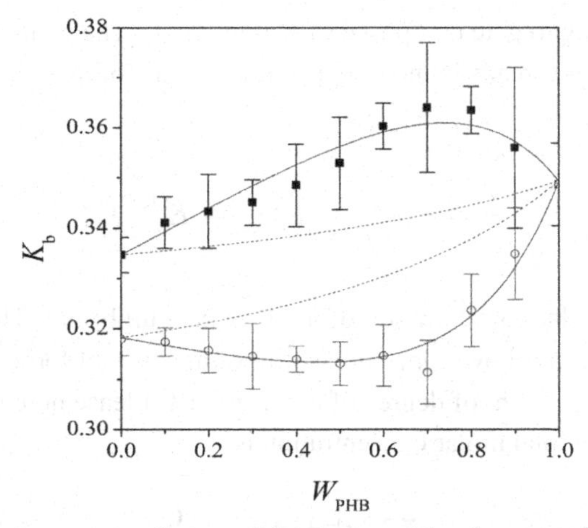

FIGURE 2 Huggins constant K_b *versus* blend composition for two ternary polymer blend solutions in chloroform, PHB blended with poly(ethylene oxide) (PEO)—(■), and poly(ε-caprolactone) (PCL)—(○), solid curves represent regression curves after Equation (22) dashed curves were calculated with $\Delta K_b = 0$.

4.3 VISCOSITY IN ASSOCIATING POLYMER SOLUTIONS

In associating polymers, we have to distinguish segment-related associations, where segments of the unimers are causing association, and molecule-related association where the number of associating sites is independent of molecule size (for example association *via* end groups). Accordingly, the association constant has to be formulated. In absence of association, ascending mass concentration leads to the same increase in molar concentration. This is not true anymore when association occurs since with increasing mass concentration the number of kinetically free moving molecules decreases and so the molar concentration. In segment–related association, the number of segments driving to association increases with molecular mass. Therefore, the equilibrium constant of association has to be given in mass concentration. In molecule-related association, the number of sites driving to association does not depend on molecular mass, we have particles with masses M_1, $M_2 = 2M_1$, ..., the equilibrium constant of association is formulated in molar concentration.

Let us discuss open multistage association in diluted polymer solutions. It might be modeled as follows:

$$A_{k-1} + A_1 \xleftrightarrow{K} A_k \quad \text{with } k-1 = 1, 2, 3, \ldots \tag{26}$$

A_k denotes an aggregate composed of k unimers A_1. The equilibrium constant (association constant) reads in molar concentrations as follows:

$$K = \frac{C_k}{C_{k-1}C_1} \Rightarrow C_k = K^{k-1}C_1^k \tag{27}$$

where C_k is the molar concentration of a k-mer in equilibrium. The second relation of Equation (27) follows since the equilibrium constant for molecule-related association is independent of degree of association k. Please note the unit of K is molar volume. The total molar concentration is:

$$C = \sum_k C_k = \sum_k K^{k-1}C_1^k = \frac{C_1}{1-KC_1} \quad \text{for } KC_1 < 1 \tag{28}$$

or

$$C_1 = \frac{C}{1+KC} \tag{29}$$

The relationship between total molar concentration C and total mass concentration c is given by:

$$C = \frac{c}{M_n} \tag{30}$$

where M_n represents the number average molecular mass. Combination of Equation (29) and (30) yields:

$$C_1^{-1} = K + \frac{M_n}{c} \tag{31}$$

For the total mass concentration, we have:

$$c = c_1 + c_2 + c_3 + \ldots = C_1 M_{n1} + 2M_{n1}KC_1^{\,2} + 3M_{n1}K^2C_1^{\,3} + \ldots$$

$$c = M_{n1}\frac{C_1}{(1-KC_1)^2} \tag{32}$$

M_{n1}—number average molecular mass of the unimer; replacing C_1 by Equation (31) gives:

$$M_n = M_{n1} + KM_{n1}\frac{c}{M_n} \tag{33}$$

Plotting M_n *versus* c/M_n gives the number average molecular mass of the unimer (intercept) and the association constant (slope). An analogous expression, we get for the weight-average molecular mass when molecule-related open association exists.

$$M_{\mathrm{w}} = M_{\mathrm{n1}} + 2KM_{\mathrm{n1}} \frac{c}{M_{\mathrm{n}}} \tag{34}$$

Equation (34) shows that for determination of M_{n1} we need M_{w} and M_{n}. With Equation (30), it follows:

$$\frac{M_{\mathrm{n}}}{M_{\mathrm{n1}}} = 1 + KC \text{ and } \frac{M_{\mathrm{w}}}{M_{\mathrm{n1}}} = 1 + 2KC \tag{35}$$

First relation of Equation (35) reads with $c = CM_{\mathrm{n}}$

$$c = M_{\mathrm{n1}}C(1 + KC) \tag{36}$$

Separation of C from Eq. (36) yields $1 + 2KC = \left(1 + 4Kc / M_{\mathrm{n1}}\right)^{1/2}$. We get from second equation (35) the relationship between true weight-average molecular mass and total mass concentration c

$$\frac{M_{\mathrm{w}}}{M_{\mathrm{n1}}} = \sqrt{1 + \frac{4K}{M_{\mathrm{n1}}} c} \tag{37}$$

The quantity M_{n1} is the molecular mass of a unimer. Equilibrium constant K results from the slope of $(M_{\mathrm{w}}/M_{\mathrm{n1}})^2$ plotted *versus* c. We recognize that molecular mass increases with concentration.

The intrinsic viscosity can be represented after Equation (8) by coil volume over number of chain segments. It reads for an aggregate of k unimers.

$$[\eta]_{\mathrm{k}} = \frac{R_{\mathrm{k}}^3}{N_{\mathrm{k}}} = \frac{k^3 R^3}{kN} = k^2 [\eta]_1 \tag{38}$$

where $[\eta]_1$ gives intrinsic viscosity of a unimer solution. With Equation (38), we can calculate the total specific viscosity of an associating solution after Equation (27). It results:

$$\frac{\eta_{\text{spec,ass}}}{c} = [\eta]_1 + 2[\eta]_1 K \frac{c}{M_n} \quad (39)$$

A plot of $\eta_{\text{spec,ass}}/c$ *versus* c/M_n yields intrinsic viscosity of the unimer $[\eta]_1$ (intercept) and association constant K (slope). Derivations of Equations (34) and (36) are given in Appendix.

One may see Equation (39) as Huggins equation. Assuming the approximation $M_n/M_{n1} = 1$, Huggins constant K_H is given by:

$$K_H = 2\frac{K}{M_{n1}[\eta]_1} \quad (40)$$

Huggins constant is here the ratio of two molar volumes. Constant K is the molar volume related to association and $M_1[\eta]$ is the molar hydrodynamic coil volume (compare Equation (8)).

APPENDIX

Derivation of Equations (34) and (39)

For weight average molecular mass of macromolecular aggregates, we get with Equation (28).

$$M_w = \frac{\sum_k n_k M_k^2}{\sum_k m_k} = \frac{M_{n1}^2}{c}\sum_k k^2 C_k = \frac{M_{n1}^2 C_1}{c}\sum_k k^2 K^{k-1} C_1^{k-1} \quad (A1)$$

As before, quantity c denotes the total mass concentration and M_1 is the molecular mass of a unimer. The integration of Equation (A1) with respect to (KC_1) gives:

$$\int d(KC_1)M_w = \frac{M_{n1}^2 C_1}{c}\sum_k k(KC_1)^k = \frac{M_{n1}^2 C_1^2 K}{c}\sum_k k(C_1 K)^{k-1} = \frac{M_{n1}^2 C_1}{c}\frac{KC_1}{(1-KC_1)^2} \quad (A2)$$

Differentiation of Equation (A2) with respect to (KC_1) yields:

$$M_{\text{w}} = \frac{M_{\text{n1}}^2 C_1}{c}\left[\frac{1}{(1-KC_1)^2} - \frac{2KC_1}{(1-KC_1)^3}\right] = \frac{M_{\text{n1}}^2 C_1}{c}\frac{1+KC_1}{(1-KC_1)^3} \quad \text{(A3)}$$

Replacing C_1 by Equation (29) gives:

$$M_{\text{w}} = M_{\text{n1}}^2 \frac{C}{c}(1+2KC)(1+KC) \quad \text{(A4)}$$

Inserting Equations (30) and (33) yields immediately Equation (34).

For the total specific viscosity, it follows:

$$\eta_{\text{spec,ass}} = \sum_{\text{k}} \eta_{\text{spec.k}} = \sum_{\text{k}} [\eta]_{\text{k}}\, c_{\text{k}} \overset{(32),\,(35)}{=} [\eta]_1 M_{\text{n1}} C_1 \sum_{\text{k}} k^2 (C_1 K)^{\text{k}-1} \quad \text{(A5)}$$

The sum in Equation (A5) agrees with the sum in Equation (A1). Hence, we get:

$$\eta_{\text{spec,ass}} = [\eta]_1 C M_{\text{n1}} (1+KC)(1+2KC) \quad \text{(A6)}$$

Equation (A6) yields Equation (39) by inserting of Equation (33).

KEYWORDS

- **Huggins equation**
- **Intrinsic viscosity**
- **Mark-Houwink-Sakurada equation**
- **Ternary solutions**
- **Viscosity of associating solutions**

REFERENCE

1. Einstein, A. *Ann. Physik*, **17**, 549–560 (1905).

CHAPTER 5

TRANSITIONS IN POLYMERS

HANS-WERNER KAMMER

CONTENTS

5.1 TRANSITIONS OF THE ISOLATED CHAIN MOLECULE

The connectivity of a huge number of chemically linked units in a chain molecule causes phase transitions in a single molecule. The relation of chain connectivity to phase transitions was discussed by Di Marzio (1999).

5.1.1 *WORMING THROUGH A MEMBRANE*

We will discuss the following situation. Two different solvents are separated by a membrane and the chain molecule, consisting of N segments, can pass segment-wise the membrane. The threading of the membrane is shown schematically in Figure 1.

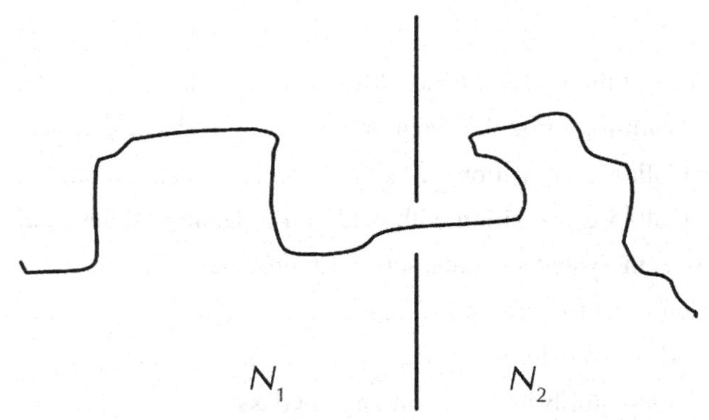

FIGURE 1 A chain molecule of N segments is crossing a membrane, left-hand N_1, right-hand N_2 segments, and one in the membrane, $N_1 + N_2 = N - 1$.

Let us discuss the distribution of N independent, not connected particles that is indistinguishable particles, in the two compartments separated by the membrane with a hole of particles size allowing for particle exchange between the phases. The partition function of the particles reads:

$$q_i = v_i \exp\left[-\frac{\varepsilon_i}{k_B T}\right] \quad i = 1,2 \tag{1}$$

The ε_i and v_i represent particle energy and volume of the compartment. Hence, it follows for the average number of particles in compartment i with $\varepsilon_i = const.$

$$\frac{\langle N_i \rangle}{N} = \frac{v_i e^{-\varepsilon/k_B T}}{\sum_i v_i e^{-\varepsilon_i/k_B T}} = \frac{V_i}{V} \tag{2}$$

Introducing the abbreviation $a \equiv q_1/q_2 > 0$, Equation (2) can be recast in:

$$\frac{\langle N_1 \rangle}{N} = \frac{a}{1+a} \tag{3}$$

The particle density is equal in both compartments, when $\varepsilon_i = const.$ According to Equation (2), no phase transition occurs in the system.

The situation changes when we turn to a chain molecule consisting of N consecutively linked segments, that is distinguishable units. The chain may worm segment-wise through the hole in the wall. One segment is in the wall, the rest is in the left and right compartment, respectively. One may see it as a chain molecule in solvents that are separated by a membrane. The quantities ε_i of Equation (1) represent then the solvent affinities towards the chain. We ask for the average number of segments in the left compartment. It results:

$$\frac{\langle N_1 \rangle}{N} = \frac{1}{N}\frac{a}{1-a} - \frac{a^N}{1-a^N} \tag{4}$$

Derivation of Equation (4) is given in the Appendix.

Equation (4) reveals that the average $<N_1>$ is a function of the intensive quantity a, which acts here as a chemical potential. We recall if any extensive quantity shows a discontinuity as a function of an intensive quantity we have a first-order transition. According to Equation (4), the chain undergoes a phase transition for large N in the region around $a = 1$.

$$\lim_{N\to\infty} \frac{\Delta\langle N_1 \rangle}{N} = \frac{\langle N_1 \rangle}{N}(1^+) - \frac{\langle N_1 \rangle}{N}(1^-) = 1 \tag{5}$$

In accordance with Equation (5), Figure 2 shows that the phase transition sharpens with increasing N.

For $N \to \infty$, the chain is just in one compartment and changes abruptly compartments at $a = 1$. It is in the right compartment when $a < 1$ whereas in the left compartment for $a > 1$. This phase transition is solely caused by the connectivity of segments in the chain molecule. Connectivity allows for marking them or we may count over the units. Therefore, phase transitions of this kind can only occur in macromolecules.

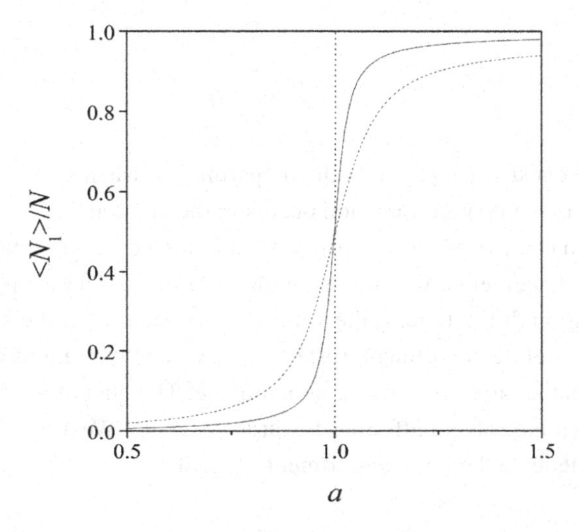

FIGURE 2 Average number of segments in compartment 1 according to Equation (4), dashed curve $N = 50$, and solid curve $N = 150$. One recognizes that the transition sharpens with increasing N.

5.1.2 COLLAPSE TRANSITION

Formally, this transition is very similar to the worming transition. Flory's approach to a real chain in a good solvent is the base of discussion (see paragraph 3.6). Flory considered in his seminal paper the balance between attraction among the monomers, wanting to collapse the polymer chain into itself, and the entropy of the chain, wanting to expand the chain. Flory's approach revealed the end-to-end distance of the swollen chain in a good solvent

$$\left\langle R^2 \right\rangle^{1/2} \propto N^{3/5} \qquad \text{(36 chapter 3)}$$

We follow here the lines of chapter 3. Accordingly, the expansion factor is introduced by:

$$\alpha^2 \equiv \frac{\langle R^2 \rangle}{\langle R^2 \rangle_o} \qquad \text{(39 chapter 3)}$$

with $\langle R^2 \rangle_o$ characterizing the size of a perfect chain. Flory formulated his approach in contributions to free energy, given in Equation (34, chapter 3) (Flory, 1949). In order to see the transition, as in Figure 2, we formulate the problem in terms of probability distributions analogous to Equation (40, chapter 3). Since it is assumed that the segments are randomly arranged in both the real chain in good solvent and the perfect chain, we approximate it by corresponding Gaussian distributions. The ratio of probabilities reads:

$$\frac{\mathrm{d}w_{\text{real}}}{\mathrm{d}w_{\text{ideal}}} = \left(\frac{\mathrm{d}\tau}{\mathrm{d}\tau_o}\right) \exp\left\{-\frac{3}{2}\frac{\langle R^2 \rangle - \langle R^2 \rangle_o}{\langle R^2 \rangle_o} - \frac{R^3}{a^3}\left[\ln\left(1-ca^3\right) + ca^3 - \chi\left(ca^3\right)^2\right]\right\} \qquad (6)$$

with $c = N/R^3$. The first term represents the ratio of the volume elements. It equals to α^3. The terms in the exponent correspond to the entropy of a swollen chain or to its elastic energy, to the excluded volume and to the interaction between solvent molecules and monomers, parameter χ, multiplied by the relative volumes, respectively. The terms in brackets (excluded volume and interaction) are formulated in the spirit of Flory-Huggins free energy. We get for Equation (6) with (Equation (39) chapter 3)

$$\frac{\mathrm{d}w_{\text{real}}}{\mathrm{d}w_{\text{ideal}}} = \exp\left[\ln \alpha^3 - \frac{3}{2}\left(\alpha^2 - 1\right) - N^{3/2}\alpha^3\left[\ln\left(1-ca^3\right) + ca^3\right] + \frac{N^{1/2}\chi}{\alpha^3}\right] \qquad (7)$$

The equilibrium value of α maximizes Equation (7). It follows:

$$\alpha^5 - \alpha^3 = -N^{3/2}\left[\ln\left(1 - N^{-1/2}\alpha^{-3}\right) + N^{-1/2}\alpha^{-3}\right] - N^{1/2}\chi \qquad (8)$$

When we develop the logarithm up to third order in $1/(N^{1/2}\alpha^3)$, we get:

$$\alpha^5 - \alpha^3 = N^{1/2}\left(\frac{1}{2} - \chi\right) + \frac{1}{6\alpha^3} \tag{9}$$

Equation (9) is a slight generalization of relationship (Equation (42) chapter 3). Expansion factor α is shown in Figure 3 for a solution of polystyrene (PS) in cyclohexane in the temperature range from 20 to 50 °C. The molecular mass of the PS was very high, $M_w = 2.7 \cdot 10^7$ g/mol. It results $N >> 1$. We observe a sharp collapse transition owing to the high value of N. Qualitatively, we find from Equation (9)

For $\alpha > 1$ and $\chi < 0.5 \Rightarrow R^5 \propto N^{1/2} R_o^5 \Rightarrow R \propto N^{3/5}$

For $\alpha = 1$ and $\chi = 0.5 \Rightarrow R = R_o \Rightarrow R \propto N^{1/2}$

For $\alpha < 1$ and $\chi > 0.5 \Rightarrow R^3 \propto \frac{R_o^3}{N^{1/2}} \Rightarrow R \propto N^{1/3}$

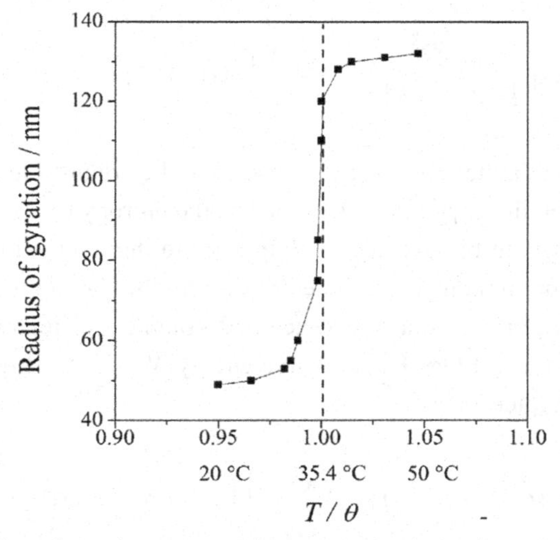

FIGURE 3 Radius of gyration *versus* T/θ for a (PS) solution in cyclohexane, data after Swislow et al. (1980).

The collapse transition for the system of Figure 3 takes place at $T = \theta$, it gives the theta temperature of the polymer in solution where the chain behaves perfectly. The phase transition takes place at $\alpha = 1$.

We throw a bridge from the single chain collapse transition to the phase transition occurring in a polymer solution. We observe the theta temperature of the system

PS/cyclohexane at $\theta = 35.4$ °C after Figure 3. For temperatures $T > \theta$, cyclohexane is a good solvent for PS, coils are swollen, the solution is homogeneous. Cyclohexane turns into a poor solvent for PS when the temperature is lower than the theta temperature, $T < \theta$. In that range, one observes phase separation. The homogeneous solution decays into two coexisting phases differing in composition, one phase is highly diluted in polymer, the other one is rich in polymer. When one heats up the system starting from the two-phase region, then the phases merge into one phase at the critical point, given by $T_c = \theta$ and $\Phi_c = \left(1+\sqrt{N}\right)^{-1}$. Therefore, this temperature is called **upper critical solution temperature** (UCST). The PS shown in Figure 3 has a very high $N \approx 26\ 000$, therefore Φ_c is very small, $\Phi_c \approx 0.006$. Precisely speaking, θ temperature equals critical temperature T_c for $N \to \infty$.

We add here that there are not only polymer solutions with UCST behavior but also with **lower critical solution temperature** (LCST). They behave in opposite way with respect to temperature as systems with UCST that is they are homogeneous at low temperatures and decay upon heating into a polymer-poor and a polymer-rich phase. Phase separation proceeds above critical temperature. Aqueous solution of poly(*N*-isopropylacrylamide) (PNIPAM), Figure 4, is an example for a system exhibiting LCST behavior. The phase diagram of an aqueous solution is depicted in Figure 5.

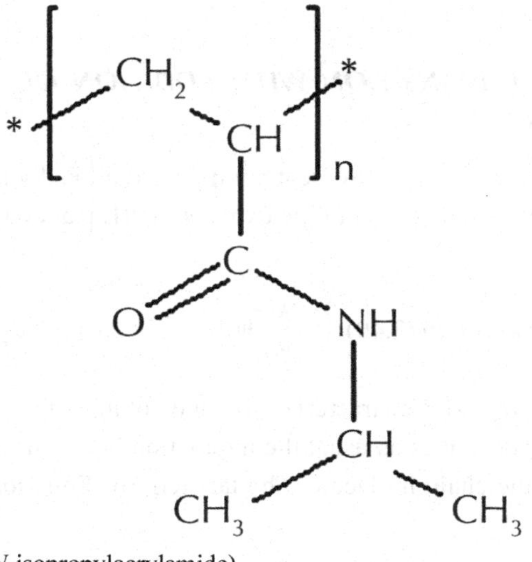

FIGURE 4 Poly(*N*-isopropylacrylamide).

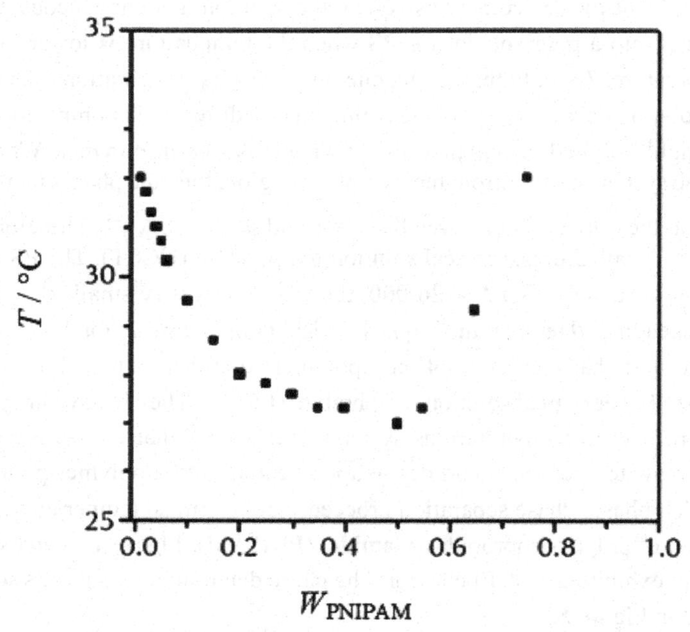

FIGURE 5 Phase diagram of an aqueous solution of PNIPAM, data after Afroze et al. (2000).

5.1.3 COLLAPSE TRANSITION WITH ADDITION OF NANOPARTICLES

When dissolved coils in a good solvent comprise additionally nanoparticles, we may reformulate the second term of the exponent in (Equation (40) chapter 3) as follows:

$$\textit{Second term of exponent in } (3.40) = -\frac{R^3}{a^3}\left[\ln\left(1-\Phi_{\text{poly}}\right)+\Phi_{\text{poly}}-\chi\Phi_{\text{poly}}^2\right]-\frac{R^3}{R_{\text{p}}^3}\Gamma\Phi_{\text{poly}}\Phi_{\text{p}} \quad (10)$$

The quantities R_p and v_p characterize size and volume of a spherical nanoparticle. Parameter χ describes as usual the interaction between solvent molecules and segments of the chain molecule. The last term of Equation (10) takes into account solely excluded volume interaction between nanoparticles and chain segments. Hence, parameter Γ is given by the corresponding ratio of excluded volumes, $\Gamma = v_p/a^3$. The quantities Φ_j in Equation (10) represent volume fractions of polymer and particles. We get with use of (Equation 39 chapter 3) and with n_p nanoparticles in the coil.

$$\Phi_{\text{poly}} = \frac{Na^3}{R^3} = ca^3 = \frac{1}{\alpha^3 N^{1/2}} \qquad \Phi_p = \frac{n_p v_p}{\alpha^3 R_o^3} \tag{11}$$

With Equations (10) and (11), the exponent of relationship Equation (7) reads:

$$\textit{exponent} =$$

$$3\ln\alpha - \frac{3}{2}\left(\alpha^2 - 1\right) - \alpha^3 N^{3/2}\left[\ln\left(1 - \Phi_{\text{poly}}\right) + \Phi_{\text{poly}}\right] + \frac{N^{1/2}}{\alpha^3}\left[\chi - \frac{v_p}{a^3}\left(\frac{n_p}{N}\right)\right] \tag{12}$$

The same procedure as applied to Equation (7) yields here instead of Equation (9):

$$\alpha^5 - \alpha^3 - \frac{1}{6\alpha^3} = N^{1/2}\left(\frac{1}{2} - \chi\right) + N^{1/2}\frac{v_p}{a^3}\left(\frac{n_p}{N}\right) \tag{13}$$

As before, we have for $\alpha > 1$ and $\chi < 0.5$ the swollen coil with $R \propto N^{3/5}$. When the particle concentration n_p/N is sufficiently high, we get the swollen coil even for $\chi > 0.5$ because the second term prevents the collapse. We may say the great excluded volume of the hard spheres hampers collapsing of the chain even under poor solvent conditions.

Let us discuss two situations, one polymer solution with UCST and one with LCST.

- The system PS dissolved in cyclohexane, sketched in Figure 3, displays an UCST. At high temperatures cyclohexane is a good solvent. It turns into a poor solvent for temperatures below 35.4 °C (Θ temperature). Expansion factor α was calculated as a function of temperature after Equation (13). Temperature dependence of interaction parameter χ might be represented by:

$$\chi = \frac{1}{2} - \frac{B}{\Theta}\left(1 - \frac{\Theta}{T}\right) \quad (\Theta = 308.6\ \text{K}) \tag{14}$$

since χ = ½ for $T = \Theta$. Parameter B is quoted for the system under discussion with B = 275.2 K (Elias, 1975). The other parameters occurring in Equation (13) were selected as follows: $N = 2.7 \times 10^4$, nanoparticles of Fe_2O_3 (density 5.24 g/cm³, M = 159.7 g/mol) with diameter of 6 nm are added to the PS with a = 0.5 nm. The ratio n_p/N varied between 0 and 7×10^{-4}. The highest ratio corresponds to mass fraction of nanoparticles in the coil of 0.37. The result is shown in Figure 6.

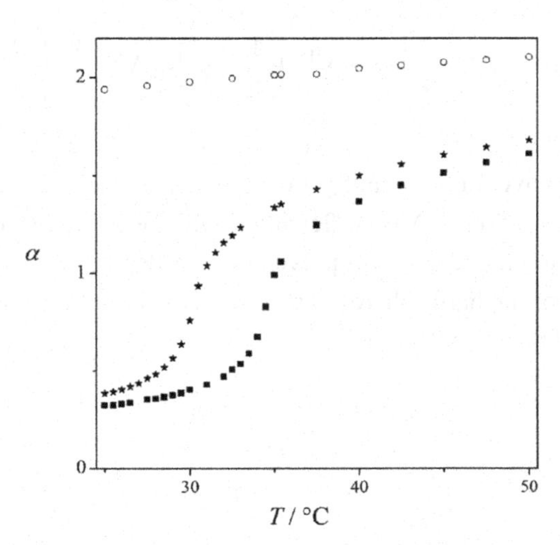

FIGURE 6 Expansion factor *versus* temperature for PS in cyclohexane calculated after Equation (13), $n_p = 0$ – solid squares, $n_p/N = 6 \times 10^{-5}$ – stars, and $n_p/N = 7 \times 10^{-4}$ – open circles.

Figure 6 reveals that with addition of nanoparticles Θ-temperature shifts to lower temperature. It shifts by $\Delta\Theta \approx -4$ K when one adds particles of ratio $n_p/N = 6 \times 10^{-5}$. Higher ratios may cause shifts of Θ-temperature where Θ-conditions are not anymore experimentally accessible.

- The PNIPAM ($M_o = 113$ g/mol) dissolved in water exhibits LCST behavior. The theta temperature is $\Theta = 30.6$ °C (Wang, 1999). Temperature dependence of interaction parameter χ can be estimated after Equation (14) with $B = 3980$ K after data published in reference (Afroze et al., 2000). Aqueous solutions of PNIPAM exhibit LCST behavior as Figure 5 shows. This is reflected in Figure 7 by the opposite variation of solvent's quality with temperature as compared to Figure 6. Water is a good solvent for $T < \Theta$ and a poor solvent for $T > \Theta$. Here, LCST shifts to higher temperature after addition of nanoparticles. The collapse of the coils is retarded by the nanoparticles.

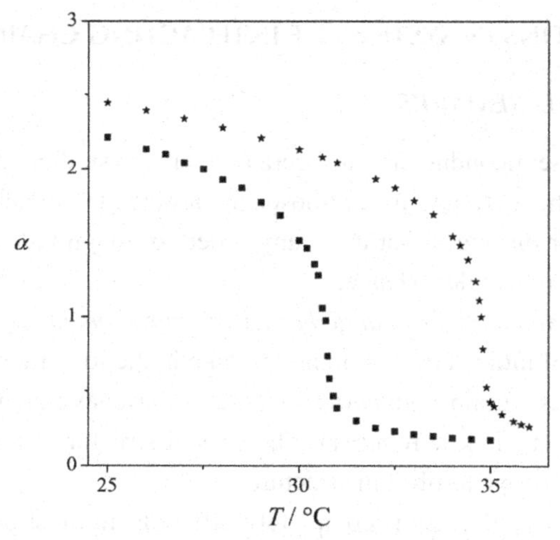

FIGURE 7 Expansion factor α *versus* temperature for PNIPAM, $N = 3 \times 10^4$, in water calculated after Equation (13), $n_p = 0$ – solid squares, and $n_p/N = 8 \times 10^{-4}$ – stars

In general, the transition becomes sharper with increasing N (see Figure 2). It shifts to lower (UCST) or higher temperatures (LCST) with ascending ratio n_p/N. The transition may disappear at all for sufficiently high concentration of nanoparticles. Figure 8 illustrates the latter point for an aqueous dispersion of PNIPAM with addition of Fe_2O_3. One may recognize qualitative agreement with Figure 7.

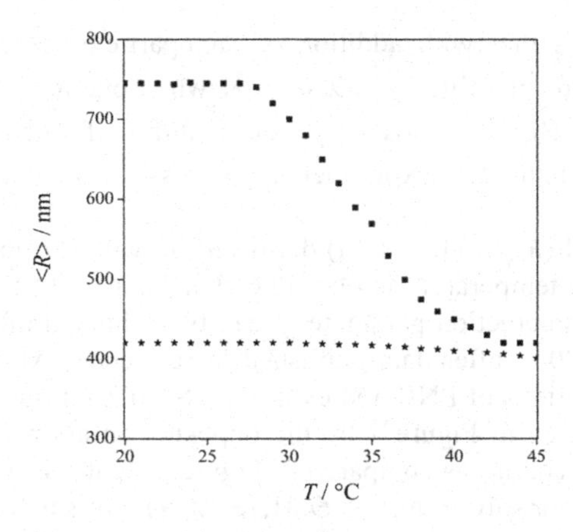

FIGURE 8 Radius of gyration for PNIPAM microgels in aqueous solution (solid squares) and after addition of mass fraction 0.4 of Fe_2O_3 nanoparticles, data after Rubio-Retama et al. (2007).

5.2 TRANSITIONS IN SYSTEMS OF INTERACTING CHAINS

5.2.1 GENERAL REMARKS

Most of metals, semiconductors, and ceramics are crystalline. Atoms and molecules of which the materials are composed are arranged with nearly perfect long ranged order over distances that are many orders of magnitude greater than the distance between molecules. Hence,

crystals are coined by long ranged order of their elements.

The fraction of molecules that do not partake in the long ranged order is very small (Nevertheless, grain boundaries or defects may have great effect on bulk properties despite their low number). The crystal structures developing in low molecular systems are equilibrium structures.

The situation in polymers is completely different. In most polymer systems, the degree of molecular order is somewhere in between the full positional order of a single crystal and the complete positional disorder of a liquid or a glass. In polymers, crystalline domains coexist with amorphous regions. Therefore, a semi-crystalline polymer is in a non-equilibrium situation.

Complete positional order $\Leftarrow$ polymer $\Rightarrow$ complete positional *disorder* (single crystal) (liquid)

5.2.2 GLASS TRANSITION

The melt solidifies to a glass when crystallization is suppressed since the rate of crystallization is smaller than the externally imposed rate of cooling. Hence, glass is a frozen in liquid. Phenomenological, the transition to the glassy state is characterized by changes of certain physical properties during isobaric cooling of a melt. Prominent examples for changes of properties are:

- Coefficient of thermal expansion
- Specific heat

These quantities are used to determine the glass transition. A schematic comparison between a first order phase transition and the glass transition is depicted in Figure 9. Figure 10 brings an example for determination of glass transition temperature T_g and melting temperature T_m from isobaric *V-T* data.

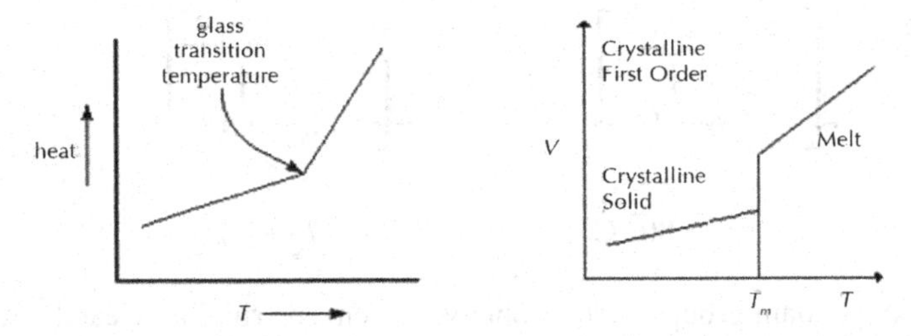

FIGURE 9 Schematic comparison of glass transition and melt-crystal transition.

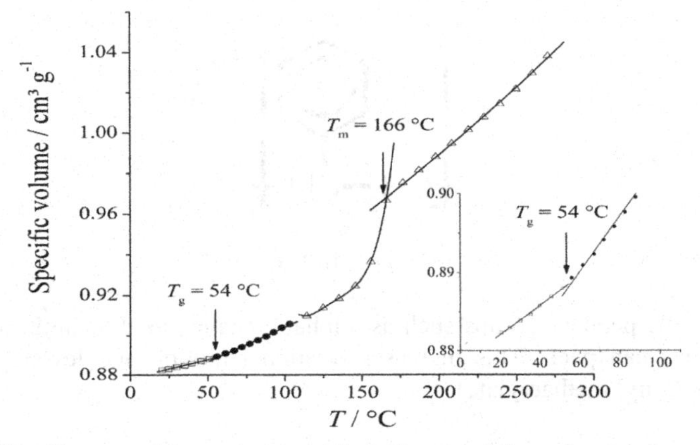

FIGURE 10 Glass transition temperature from change of thermal expansion coefficient in isobaric *V-T* diagram for random copolymer poly(ethylene-*co*-vinyl alcohol) with 44 mol% of ethylene, data after Funke et al. (2007).

What Properties Influence Glass Transition Temperature T_g of Polymers?

- Reduced flexibility of chains causes higher T_g of poly(ethylene terephthalate) (PET) as compared to poly(ethylene adipate).

$$\left[-O-CH_2CH_2-O-\overset{O}{\overset{\parallel}{C}}-C_6H_4-\overset{O}{\overset{\parallel}{C}}-\right]_n \qquad \left[-O-CH_2CH_2-O-\overset{O}{\overset{\parallel}{C}}-CH_2CH_2CH_2CH_2-\overset{O}{\overset{\parallel}{C}}-\right]_n$$

$T_g = 70\ °C$ $\qquad$ $T_g = -70\ °C$

- stronger intermolecular forces lead to higher T_g of atactic poly(vinyl chloride) with dipole-dipole interaction of C–Cl bonds as atactic polypropylene.

$$\left[-CH_2-\underset{}{\overset{Cl}{\overset{|}{CH}}}-\right]_n \qquad \left[-CH_2-\overset{CH_3}{\overset{|}{CH}}-\right]_n$$

$T_g = 80\ °C$ $\qquad$ $T_g = -20\ °C$

- bulky pendant groups such as phenyl or benzene ring, may catch on neighboring chains like a fish hook and restrict rotational freedom. This increases T_g of atactic PS as compared to polypropylene.

$$\left[-CH_2-\overset{C_6H_5}{\overset{|}{CH}}-\right]_n$$

$T_g = 100\ °C$

- flexible pendant groups such as aliphatic chains, tend to limit how close chains can pack. This increases rotational motion and lowers T_g as in poly(butyl methacrylate).

$$\left[-CH_2 - \underset{COO(CH_2)_3CH_3}{\overset{CH_3}{C}} - \right]_n$$

$T_g = 20\ °C$

In Figure 11, we illustrate schematically glass transition in an isobaric volume temperature diagram, specific volume $V = V(T,P)$. The volume does not display a jump at the glass transition, as it does during melting (see right diagram of Figure 9). We may state:

Glass transition depends on cooling rate, the slower the cooling rate, the lower the temperature where thermal expansion coefficients α_{glass} and $\alpha_{crystal}$ (the slope) changes.

The coefficient of expansion is not very different in glassy and crystalline state.

In the glassy state, the long ranged order of crystals does not occur-glasses are frozen-in liquids. The transition temperature depends on cooling rate.

Glass is a non-equilibrium situation with very low relaxation rates towards equilibrium.

The relationship between relaxation time constant τ of segments and characteristic time t imposed by cooling rate, $t = \frac{\Delta T}{T}$, governs which glassy state develops. We may summarize it as follows:

$\tau << t$ melt in equilibrium, segments can follow any temperature change

$\tau \approx t$ glass transition

$\tau >> t$ glass in non-equilibrium, segments cannot follow temperature change

5.2.3 NON-EQUILIBRIUM PHENOMENA IN GLASSES

In thermodynamic sense, glass transition is characterized by:

$$dV_l(T_g,P) = dV_g(T_g,P,\zeta) \tag{15}$$

where ζ symbolizes an internal variable (see Chapter 7, Polymers in Non-equilibrium). Specific volume in the glassy state is also characterized, apart from T and P, by an internal variable reflecting the non-equilibrium situation. As Figure 11 shows only the slope displays a jump not the volume itself. Therefore, we ignore the index on V. Thermal expansion coefficient and compressibility of the glass are defined as follows:

$$\alpha_\zeta = \frac{1}{V}\left(\frac{\partial V}{\partial T}\right)_{P,\zeta} \qquad \kappa_\zeta = -\frac{1}{V}\left(\frac{\partial V}{\partial P}\right)_{T,\zeta}$$

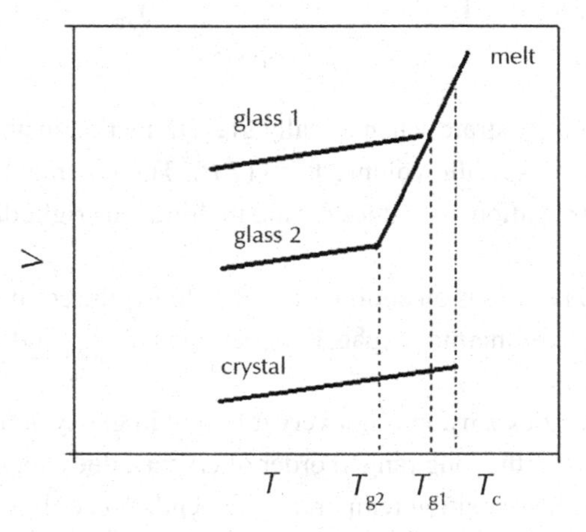

FIGURE 11 Glass transition in *V-T* diagram, a high cooling rate leads to glass 1 and a lower to glass 2.

For the liquid state, these quantities are defined in the usual way. Equation (15) yields:

$$V\alpha_l \mathrm{d}T_g - V\kappa_l \mathrm{d}P = V\alpha_\zeta \mathrm{d}T_g - V\kappa_\zeta \mathrm{d}P + \left(\frac{\partial V}{\partial \zeta}\right)_{P,T} \mathrm{d}\zeta \quad (16)$$

Neglecting the last term on the right-hand side leads to the famous Ehrenfest equation for the pressure dependence of transition temperature in a second order transition at equilibrium.

$$\frac{\mathrm{d}T_{trans}}{\mathrm{d}P} = \frac{\Delta\kappa}{\Delta\alpha} \text{ with } \Delta\kappa \equiv \kappa_l - \kappa_\zeta \text{ (analogously for } \Delta\alpha) \quad (17)$$

Applying Equation (16) results in the following pressure dependence of glass transition temperature.

$$\frac{\mathrm{d}T_{\mathrm{g}}}{\mathrm{d}P} = \frac{\Delta\kappa_{\zeta} - \frac{1}{V}\left(\frac{\partial V}{\partial \zeta}\right)\frac{\mathrm{d}\zeta}{\mathrm{d}P}}{\Delta\alpha_{\zeta}} \quad \text{with } \Delta\kappa_{\zeta} \equiv \kappa_{1} - \kappa_{\zeta} \tag{18}$$

We test these equations for glass transition using *PVT* data found for polymers. Two different pathways were traversed in a melt-glass system, one at normal pressure, the other one at high pressure. We sketch them schematically in the *V-T* plane (see Figure 12). The accompanying change in volume is given after Equation (18) by:

$$\frac{\mathrm{d}\Delta v}{v} = \Delta\alpha\left[\frac{\Delta\kappa_{\zeta}}{\Delta\alpha_{\zeta}} - \frac{\mathrm{d}T_{g}}{\mathrm{d}P}\right]\mathrm{d}P \tag{19}$$

where v represents the constant initial volume. We consider (19) along the two pathways indicated in Figure 12. It follows:

$$\frac{|\Delta v|_{\mathrm{II}}}{v} - \frac{|\Delta v|_{\mathrm{I}}}{v} = \Delta\alpha\left(P_{2} - P_{1}\right)\left[\frac{\Delta\kappa_{\zeta}}{\Delta\alpha_{\zeta}} - \frac{\mathrm{d}T_{\mathrm{g}}}{\mathrm{d}P}\right] \tag{20}$$

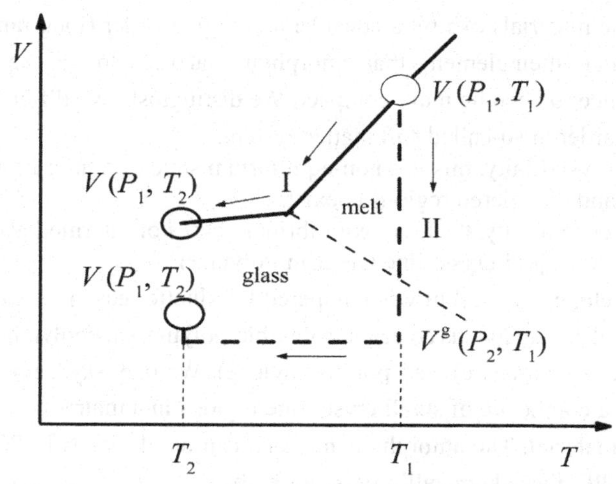

FIGURE 12 Cooling down a melt to a glass along different pathways, I) cooling at normal pressure and II) cooling at enhanced pressure and adjusting the final volume at the same pressure and temperature as in I).

If the glass transition would be a true 2nd order phase transition, we would find zero for the volume difference $\Delta v_{\mathrm{I}} = v_{\mathrm{II}}$, in agreement with Ehrenfest Equation (17). Precise measurements by Rehage and Breuer (Rehage and Breuer, 1967a; Rehage and Breuer, 1967b) yielded for atactic PS.

$$\frac{\Delta\kappa_{\zeta}}{\Delta\alpha_{\zeta}} = 0.5\frac{\mathrm{K}}{\mathrm{MPa}} \quad \text{and} \quad \frac{\mathrm{d}T_{\mathrm{g}}}{\mathrm{d}P} = 0.25\frac{\mathrm{K}}{\mathrm{MPa}} \tag{21}$$

$$\Rightarrow |\Delta v|_{\mathrm{II}} > |\Delta v|_{\mathrm{I}}$$

A glass formed at high pressure is denser as a glass formed under normal pressure. The glass transition is not a genuine phase transition of second order after Ehrenfest but a relaxation transition.

5.2.4 CRYSTALLIZATION OF POLYMERS

Some Basics

We focus solely on some basics. The Crystallization and Melting of Polymers are extensively discussed in chapter 6 (see chapter 6, Crystallization and Melting of Polymers).

Crystalline materials exhibit a considerably higher order (long ranged order) in arrangement of their elements than amorphous materials do. This causes a crystalline substance to become more compact. We distinguish two different types of intermediate order in so-called soft matter systems.

- Partial crystallinity, this is a non-equilibrium state of matter in which crystalline and disordered regions coexist.
- Liquid crystallinity, these are equilibrium phases of intermediate order (see chapter 8, Liquid-crystalline Order in Polymer).

Fully developed crystallinity is hampered by kinetic reasons. Partial crystallinity is typical for many stereo regular flexible polymers as polyethylene (PE), poly(ethylene terephthalate), and poly(propylene). We may say, semi-crystalline polymers are a composite of small crystalline regions in a matrix of dramatically less ordered material. The amorphous material can be glassy as in PET or more rubbery as in PE. Partial crystallinity is caused by

- Slow kinetics, even quite slow cooling rates lead to glass (disentanglement does not constitute a barrier for crystallization).

- Kinetics is also ruled by stereo-regularity. When a chain has many branches, it makes it more difficult to pack the chains into regular crystals.

Low-density PE, which is branched, is less crystalline than high-density PE, which is strictly linear. A polymer comprising irregular chain structure does not crystallize.

Mutual positioning of structural elements is affected by

- Chemical architecture, length and flexibility of the chain
- Relaxation behavior of the macromolecules.

Hence, types of crystalline structures and the laws of their appearance are ruled by internal parameters and external conditions of crystallization.

Crystallization means establishment of order of elements. For chain molecules, we have to specify what element means. It could be segments of the chain as well as the whole chain. Therefore, three types of order, we have to distinguish:

1. order of both segments and chains
2. order of segments only
3. order of chains only

All of these ordered arrangements, we call crystalline. The last arrangement, order of chains only, is observed in colloidal solutions of chain molecules in a poor solvent. Chains are collapsed and exert attractions to each other that may lead to colloidal crystals. It is order of chains but not of segments. These phenomena are outside the scope of the following discussion.

Different types of crystalline structures develop: Fibrillar or extended-chain crystals and lamellar crystallites. Development of the first mentioned structures is an equilibrium process. We do not have an amorphous phase since both segments and chains are ordered. Extended-chain crystals develop only in polymers of low molecular mass ($M_w \leq 10$ kg/mol) and in the range of low undercoolings, that is $\Delta T_c = T_m - T_c$ is small.

In most cases, we observe formation of lamellae. This represents a more modest arrangement as compared to fibrillar crystallization (Lauritzen Jr. and Hoffman, 1960; Frank and Tosi, 1961, Keller, 1957; Keller, 1968; Strobl, 2000). Only parts of chains assume an ordered arrangement. In other words, only segments have long ranged order. Chains fold back and forth on themselves to form crystalline platelets or lamellae with a thickness in chain direction between 80 and 300 Å and lateral dimensions extending up to several microns. These lamellae grow usually at large undercoolings. We recognize that they have a highly disordered fold surface with a remarkable surface tension that is lamellae are non-equilibrium structures. Formation of lamellae shows directly semi-crystallinity of polymers or coexistence of crystalline domains surrounded by amorphous regions. The crystallinity is always less than 100 %. An example for variation of crystallinity with composition in a polymer blend is shown in Figure 13. Degree of crystallinity is defined by $X = \Delta H/\Delta H_{ref}$ where ΔH_{ref} represents the melting enthalpy of 100 % crystalline material. The two constituents, poly(3-hydroxy butyrate-*co*-3-hydroxy valerate) (PHBV), comprising 12 mol% valerate, and poly(ethylene oxide) (PEO), are miscible in the molten state and crystallize separately at different temperatures. The PHBV crystallizes at a temperature (at around 100 °C) where PEO is unable to do so. The PEO starts crystallizing at around 50 °C. Thus, the two miscible constituents in the melt are unable to form mixed crystals. Figure 13 nicely demonstrates that crystallization of PHBV is not influenced by molten PEO. On the other hand, PEO was crystallized when PHBV crystallites existed (at T_c = 44 °C). One recognizes that crystallization of PEO is hampered by existence of crystalline PHBV.

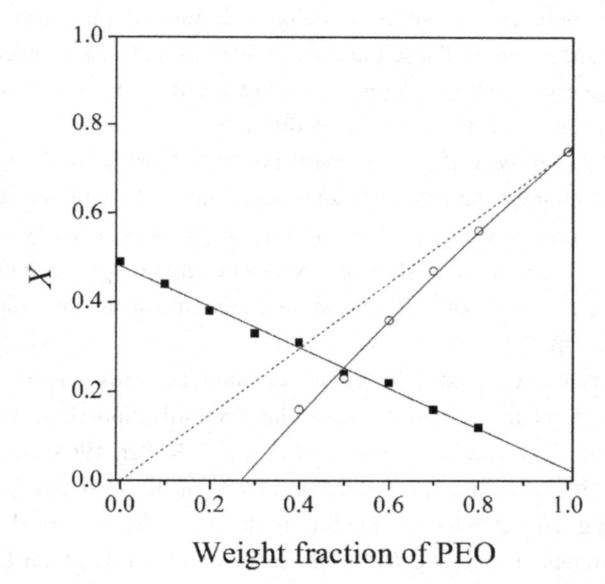

FIGURE 13 Degree of crystallinity in blends of PHBV and PEO, solid squares for crystallinity of PHBV, crystallized at 104 °C, and open circles for PEO, crystallized at 44 °C, data after Tan et al. (2006).

HIERARCHY OF STRUCTURES

There are different structures of different length scales. The basic unit is the chain folded lamella (see Figure 14). It turns out that lamellar thickness l is:

- independent of molecular mass.
- function of supercooling when the crystal was formed, $l \propto (\Delta T_c)^{-1}$.

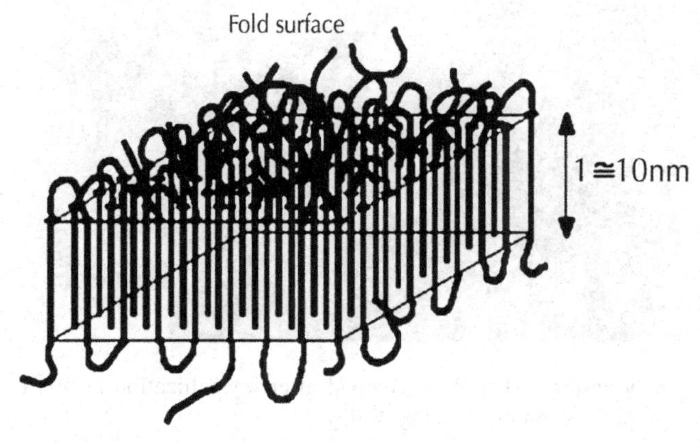

FIGURE 14 The basic unit of semi-crystalline polymers is the chain folded lamella of thickness l (in the order of 10 nm) and a fold surface having surface tension σ against the melt.

Lamellar thickness is in the order of 10 nm. Lamellae are separated by amorphous regions. Individual chains may be involved in more than one lamella as well as in the amorphous region in between. The chain folded lamellae are themselves organized in larger scale structures, spherulites. These structures, which may be many microns in space, consist of sheaves of individual lamellae, which grow out from a central nucleus until finally the space is filled by these structures separated by amorphous regions. Figure 15 presents a spherulite grown in PHBV after crystallization for 24 hr at 50 °C that is undercooling $\Delta T_c = 140$ K (Chan and Kammer, 2009).

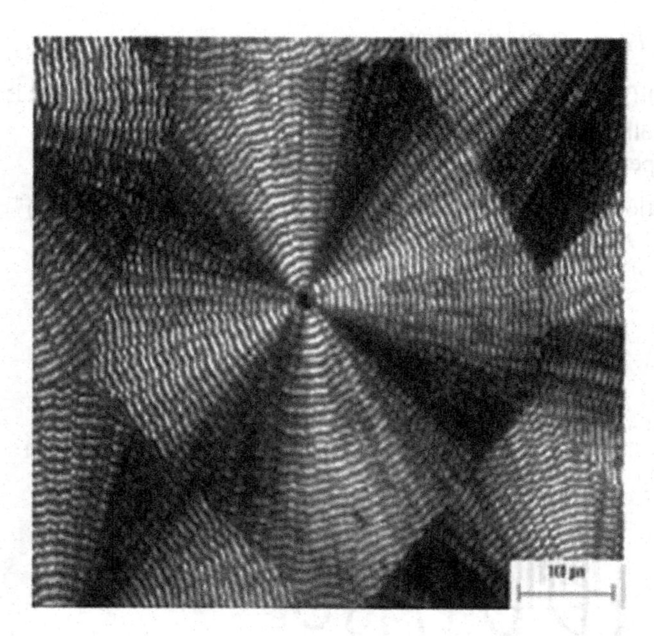

FIGURE 15 Spherulites of PHBV developed after crystallization at 50 °C for 1 day (Chan et al. 2004, permission provided by Wiley).

5.2.5 LIQUID-LIQUID PHASE SEPARATION IN POLYMER BLENDS

The phase behavior of polymer blends that is mixtures of two chemically different polymers is experimentally well accessible in a window which is bounded at high temperatures by thermal decomposition temperature, T_d, of the polymer components and at low temperatures by the glass transition temperature, T_g, of the system. This is shown schematically, in Figure 16. When LCST and UCST merge, we have immiscibility or heterogeneity of the blend (see Figure 17).

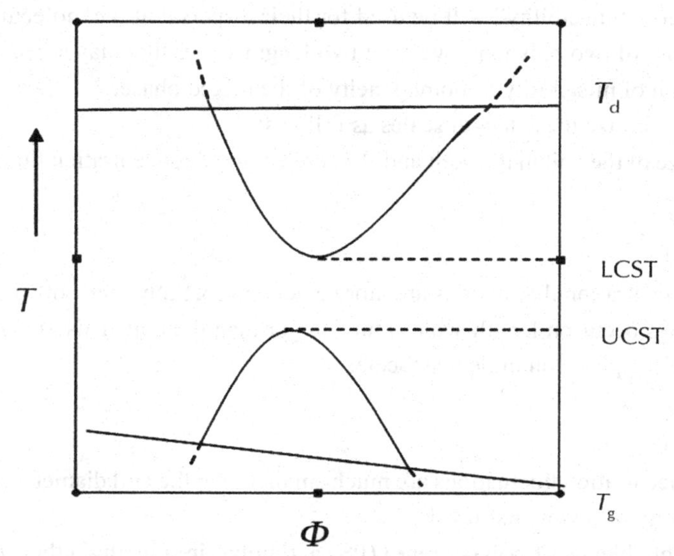

FIGURE 16 General phase diagram of polymer blends, phase behavior can be monitored in the window between T_g and T_d, areas enclosed by the curves are regions of phase instability.

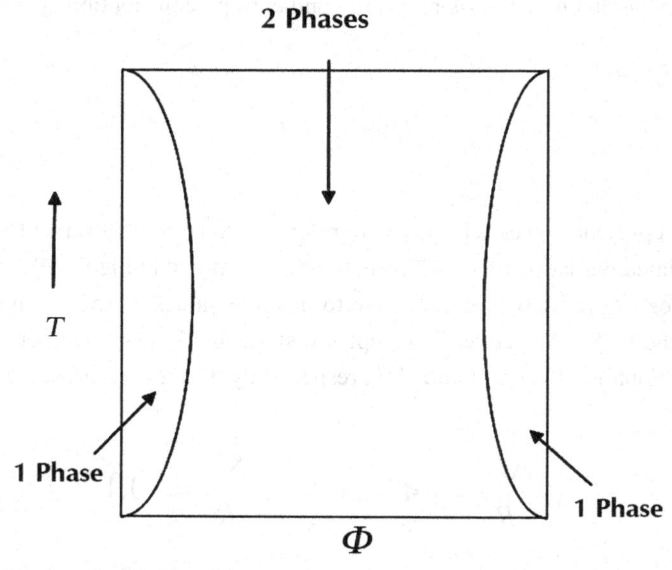

FIGURE 17 Immiscible blends have only marginal regions of miscibility.

The term "miscibility" will be used for their dispersal at the molecular level. In a mixture of two polymers, we have two length scales that may serve for characterization of miscibility or homogeneity of the mixed phase.

We symbolize these length scales as follows:

R_o- size of the coil in the melt and Λ-wavelength of concentration fluctuations.

I. $\frac{R_o}{\Lambda} << 1$

—concentration fluctuations are more extended spatially than coil size or heterogeneities occur on length scales much larger than the coil. This means homogeneity of the phase on molecular scale.

II. $\frac{R_o}{\Lambda} >> 1$

—concentration fluctuations are much smaller than the coil diameter. It means homogeneity on segmental level.

Miscible blends of poly(styrene) (PS) and poly(vinyl methyl ether) (PVME) show LCST behavior as presented in Figure 18. Phase separation occurs above 152 °C. This liquid-liquid phase separation, we may discuss in terms of Flory-Huggins parameter given as a function of temperature. We see parameter χ as a free-energy parameter comprising energy and entropy contribution, $\chi = \chi_U + \chi_S$.

$$\chi T = \frac{U}{R} - T\frac{S^E}{R} \tag{22}$$

where S^E represents excess entropy with reference to combinatorial entropy. One observes linear variation of χT with temperature as shown in Figure 19. The quantity χT crosses precisely from negative to positive values at 152 °C marking in that way the LCST. Moreover, intercept and slope of χT *versus* temperature provide after Equation (22) U/R and S^E/R, respectively. From Figure 19, we get:

$$\frac{\Delta U}{R} = -43K \qquad \frac{S^E}{R} = -0.1 \tag{23}$$

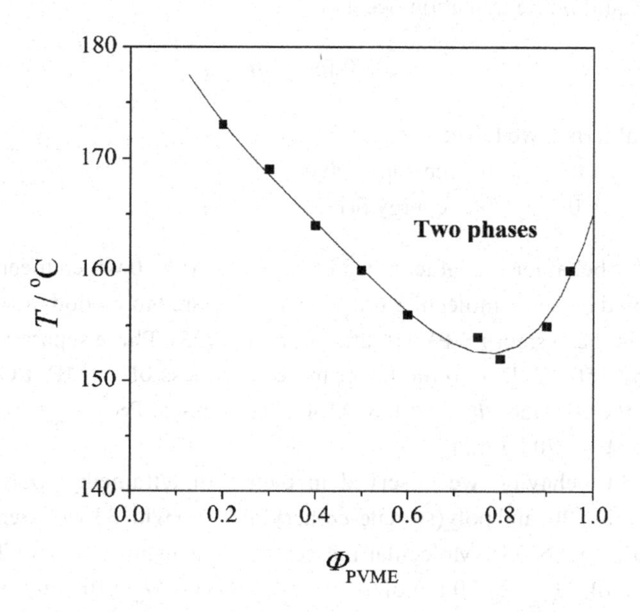

FIGURE 18 The phase diagram of PS/PVME blend (data after Troeger et al., 1992).

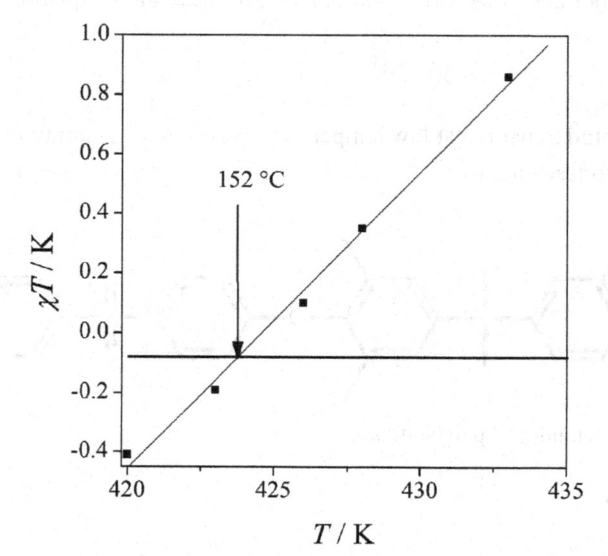

FIGURE 19 Interaction parameter χT *versus* temperature for PS/PVME 20/80 blends.

Liquid-liquid phase transition occurs at:

$$\chi = 0 \text{ for } N \rightarrow \infty \tag{24}$$

In general terms, we have:

$$\begin{array}{llll} \text{LCST} & \chi_U < 0 & \chi_S > 0 & \text{entropy driven} \\ \text{UCST} & \chi_U > 0 & \chi_S < 0 & \text{energy driven} \end{array} \tag{25}$$

The LCST behavior is characterized by $\chi_S > 0$ or $\Delta S^E < 0$ which means mixing is accompanied by some molecular ordering as compared to random mixing. This can be seen in the system PS/PVME after Equation (23). Phase separation occurs above around 150 °C. Due to the lower molecular mass of PVME, LCST shifts towards the PVME side. $\Phi_{PVME} = 0.8$ (Molecular masses: PS – $M_w = 500$ kg/mol and PVME – $M_w = 70$ kg/mol).

The UCST behavior, we observed in blends of tetramethyl poly(sulfone) (TMPSF), Figure 20, and poly(styrene-*co*-acrylonitrile) with 33 wt% acrylonitrile (symbolized by SAN-33). Molecular masses of the constituents were TMPSF – $M_w = 2.5$ kg/mol, $M_{mono} = 470$ g/mol $\rightarrow N \approx 5$ SAN-33 – $M_w = 100$kg/mol, $M_{mono} =$ 79 g/mol$\rightarrow N \approx 1270$. Figure 21 shows that TMPSF behaves like a low molecular solvent for SAN, actually a poor solvent. We observe the typical phase separation into a SAN-poor and a SAN-rich phase. Here, we have after Equation (25)

$$\chi_U > 0 \qquad \chi_S < 0$$

Liquid-liquid transition at low temperature is driven by an unfavorable energy contribution to free energy.

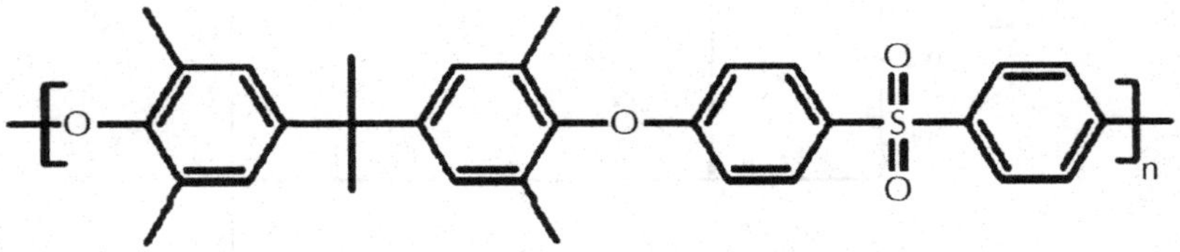

FIGURE 20 Tetramethyl poly(sulfone).

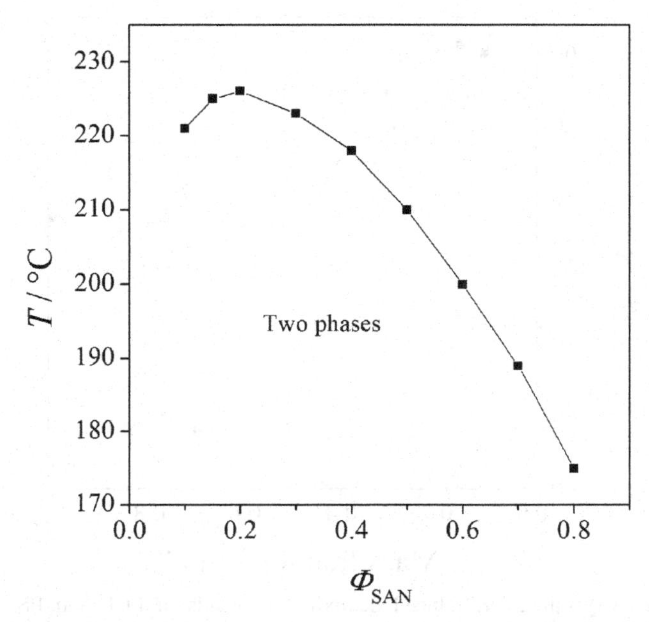

FIGURE 21 The phase diagram of TMPSF and SAN-33, data after Pfefferkorn et al. (2012).

SUPERPOSITION OF LIQUID-LIQUID PHASE SEPARATION AND CRYSTALLIZATION

The studies on superposition of phase transitions in polymers are rare. Most studies are devoted to competition of liquid-liquid phase transition and crystallization. We may distinguish:

1. Crystallization-induced demixing.
2. Demixing-induced crystallization.

The first situation is observed in blends of low molecular PS and poly(ε-caprolactone) (PCL). Figure 22 gives the phase diagram. The glass transition temperature varies between 19 °C (PS) and –69 °C (PCL). A UCST is observed at 166 °C and the melting temperature of PCL at 67 °C. Only tiny melting point depression occurs. In the range 0.6 $W_{PCL} \leq 1$ below melting temperature, it coexists crystalline PCL and an amorphous phase of varying composition but rich in PCL. For $W_{PCL} \leq 0.6$ and above T_m, we observe coexistence of SAN-rich and SAN-poor phases.

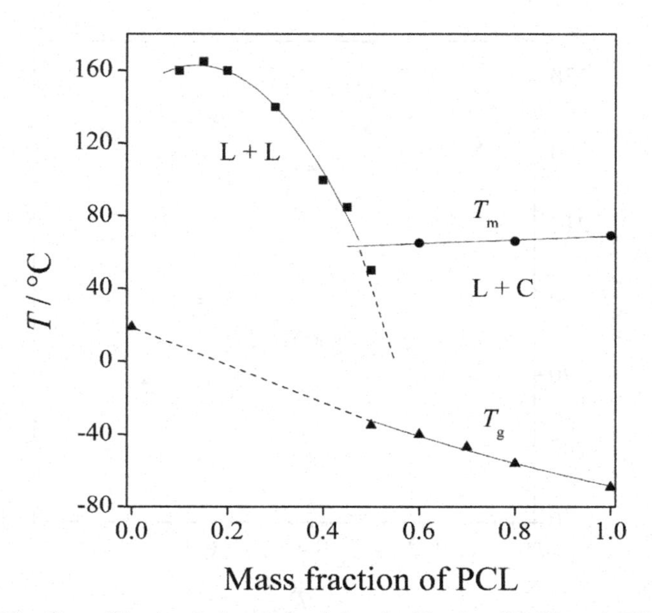

FIGURE 22 Crystallization-induced demixing in blends of PCL and PS, data after Pfefferkorn (2008).

Demixing-induced crystallization, we observed in PCL (M_w = 40.4 kg/mol) mixed with SAN-26.4 (M_w = 168 kg/mol). An LCST occurred at 165 °C, Figure 23. Glass transition temperature of SAN was observed at 110 °C. Equilibrium melting temperature of PCL was found at 83 °C exhibiting tiny melting point depression only. In addition, we observed non-equilibrium UCST at around 35 °C. Polarized light scattering revealed that concentration fluctuations causing development of two isotropic phases precede the evolution of optical anisotropy. The V_v scattering intensity started to increase dramatically indicating liquid-liquid phase separation whereas the H_v scattering profile did not change. It means no anisotropy develops. V_v changed from 0 to 85 s significantly whereas H_v was constant equal zero until 100 s , afterwards it increased. In other words, before crystallization starts, liquid-liquid phase separation occurs, generating PCL-rich and PCL-poor phase. Then, PCL crystallizes below UCST in the two-phase region. One recognizes that this UCST is a transient non-equilibrium situation.

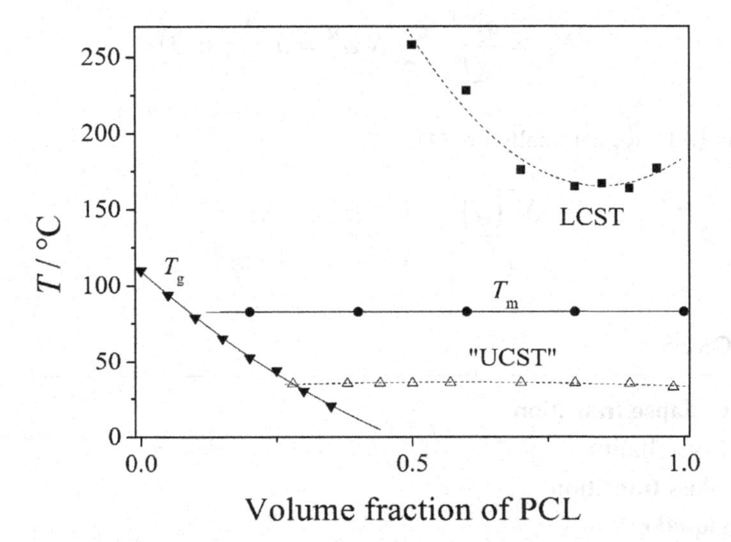

FIGURE 23 Demixing-induced crystallization in PCL/SAN blends, data after Svoboda et al. (1994).

APPENDIX

DERIVATION OF EQUATION (4)

The partition function for the chain molecule, consisting of freely rotating but distinguishable segments N, is in contrast to Equation (1).

$$Q = \sum_{N_1=0}^{N-1} q_1^{N_1} q_2^{N-1-N_1} = q_2^{N-1} \sum_{N_1=0}^{N-1} a^{N_1} = q_2^{N-1} \frac{1-a^N}{1-a} \quad \text{(A1)}$$

The ratio of the segment partition functions has been denoted by a, $a \equiv q_1/q_2$. Please note, the free energy $F = -k_B T \ln Q$ is proportional to N after Equation (2) for the system of N independent particles. This is so since free energy is an extensive quantity. The free energy after Equation (A1) cannot be an extensive quantity with respect to degree of polymerization because the segments are not independent of each other. For the average segment number in compartment 1, it follows:

$$\langle N_1 \rangle = \frac{q_2^{N-1}}{Q} \sum_{N_1=0}^{N-1} N_1 a^N = a \frac{\partial}{\partial a}(\ln Q)$$

With (A1), we get finally Eq. (4).

$$\frac{\overline{N_1}(a)}{N} = \frac{1}{N} \frac{a}{1-a} - \frac{a^N}{1-a^N}$$

KEYWORDS

- **Collapse transition**
- **Crystallinity**
- **Glass transition**
- **Liquid crystals**
- **Liquid-liquid phase separation in blends**

REFERENCES

1. Afroze, F., Nies, E., and Berghmans, H. *J. Mol. Structure*, **554**, 55 (2000).
2. Chan, C. H., Ismail, J., and Kammer, H. W. *Polymer Degradation and Stability*, **85**, 947 (2004).
3. Chan, C. H. and Kammer, H. W. *Polymer Bull.*, **63**, 673 (2009).
4. Di Marzio, E. A. *Progr. Polym. Sci.*, **24**, 329–377 (1999).
5. Elias, H. G. *Makromoleküle;* Hüthig & Wepf: Basel, p. 202 (1975).
6. Flory, J. P. *J. Chem. Phys.*, **17**, 303 (1949).
7. Frank, F. C. and Tosi, M. *Proc. Roy. Soc.* London, **A263**, 323 (1961).
8. Funke, Z., Hotani, Y., Ougizawa, T., Kressler, J., and Kammer, H. W. *Europ. Polym. J.*, **43**, 2371 (2007).
9. Keller, A. *Phil. Mag.*, **2**, 1171 (1957).
10. Keller, A. *Rep. Progr. Phys.*, **31**(2), 623 (1968).
11. Lauritzen Jr., J. I. and Hoffman, J. D. *J. Res. Natl. Bur. Stand.*, **64A**, 73 (1960).
12. Pfefferkorn, D., Browarzik, D., Steininger, H., Weber, M., Gibon, C., Kammer, H. W., and Kressler, J. *Europ. Polym. J.*, **48**, 200 (2012).
13. Pfefferkorn, D. Master Thesis, Martin-Luther-University Halle, Germany (2008).
14. Rehage, G. and Breuer, H. *J. Polym. Sci.*, **C16**, 2299 (1967a).
15. Rehage, G. and Breuer, H. *In the glassy state of polymers*; Springer: Berlin (1967b).
16. Rubio-Retama, J., Zafeiropoulus, N. E., Serafinelli, C., Roja-Reyna, R., Voit, B., Cabarcos, E. L., and Stamm, M. *Langmuir*, **23**, 10280 (2007).
17. Strobl, G. *Eur. Phys. J.*, **E3**, 165 (2000).

18. Svoboda, P., Kressler, J., Chiba, T., Inoue, T., and Kammer, H. W. *Macromolecules*, **27**, 1154 (1994).
19. Swislow, G., Sun, S. T., Nishio, I., and Tanaka, T. *Phys. Rev. Lett.*, **44**, 796 (1980).
20. Tan, S. M., Ismail, J., Kummerlowe, C. and Kammer, H. W. *J. Appl. Poly. Sci.*, **101**(5), 2776 (2006).
21. Troeger, J. and Kammer, H. W. *Acta Polymerica*, **43**, 331 (1992).
22. Wang, X. H. and Wu, C. *Macromolecules*, **32**, 4299 (1999).

CHAPTER 6

CRYSTALLIZATION AND MELTING OF POLYMERS

HANS-WERNER KAMMER

CONTENTS

6.1 CRYSTALLIZATION

Is crystallization from the melt controlled by entanglement? Chain molecules are entangled in the melt above the entanglement molecular mass M_e. For poly(ethylene) this molecular mass amounts to $M_e = 3800$ g/mol. One may estimate the relaxation time for disentanglement by estimating the time a chain needs diffusing across the distance corresponding to its end-to-end distance. The chain may have disentangled itself after traversing this distance. It is:

$$\text{Rate of disentanglement} = \frac{D}{\langle R_o^2 \rangle} = \frac{10^{-14}\ \text{cm}^2/\text{s}}{100\ \text{nm}^2} = O\left(10^{-2}\,\text{s}^{-1}\right)$$

The rate of crystallization for example of poly(ethylene oxide) (PEO) or poly(ε-caprolactone) (PCL) was found at undercooling $\Delta T_c = 30$ K:

Rate of crystallization = $O\ (10^{-3}\ \text{s}^{-1})$

Comparison shows:

$$R_{\text{disentanglement}} > R_{\text{crystallization}}$$

Hence, entanglement does not constitute a barrier for crystallization. The unmixing of crystallizable and non-crystallizable parts of the chains proceeds without entanglement barrier. Experimental evidence for disentanglement during crystallization is given by:

- Fractionation of chains during crystallization in samples with broad molecular mass distribution
- Phase separation during crystallization in miscible amorphous/crystalline blends. In the melt, different species are entangled. After crystallization, we have neat crystallites of the respective constituent.

6.2 DYNAMICS OF CRYSTALLIZATION

We start the discussion with crystallization under isothermal conditions. When the temperature drops below the melting temperature, density fluctuations increase to critical size. Below critical size, fluctuations decay and above critical size, they become crystallization nuclei that are able to grow to the new phase by attaching stems of adjacent chains (so-called nucleation and growth mechanism). The dynamics of crystallization is governed by the dynamic law (see Chapter 7, Equation (9)):

Flow ∝ thermodynamic driving force

$$\frac{dX}{dt} = -L\left(\frac{\partial g}{\partial X}\right) \tag{1}$$

where X represents the degree of crystallinity. The most general solution of Equation (1) is given by the stretched exponential. The amorphous part descends as follows:

$$1 - X = \exp\left[-Kt^{n}\right] \tag{2}$$

The Equation (2) is frequently called Avrami equation (Avrami, 1939). The rate of crystallization can be expressed by reciprocal half time.

$$t_{0.5}^{-1} = \left(\frac{K}{\ln 2}\right)^{1/n} \tag{3}$$

From Equation (3), we learn that for $K = const$ the half time decreases or the rate increases with increasing n. For $n = const$, the rate increases with K. This is illustrated in Figure 1. Using Equation (3), we may reformulate the stretched exponential:

$$1 - X = \exp\left[-\ln 2\left(\frac{t}{t_{0.5}}\right)^{n}\right] \tag{4}$$

This function employs two parameters that might be illustrated as follows:

- $t_{0.5}$ sets the time scale, after $t = t_{0.5}$ half of the material is transformed
- Equation (4) is increasingly compressed with ascending n
- $(t_{0.5})^{n}$ provides a measure for duration of the transformation process

As stated before the greater $(t_{0.5})^{n}$, the slower the conversion process.

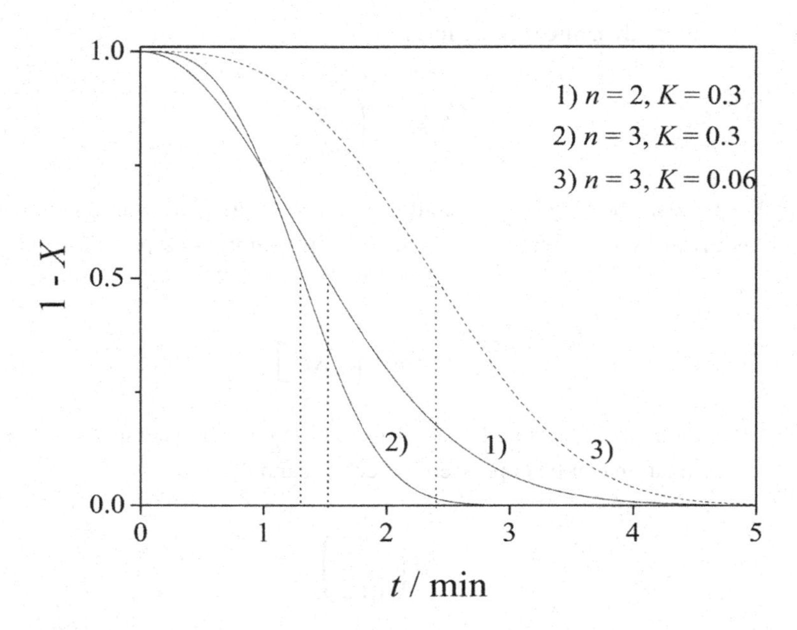

FIGURE 1 Stretched Equation (2) for $n = const$ and $K = const$.

If data fit to Equation (2), double-logarithmic plots serve to determine parameters $K(T)$ (intercept) and n (slope). Examples for Avrami plots after Equation (2) are shown in Figure 2. Deviations from Avrami occur at $X \approx 0.7$ due to impingement of spherulites. Polyhydroxybutyrate (PHB) was mixed with a low-molecular PHB. One may say, the low-molecular component acted as a solvent for high-molecular mass PHB. It crystallized at T_c = 122 °C, where the low-molecular component cannot crystallize. The rates of crystallization and Avrami parameter n are listed for different combinations of high-molecular and low-molecular mass polymer in Table 1. It is interesting to note that the rate of crystallization slows down with increasing amount of the low-molecular constituent. One may argue, addition of the low-molecular constituent causes melting point depression of PHB. Hence, the temperature of crystallization is increasingly closer to melting temperature. This causes descending rate of crystallization, which is accompanied by increasing parameter n.

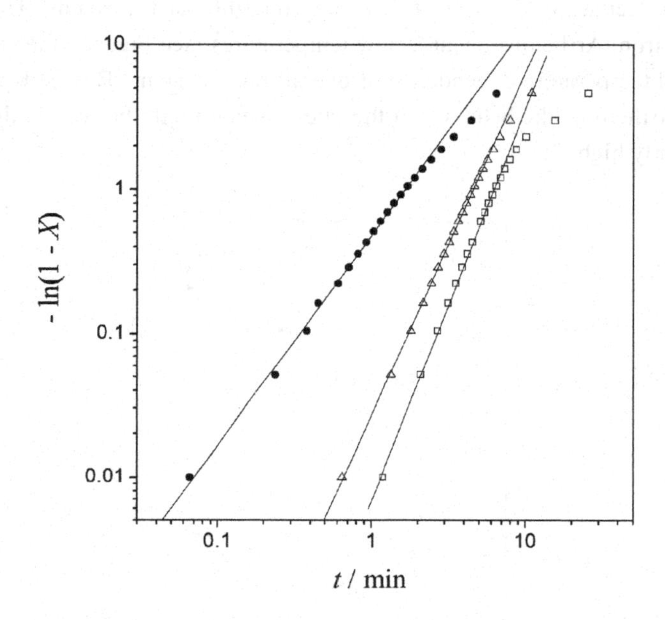

FIGURE 2 Dynamics of crystallization of PHB at T_c = 122 °C, the PHB (M_w = 350 kg/mol) was mixed with a low-molecular PHB (M_w = 4 kg/mol), which acted as a solvent, • neat PHB, Δ 90 % PHB, Δ 50 % PHB; data after Chan et al. (2004).

TABLE 1 Characteristic quantities for crystallization of PHB in solutions with low-molecular PHB, T_c = 122 °C

PHB/low-*M* PHB	$K^{1/n}$ / min^{-1}	$t_{0.5}^{-1}$ / min^{-1}	*n*
100/0	0.60	0.77	1.46
90/10	0.22	0.25	2.38
50/50	0.16	0.19	2.72

The temperature dependence of the rate constant is ruled by two processes:

1. Nucleation and growth of fluctuations
2. Transport of chains to the site of growing crystallites

These processes have different temperature dependencies. The first one, describing the birth and growth of nuclei, increases with decreasing crystallization temperature T_c. The transport, responsible for the growth of crystallites slows down with decreasing T_c due to descending mobility of the chains. Therefore, deviations from Arrhenius occur at low temperatures (see Figure 3) leading to a bell shaped temperature dependence of overall rate constant. This is depicted in Figure 4. Arrhenius-like behavior of the rate constant might be seen in the range of sufficiently high T_c.

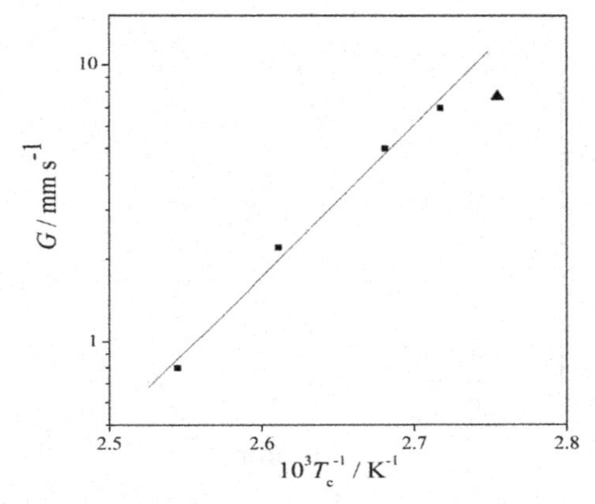

FIGURE 3 Arrhenius plot for spherulite growth rate of poly(hydroxyl butyrate–*co*–valerate) PHBV; data after Chan et al. (2011).

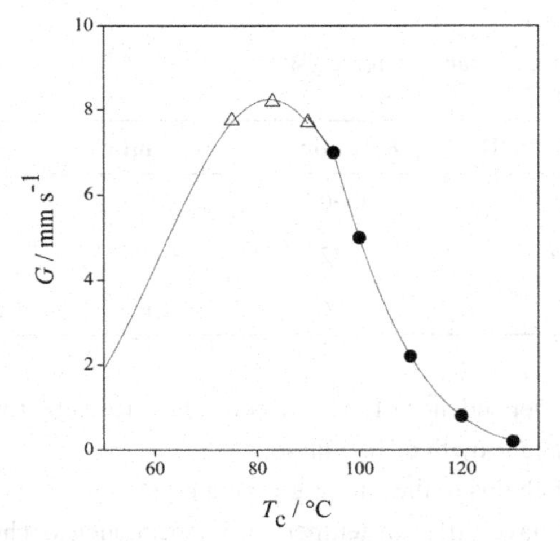

FIGURE 4 The rate of crystallization as a function of crystallization temperature for PHBV (T_g = 0 °C, T_m° = 146 °C), the solid circles correspond to Arrhenius equation, the open triangles deviate from this equation.

As mentioned in chapter 5, the basic unit during process of crystallization is the chain-folded lamella (see chapter 5 Figure 14). Growing lamellae are not equilibrium structures; the fold surface is considerably disordered and there is a substantial interfacial energy σ associated with it. We anticipate fully extended chain crystals in equilibrium. The finite thickness of lamellae is ruled by kinetic effects not governed by equilibrium thermodynamics (Chan et al., 2004; Marand et al., 1998). Thus, polymers form lamellae of well-defined thickness because crystals with this thickness grow the fastest. Where does the length scale selection come from? It is competition of two rate processes. The two rate processes are:

1. The rate of formation of a lamella with thickness l, it is proportional to $(1-\exp[\Delta g/k_B T])$
2. The rate, which generates the degree of stretching or uncoiling of a chain in the melt, $\exp(\Delta S/k_B)$

The second rate, we approximate by the $\exp(-l/l_o)$ where length l_o refers to the length when Δg disappears. Combining these two rates, we have:

$$\frac{1}{\tau}=\frac{1}{\tau_o}\left(1-e^{\Delta g/k_B T}\right)\exp\left[-\frac{l}{l_o}\right]\approx\frac{1}{\tau_o}\left(-\frac{\Delta g}{k_B T}\right)\exp\left[-\frac{l}{l_o}\right] \quad (5)$$

The quantity Δg symbolizes the free energy of formation for a lamella. It reads for a lamella with thickness l and lateral dimensions αl and βl (we set $k_B T = 1$):

$$\Delta g(l)=-\alpha\beta\frac{\Delta\mu_c}{V_c}l^3+2\alpha\beta l^2\sigma \quad (6)$$

where V_c is the molar volume of the crystallite and σ represents the interfacial tension between fold surface and amorphous surrounding. The difference in chemical potentials, $-\Delta\mu_c=\mu_c-\mu_1$, is the driving force towards crystallization. The free energy (Equation (6)) as a function of thickness l does not have a minimum. The lowest possible lamellar thickness reads:

$$l_o=\frac{2\sigma V_c}{\Delta\mu_c} \quad (7)$$

Inserting this in Equation (6) and defining $A\equiv\alpha\beta l^2$, we arrive at:

$$\Delta g(l) = \frac{2A\sigma}{k_B T}\left(1 - \frac{l}{l_o}\right) \tag{8a}$$

The crystalline phase is more stable for $T < T_m^o$. It follows from Equation (8a):

$$\Delta g < 0 \text{ if } l > \frac{2\sigma V_c}{\Delta \mu_c} \tag{8b}$$

Hence, the most stable lamella is characterized by $l \rightarrow \infty$. The opposite limit, the lowest possible lamella thickness, is given by Equation (7).

With Equation (8), we get for the rate of crystallization, Equation (5):

$$\frac{1}{\tau} = \frac{2A\sigma}{\tau_o k_B T}\left(\frac{l}{l_o} - 1\right)\exp\left[-\frac{l}{l_o}\right] \tag{9}$$

One easily verifies that we have rate = 0 when $l = l_o$ and maximal growth rate at $l_{max} = 2l_o$. The Equation (9) is illustrated in Figure 5.

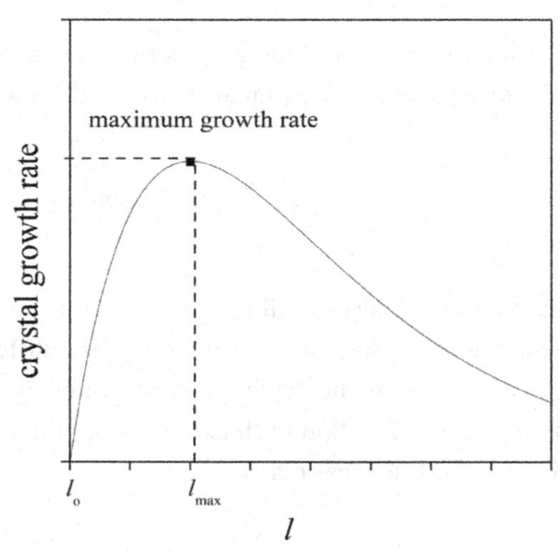

FIGURE 5 Crystal growth rate as a function of thickness l; crystals of thickness l_{max} are growing with the maximal growth rate and dominate the morphology, whereas crystals of thickness l_o are in equilibrium with the melt and do not grow.

Thermodynamic standard procedures yield for the driving force of crystallization (see Appendix 6):

$$\Delta\mu_{\mathrm{c}} = \frac{\Delta H_{\mathrm{m}}}{T_{\mathrm{m}}^{\circ}} \Delta T_{\mathrm{c}} \tag{10}$$

where ΔH_{m} and T_{m}° represent molar melting enthalpy and equilibrium melting point. $\Delta T_{\mathrm{c}} \equiv T_{\mathrm{m}}^{\circ} - T_{\mathrm{c}}$. Free enthalpy of formation of a critical nucleus of thickness L follows from Equation (6) with Equation (10):

$$\Delta g^{*} = L\alpha\beta \frac{\sigma^{2} T_{\mathrm{m}}^{\circ}}{\Delta \tilde{H}_{\mathrm{m}} \rho \Delta T_{\mathrm{c}}} \tag{11}$$

where $\Delta \tilde{H}_{\mathrm{m}}$ represents specific melting enthalpy. We may formulate an Arrhenius-like relationship with Equation (11) for the reciprocal half time representing the overall rate constant $t_{0.5}^{-1}$ of crystallization at T_{c}:

$$\frac{1}{t_{0.5}} = \frac{1}{\tau} \exp\left[-\frac{\Delta g^{*}}{k_{\mathrm{B}} T_{\mathrm{c}}}\right] \tag{12}$$

where τ is a constant that does not depend on temperature. With Equation (11) and abbreviation of all temperature independent quantities by B, we arrive at:

$$\frac{1}{t_{0.5}} = \frac{1}{\tau} \exp\left[-\frac{B T_{\mathrm{m}}^{\circ}}{T_{\mathrm{c}} \Delta T_{\mathrm{c}}}\right] \approx \frac{1}{\tau} \exp\left[-\frac{B}{\Delta T_{\mathrm{c}}}\right] \tag{13}$$

The last relationship follows with:

$$\Delta T_{\mathrm{c}} \left(1 - \frac{\Delta T_{\mathrm{c}}}{T_{\mathrm{m}}^{\circ}}\right) \approx \Delta T_{\mathrm{c}}$$

The rates of crystallization for three polymers follow Equation (13) as Figure 6 shows for PHBV, poly(ethylene terephthalate) (PET), and PEO (Chan et al., 2011). The relevant data are listed in Table 2. The comparison of Equation (11) and Equation (13) shows that $k_B B/\Delta H_m$ is proportional to the ratio of surface and bulk energy of the critical nucleus.

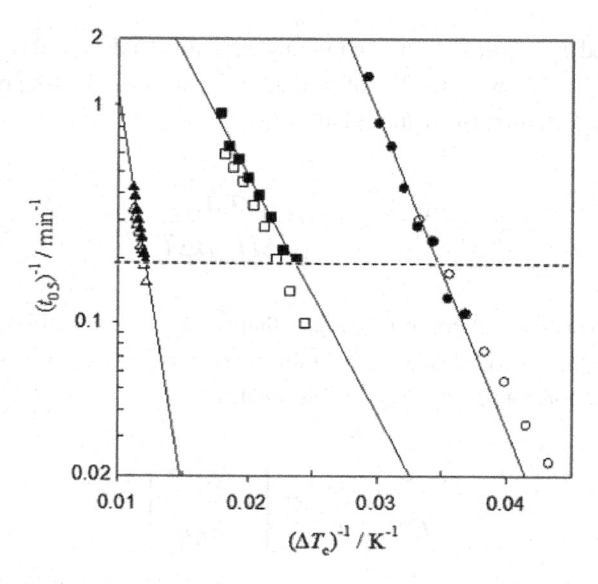

FIGURE 6 The rate of crystallization *versus* reciprocal undercooling after Equation (13), ▲PHBV, ■ PET, and ● PEO, the dashed line indicates constant rate of crystallization, the open symbols refer to 50/50 blends with epoxidized natural rubber (ENR) (Chan et al. 2011; permission provided by Wiley).

TABLE 2 Data characterizing rate of crystallization of a few polymers

	PHBV	PEO	PET
ΔT_c / K for 0.2 min^{-1}	83	29	46
B / K	830	337	253
ΔH_m / kJ mol^{-1}	9.8	5.2	32.3
$RB/\Delta H_m$	0.71	0.54	0.07

PHBV exhibits the highest ratio and PET the lowest. We have qualitatively:

$$\left(\frac{RB}{\Delta H_{\mathrm{m}}}\right)_{\mathrm{PHBV}} \approx \left(\frac{RB}{\Delta H_{\mathrm{m}}}\right)_{\mathrm{PEO}} > \left(\frac{RB}{\Delta H_{\mathrm{m}}}\right)_{\mathrm{PET}} \tag{14}$$

Morphologies that developed at the indicated undercooling's in symmetric blends of PET and PHBV with ENR are shown in Figure 7. Undercooling's were selected in a way that all systems crystallized with the same rate (see Figure 6). We recognize very small spherulites in PET blends owing to the low ratio of surface to bulk energy and large spherulites of PHBV due to the high ratio.

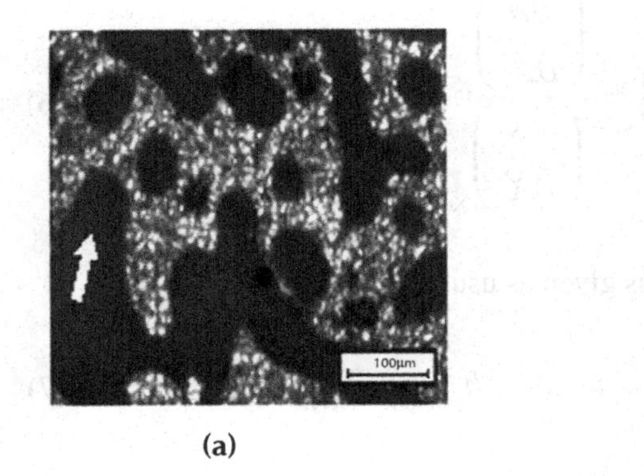

(a)

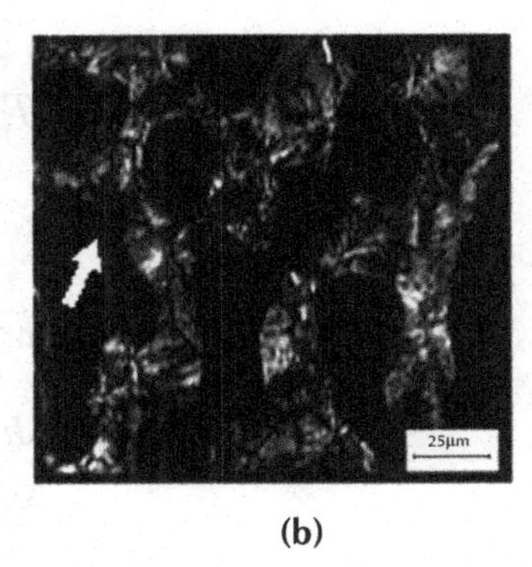

(b)

FIGURE 7 Morphologies in symmetric blends of PET (a) and PHBV (b) with ENR, the arrows mark ENR domains (Chan et al. 2011; permission by Wiley).

6.3 MELTING OF SEMICRYSTALLINE POLYMERS

In contrast to low-molecular substances, crystallization and melting of polymers proceed under non-equilibrium conditions but near to equilibrium. As mentioned in chapter 5, during crystallization lamellae and spherulites develop. These are non-equilibrium structures where crystalline and amorphous regions coexist. In those systems, the degree of crystallinity X acts as natural internal variable. In equilibrium, degree of crystallinity becomes a function of P and T, $X_{\mathrm{eq}} = X(P, T)$, it is not anymore an independent variable. Outside equilibrium, we have analogously to equation (Equation (3) chapter 7).

$$g = g(P, T, X)$$

Equilibrium and stability conditions read analogously to equation (see Equation (7) chapter 7). It follows for equilibrium with internal variable X:

$$\left(\frac{\partial h}{\partial X}\right)_{P,T} - T\left(\frac{\partial s}{\partial X}\right)_{P,T} = 0 \tag{15}$$

We define melting temperature as the temperature where crystallinity disappears, $X = 0$. The melting temperature reads:

$$T_m = \frac{\left(\frac{\partial h}{\partial X}\right)_{X=0}}{\left(\frac{\partial s}{\partial X}\right)_{X=0}} \tag{16}$$

The equilibrium melting point is given as usual by:

$$T_m^o = \frac{\Delta h}{\Delta s} \text{ with } \Delta h = h_{amorph} - h_{cryst} \tag{17}$$

We approximate the semicrystalline system by a two-phase approximation

$$\Delta h = h_a(1 - X) + h_c X$$

$$\Delta s = s_a(X)(1 - X) + s_c X \tag{18}$$

For the enthalpy, we use a linear approximation, but the entropy varies nonlinearly with degree of crystallinity since crystallization affects also the order (or disorder) in amorphous regions. The combination of Equation (16) and Equation (18) yields for the melting temperature of a material with degree of crystallinity X (Kammer, 2009):

$$T_{\mathrm{m}} = \frac{T_{\mathrm{m}}^{\mathrm{o}}}{1 - \frac{1}{\Delta s}\left(\frac{\partial s_{\mathrm{a}}}{\partial X}\right)_{\mathrm{o}}} \tag{19}$$

Quantity, $-\frac{1}{\Delta s}\left(\frac{\partial s_{\mathrm{a}}}{\partial X}\right)_{\mathrm{o}}$ is the measure for distance to equilibrium.

Its magnitude is usually small as compared to unity. One gets with $\Delta T_{\mathrm{m}} \equiv T_{\mathrm{m}}^{\mathrm{o}} - T_{\mathrm{m}}$ from Equation (19):

$$\frac{\Delta T_{\mathrm{m}}}{T_{\mathrm{m}}^{\mathrm{o}}} = -\frac{1}{\Delta s_{\mathrm{m}}}\left(\frac{\partial s_{\mathrm{a}}}{\partial X}\right)_{X=0} \tag{20}$$

Since $T_{\mathrm{m}} < T_{\mathrm{m}}^{\mathrm{o}}$, it follows:

$$\left(\frac{\partial s_{\mathrm{a}}}{\partial X}\right)_{\mathrm{o}} < 0 \tag{21}$$

6.3.1 *HOFFMAN–WEEKS APPROXIMATION*

It is a well-established experimental fact that melting temperature (Equation (16)) of polymers is a function of crystallization temperature:

$$T_{\mathrm{m}} = T_{\mathrm{m}}(T_{\mathrm{c}}) \tag{22}$$

It is also manifold experimentally confirmed that melting temperature depends linearly on crystallization temperature towards approach to equilibrium. In phenomenological terms, it follows:

$$T_{\mathrm{m}} = T_{\mathrm{m}}^{\mathrm{o}} - \frac{\mathrm{d}T_{\mathrm{m}}}{\mathrm{d}T_{\mathrm{c}}}\left(T_{\mathrm{m}}^{\mathrm{o}} - T_{\mathrm{c}}\right) \tag{23a}$$

With $\frac{dT_m}{dT_c} = const$, we get immediately the famous Hoffman–Weeks equation (Hoffman and Weeks, 1962a; Hoffman and Weeks, 1962b):

$$T_m = \alpha T_c + (1 - \alpha) T_m^o \qquad (23b)$$

Moreover, it follows in symmetric way to Equation (20):

$$\alpha \frac{\Delta T_c}{T_m^o} = -\frac{1}{\Delta s_m} \left(\frac{\partial s_a}{\partial X} \right)_{X=0} \qquad (24)$$

where difference ΔT_c is analogously defined to ΔT_m. The ratio of entropies $-\frac{1}{\Delta s} \left(\frac{\partial s_a}{\partial X} \right)_o$ gives in relationship Equation (20) deviation from equilibrium and in Equation (24) subdivided by parameter α the driving force for crystallization. This is quite consistent, because for crystallization and melting at equilibrium, it becomes $\alpha = 1$ and deviation from equilibrium and driving force to crystallization are equal.

Hoffman-Weeks approximation is illustrated in Figure 8. Lamellar crystallites develop in the high-molecular mass PCL, but not so in the low-molecular mass PEO. In the latter, it observes $\alpha = 0$ indicating formation of extended-chain crystals. This is consistent with Equation (24) since formation of extended-chain crystals is not accompanied by an amorphous phase.

In addition, one observes that extended-chain crystals develop under equilibrium conditions. The data as in Figure 8 allow estimation of distance to equilibrium. For PCL, one follows with $T_m^o = 339$ K and $\alpha = 0.3$ at supercooling of 20 K $-\frac{1}{\Delta s} \left(\frac{\partial s_a}{\partial X} \right)_o \approx 0.02$. The same value was found for poly(hydroxybutyrate) (PHB) ($T_m^o = 462$ K, $\alpha = 0.18$) but at undercooling $\Delta T_c = 50$ K (Chan et al., 2004). These data show indeed that the ratio of entropies or deviations from equilibrium are small at the indicated undercoolings.

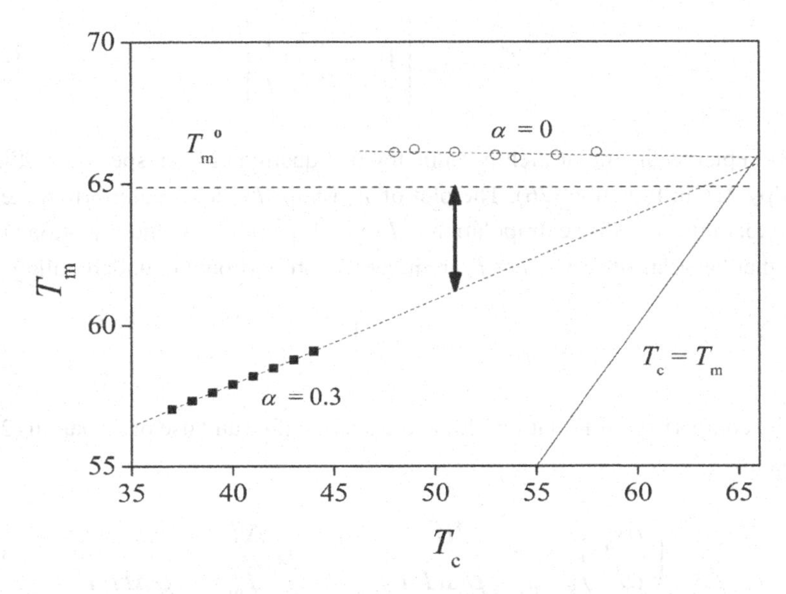

FIGURE 8 Hoffman-Weeks plots for PCL of M_w = 80 000 g/mol (■), data after Chong et al. (2004), and PEO of M_w = 13 000 g/mol (o), data after Alfonso and Russell (1986), the arrow marks the distance to equilibrium after Equation (20).

6.3.2 GIBBS–THOMSON EQUATION

The equilibrium melting temperature may also be determined by extrapolation of lamellar thickness *l* (Keller, 1968; Strobl, 2000; Al-Hussein and Strobl, 2002). In contrast to extended-chain crystals, lamellae grow at large undercoolings.

We want to formulate an analogous equation like Equation (19) but expressed by quantities characterizing the lamella. In discussing of melting, we adopt Equation (6). The quantity $\Delta\mu$ gives then the change of chemical potential accompanying the transition from crystallite to melt. Accordingly, we have $\Delta\mu_m$ instead of Equation (10). It is given by:

$$\Delta\mu_m = \frac{\Delta H_m}{T_m^o}\Delta T_m \tag{25}$$

We define, analogously to Equation (16), melting temperature T_m of a lamella with thickness *l* as the temperature where free enthalpy (Equation (6)) disappears. It follows the Gibbs–Thomson equation:

$$T_m = T_m^o \left[1 - \frac{2\sigma}{\rho \Delta \tilde{H}} \frac{1}{l} \right] \tag{26}$$

We replaced the molar melting enthalpy of Equation (25) by specific melting enthalpy $\Delta \tilde{H}$ in Equation (26). The plot of T_m *versus* l^{-1} yields equilibrium melting temperature $T_m{}^o$ after extrapolation to $l \to \infty$. Equation (8b) and Equation (10) show that lamellae grown at $T = T_c$ are inversely proportional to undercooling.

$$l_c \propto \frac{1}{\Delta T_c} \tag{27}$$

The comparison of Equation (20) and Equation (26) and use of Equation (24) results in:

$$-\frac{1}{\Delta s_m} \left(\frac{\partial s_a}{\partial X} \right)_{X=0} = \frac{2\sigma}{\rho \Delta \tilde{H}} \frac{1}{l} \qquad \alpha \frac{\Delta T_c}{T_m^o} = \frac{2\sigma}{\rho \Delta \tilde{H}} \frac{1}{l} \tag{28}$$

It demonstrates that Gibbs–Thomson and Hoffman–Weeks approach are completely equivalent. The extrapolation of $l \to \infty$ equals in thermodynamic terms extrapolation of $T_c \to T_m{}^o$ after Equation (28). Use of Equation (7) in Equation (28) results in:

$$\alpha = \frac{l_o}{l} \tag{29}$$

With the value calculated for PHB, $-\frac{1}{\Delta s_m} \left(\frac{\partial s_a}{\partial X} \right)_{X=0} \approx 0.02$, we may estimate surface energy density σ/l from Equation (28). Assuming for PHB $\rho = 1.15$ g/cm^3 and $\Delta \tilde{H} = 146$ J/g, one obtains:

$$\frac{\sigma}{l} = 1.7 \cdot 10^9 \text{ mN/m}^2 \tag{30}$$

For a lamella with $l = 10$ nm, the interfacial tension between crystallite and amorphous surrounding comes to 17 mN/m. It is a quite realistic value.

An observation deviating from results shown in Figure 8 is presented in Figure 9. The variation of $T_m(T_c)$ as given in Figure 9 advanced speculations about failing of Hoffman–Weeks (Marand et al., 1998). There is no reason for such a conclusion. Close to equilibrium melting point T_m^o, Equation (23) holds true, that is one observes linear variation of T_m with T_c as shown in Figure 9. The melting points at greater distances from equilibrium melting point vary also linearly with T_c to a good approximation, but with slope α different to that one in the range close to T_m^o. One may approach function $T_m(T_c)$ by two linear approximations (Equation (23)) holding true in different ranges of temperature. The crossing point occurs here at $T_c^* = 362$ K, and we have the range $T_c \geq T_c^*$ close to T_m^o, and the other one, $T_c \leq T^*$, more distant to T_m^o. Formally, we may design two equilibrium melting temperatures in terms of Hoffman–Weeks, but only the upper melting point is experimentally accessible, the lower one might be seen as a virtual melting point of lamellae growing fastest in the temperature range below T^*.

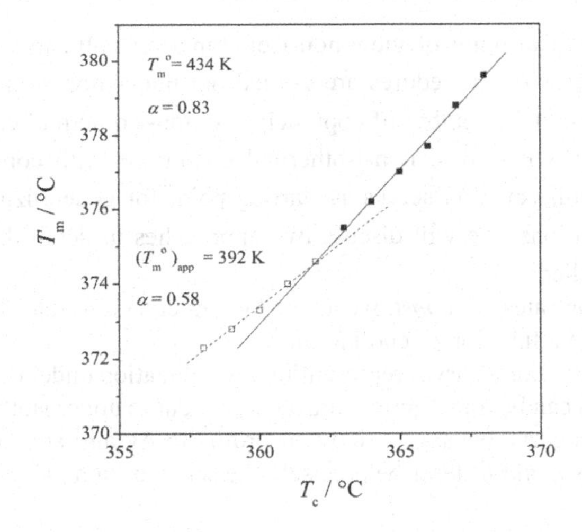

FIGURE 9 Hoffman–Weeks plot for poly(tetramethylene succinate) (PTMS), data after Chong et al. (2004).

We discuss this jump of slopes at T^* in terms of thermodynamics under assumption that the change is solely caused by entropy effects. The molar melting enthalpy of 100 % crystalline PTMS amounts to 19.6 kJ/mol (Chong et al.,

2004) and T_m* = 374.6 K after Figure 9. Superscripts u and l refer to "upper" and "lower" curve. It results:

$$\left[-\frac{1}{\Delta s_m}\left(\frac{\partial s_a}{\partial X}\right)_o\right]^u - \left[-\frac{1}{\Delta s_m}\left(\frac{\partial s_a}{\partial X}\right)_o\right]^l = 0.09 > 0 \quad (31)$$

The change in α reflects a considerably higher order in the amorphous phase of the newly developing lamella above T_c^*. If we suppose that only entropy changes rule the change in α then Equation (31) and Equation (27) tell us:

$$l^l - l^u > 0 \quad \text{at } T_c = T^* \quad (32)$$

This is consistent with Equation (28).

6.4 NON-ISOTHERMAL CRYSTALLIZATION

From a technological point of view non-isothermal crystallization is important since many processing procedures are carried out under non-isothermal conditions. There are quite a number of approaches to non-isothermal crystallization. We restrict the discussion to non-isothermal conditions with constant cooling rate s. The Avrami equation serves as starting point for generalizations to non-isothermal conditions. We will discuss two approaches in detail that have been successfully applied.

- For cooling rates $s = const$, we adopt Avrami equation; rate constant K becomes then a function of cooling rate s.
- A series of cooling curves representing crystallization under non-isothermal conditions can be transformed into a cooling curve under isothermal conditions. This curve behaves strictly according to Avrami and corresponding parameters n and K describe precisely the series of non-isothermal cooling curves.

For many polymers, reliable values for the two Avrami parameters could be determined by applying the indicated procedures.

We focus on crystallization under non-isothermal conditions with cooling rate $s = const$:

$$s = \frac{\mathrm{d}T}{\mathrm{d}t} = const \quad \Rightarrow \quad t = \frac{\Delta T}{s} \quad (33)$$

We may then write Avrami-equation (Equation (2)) as follows:

$$-\ln\left(1-X\right)=K^{+}\left(s\right)\left(\frac{\Delta T}{s}\right)^{n} \quad s=const \qquad \Delta T=T_{onset}-T,\ \ T<T_{onset} \tag{34}$$

As Figure 10 and Figure 11 show relationship Equation (34) is an acceptable approximation even for low conversions when cooling rates are low. Some care is needed for high cooling rates. In some cases, only conversions exceeding $X \approx 0.5$ obey the modified Avrami equation (Equation (34)). In PHBV deviations are less pronounced as in PTMS even at high cooling rates (see Figure 11).

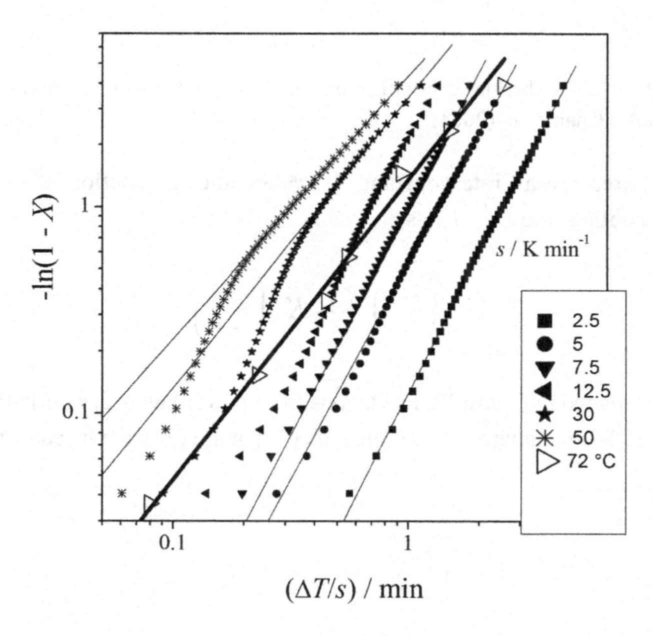

FIGURE 10 Non-isothermal crystallization of PTMS under various cooling rates $s = const$, data after Chong et al. (2004), to the curve referring to $T_c = 72$ °C see below.

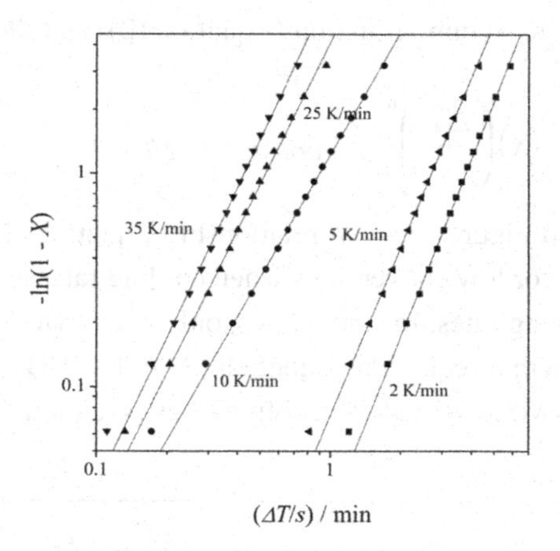

FIGURE 11 Non-isothermal crystallization of PHBV under various cooling rates s = *const*, data after Chan et al. (2004).

The Figures reveal, rate constant $(K^{+})^{1/n}$ becomes a function of parameter s (reference cooling rate $s^* = 1$ K/s):

$$\left(K^{+}\right)^{1/n} = K_{o}\left(\frac{s}{s^{*}}\right)^{\alpha} \tag{35}$$

This is depicted in Figure 12. Deviations from power law occur for PHBV at low cooling rates. The parameters determined after Equation (34), are listed in Table 3.

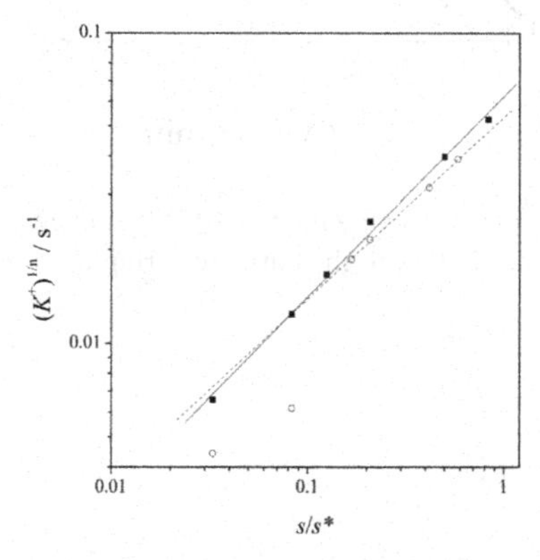

FIGURE 12 The variation of rate constant $K^{+\,1/n}$ with reduced cooling rate for PTMS and PHBV (o); $s^* = 1$ K/s.

TABLE 3 Parameters characterizing non-isothermal crystallization after Eq. (34)

Polymer	s / K min^{-1}	$K^{+1/n}$ / min^{-1}	n
PTMS	2.5	0.40	2.2
PTMS	30	2.4	1.4
PHBV	2	0.27	2.8
PHBV	25	1.9	2.2
PTMS	$\left(K^{+}\right)^{1/n}/\mathrm{s}^{-1} = 0.06(s/s^{*})^{0.65}$ (s > 5 K/min)		
PHBV	$\left(K^{+}\right)^{1/n}/\mathrm{s}^{-1} = 0.05(s/s^{*})^{0.59}$		

Overall rate constants $t_{0.5}^{-1}$ for isothermal crystallization is given by Equation (3). In analogous way, one may define a rate constant for non-isothermal crystallization on base of Equation (34). This is done by replacing half time under T = *const* by half time of crystallization under condition s = *const*, $(\Delta T/s)_{0.5}$. It follows:

$$\left(\frac{\Delta T}{s}\right)_{0.5}^{-1} = \left(\frac{K^{+}}{\ln 2}\right)^{1/n} \tag{36}$$

The relationship (Equation (36)) is seen as analogue to Equation (13). After Equation (34) undercooling ΔT_c is given by $\Delta T = T_{onset} - T$. Plots after Equation (13) and Equation (36) are given in Figure 13 for PTMS. Crystallization temperatures were selected between 92 °C and 85 °C. Therefore, the lower equilibrium melting temperature of Figure 9 was selected as reference for undercooling under isothermal conditions. For non-isothermal crystallization, cooling rates were changed between 2.5 K/min and 50 K/min. It results a quite extended range of corresponding ΔT^{-1}, approximately $0.1 \leq \Delta T^{-1}$ / K$^{-1} \leq 0.2$. The corresponding range for undercooling under isothermal conditions is by one order of magnitude smaller, $0.03 \leq \Delta T_c^{-1}$ / K$^{-1} \leq 0.04$. The rates of crystallization for isothermal and non-isothermal crystallization are only comparable for very slow cooling rates. It might be called the quasi-isothermal range. High undercoolings under isothermal conditions are needed for achieving rates of crystallization at high cooling rates. As Figure 13 shows, one needs for changes of crystallization rates by one order of magnitude under non-isothermal conditions much higher changes of ΔT_c^{-1} as under isothermal regime.

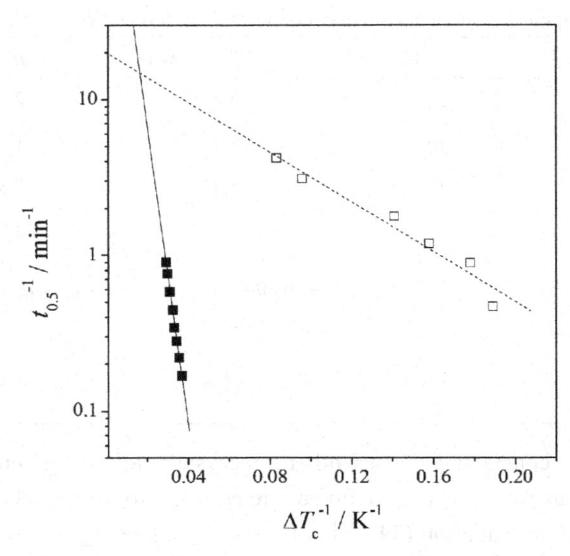

FIGURE 13 The rate of crystallization *versus* undercooling for PTMS under isothermal and non-isothermal conditions (open squares) after Equation (2) and Equation (34), data after Chong et al. (2004); compare text.

The comparison of curves in Figure 13 with Equation (13) reveals:

$$B_{\text{isotherm}} > B_{\text{non-isotherm}} \tag{37}$$

This is a direct consequence of the more extended range of ΔT_c^{-1} covered under non-isothermal conditions to achieve the same change of rates as under isothermal conditions. The corresponding energies *RB* differ by one order of magnitude:

$$RB_{\text{isotherm}} = 1.84 \text{ kJ/mol} \qquad RB_{\text{non-isotherm}} = 0.15 \text{ kJ/mol} \tag{38}$$

The values (Equation (38)) clearly demonstrate that crystallites grown under isothermal conditions exhibit higher interfacial energy towards surrounding amorphous phase as crystallites developed in non-isothermal regime.

6.4.1 QUASI-ISOTHERMAL APPROACH

The kinetic parameter (Equation (36)) results from Equation (34) for a fixed conversion, $X = 0.5$. Now, we allow for variation of X but fix the temperature.

Depending on rate of cooling, crystallization starts at T_{onset} and we select the conversion at temperature $T = const$ but originating from histories with different cooling rates s_1, s_2, s_3,... and so on. In terms of Equation (34), it reads:

$$-\ln\left(1-X_{\mathrm{T}}\right)=K(\mathrm{T})\left[\frac{\left(T_{\mathrm{ons}}-T\right)}{s}\right]^{\mathrm{n}} \tag{39}$$

A series of non-isotherms with $s = const$ is in that way characterized by an isothermal equivalent. There is a different law of conversion for each period of time. An example is given in Figure 10 for $T = 72$ °C. Only a selected number of non-isotherms contributes to the Avrami-like relationship, from $s = 7.5$ to 30 K/min. As in isothermal Avrami plots (see Figure 2), deviations occur at high conversions. The quasi-isothermal approach is depicted in Figure 14 for non-isothermal crystallization of PHBV.

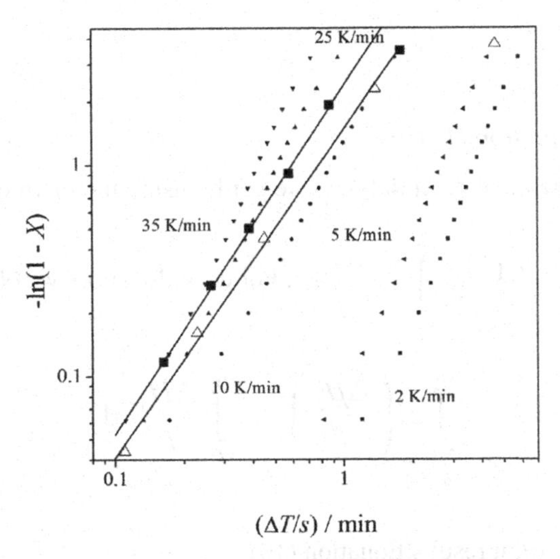

FIGURE 14 Non-isotherms with $s = const$ for PHBV and quasi-isotherms after Equation (39),

■ $T_c = 91$ °C, non-isotherms to $s = 15$–35 K/min contribute the quasi-isotherm;

Δ $T_c = 95$ °C, non-isotherms to $s = 10$–30 K/min contribute to the quasi-isotherm.

We observe that quasi-isotherms shift downwards when the temperature T_c increases, it occurs as a consequence of slowing down of rate of crystallization. One may state, isotherms and quasi-isotherms show deviations from Avrami at high conversions whereas, non-isotherms with $s = const$ display deviations from Equation (34) at low conversions. The comparison of relevant parameters for isotherms and quasi-isotherms reveal that rates are lower and Avrami exponents are higher for isotherms as compared to quasi-isotherms. An example is given in Table 4.

TABLE 4 The rates and exponents n for isotherms and quasi-isotherms of PHBV at $T_c = 91$ °C

	$K^{1/n}$ / min^{-1}	n
Isotherm	0.1	2.3
Quasi-isotherm	1.7	1.7

APPENDIX

Derivation of Equation (10)

The change in chemical potential $\Delta\mu_c$, we get by using the thermodynamic standard relationship $\frac{\partial}{\partial T}\left(\frac{\Delta\mu_c}{T}\right) = -\frac{\Delta H_m}{T^2}$. For crystallization, we obtain:

$$\int_{\frac{\Delta\mu_c}{T_c}}^{0} d\left(\frac{\Delta\mu_c}{T}\right) = -\int_{T_c}^{T_m^o} \frac{\Delta H_m}{T^2} dT$$

For $\Delta H_m = const$, it results Equation (10)

$$\Delta\mu_c = \Delta H_m \frac{\Delta T_c}{T_m^o}$$

KEYWORDS

- **Gibbs-Thomson equation**
- **Hoffman-Weeks approach**
- **Isothermal crystallization**
- **Non-isothermal crystallization**
- **Rates of crystallization**

REFERENCES

1. Alfonso, G. C. and Russell, P. T. *Macromolecules*, **19**, 1143 (1986).
2. Al-Hussein, M. and Strobl, G. *Macromolecules*, **35**, 1672 (2002).
3. Avrami, M. *J. Chem. Phys.*, **7**, 1103-1112 (1939).
4. Chan, C. H., Sulaiman, S. F., Kammer, H. W., Sim, L. H., and Harun, M. K. *J. Appl. Polym. Sci.*, **120**, 1774-1781 (2011).
5. Chan, C. H., Kummerlöwe, C., and Kammer, H. W. *Macromol. Chem. Phys.*, **205**, 664-675 (2004).
6. Chong, K. F., Schmidt, H., Kummerloewe, C., and Kammer, H. W. *J. Appl. Polym. Sci.*, **92**, 149 (2004).
7. Hoffman, J. D. and Weeks, J. J. *J. Res. Natl. Bur. Stand.* (A), **66**, 13 (1962a).
8. Hoffman, J. D. and Weeks, J. J. *J. Chem. Phys.*, **37**, 1723 (1962b).
9. Kammer, H. W. *On Crystallization and Melting of Polymers*, PERFIK ATP, Conference Proceedings (2009).
10. Keller, A. *Rep. Progr. Phys.*, **31**(2), 623 (1968).
11. Marand, H., Xu, J., and Srinivas, S. *Macromolecules*, **31**, 8219 (1998).
12. Strobl, G. *Eur. Phys. J.*, **E3**, 165 (2000).

CHAPTER 7

POLYMERS IN NON-EQUILIBRIUM

HANS-WERNER KAMMER

CONTENTS

7.1 THERMODYNAMICS OF NON-EQUILIBRIUM PROCESSES

Thermodynamics of equilibrium is extended to processes taking place near to equilibrium. We recapitulate very briefly, thermodynamics of equilibrium. In a system consisting of one component, Gibbs free energy is just a function of pressure P and temperature T [1]

$$g = g(P,T)$$

$$g = h - Ts$$

$$\mathrm{d}g = -s\mathrm{d}T + v\mathrm{d}P \tag{1}$$

Under isothermal-isobaric conditions, it follows:

$$\mathrm{d}g = -T\mathrm{d}s \leq 0 \tag{2}$$

The Equation (2) shows, Gibbs free energy is minimum in equilibrium. This minimum principle is a consequence of 2nd Law. More precisely, we write Equation (2) as:

$$\mathrm{d}g = -T\mathrm{d}_i s \leq 0 \tag{2a}$$

where $\mathrm{d}_i s$ represents the internally produced entropy in the system with:

$$\mathrm{d}_i s \geq 0$$

positive for irreversible processes and zero for reversible ones.

For non-equilibrium situations, but situations near to equilibrium, we assume that all thermodynamic equations are also valid. The only generalization, we have to make: There are more than two independent variables since non-equilibrium situations are characterized by additional variables. For example, a system not in equilibrium with respect to temperature does not have one (constant) temperature. It is characterized by many different temperatures (a gradient of temperature) leading to conductance of heat. In what follows, we will not discuss those inho-

[1] Note extensive quantities are symbolized by letters, intensive quantities by capitals.

mogeneities with respect to external conditions. We will focus on additional internal degrees of freedom characteristic for polymers and homogeneity of external conditions as pressure and temperature. In polymer systems, internal variables are related to a reduced number of conformations owing to frozen in degrees of translation and rotation. To keep it tractable, we suppose the existence of just one internal variable. We call the phenomenological internal variable ζ. It corresponds to the internal degree of freedom. Then, in generalization of Equation (1), Gibbs free energy reads:

$$g = g(P, T, \zeta) \tag{3a}$$

$$\mathrm{d}g = -s\mathrm{d}T + v\mathrm{d}P + \left(\frac{\partial g}{\partial \zeta}\right)\mathrm{d}\zeta \tag{3b}$$

The differential belonging to ζ is called affinity A. It represents the driving force for the process towards equilibrium:

$$\left(\frac{\partial g}{\partial \zeta}\right) = -A \tag{4}$$

In equilibrium internal variable becomes a function of pressure and temperature, it is not anymore an independent variable.

$$\zeta_{eq} = \zeta(P, T) \tag{5}$$

It is supposed that Equation (3) is valid in any time interval dt. It follows when *T,P = constant*

$$\frac{\mathrm{d}g}{\mathrm{d}t} = \left(\frac{\partial g}{\partial \zeta}\right)\frac{\mathrm{d}\zeta}{\mathrm{d}t} \tag{6}$$

Variation of Equation (3) around minimum results in:

$$\delta g = \left(\frac{\partial g}{\partial \zeta} \right) \delta \zeta = 0 \quad \Rightarrow \quad \left(\frac{\partial g}{\partial \zeta} \right)_{\text{eq}} = 0$$

Hence, it follows for equilibrium and stability

$$\left(\frac{\partial g}{\partial \zeta} \right)_{\text{eq}} = 0 \qquad \left(\frac{\partial^2 g}{\partial \zeta^2} \right)_{\text{eq}} \geq 0 \tag{7}$$

So far we argued only in a static way. Actually, non-equilibrium states are characterized by some dynamic behavior ruling relaxation towards equilibrium. Frequently, we have competition of two rate processes, an external and an internal:

- Rate of change of external conditions, R_{change}
- Rate of system's approach to equilibrium or steady state, R_{relax}.

Two limiting situations exist:

$$R_{\text{change}} >> R_{\text{relax}} \qquad \text{and} \qquad R_{\text{change}} << R_{\text{relax}} \tag{8}$$

In the first case, the change of external conditions proceeds so rapidly that the system cannot follow it. Therefore, it is in a "frustrated" situation (actually non-equilibrium) and equilibrium (or steady state) cannot be reached at all. In contrast, in the second case, the system follows immediately external changes and is always in equilibrium (steady state).

Relaxation phenomena can be detected somewhere in the middle between these two extremes. For polymers, it is important to be aware of their relaxation behavior. We mention two typical relaxation phenomena, glass transition and high elasticity.

Thus, in non-equilibrium thermodynamics, we need a dynamic law ruling relaxation towards equilibrium. In linear non-equilibrium thermodynamics, the rate of the process is just proportional to affinity.

$$\frac{\mathrm{d}\zeta}{\mathrm{d}t} = -L \left(\frac{\partial g}{\partial \zeta} \right) \tag{9}$$

The L is called phenomenological coefficient, $L > 0$. Generally, it is a function of external conditions, T, P. Here, we take $L = constant$. We obtain a differential equation for ζ from Equation (9) just by differentiation with respect to time:

$$\frac{\mathrm{d}^2\zeta}{\mathrm{d}t^2} + L\left(\frac{\partial^2 g}{\partial \zeta^2}\right)\frac{\mathrm{d}\zeta}{\mathrm{d}t} = 0 \tag{10}$$

One immediately verifies:

$$\tau^{-1} \equiv L\left(\frac{\partial^2 g}{\partial \zeta^2}\right) \tag{11}$$

Quantity τ is the relaxation time constant governing relaxation towards equilibrium. The simplest solution of Equation (9), when only one exists, is the saturation curve:

$$\zeta(t) = \left[\zeta(0) - \zeta_{eq}\right]e^{-t/\tau} + \zeta_{eq} \tag{12}$$

$$\zeta(t) = \zeta_{eq} \qquad \text{for} \quad t >> \tau$$

$$\zeta(t) = \zeta(0) \qquad \text{for} \quad t << \tau$$

The last line represents quasi-equilibrium since $\zeta(0) \neq \zeta_{eq}$. Hence, in the limiting situations, we have equilibrium or almost so whereas non-equilibrium thermodynamics becomes operative in the region $t \approx \tau$.

When we have more than one relaxation time constant, the time dependence of internal variables is more adequately described by a stretched exponential instead of Equation (12).

$$\Phi(t) = \exp\left[-\left(\frac{t}{\tau}\right)^{\beta}\right] \tag{13}$$

This is so because Equation (13) can be formulated in a Fourier-series:

$$\Phi(t)=\sum_{i} w_{i} \exp\left[-\frac{t}{\tau_{i}}\right] \text{ with } \sum_{i} w_{i}=1$$

The mean value of τ over the distribution of relaxation constants is given by:

$$\langle\tau\rangle=\tau\Gamma\left(1+\frac{1}{\beta}\right)$$

with Γ being the gamma-function.

7.2 AGING OF POLYMERS - ENTHALPY RELAXATION

Enthalpy relaxation reflects physical aging of polymers. The melt in equilibrium has a certain number of holes. When cooled down below glass transition temperature T_g, the fraction of holes is not anymore in equilibrium due to the loss of conformations. The fraction of holes η is here the internal variable and rules the driving force, $\left(\frac{\partial g}{\partial \eta}\right)$, towards equilibrium. This is reflected in an enthalpy relaxation or change in heat capacity, which we can monitor.

For detecting aging, samples are subjected to the following thermal history:

- Annealing for a certain time in the melt to erase all history,
- Quenching of the sample to $T_{annealing} < T_{g,}$
- Annealing there and monitoring the relaxation towards equilibrium.

This is schematically shown in Figure 1. Figure 2 presents an example for enthalpy relaxation of a random copolymer.

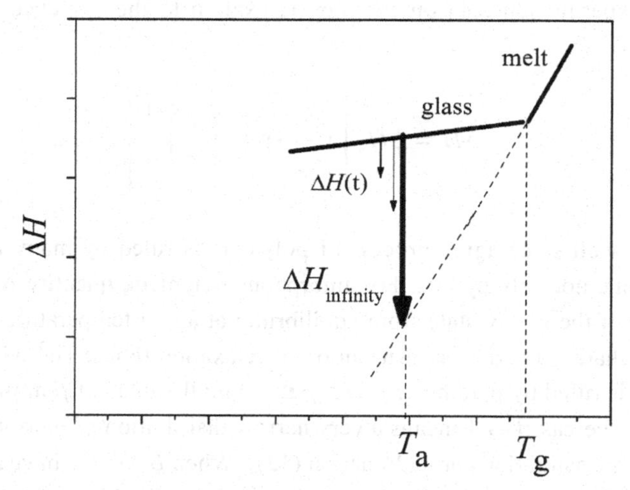

FIGURE 1 Schematic representation of aging at temperature T_a; the enthalpy grows up to $\Delta H_{infinity}$ corresponding to deviation of the glass from equilibrium.

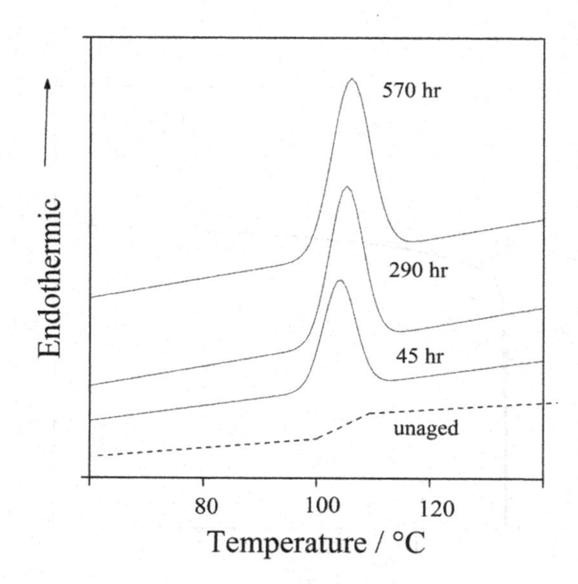

FIGURE 2 Enthalpy changes reflecting aging of poly(styrene-*co*-acrylonitrile) with 11 wt% of AN; sample with T_g = 105 °C was aged at 85 °C for the indicated periods of time, data after Pauly and Kammer (1994).

We fit experimental data on enthalpy relaxation to the stretched potential, Equation (13).

$$\Delta H = \Delta H_{\infty}\left[1 - \exp\left\{-\left(\frac{t}{\tau}\right)^{\beta}\right\}\right] \tag{14}$$

It works well since aging process of polymers is ruled by many relaxation time constants, not only by one. For amorphous polymers, quantity ΔH_{∞} marks the distance of the glassy state from equilibrium at aging temperature T_a. Equation (13) is characterized by a distribution of relaxation times. The width of the distribution is ruled by parameter β, $0 \leq \beta \leq 1$. Small values of β imply a broad distribution whereas $\beta \rightarrow 1$ means a very narrow distribution of relaxation times ($\beta = 1$ gives the saturation curve, Equation (12)). When $\beta = 0$, we have a constant function which decays over an infinite period of time. The Figure 3 presents the function $(1 - \Phi)$ for different values of parameter β. With decreasing β the system needs much more time, expressed in relaxation times τ, for entering the plateau region.

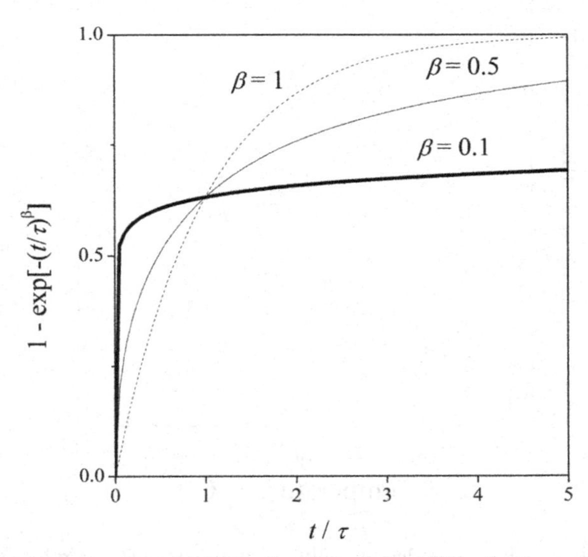

FIGURE 3 Representation of the function $(1 - \Phi)$ for different values of exponent β; the dotted curve gives the saturation curve (Equation (11)), the solid bold curve represents a broad distribution of relaxation times due to $\beta = 0.1$.

We discuss here as an example aging of poly(vinyl methyl ether) (PVME), polystyrene (PS) and their 50/50 blend. The data were adopted from reference (Cowie and Ferguson, 1989). Polymers have molar masses of around 10^5 g/mol. The phase diagram of the system is very similar to the diagram shown in Figure 18 of chapter 5 (see Chapter 5, Transitions in Polymers). We recognize lower critical solution temperature (LCST) behavior with phase separation slightly above 150 °C. Dependence of glass transition on blend composition is presented in Figure 4 (Cowie and Ferguson, 1989). The glass transition of PVME was observed at T_g = 256 K. Parameters in Equation (13) were determined in the same way as well-known from Avrami-like plots. A few data about aging of PVME are listed in Table 1. The distance to equilibrium increases with decreasing aging temperature. Consequently, the relaxation time constant increases. At aging temperature closest to T_g, PVME approaches 99 % of ΔH_∞ after around 1.5 hr. It needs 130 hr for the same distance to saturation at T_a = 240 K and 235 K.

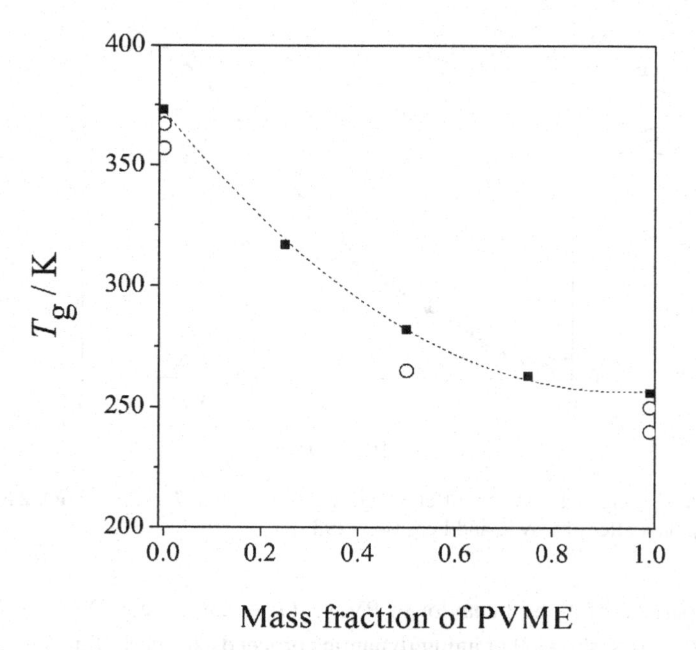

FIGURE 4 Glass transition temperature T_g of PVME/PS blend *versus* composition; the open circles indicate annealing temperatures.

TABLE 1 Aging data for PVME at different aging temperatures (Cowie and Ferguson, 1989) calculated after Equation (13)

T_a / K	**250**	**240**	**235**
τ / min	17	348	683
ΔH_∞ / J g^{-1}	1.1	3.2	3.8
β	0.92	0.49	0.63

The PVME displays here the widest distribution of relaxation time constants at $T_a = 240$ K. Time dependence of aging, observed for PVME, is plotted in Figure 5.

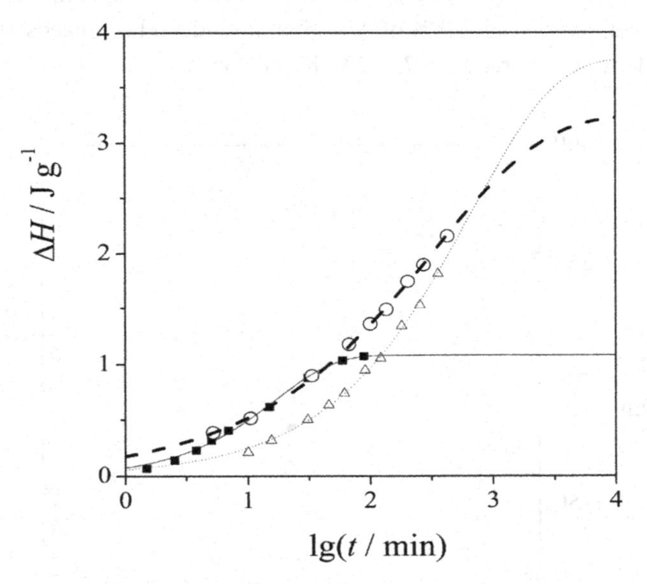

FIGURE 5 Aging of PVME at different aging temperatures: $T_a = 250$ K (■), 240 K (○), 235 K (Δ), data after of Cowie and Ferguson (1989).

Comparison of aging behavior of PVME ($T_g = 256$ K) and PS ($T_g = 373$ K) at $(T_g - T_a) = 6$ K shows that aging dynamics proceeds similarly for the polymers in the same distance from equilibrium (*see* Figure 6). Relevant data are listed in Table 2.

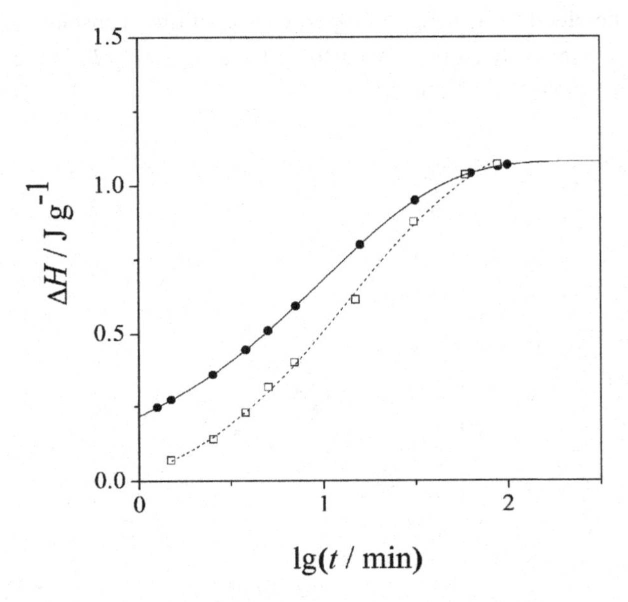

FIGURE 6 Aging of PS (solid curve) and PVME (dotted curve) at $(T_g - T_a) = 6$ K, data after Cowie and Ferguson (1989).

TABLE 2 Aging data for PVME and PS at $(T_g - T_a) = 6$ K, data after Cowie and Ferguson (1989)

	ΔH_∞ / J g^{-1}	τ / min	β
PVME	1.1	17	0.92
PS	1.1	10	0.65

Let us turn to aging of the blend. The polymers are completely miscible at low temperatures as revealed by phase diagram (Figure 18, chapter 5) and dependence of glass transition temperature on composition (see Figure 4). Aging of the 50/50 blend can only be observed in sufficient distance from its $T_g = 282$ K. The Figure 7 presents the aging behavior at $(T_g - T_a) = 16$ K and Table 3 summarizes the data. The two parent polymers display again very similar aging dynamics with roughly the same distance to equilibrium. The blend is closer to equilibrium than the constituents. However, the aging dynamics of the blend does not change markedly as compared to that of the constituents. One easily verifies that the relaxation time constant of the blend is the weighed superposition of time constants of the parent polymers. The thermodynamic driving force for aging, $\Delta H_\infty/T_a$, is the highest for PVME and the lowest for the blend.

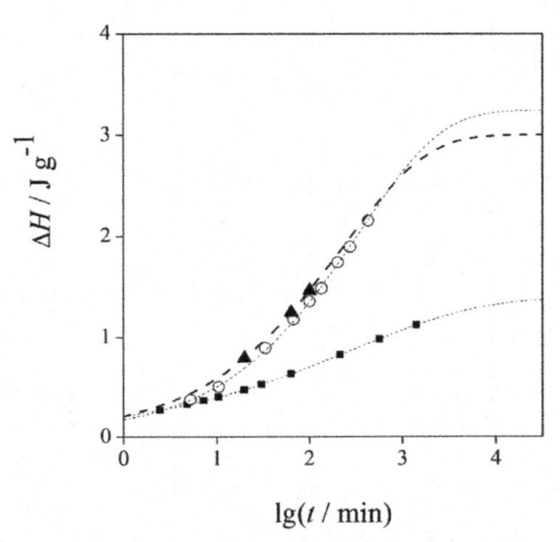

FIGURE 7 Aging of PVME (○), PS (▲), and 50/50 blend (■) at $(T_g - T_a) = 16$ K, data after Cowie and Ferguson (1989).

TABLE 3 Aging data for PVME, PS and the 50/50 blend at $(T_g - T_a) = 16$ K, calculated after results of Cowie and Ferguson (1989)

	T_a / K	ΔH_∞ / J g^{-1}	τ / min	β
PVME	240	3.2	348	0.49
PS	357	3.0	229	0.48
50/50	266	1.4	288	0.32

7.3 LINEAR VISCOELASTIC BEHAVIOR OF POLYMERS-STRESS RELAXATION

7.3.1 PRELIMINARIES

In the study on enthalpy relaxation, we studied the response of polymers subjected to aging in a non-equilibrium (glassy) state. We monitored relaxation of the system towards equilibrium under static external conditions. This approach may be described by Equation (9). Here, we turn to systems under flow. Again relaxation will occur when these systems are subjected to change of external conditions. They will respond by approaching a new steady state, which is compatible with the new conditions. The basic equations for discussion of linear viscoelasticity are Hooke's (deformation) and Newton's law (flow):

Stress $\propto$ deformation stress $\propto$ rate of deformation

For shear deformation γ and rate of shear $\dot{\gamma}$, these laws read:

$$\sigma_s = G\gamma \qquad \sigma_s = \eta\dot{\gamma} \tag{15}$$

where G and η are shear modulus and shear viscosity. These quantities are material properties and in simplest case constants.

7.3.2 LINEAR VISCOELASTICITY

When an external stress is imposed to the system and the state of the system is characterized by change of stress in the course of time, we call it stress relaxation. Usually, we assume a linear relationship of type Equation (9). As long as the total deformation is small, we may assume:

rate of approaching steady state (or equilibrium) $\propto$ *deviation of the system from steady state*

In mathematical terms, we postulate for stress relaxation:

$$\frac{\mathrm{d}\sigma}{\mathrm{d}t} = -\frac{1}{\tau}\sigma \tag{16}$$

Proportionality factor τ^{-1} is a material-specific quantity; τ is called relaxation time constant. It follows:

$$\sigma(t) = \sigma_o e^{-t/\tau} \tag{17}$$

The initial stress diminishes ("relaxes") in the course of time. When we say, t is the time, an external force acts on the specimen then inequalities (Equation (8)) read in terms of t and τ:

$R_{\text{change}} >> R_{\text{relax}}$ corresponds to $\frac{\tau}{t} >> 1$ or after Equation (17) $\sigma \approx \sigma_o$

that is, the sample remains in the frustrated situation. The other limit:

$R_{change} << R_{relax}$ corresponds to $\frac{\tau}{t} << 1$ or after Equation (17) $\sigma \to 0$

that is, the sample is always in equilibrium.

The ratio of times is sometimes called Debora number D:

$$D = \frac{\tau}{t} \tag{18}$$

It is experimentally well established that properties of polymers are determined by the ratio of relaxation time constant τ and duration of external force action t. The smaller quantity D the more rapidly the system relaxes towards equilibrium (steady state). For low-molecular liquids, D is very small. The same situation turns up if a polymer is subjected to stress for a very long time (low-frequency range). Then, it behaves like a liquid that is exhibits flow. This can be observed for example with butyl rubber when stored for sufficiently long period of time; it displays "cold" flow. On the other side, polymer composites turn in the low-frequency range from liquid-like to solid-like behavior, compare later. It follows:

$$\begin{array}{ll} D << 1 & \text{polymers behave like liquids} \\ D \approx 1 & \text{viscoelastic behavior of polymers} \\ D >> 1 & \text{polymers behave like solids} \end{array} \tag{19}$$

We see, properties of polymers depend not only on their chemical structure but also on duration external loads act. In the state of high elasticity, many relaxation phenomena of polymers can be detected noticeably.

Let us discuss as an introduction stress relaxation and stress-strain curve. In stress relaxation, a sample is rapidly deformed to a preset strain value ε_o and is kept in the deformed state while monitoring the time decay of stress, $\sigma = \sigma(t)$:

$$\varepsilon_o = constant \Rightarrow \sigma = \sigma(t)$$

In a Hookean body, we would have:

$$\varepsilon_o = constant \Rightarrow \sigma = E\varepsilon_o = constant$$

But in a viscoelastic material, we observe $\sigma(t)$. In generalization of Hooke, we have:

$$\sigma(t) = E(t)\varepsilon_o \tag{20}$$

The solution of Equation (16) reads:

$$\sigma(t) = \sigma_\infty + (\sigma_o - \sigma_\infty) e^{-t/\tau} \tag{21}$$

When experimental data can be fitted by Equation (21), the relaxation time constant of the process can be determined. The stress relaxation experiment is schematically depicted in Figure 8.

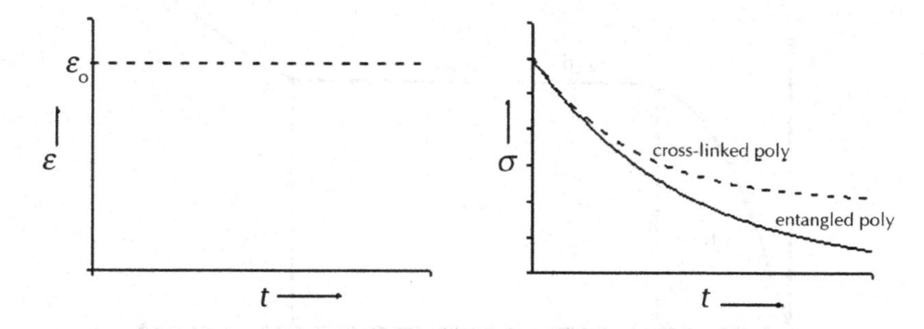

FIGURE 8 Stress relaxation experiment; for cross-linked polymers σ_∞ is finite.

The quantity σ_∞ is the equilibrium stress. If a sample behaves like plotted in curve "cross-linked poly", flow was initiated in the specimen. When the sample is unloaded from the preset strain, it will remain in its final form and no contraction occurs. The elastic strain, which was initially imposed to the sample, was transformed completely into flow (or was irreversibly dissipated into flow).

Generalization of Hooke's law shows that in the range of linear viscoelasticity, material parameters become a function of time:

$$E(t) = \frac{\sigma(t)}{\varepsilon_o} \qquad G(t) = \frac{\sigma_s(t)}{\gamma_o} \tag{22}$$

We may also make the opposite experiment to stress relaxation. In the creep experiment, the sample is subjected to a preset stress σ_o = *constant* and we monitor deformation as a function of time, $\varepsilon = \varepsilon(t)$. The result is depicted in Figure 9. Let us discuss deformation $\varepsilon(t)$ in more detail. We introduce the terminal relaxation time τ_t. It is the longest relaxation time that can be observed in measurements of mechanical properties. For very short times as compared to τ_t chains cannot follow external changes, they are frozen in. One observes behavior typical for a glass. For longer times, but still $t < \tau_t$, conformations may change, however, chains remain entangled since the time for disentanglement is longer as t or $D > 1$. Hence, we observe behavior typical for a rubber network. No slippage or disentanglement occurs.

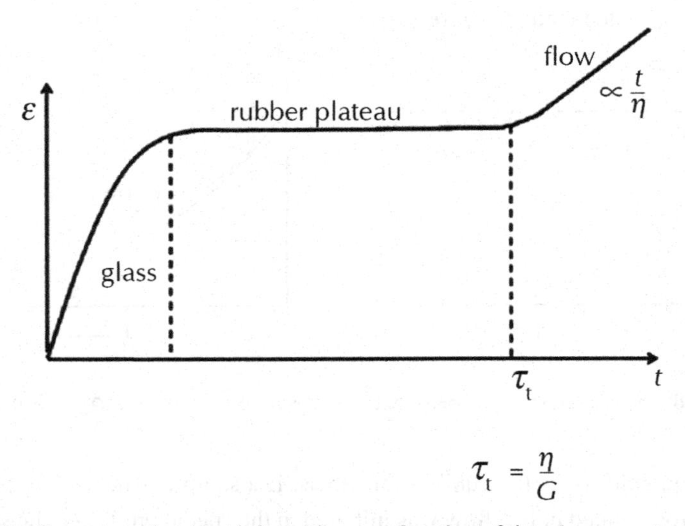

FIGURE 9 Creep experiment – deformation as a function of time (compare text).

At very long times of stress action (σ_o), disentanglement occurs which means flow. The deformation ε (t) reads:

$$\varepsilon = \frac{\sigma_o}{\eta} t \qquad \text{for } t > \tau_t \tag{23}$$

The Equation (22) yields immediately the relationship among plateau modulus, viscosity, and terminal relaxation time:

$$\eta = E\tau_t \tag{24}$$

We will see that relationship (Equation (24)) determines the frequency where the loss modulus has its maximum.

7.3.3 CYCLIC DEFORMATIONS

In a dynamic mechanical analyzer (DMA), a sinusoidal changing stress is imposed to the sample and will produce a sinusoidal change in strain (or vice versa).

Elastic sample after Equation (15):

$$\varepsilon = \varepsilon_o \sin \omega t \qquad \sigma = E\varepsilon_o \sin \omega t \tag{25}$$

For the elastic body stress and strain are "in phase", no lag occurs. The elastic body responds instantaneously to an external change. There is no phase shift angle; it is zero.

Ideal liquid: Applying sinusoidal varying stress, $\sigma_s = \sigma_{so} \sin \omega t$, we have with $\sigma_s = \eta \dot{\gamma}$

$$\frac{d\gamma}{dt} = \frac{\sigma_{so}}{\eta} \sin \omega t$$

It results:

$$\gamma(t) = -\frac{\sigma_{so}}{\eta\omega} \cos \omega t = \frac{\sigma_{so}}{\eta\omega} \sin\left(\omega t - \frac{\pi}{2}\right) \tag{26}$$

The relationship (Equation (26)) reveals that in a perfect liquid the phase shift angle of stress and strain is $\pi/2$. Figure 10 illustrates the phase shift δ for a visco-elastic body. It becomes obvious that zero strain amplitude coincides with maximum of stress for a perfect liquid. In other words, in a perfect liquid, stress and rate of strain are in phase rather than stress and strain. Strain lags behind stress by $\pi/2$. It follows for the phase shift angle δ:

Elastic body	$\delta = 0$	
Viscous body	$\delta = \pi/2$	(27)
Visco-elastic body	$0 < \delta < \pi/2$	

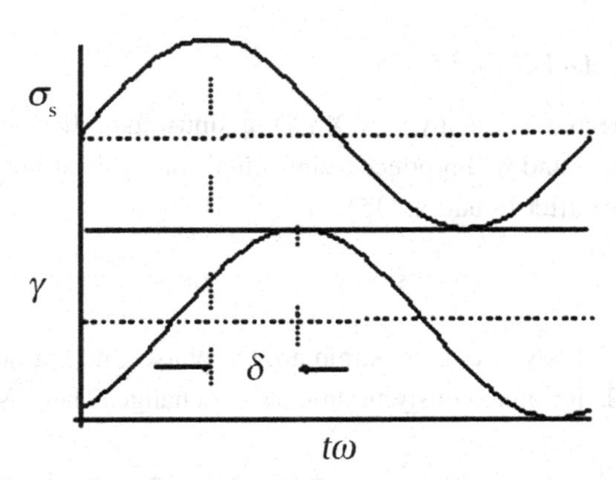

FIGURE 10 Phase shift for a viscoelastic body.

The lag of the response behind the preset value of stress or strain reflects relaxation processes. For each preset value of stress or strain, the system needs time for response. The response depends on the polymer and its relaxation time constants. Imposing a preset periodic shear strain, we get analogous to (26) for a viscoelastic body.

$$\gamma(t) = \gamma_o \sin \omega t \qquad \sigma_s = \sigma_{so} \sin(\omega t - \delta) \tag{28}$$

These periodic variations, given by Equation (26) and Equation (28), inspire to a more compact notation for viscoelastic properties. We see the shear stress for example as a complex quantity having elastic and viscous component. In that way, relationship (Equation 15) is generalized:

$$\sigma_s{}^* = G^* \gamma \tag{29a}$$

The complex quantities read:

$$\sigma_s{}^* = \sigma_s{}' + i\sigma_s{}'' \qquad G^* = G' + iG'' \tag{29b}$$

where X' and X'' are called real and imaginary part of complex quantity X^*. Subdividing $\sigma_s{}^*$ of Equation (30) by γ tells us that G' is related to elastic shear modulus after first Equation (15) and G'' reflects viscosity after second Equation (15), η/τ (compare Equation (24)). Therefore, they are called storage and loss modulus, respectively. The ratio of the two modules gives the phase shift:

$$\tan \delta = \frac{G''}{G'} \quad (30)$$

The situation is illustrated in the complex plane, Figure 11.

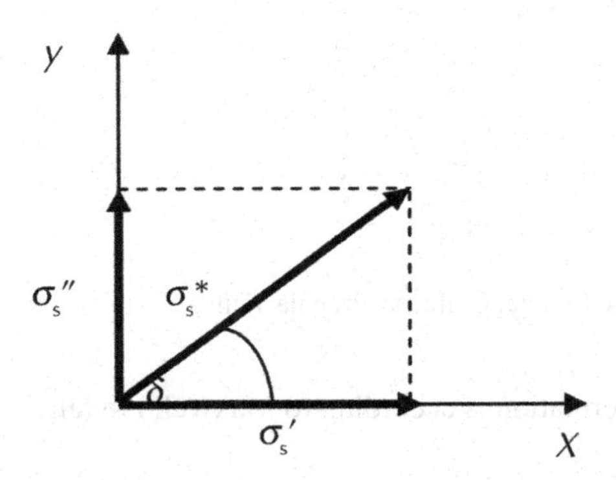

FIGURE 11 Complex shear stress and its components in the complex plane.

The stress displayed under sinusoidal strain by a viscoelastic body might be represented by a complex function:

$$\sigma_s{}^* = \sigma_{so}\exp[-i(\omega t - \delta)]$$

The modulus becomes also a complex quantity, G^*:

$$G' = \frac{\sigma_s{}'}{\gamma} = \frac{\sigma_{so}}{\gamma}\cos\delta \qquad G'' = G'\tan\delta = \frac{\sigma_{so}}{\gamma}\sin\delta \quad (31)$$

Maxwell's model on linear viscoelasticity provides insights in frequency dependence of complex modulus G^*. It discusses deformation in a system consisting of a spring with spring constant G in series with a dashpot comprising a liquid with viscosity η (see Figure 12). The dashpot resists deformation *via* viscous friction.

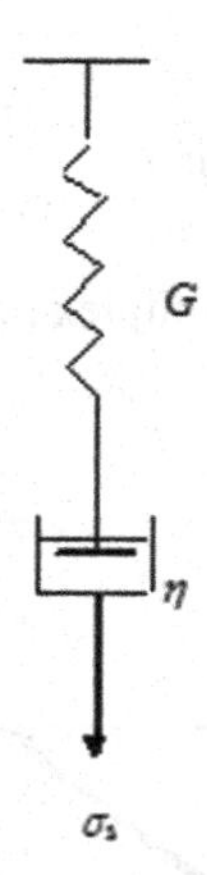

FIGURE 12 Maxwell model of linear viscoelasticity.

The rate of deformation is according to Maxwell model:

$$\frac{d\gamma}{dt} = \frac{d\gamma_{el}}{dt} + \frac{d\gamma_{visc}}{dt}$$

With Eq. (15), $\sigma_s = G\gamma$ and $\sigma_s = \eta\dot{\gamma}$, we get

$$\frac{d\gamma}{dt} = \frac{1}{G}\frac{d\sigma_s}{dt} + \frac{\sigma_s}{\eta}$$

Periodic variation of deformation and rate of deformation, $\gamma = \gamma_o e^{i\omega t}$ and $\dot{\gamma} = i\omega\gamma$, gives a linear differential equation for σ_s

$$\frac{d\sigma_s}{dt} + \frac{G}{\eta}\sigma_s = i\omega\gamma G \tag{32}$$

We use the following ansatz for solving Equation (32):

$$\sigma_s(t) = G^*\gamma(t) \tag{33}$$

It follows:

$$G^* = \frac{i\omega G\eta}{G + i\omega\eta}\frac{G - i\omega\eta}{G - i\omega\eta} = \frac{\omega^2\eta^2 G + i\omega\eta G^2}{G^2 + \eta^2\omega^2} \tag{34}$$

$$G' = G\frac{\frac{\eta^2}{G^2}\omega^2}{1 + \frac{\eta^2}{G^2}\omega^2} \quad \text{and} \quad G'' = \frac{\eta\omega}{1 + \frac{\eta^2}{G^2}\omega^2} \tag{35}$$

The ratio η/G equals after Equation (24) relaxation time constant τ, $\eta/G = \tau$. We may write for Equation (35):

$$G' = G\frac{(\tau\omega)^2}{1 + (\tau\omega)^2} \overset{\omega\to 0}{\propto} \omega^2 \qquad G' \overset{\omega\to\infty}{=} G$$

and

$$G'' = G\frac{\omega\tau}{1 + (\omega\tau)^2} \overset{\omega\to 0}{\propto} \omega \qquad G'' \overset{\omega\to\infty}{=} 0 \tag{36}$$

It becomes obvious that G' reflects the elastic part of deformation. We may say that G' shows how much energy a polymer accumulates at a given strain, and then returns when the load is removed. The second relationship of Equation (35) reflects work lost irreversibly in one cycle. It refers to the viscosity that is the flow occurring during deformation. Therefore, G'' is called loss modulus.

One easily proofs that G'' has a maximum at $(\omega\tau) = 1$

$$\omega_{\max} = \frac{G}{\eta} = \tau^{-1} \qquad G''_{\max} = \frac{G}{2} = \frac{\eta}{2\tau} \tag{37}$$

The maximal energy loss corresponds after Equation (18) to:

$$D = 1$$

where characteristic external time is equal to characteristic internal time. In other words, we have $D = 1$ when duration of action of the force (determined by frequency ω) equals the relaxation time constant of the polymer. The frequency dependence of G' and G'' is depicted in Figure 13.

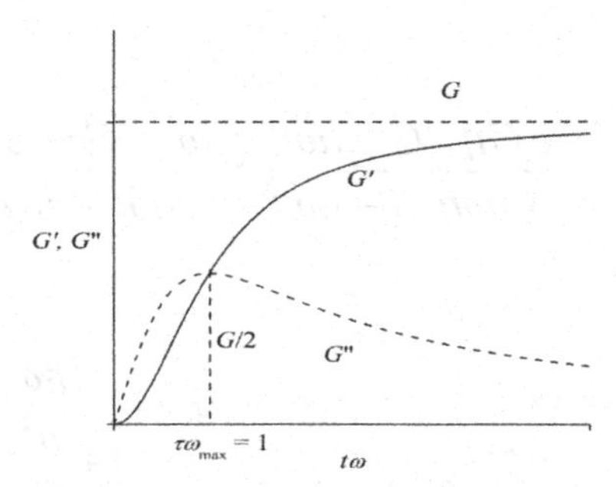

FIGURE 13 Frequency dependence of real and imaginary part of G^* after Maxwell model.

For illustration, we mention here linear viscoelastic behavior of nanocomposites consisting of poly(methyl methacrylate) (PMMA) and nanotubes (Du et al., 2004). Nanotubes were combined with PMMA of $M_w = 10^5$ g/mol. The highest mass fraction of nanotubes amounted to 0.02. Rheological measurements were performed at 200 °C and 0.5 % preset strain. As Figure 14 shows, nanotubes have dramatic effect on rheological behavior in the range of low frequencies. As the loading increases both G' and G'' increase especially at low frequencies. At 200 °C and low ω, the PMMA chains are fully relaxed, G' and G'' vary according to Equations (36):

$$G' \propto \omega^2 \text{ and } G'' \propto \omega$$

Due to polydispersity, slopes are slightly lower, 1.95 and 0.96, respectively.

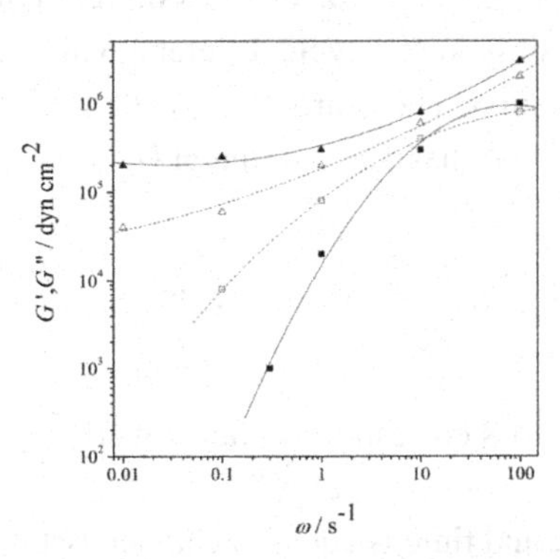

FIGURE 14 The variation of G' (solid markers) and G'' (open markers) with frequency PMMA – squares, PMMA + 2 wt% nanotubes – triangles; graph after data of Du et al. (2004).

At nanotube loadings higher than 0.2 wt% this terminal behavior disappears and the dependence of G' and G'' on ω at lower frequencies becomes weak (at 0.2 wt% loading slope of G' 0.09 and of G'' 0.21; compare Figure 14). Thus, large-scale polymer relaxations in the nanocomposites are effectively restrained by the presence of nanotubes. The fact that G' is almost independent of ω at low frequencies, when the nanotube content is higher than 0.2 wt%, is indicative of a transition from liquid-like to solid-like viscoelastic behavior. We also observe that G' exceeds G'' even at low ω, which demonstrates solid-like behavior at high nanotube loadings. This non-terminal behavior may be attributed to a nanotube network, which restrains the long-range motion of polymer chains. At high frequencies, the effect of nanotubes on rheological behavior is weak. This behavior suggests that nanotubes do not significantly influence the short-range dynamics of polymer chains, particularly on length scales comparable to the entanglement length. Nanotubes do not vary T_g. Hence, presence of nanotubes has a substantial influence on polymer chain relaxations, but has little effect on polymer motion at length scales comparable to entanglement length.

We may also define a complex viscosity:

$$\eta^* = \eta' - i\eta'' \tag{38}$$

We proceed analogously to Equation (28). It follows with strain $\gamma = \gamma_o e^{i\omega t}$ after second Equation 15:

$$\sigma_s^* = \eta^* \dot{\gamma} = \eta^* i\omega\gamma \tag{39}$$

It follows after Equation 29:

$$\eta^* = \frac{G^*}{i\omega} \tag{40}$$

Using (Equation 31), we get:

$$\eta' = \frac{\sigma_{so}}{\omega\gamma}\sin\delta \qquad \eta'' = \frac{\sigma_{so}}{\omega\gamma}\cos\delta = |\eta^*|\cos\delta$$

Inserting Equation 34, we get from Equation 40:

$$\eta^* = \frac{G^2\eta}{G^2+\omega^2\eta^2} - i\frac{\omega\eta^2 G}{G^2+\omega^2\eta^2}$$

The real and imaginary part of complex viscosity read:

$$\eta' = \eta\frac{1}{1+(\omega\tau)^2} \qquad \eta'' = \eta\frac{\omega\tau}{1+(\omega\tau)^2} \tag{41}$$

For the limits in frequency, it follows:

$$\eta' \to \eta \text{ for } \omega \to 0 \qquad \eta'' \to 0 \text{ for } \omega \to 0$$

$$\eta', \eta'' \to 0 \text{ for } \omega \to \infty$$

The real part η' approaches for low frequencies the steady-state flow viscosity. This tendency can be used to determine flow viscosity of systems with an extremely high viscosity. The magnitude of complex viscosity is given by:

$$|\eta^*| = \sqrt{\eta'^2 + \eta''^2} \tag{42}$$

The Figure 15 gives real and imaginary part of complex viscosity according to Maxwell model, Equations (40) and (41). At high frequencies, frequency dependencies are similar to the real and imaginary part of G^*.

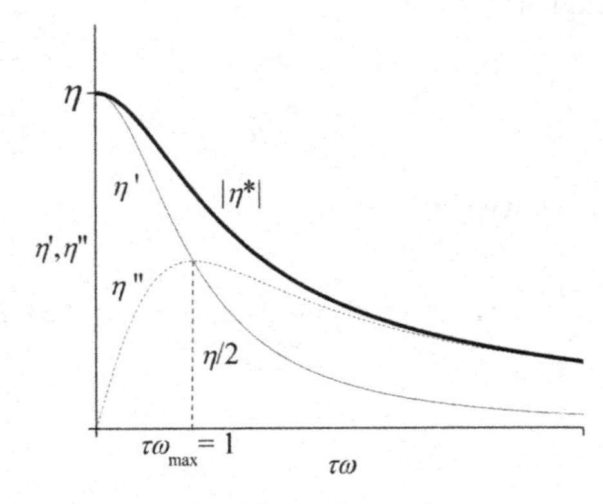

FIGURE 15 Complex viscosity after Maxwell model.

The Figure 16 presents complex viscosity of nanocomposites. The composites are dispersions of silicate layers in a matrix of polylactide (PLA). Viscosity measurements were carried out at 175 °C and a preset strain of 5 %. We recognize at low $\omega < 10\ s^{-1}$ that pure PLA exhibits Newtonian behavior whereas all composites show shear thinning tendency. The high viscosity of the composites might be explained by flow restrictions of polymer chains in the molten state arising from the presence of nanoparticles (silicate layers). Perhaps also shear-induced alignment of dispersed nanoparticles occurs. At high frequencies no significant difference between polymer PLA and the composites can be seen.

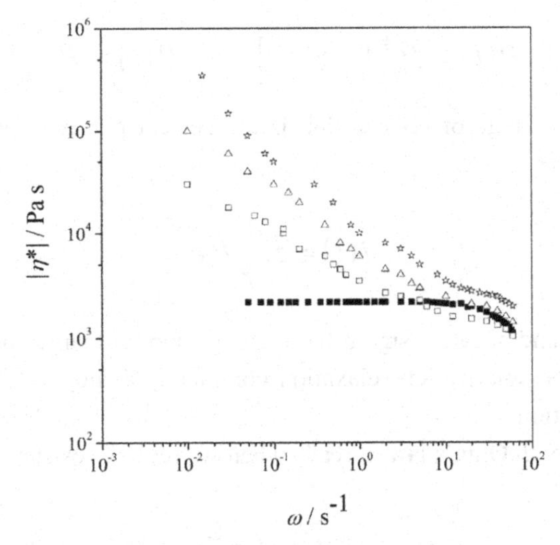

FIGURE 16 Complex viscosity of PLA-silicate nanocomposites, graph after data of Sinha and Okamoto (2003) • PLA, ▫ PLA + 2 wt%, Δ PLA + 3 wt%, ☆ PLA + 5 wt% of silicate.

7.4 DIELECTRIC POLARISATION IN POLYMER-SALT SYSTEMS

In this paragraph, electric oscillations are our concern. The experimental tool is impedance spectroscopy. It provides information on electric properties and electrochemical processes proceeding in materials. We are focusing in the following on electric properties of polymer electrolytes that form dielectrics.

The impedance is a generalized resistance. It refers to the ability of a material to resist flow of an electric current. We immediately recognize that impedance in the linear regime must display similar properties with respect to electric current as complex viscosity (Equation 38) with respect to flow of matter.

We may also argue analogous to stress relaxation (Equation 16). Here, we discuss polarization relaxation $P(t)$:

$$\frac{\mathrm{d}P}{\mathrm{d}t} = -\frac{1}{\tau} P(t) \tag{43}$$

Solutions of Equation (43) read under conditions of charge and discharge, respectively, as follows:

$$P(t) = P_o\left[1 - e^{-t/\tau}\right] \quad \text{or} \quad P(t) = P_o e^{-t/\tau} \tag{44}$$

The rate of change of polarization is the current per area that flows in the external circuit:

$$i(t) = \pm\frac{1}{\tau} P_o e^{-t/\tau} \tag{45}$$

where, positive and negative sign refer to charge and discharge conditions. Thus, we formulate for polarization relaxation completely analogous relationships as for stress relaxation.

Let us start with Ohm's law. It reads when we replace resistance R by electric conductivity σ:

$$I = A\sigma\frac{U}{L} \tag{46}$$

where A and L are area and thickness of the conducting sample, U/L symbolizes the electric field. After Equation (45), we have also:

$$I = A\dot{P} \tag{47}$$

The polarization P is given in phenomenological terms by:

$$P = \varepsilon_o \chi \frac{U}{L} \tag{48}$$

with $\varepsilon_o = 8.85 \cdot 10^{-12}$ A s/V m, permittivity of the empty space, and the material parameter dielectric susceptibility χ. The polarization varies under a periodic electric field in the same way as deformation (compare Equation (32)). It follows with $\dot{P} = i\omega P$ and use of Equations (46) – (48) for complex conductivity:

$$\sigma^* = i\omega\varepsilon_o\chi^*(\omega) \tag{49}$$

For the electrical power, $N = UI = \sigma' U^2 A/L$, we get with the real part of Equation (49):

$$N_1 = \chi'' \omega\varepsilon_o U^2 A/L \tag{50}$$

Equation (50) represents the loss of energy per second in the conducting system under voltage U.

Suppose, we apply a sinusoidal electrical potential. In linear systems, the response to this potential will be a sinusoidal current having the same frequency but shifted in phase. We get exactly the same periodicities as in Figure 10 when we replace σ_s by electric potential U and deformation γ by current I. Resistance R is defined as the ratio of voltage to current by Ohm's law:

$$R = \frac{U(t)}{I(t)} \tag{51}$$

Quantity R does not depend on frequency. Analogous to elasticity (compare Equation (27)), alternating current (AC) and voltage over a resistor are in phase with each other. The direct current (DC) knows only a resistor, but an AC current feels also resistance towards a capacitor C and an inductor L. Table 4 lists the electric circuit elements, the corresponding versions of Ohm's law and the impedance relations. Resistance R does not have an imaginary part. In contrast, capacitor and conductor do not have real parts. As a result, the current through a capacitor is shifted by –90 degrees with respect to voltage whereas that through an inductor is shifted by +90 degrees compared to voltage.

TABLE 4 The elements of complex impedance

Element	$U-I$ relationship	Impedance
resistor	$U=RI$	R
capacitor	$\frac{dU}{dt}=\frac{I}{C}$	$\frac{1}{i\omega C}$
inductor	$U=L\frac{dI}{dt}$	$i\omega L$

The periodic variation of voltage leads to a similar differential equation as for variation of deformation, Equation (32). Currents passing a circuit with resistor and capacitor read with complex voltage U.

$$I=C\dot{U}+\frac{U}{R} \tag{52}$$

The complex voltage is related to complex impedance by $U=ZI$. Current varies periodically as deformation in Equation (32). Hence, we have for $\dot{U}$, $\dot{U}=i\omega U$. With $\tau=CR$, complex impedance has the same structure as complex viscosity, Equation (41).

$$Z=\frac{R}{1+i\omega\tau} \tag{53}$$

It means real and imaginary part of complex impedance show the same frequency dependence as the parts that constitute complex viscosity. This frequency dependence is depicted in Figure 15. The relation (Equation 53) was firstly derived by Debye (1928). We get the same expression for complex impedance if we employ the circuit sketched in Figure 17. It is the equivalent circuit with Ohm resistance coupled parallel to a capacitor. For parallel arrangement of R and C, we get after Table 4:

$$\frac{1}{Z}=\frac{1}{R}+i\omega C$$

which is identical with Equation (53). Conductivity equals specific impedance:

$$\sigma = \frac{L}{AZ} \tag{54}$$

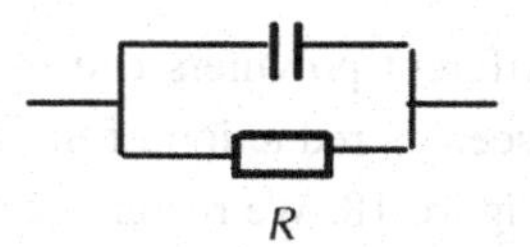

FIGURE 17 Parallel arrangement of resistor and capacitor.

The geometric factor is given as in Equation (46). The unit of conductivity is usually (cm Ω)$^{-1}$.

We note here, frequently capacitors do not behave ideally as listed in Table 4. In experiments, it is observed instead:

$$Z = \frac{R}{1+(i\omega\tau)^{n}} \text{ with } n < 1 \tag{55}$$

The empirical impedance contribution is sometimes called constant phase element. According to Figure 15, real and imaginary parts of complex viscosity are crossing each other in the maximum of η". This is also true for real and imaginary part of impedance as long as Equation (53) is obeyed. When in complex impedance a contribution like Equation (55) occurs then the two curves do not cross precisely in the maximum of Z". Crossing is shifted to higher frequency.

In $Z' - Z''$ plane, the impedance after Equation (53) forms a circle with center at $R/2$ and radius $R/2$:

$$|Z - \frac{R}{2}|^2 = \left(\frac{R}{2}\right)^2 \tag{56}$$

The magnitude of complex impedance is given by:

$$|Z|^2 = (Z'^2 + Z''^2) \tag{57}$$

7.4.1 GLASS TRANSITION TEMPERATURE OF POLYMER ELECTROLYTES

The solid polymer electrolytes are mixtures of an inorganic salt and an organic polymer, preferably polyether like poly(ethylene oxide) (PEO). Relaxation and diffusion phenomena in context of ionic conductivity in those systems are our concern in this paragraph.

Lithium salt added to different polymers causes changes of glass transition temperature. This might be seen as reduction of the free volume with increasing salt content. It is shown in Figure 18. We recognize that addition of salt leads to dramatic increase in T_g of PEO whereas the T_g of epoxidized natural rubber (ENR-25 with 25 mol% of epoxy) exhibits a negligible shift. The T_g of PMMA is not influenced by nanotubes (Figure 18(b)), data after Du et al. (2004). In more general terms, we may state salt restricts chain mobility in PEO on lengths scales of entanglement length. But there is no effect on local motions in ENR or in PMMA.

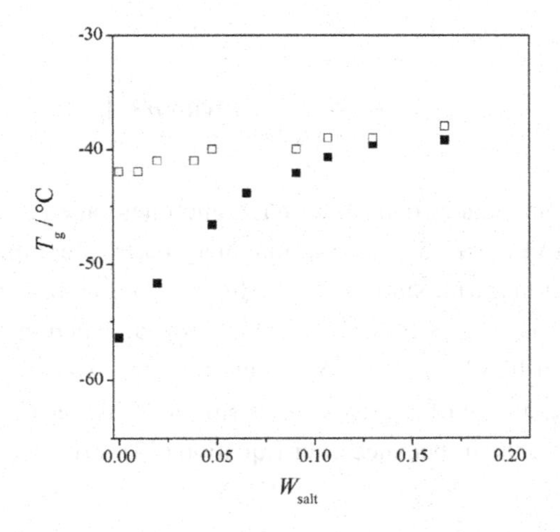

FIGURE 18 *(Continued)*

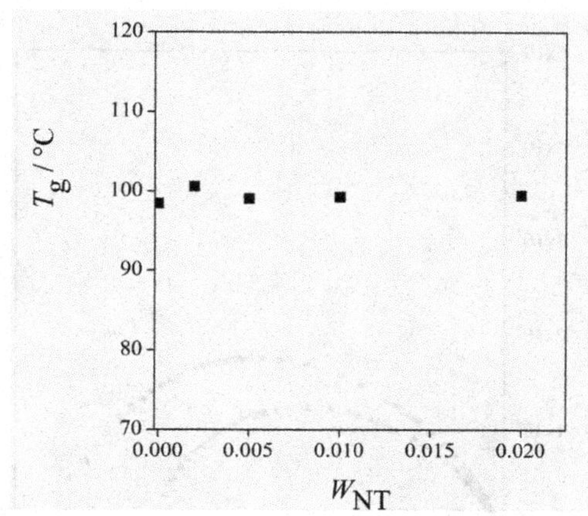

FIGURE 18 Glass transition temperature as a function of (a) $LiClO_4$ content and (b) nanotube content. a) PEO – solid squares, ENR-25 – open squares and b) PMMA + nanotubes.

7.4.2 COMPLEX IMPEDANCE

The turning to impedance spectroscopy, polymer-salt systems should exhibit semicircles in the Z'- Z'' plane. Figures 19 and 20 provide examples that refer to mixtures of polymer and $LiClO_4$. We recognize that in the range of low frequencies deviations from the circle occur.

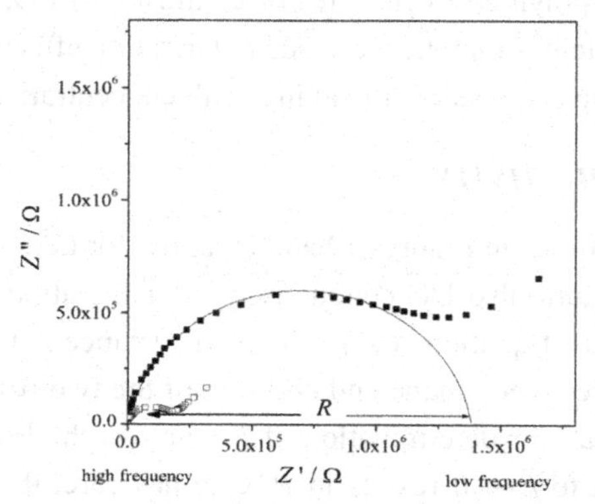

FIGURE 19 Semicircles for PEO; solid squares PEO + 2% salt[1]; the semicircle was calculated with $R = 1.4\times10^6\ \Omega$ after Equation (56); the center of the circle is around $10^5\ \Omega$ below $Z'' = 0$; open squares PEO + 5% salt. Figure is adapted from reference (8).

[1]The concentration measure here used is Y = (mass of salt)/(mass of polymer); it follows for mass fraction of salt $W_s = Y/(Y + 1)$

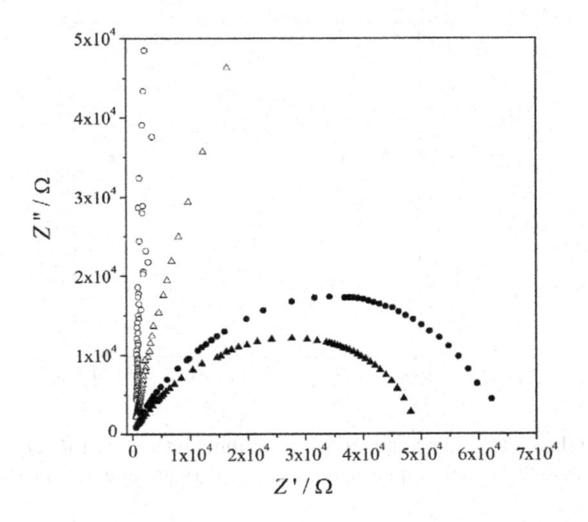

FIGURE 20 Semicircles for ENR-25; open circles ENR + 5% salt, open triangles ENR + 20% salt, solid circles ENR + 25%, solid triangles ENR + 30% salt. Figure is adapted from reference (8).

These deviations are caused by electrode polarization owing to double layer formation in electrode-electrolyte interface. Only the semicircle reflects the bulk conductivity of the polymer electrolyte. In contrast to PEO, ENR-25 displays even at high salt contents dramatic electrode polarization effects. This dominating electrode polarization disappears only at high salt concentrations, $Y > 0.2$.

7.4.3 DC CONDUCTIVITY

The semicircles as shown in Figures 19 and 20 allow for determination of Ohm's resistance R and in turn also DC conductivity after Equation (54). We mention here in anticipation of Equations (57)—Ohm's resistance R follows from maximum of Z'' in Z''–frequency plane and crossing of the two parts of complex impedance. Another way for determination of R employs the low frequency range of Z'. Extrapolation to $f \to 0$ results in R. One has carefully to select the low frequency part that does not comprise electrode polarization at lowest frequencies (compare Figure 19). All these procedures yield approximately the same result.

In the range of low concentrations of $LiClO_4$ in ENR, we observe variation of DC conductivity with salt concentration in a way as it is typical for weak electrolytes. A graph of equivalent conductivity, $\Lambda = \sigma_{DC}/c$, as a function of root of molar concentration is given in Figure 21. The concentration range refers to $Y = 0.01$ up to 0.2 which corresponds to $c = 1\times10^{-4}$ up to $\approx 2\times10^{-3}$ mol/cm³. It follows that equivalent conductivity obeys in the range of low concentration Kohlrausch's law.

$$\Lambda = \Lambda_o - Ac^{1/2}$$

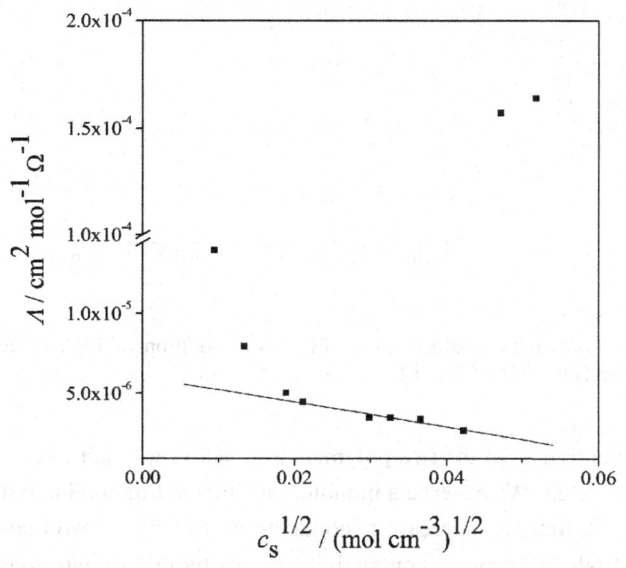

FIGURE 21 Equivalent conductivity of ENR-25 with addition of $LiClO_4$, the concentration range corresponds to $0.01 \leq Y \leq 0.3$.

Deviations occur at lowest concentrations, which means here below $Y = 0.05$ corresponding to $c \approx 4{\cdot}10^{-4}$ mol/cm³. We also recognize that conductivity jumps above $Y > 0.2$ by two orders of magnitude.

Completely different behavior displays PEO at low concentrations of salt (Figure 22). The concentration covers the range $c_s = 1{\cdot}10^{-4}$ to $1{\cdot}10^{-3}$ mol/cm³. In contrast to ENR-25, equivalent conductivity increases linearly with $c^{1/2}$.

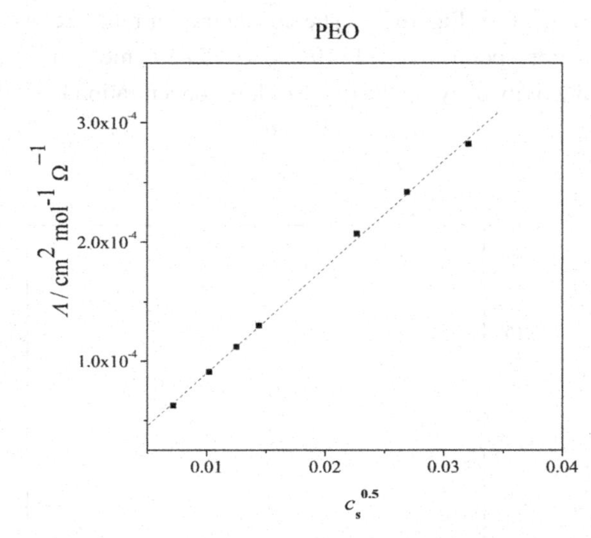

FIGURE 22 Equivalent conductivity of PEO with addition of $LiClO_4$, concentration range corresponds to $0.005 \leq Y \leq 0.1$

The conductivities of the two polymers plus salt *versus* salt concentration are shown in Figure 23. We observe a monotonous increase of conductivity with salt content in PEO. Actually, increase of conductivity follows a power law (compare Figure 24). ENR-25 displays conductivity values by one to two orders of magnitude lower as compared to PEO. A saturation value occurs at relatively low salt concentration. However, we observe a jump in conductivity by two orders of magnitude at quite high salt concentration. This transition insulator-conductor reminds of formation of a percolation network. Good electrical conductivity in polymer-salt systems seems to be related to heterogeneities or formation of a random network of density fluctuations in the matrix polymer. The results just discussed indicate that network-like structures are formed differently in the polymers under consideration. In PEO, they already exist in the neat polymer due to associations between chains. Interactions between salt and chain molecules, as the T_g variation shows, stabilize these heterogeneities. Only weak interactions exist between salt and ENR-25 and no associations between chains. Therefore, heterogeneities are only formed at high salt content leading to the jump in conductivity shown in Figure 23. We will return to this important point in the context of polarization relaxations.

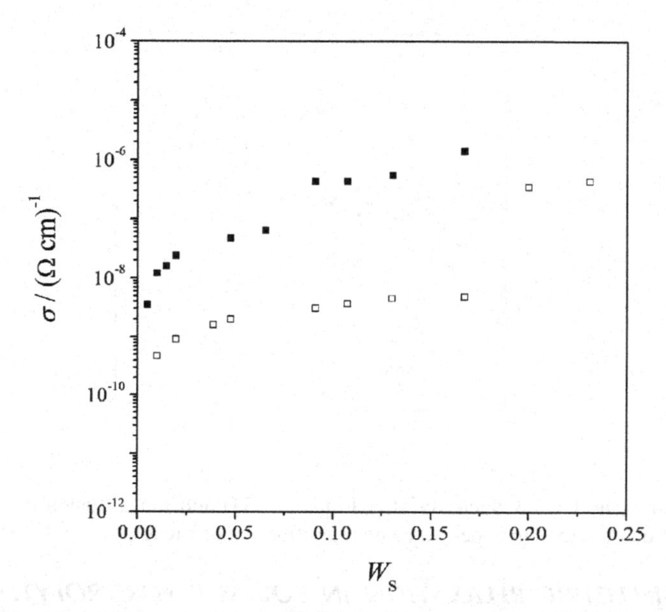

FIGURE 23 The DC conductivity of PEO (solid squares) and ENR-25 *versus* mass fraction of $LiClO_4$ at 300 K.

Following Nernst–Einstein equation, one may relate DC equivalent conductivity to ion diffusion coefficient:

$$D_{ion} = \frac{RT}{F^2} \Lambda \tag{58}$$

The quantity F is Faraday's constant (96485 A s/mol). Figure 24 presents plots of conductivity as well as diffusion coefficient *versus* mass fraction of salt. We observe power law dependencies on salt content:

$$\sigma / (\Omega\ cm)^{-1} \propto W_s^{1.6} \qquad D_{ion} / cm^2\ s^{-1} \propto W_s^{0.54} \tag{59}$$

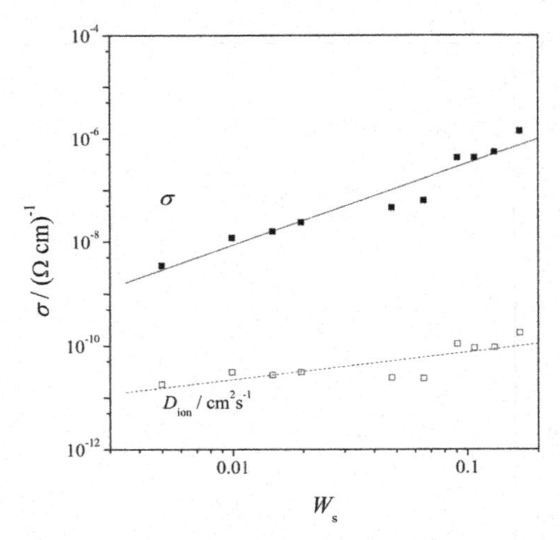

FIGURE 24 The DC conductivity after Equation (54) and ion diffusion coefficient in PEO after Equation (58) in dependence on mass fraction of $LiClO_4$.

7.4.4 DIELECTRIC RELAXATION IN POLYMER ELECTROLYTES

The addition of salt to a polymer may lead to formation of dipoles between ions and chain molecules. These dipoles have a (longitudinal) component along the chains as it is sketched in Figure 25. The impedance measurements reflect relaxation of these longitudinal dipole components since they relax sufficiently slowly. Dipole relaxations occur as maxima in plots of Z'' as a function of frequency. It is analogous to complex viscosity sketched in Figure 15. Real and imaginary part of complex impedance (Equation 53) read:

$$Z'(\omega) = \frac{R}{1+(\omega\tau)^2} \qquad Z''(\omega) = R\frac{\omega\tau}{1+(\omega\tau)^2} \tag{60}$$

The angular frequency is given by $\omega = 2\pi f$.

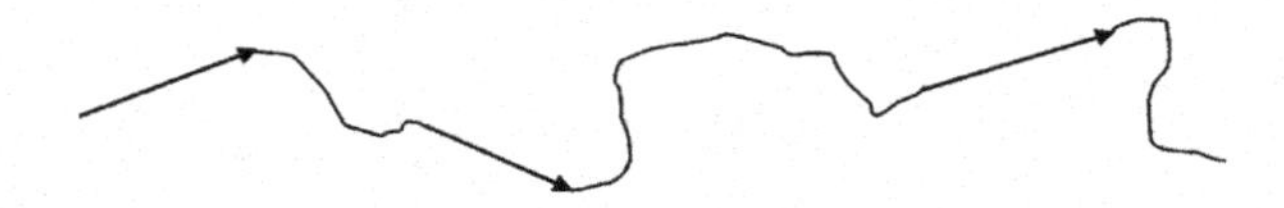

FIGURE 25 Dipole components along a chain molecule.

One easily verifies that Z'' has a maximum at $\omega_0 \tau = 1$. We denote quantity τ as fundamental relaxation time. The real and imaginary parts of Z *versus* frequency f are shown for PEO in Figure 26. Polarization relaxations, reflected by constancy of Z' at low frequencies and a maximum of Z'', occur in PEO even at very low salt concentrations, but they do not occur in neat PEO. Salt concentration $Y = 0.01$ means that one salt molecule occurs among around 240 monomer units of PEO. We observe that the two curves cross very close to the maximum of Z''. It means in the low-frequency range we have Debye relaxation after Equation (60) to a good approximation. Moreover, we recognize crossing shifts to higher frequencies with increasing salt content or fundamental relaxation time shortens. The relaxation of polarization towards equilibrium becomes faster with increasing salt concentration.

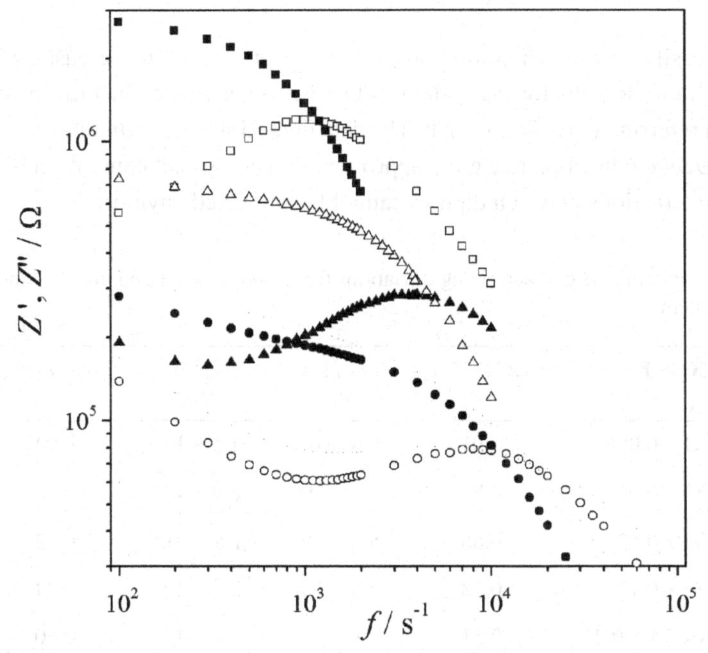

FIGURE 26 The real part (solid symbols) and imaginary part of impedance of PEO *versus* frequency for different salt concentrations at 300 K; Y: ■ – 0.005, ▲ – 0.01, ● – 0.02.

It might be interesting to have a closer look to crossing of the two impedance constituents. The Equation (53) after Debye describes relaxations of non-interacting dipoles. In such an ideal system, real and imaginary part of impedance cross at the maximum of $Z''(\omega)$. Interactions between dipoles become more important with increasing salt content. This is taken empirically into account by inserting contributions (Equation 55) into Equation (53). It follows:

$$Z = \frac{R}{1+(i\omega\tau)^{n}} \quad \text{with } n \leq 1 \tag{61}$$

The corresponding expressions to Equation (60) are given in Appendix. The ratio of interest here is:

$$\frac{Z''(\omega\tau_o = 1)}{Z'(\omega\tau_o = 1)} = \frac{\sin(n\pi/2)}{1+\cos(n\pi/2)} \tag{62}$$

One easily verifies, the ratio equals unity for $n = 1$ and n decreases with descending ratio. Results for the system PEO + $LiClO_4$ are listed in Table 5. We add also data referring to ENR-25 + salt. The data in the Table confirm that we observe in PEO Debye relaxation to a good approximation at low salt content. At high salt content interactions between dipoles cannot be neglected anymore.

TABLE 5 Parameters characterizing deviations from Debye relaxation in PEO- and ENR-25-salt system

PEO + *Y*	$(Z''/Z')_{\omega\tau=1}$	ω_o / s^{-1}	τ / s	n
PEO + 0.005	0.91	6.6×10^{3}	1.5×10^{-4}	0.94
PEO + 0.01	0.88	2.6×10^{4}	3.9×10^{-5}	0.92
PEO + 0.02	0.88	5.6×10^{4}	1.8×10^{-5}	0.92
PEO + 0.1	0.78	4.6×10^{5}	2.2×10^{-6}	0.84
ENR-25 + 0.25	0.51	8.1×10^{3}	7.7×10^{-4}	0.60
ENR-25 + 0.3	0.49	1.6×10^{4}	3.8×10^{-4}	0.58

The relaxation frequency ω_o of PEO varies with salt concentration according to a power law:

$$\omega_0 \cong 10^7 W_s^{1.4} s^{-1l} \quad \text{for} \quad 0.005 \leq Y \leq 0.2 \tag{63}$$

The Figure 27 presents responses of ENR-25 to periodic voltage excitations. No polarization relaxation can be observed up to $Y = 0.2$, Z'' exceeds Z' and therefore, no crossing can be monitored. This behavior is consistent with negligible changes in glass transition temperature of ENR-25 under influence of salt (compare Figure 18), indicating weak interaction between salt molecules and chains. Polarization relaxation starts only at high salt content, $Y \approx 0.25$. One observes pronounced shift of crossing towards high frequencies as it is manifested in data of Table 5. It leads to small values of exponent n in relationship (Equation 61) indicating deviation from Debye relaxation.

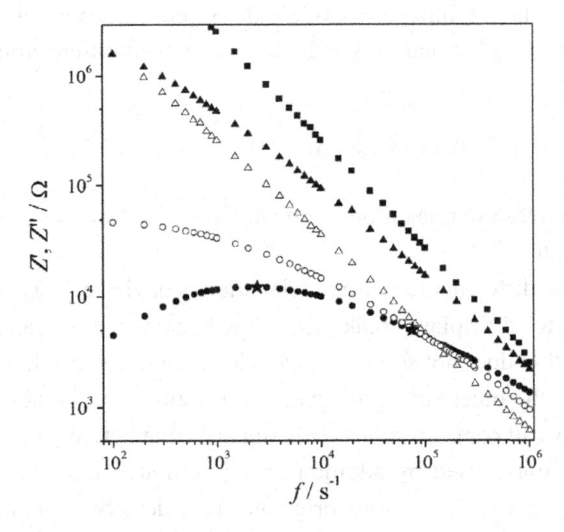

FIGURE 27 The real part and imaginary part (solid symbols) of impedance of ENR-25 *versus* frequency for different salt concentrations at 300 K, Z''-solid symbols; Y: ■ – 0.1; ▲ – 0.2, ● – 0.3; stars indicate maximum of Z'' and crossing of Z' and Z''.

After Table 5, one observes nearly perfect behavior in PEO at salt concentration $Y = 0.02$. The ratio of (ethylene oxide) units and salt molecules amounts to $N_{EO}/N_{LiClO4} \approx 120$ for this concentration Y. Hence, dipoles are highly diluted and behave perfectly. The ratio reduces to $N_{unit(ENR)}/N_{LiClO4} \approx 6$ for ENR-25 with salt concentration $Y = 0.25$. It becomes understandable that interactions between dipoles are not anymore negligible.

One may get an impression about the size of the relaxing unit by applying Einstein's relationship. The relaxing dipole traverses with diffusion coefficient D_{ion} distance R_{d}. It is:

$$R_{\text{d}}^2 = \tau D_{\text{ion}} \tag{64}$$

Using values given in Figure 24 and Table 5, we have $D_{\text{ion}} = 1.1 \times 10^{-10}$ cm^2 s^{-1} and $\tau = 2.2 \times 10^{-6}$ s for PEO with $Y = 0.1$. We get for the size of the relaxing unit:

$$R_{\text{d}} = 0.16 \text{ nm}$$

This size reduces slightly with salt content. We get $R_{\text{d}} = 0.10$ nm for $Y = 0.2$. The size of the relaxing unit in ENR-25-salt system is one order of magnitude greater than for PEO. Estimation yields for two concentrations after percolation threshold:

$$R_{\text{d}}\,(Y = 0.25) = 1.8 \text{ nm} \qquad R_{\text{d}}\,(Y = 0.3) = 1.3 \text{ nm}$$

Owing to increase in relaxation frequency, size R_{d} decreases slightly with increasing salt content.

We note quite different electric and relaxation behavior in PEO- and ENR-salt system. The ENR-25 displays weak electrolyte behavior in the range of low salt concentration. PEO does not show this behavior even not at still lower concentrations of salt. On the other side, polarization relaxations can be observed in PEO at extremely low salt concentrations. This suggests that heterogeneities or density fluctuations are not caused by addition of salt although they are stabilized by salt. The concentration fluctuations originate from density fluctuations in PEO. Therefore, we cannot clearly identify a percolation threshold in PEO in contrast to ENR-25 (compare Figure 23). This conclusion is strongly supported by dynamic light scattering experiments (Walter et al., 2002). The scattering experiments revealed that PEO in the molten state does not form a homogeneous system. There exist local density fluctuations caused by associations between chains. Addition of salt leads to stabilization of those fluctuations. Therefore, the system displays even at very low salt concentrations dipole relaxations as shown in Figure 26. Thus in PEO, interactions between both polymer chains as well as salt molecules and monomer units, lead to formation of random networks that generate dielectric response. Heterogeneities in ENR-25 originate dominantly from salt-monomer interactions. These couplings lead to local concentration fluctuations which are equivalent to random networks above a critical salt concentration. As a result, one observes percolation threshold.

7.4.5 AC CONDUCTIVITY

The real part of complex conductivity in ion conducting heterogeneous materials approaches at low frequencies, $\omega \rightarrow 0$, DC conductivity in absence of electrode polarization. In the limit of high frequencies, the AC conductivity turns to a power law dependence on frequency, $\sigma_{AC} \propto \omega^{n}$. One may combine these dependencies on frequency to the following relationship:

$$\sigma_{ac} = \sigma_{dc}\left[1+\left(\frac{\omega}{\omega_{o}}\right)^{n}\right] \tag{65}$$

The frequency ω_{o} terminates the DC conductivity plateau. For $\omega > \omega_{o}$, AC conductivity becomes dominant, $\sigma_{AC}(\omega) > 2\sigma_{DC}$.

The simulations of conductivity in random resistor-capacitor (RC) networks confirmed frequency dependence of Equation (65) (Panteny et al., 2005). In random networks percolated resistors lead to the DC plateau at low frequencies. In contrast, non-percolated resistors generate negative deviation of conductivity from DC plateau and non-monotonous decrease of conductivity at low frequencies. These regular or irregular deviations from DC plateau are in polymer-salt systems due to electrode polarization. At high frequencies random network simulations recovered the power-law dependency of conductivity as in Equation (65). Generally, one may say conductivity in random RC networks will be preferably determined by resistors at low frequencies and by capacitors at high frequencies.

The Figures 28 and 29 present curves $\sigma_{AC}(f)$ after Equation (65). The AC conductivity of the heterogeneous PEO-salt system is coined by percolated resistors at low frequency. Almost no deviation from DC plateau displays the system with the lowest salt content. Electrode polarization is negligible. It becomes remarkable at higher salt concentration (compare Figure 19).

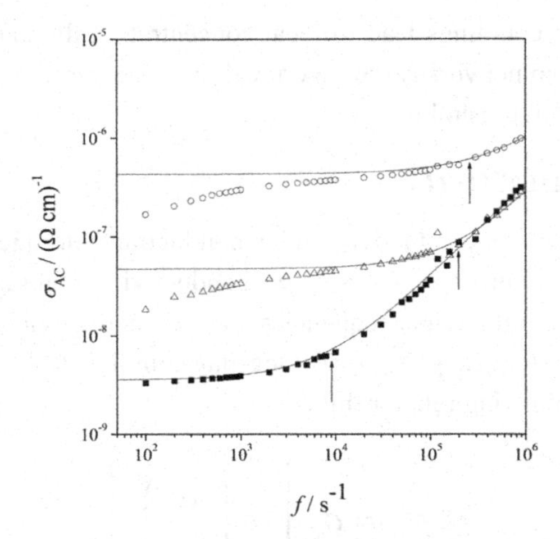

FIGURE 28 The AC conductivity of PEO as a function of frequency; arrows mark frequency f_o; Y: ■ – 0.005, Δ – 0.05, o – 0.1; solid curves according to Equation (65) with (f_o; n) = (8.7×10^3 s^{-1}; 0.96), (2.1×10^5 s^{-1}; 1.01), (7.0×10^5 s^{-1}; 0.91), respectively.

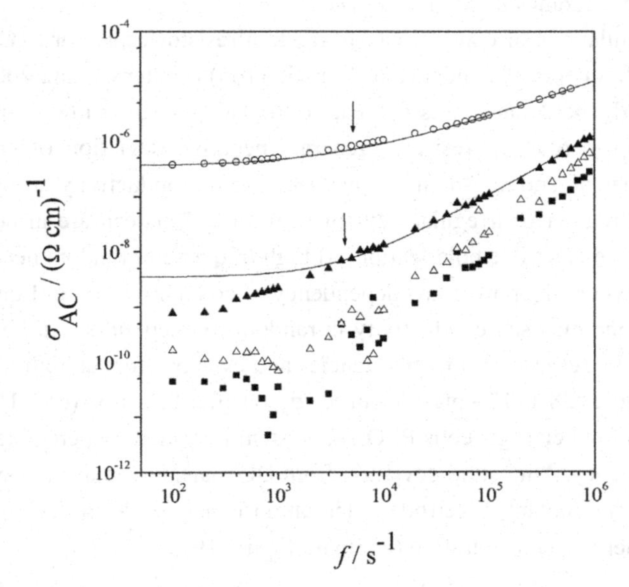

FIGURE 29 The AC conductivity of ENR-25 as a function of frequency after Equation (65); arrows mark frequency f_o; Y: ■ – 0.01, Δ – 0.05, ▲ – 0.1, o – 0.25; solid curves according to Equation (65) with (f_o; n) = (4.1×10^3 s^{-1}; 0.98), (5.0×10^3 s^{-1}; 0.67), respectively.

The behavior of ENR-25 at low salt concentration is different from PEO. Non-percolated resistors determine irregularities occurring in the low-frequency range. At high frequencies, one may recognize power-law dependency even at low salt concentration. Heterogeneities at low frequency transform in ENR-25 to percolated resistors for salt concentration $Y \geq 0.1$. Frequency f_o characterizes onset of AC conductivity. The Figures 28 and 29 reveal that DC plateau is more extended with increasing salt content. ENR-25 displays quite low frequencies f_o as compared to PEO indicating that the DC plateau is restricted to smaller frequency range as in PEO.

APPENDIX

The complex impedance (Equation 48) might be generalized as indicated in Equation (49). It reads:

$$Z(\omega) = \frac{R}{1 + (i\omega\tau)^{n}} \tag{A1}$$

Equation (48) describes non-interacting dipoles. Interaction between dipoles is taken into account by the empirical generalization (Equation (A1)). The real and imaginary part of Equation (A1) might be given by:

$$Z' = R\frac{1 + (\omega\tau)^{n} \cos\left(n\frac{\pi}{2}\right)}{N} \qquad Z'' = R\frac{(\omega\tau)^{n} \sin\left(n\frac{\pi}{2}\right)}{N} \tag{A2}$$

With,

$$N \equiv 1 + 2(\omega\tau)^{n} \cos\left(n\frac{\pi}{2}\right) + (\omega\tau)^{2}$$

KEYWORDS

- **Debora number**
- **Dielectric relaxation**
- **Enthalpy relaxation**
- **Linear viscoelasticity**
- **Non-equilibrium thermodynamics**

REFERENCES

1. Cowie, J. M. G. and Ferguson, R. *Macromolecules*, **22**, 2307 (1989).
2. Du, F. M., Scogna, R. C., Zhou, W., Brand, S., Fischer, J. E., and Winey, K. I. *Macromolecules*, **37**, 9048 (2004).
3. Panteny, S., Stevens, R., and Bowen, R. *Ferroelectrics*, **319**, 199 (2005).
4. Pauly, S. and Kammer, H. W. *Polym. Networks Blends*, **4**, 93 (1994).
5. Sinha, S. and Okamoto, M. *Macromol. Mater. Eng.*, **288**, 936 (2003).
6. Troeger, J. and Kammer, H. W. *Acta Polymerica*, **43**, 331849 (1992).
7. Walter, R., Selser, J. C., Smith, M., Bogoslovov, R., Piet, G. *J. Chem. Phys.*, **117**, 427 (2002).
8. Chan, C. H., Kammer, H. W., Sim, L. H., Mohd Yusoff, S. N. H., Hashifudin, A., and Winie, T. *Ionics* (accepted) doi:10.1007/s11581-013-0961-7 (2013).

CHAPTER 8

LIQUID CRYSTALLINE ORDER IN POLYMERS

HANS-WERNER KAMMER

CONTENTS

8.1 LIQUID CRYSTALLINE ORDER IN POLYMERS

In the crystalline state there is a long ranged order in position and orientation of the molecules. There are three possible transitions from crystalline order to states having lower order. Firstly, all order disappears that means we are entering the liquid state. Secondly, only orientation order disappears during the transition, positional order is kept original. This is transition to the phase of plastic crystal which is outside the scope of this chapter. Thirdly, melting may lead to disappearance of positional order; we are entering the liquid crystalline (LC) state. Hence, order in LC state is intermediate between crystalline solid and isotropic liquid state (Figure 1). There is no long ranged spatial order, the same as in liquids but we observe long ranged orientational order. As a consequence, liquid crystals are anisotropic. Therefore, liquid crystals develop in systems of rod-like molecules. The anisotropy may arise for solely steric reasons since isotropic arrangement of rod-like particles in densely packed systems becomes unfavorable from an entropy point of view.

FIGURE 1 Order in different states of matter.

More than 80,000 LC compounds have been found. One observes LC phases in:

- Organic compounds consisting of highly anisotropic molecules of rod-like conformation
- Polymers with high degree of rigidity in the backbone or side-chains
- Polymers or associated molecules that form rigid-rod like structures

The molecules or groups of molecules that display LC order are called mesogens or mesogenic compounds. The first mesogen was found by Reinitzer (Austria) and Lehmann (Germany) in 1888, cholesteryl nonanoate (ester of cholesterol and nonanoic acid) (Figure 2). This compound displays two melting points:

$$T_{m1} = 145.5\ °C \Rightarrow \text{cloudy liquid}$$

$$T_{m2} = 178.5\ °C \Rightarrow \text{clear liquid}$$

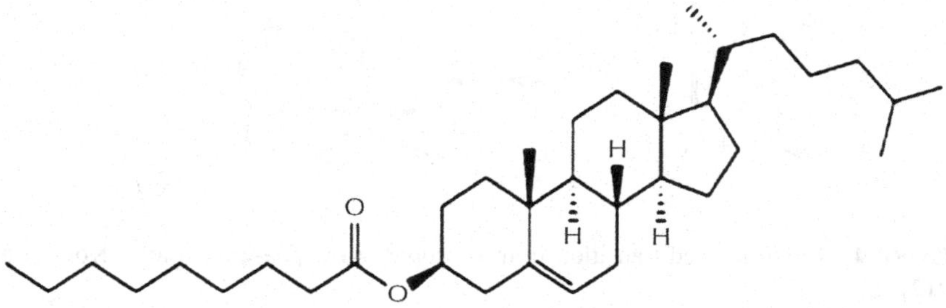

FIGURE 2 Cholesteryl nonanoate.

The organic compound 4-methyloxybenzylidene – 4'-butylaniline (MBBA) (Figure 3) was the first compound found that displayed liquid crystallinity at room temperature (Kelker and Scheurle, 1969).

$$\text{crystal} \overset{20\ °C}{\leftrightarrow} \text{nematic phase} \overset{47\ °C}{\leftrightarrow} \text{isotropic}$$

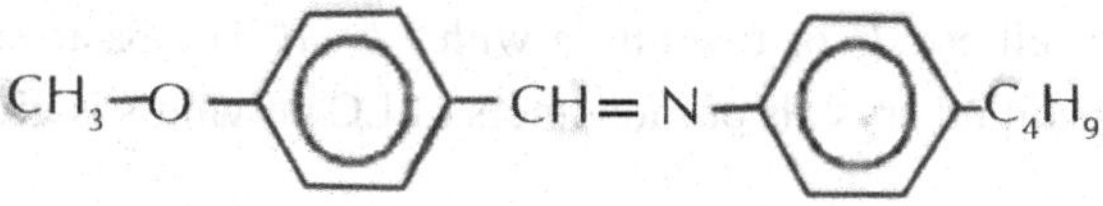

CH_3–O–C$_6$H$_4$–CH=N–C$_6$H$_4$–C_4H_9

FIGURE 3 Mesogenic MBBA (*n*-*p*-methoxy-benzylidene-*p*-butylaniline).

The naphthopyran, as shown in Figure 4, undergoes a reversible photo induced conformational change by ring-opening (Kosa et al., 2012). It displays increase in order under light exposure.

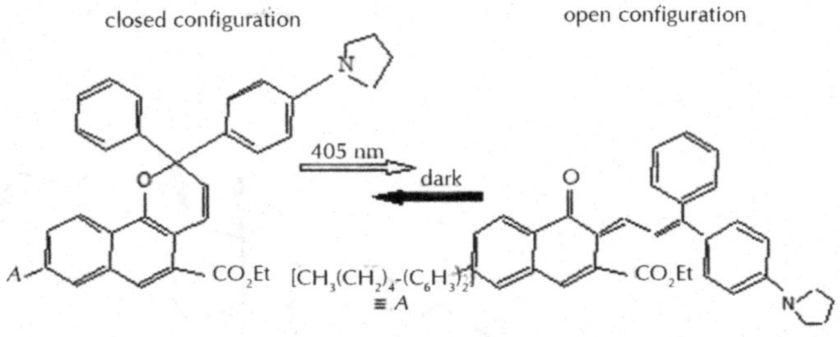

FIGURE 4 Light-induced transition from isotropic to LC phase; data after Kosa et al. (2012).

The LC phases, illustrated by examples in Figures 2 to 4, originate from different driving forces. In the first example, heat transfer causes transition from isotropic to LC state. The order parameter in the second example is ruled by concentration of MBBA in solution. In the last example the mesophase is formed by light exposure of the material. We see, liquid crystals develop either under thermotropic or lyotropic conditions depending on the driving force for organization of the system.

8.2 MOLECULAR STRUCTURE

The molecules displaying LC phases are highly anisotropic. They might be seen as rigid rods or ellipsoids of revolution with $l >> d$. The basic structure of low molecular mass liquid crystals or monomers of LC polymers is given in Figure 5.

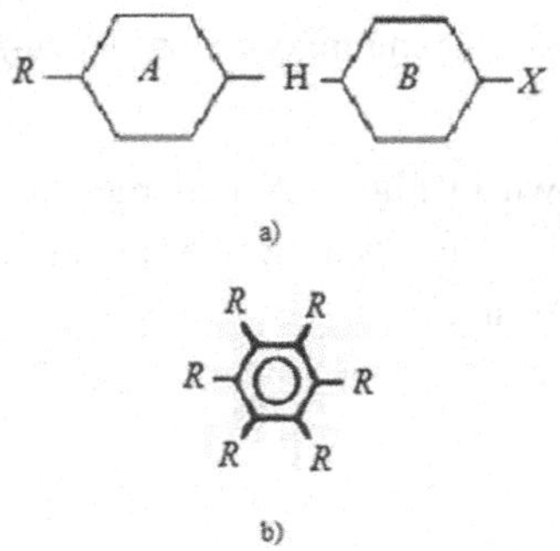

FIGURE 5 Basic molecular structures of LC molecules: a) rod-like with A, B aromatic rings, H – linking group, R terminal group and X side group; b) discotic.

The disk-like molecules consist of a flat aromatic core connected with a number of flexible side chains. The discotic molecules exhibiting LC phases are about 1 nm in thickness and a few nanometers in diameter.

The solutions of amphiphilic molecules form another class of the LC family. The amphiphiles are hydrophilic/hydrophobic molecules. One end of the molecule consists of a polar hydrophilic group and the other end is formed by a nonpolar hydrophobic tail. Soap forms a LC phase in solution. In Figure 6, sodium laureate is shown as an example for an amphiphile, $-COO - Na^+$ is the polar group and soluble in water whereas the hydrocarbon paraffin group, $CH_3(CH_2)_{14}$ – , is hydrophobic.

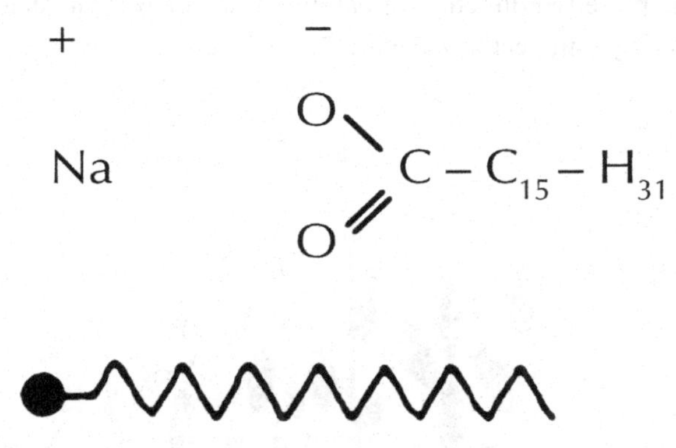

FIGURE 6 Amphiphilic sodium laureate.

The different driving forces lead to LC phases in these low molecular compounds. Rod-like systems display LC behaviour in a certain range of temperature; therefore, they are thermotropic. Amphiphiles are lyotropic LCs; they show LC behaviour in a certain range of concentration.

8.3 CLASSIFICATION OF LC PHASES AND THEIR STRUCTURAL FEATURES

We distinguish two fundamental molecular orderings in macroscopic systems:

Positional and orientational order

At low temperatures, we have the crystalline state characterized by long ranged order with both regular position and orientation of the elements. The liquid state without any long ranged order develops at high temperatures. When cooled down, matter will turn in two ways in semi-disordered state:

- By losing orientational order and retaining translational order, plastic crystals or semi-crystalline materials develop,
- By losing translational order and keeping orientational order, liquid crystals are formed.

All LC systems are characterized by their orientational order but they display varying degrees of translational order as well, except nematics, they do not have positional order.

I. A stable nematic phase is observed between isotropic melt and crystalline solid. Nematics exhibit orientational order between rod-like molecules, but no positional order. They are the most important members of the LC family. Nematics are widely used in display industry. In those phases, molecules tend to parallel to each other. The preferred direction of parallel orientation is characterized by the *director* denoted by unit vector n (Figure 7).

FIGURE 7 Orientational order in nematic phase by aligning along director n.

II. Smectic A and C phases have also one-dimensional translational order as shown in Figure 8. As the temperature is further cooled, molecules begin to segregate into planes giving rise to smectic A or smectic C phase. In a smectic phase, molecules arrange themselves in sheets. They are arranged in layers but display no positional short range order. The transition nematic $\Rightarrow$ smectic means a transition from no positional order to long ranged positional order in one dimension only. In smectic A, the director is parallel to the normal of the layer. Thus, in addition to orientational order of the nematic phase, smectic A and C phases exhibit one-dimensional translational order and form layered structures. There is a liquid-like motion of rods in each layer and no correlation of the molecular positions from one layer to the next. In each layer the mass centers of the molecules are randomly distributed as in isotropic liquids. The layers can easily slide. In smectic A phase, molecules tend to be perpendicular to smectic layers. The layer thickness d is approximately the same as the molecular length l.

$$\text{smectic A} \quad d \approx l$$

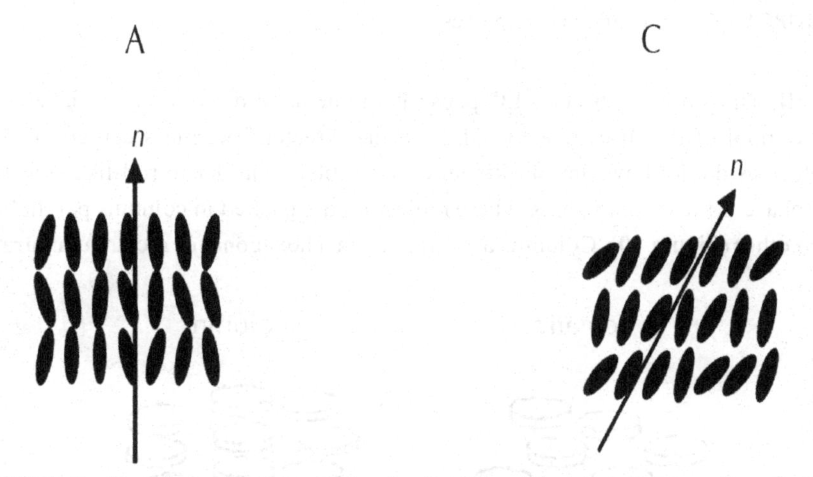

FIGURE 8 One-dimensional translational order in smectic A and C phases; in smectic A the thickness of the layer is close to monomer length l, whereas in smectic C it is $l \cos \theta$ with θ being the tilt angle to normal.

In a polymer LC, length d equals approximately monomer length. In the smectic C phase, molecules in the layers are parallel and tilted in arrangement with respect to the normal of the layers by a tilt angle θ.

$$\text{smectic C} \quad d \approx l \cos \theta$$

The orderings in smectic A and C phases are higher than in the nematic phase. They appear at lower temperatures as the nematics. In systems that show smectic A and C, phase A appears first as the temperature decreases.

In summary, ordering in LC phases is depicted in Figure 9.

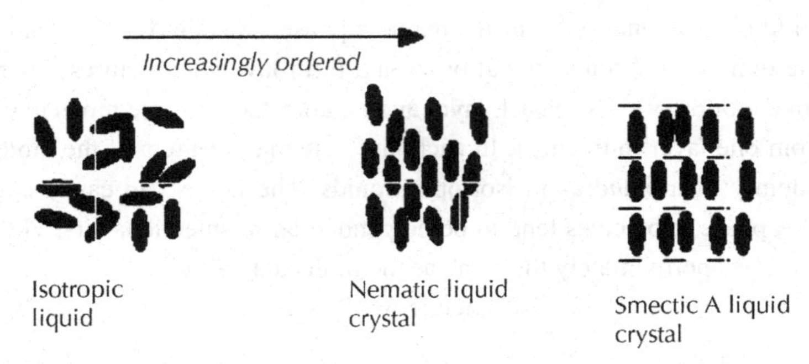

FIGURE 9 Order in different LC phases.

III. The simplest discotic LC phase is the nematic discotic phase, in which the normal of the discs tends to align to the director. The mass centers of the molecules do not have any positional order. A phase similar to rod-like smectic LC phase is the column phase where molecules are packed in columns parallel to each other (Figure 10). Columns are arranged in a hexagonal or rectangular array.

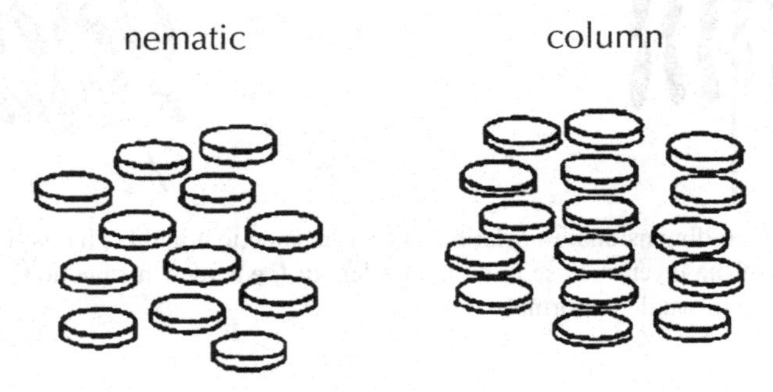

FIGURE 10 Discotic LC phases.

IV. As mentioned before LC phases appear also in response to changes in concentration of water, oil or surfactants in mixtures. These phases are called lyotropic LCs. When amphiphiles are dissolved in water, their polar heads tend to be close to each other while the nonpolar hydrophobic tails are as far away from water as possible.

Depending on concentration, the amphiphilic molecules aggregate to form spheres columns or layer structures. One example is shown in Figure 11.

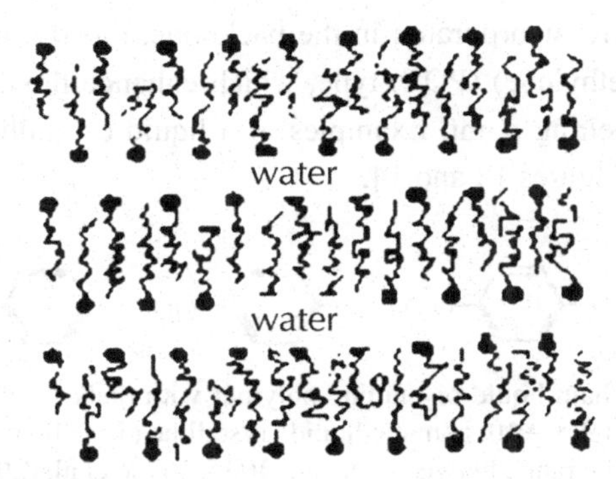

FIGURE 11 Amphiphilic liquid crystal.

8.4 LIQUID CRYSTALLINE POLYMERS

The LC polymers became popular in the 1970 of last century due to their potential in forming of high modulus fibers. Around 3,000 LC polymers have been synthesized and characterized until 2010.

In rigid chain macromolecules, the length of a segment l is much greater than its thickness d, $l/d >> 1$. Therefore, LC phases are not only found in polymer melts but also in concentrated solutions. The former represent thermotropic liquid crystals and the latter lyotropic liquid crystals.

The LC polymers comprise mesogenic groups either in the backbone or in side chains, as indicated in Figure 12. They form a stable state of matter. These polymers are attractive because they combine typical properties of low molecular liquid crystals with mechanical and thermal properties of macromolecular materials, that is they can be processed to fibres or films and we can freeze in the LC state.

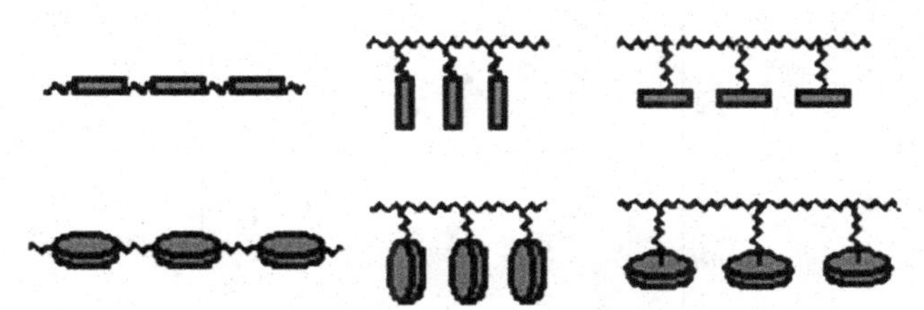

FIGURE 12 Structure of main chain and side chain LC polymers.

Let us discuss briefly main chain LC polymers. Linking of mesogenic units leads to rod-like chains with very high melting point. Therefore, flexible groups, so-called spacers, are incorporated in the backbone, mostly of poly(methylene) (PM)- or poly(oxyethylene) (POE)-type, which enhance flexibility of the backbone and reduce melting point. Examples of a liquid crystalline polymer and a dimer are given in Figures 13 and 14.

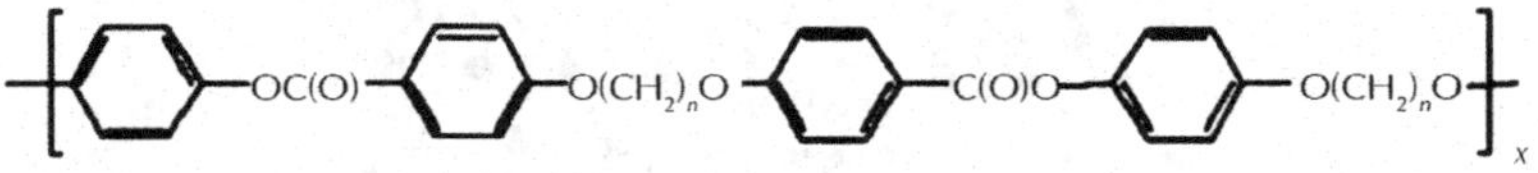

FIGURE 13 Main chain liquid crystalline polymer with dialkoxyalkane spacer; with n varying between 2 and 9, $x = 10$; nematic liquid crystalline phase between crystalline solid and isotropic melt in the range between 100 and 300 °C, Abe et al. (2005).

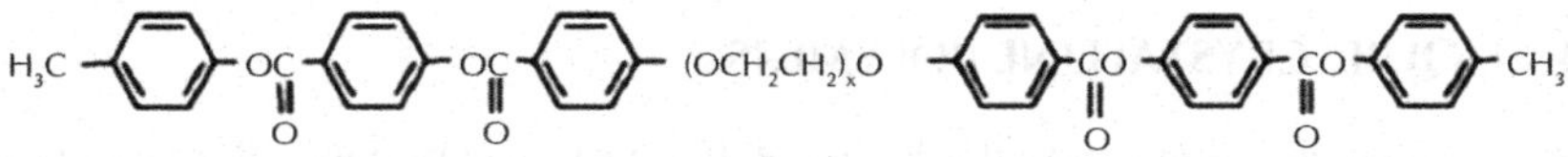

FIGURE 14 Dimer MBBE with POE-type spacers; $x = 2$ to 9, Hiejima et al. (2003).

The phase diagram is illustrated in Figure 15 for the dimer with varying spacer length. A stable nematic phase is observed between isotropic melt and crystalline solid. The phase diagram is illustrated in Figure 15 for the dimer with varying spacer length. A stable nematic phase is observed between isotropic melt and crystalline solid. Corresponding chain conformations are sketched schematically in Figure 16.

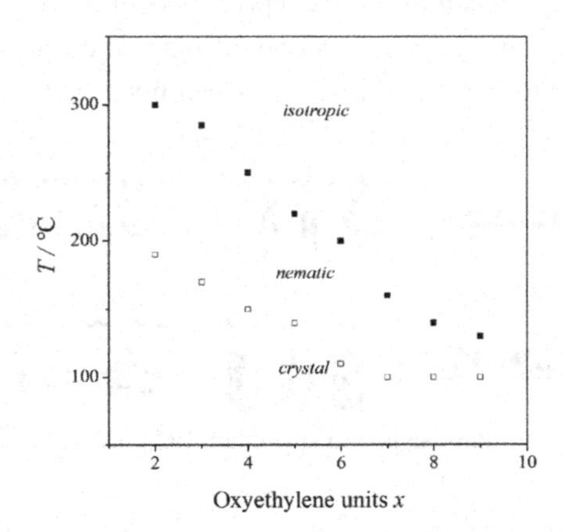

FIGURE 15 Phase diagram of MBBE with oxyethylene spacer of different length; data after Abe et al. (2005); (■) T_{NI}, (□) T_{NC}.

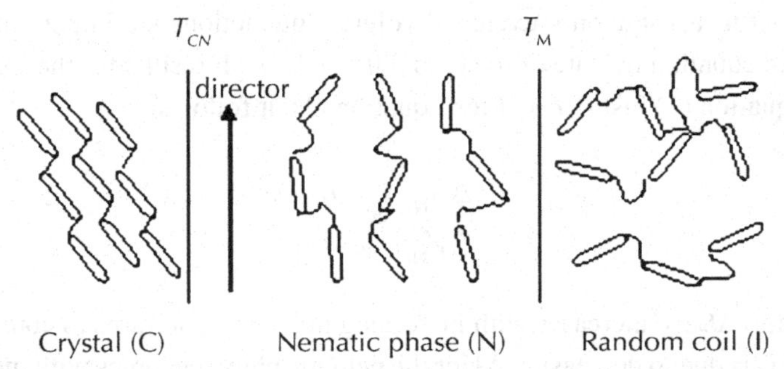

FIGURE 16 Transition isotropic liquid phase (neither positional nor orientation order) → nematic phase (orientation order but still no positional order) → ordered crystalline phase.

From a thermodynamic point of view, the LC polymer is in an unfavorable situation similar to a block copolymer comprising immiscible blocks. Covalent links between mesogenic units and spacers prevent phase separation driven by competing entropy effects. Flexible coils dissolved in a nematic solvent would segregate from the solvent because mixing of the constituents would lead to conformational restrictions for the flexible chains or stretching of the coils would reduce their entropy. On the other side, solubility of the constituents would require for the nematic component random arrangement which is above a certain concentration not anymore sterically possible. This conflicting situation is inherent in the molecular architecture. We recognize that this segregating force descends with increasing spacer length.

As in low molecular systems, the nematic phase exhibits orientation order between rod-like units, but no positional order. Due to orientation of mesogenic units also spacers are oriented by stretching. Molecules are oriented with respect to director. Thus, the isotropic – nematic transition is a disorder-order transition from a phase without long range order to a phase with long range orientation order but still liquid-like. It is an equilibrium transition since there is no undercooling. This transition is characterized by $\Delta S_{NI} \equiv S_N - S_I < 0$ and $\Delta V_{NI} < 0$. We may relate it to equation of Clausius–Clapeyron. It follows:

$$\frac{\Delta S_{NI}}{\Delta V_{NI}} = \frac{dP}{dT} \tag{1}$$

Now, let us assume only excluded volume interactions are important, that is we have an equation of state of the kind $\tilde{P}(\tilde{V}-\tilde{B})=\tilde{T}$. It results for the right-hand side of Equation (1) just P/T and for Equation (1), it follows:

$$\frac{\Delta S_{\mathrm{NI}}}{\Delta V_{\mathrm{NI}}}=\frac{const}{T_{\mathrm{NI}}} \tag{2}$$

The ratio $\Delta S/\Delta V$ increases with increasing number n and turns to constancy at high n. This is due to decreasing ΔV or the nematic phase becomes with increasing n more and more liquid-like.

The transition from isotropic to nematic state is coined by an entropy penalty outweighed by a dramatic decrease in excluded volume, as mentioned before, and additionally, by favorable interactions in the aligned state. The positive contribution to the free energy arising from the loss of conformational entropy must be compensated by shrinking excluded volume interaction and by favorable attractive van-der-Waals interactions; both effects will be maximized when molecules are aligned. The total free energy change per molecule reads qualitatively for transformation from the isotropic to the nematic state.

$$\Delta F_{I\rightarrow \mathrm{N}}(\Sigma)=-u\Sigma^2-T\Delta S_{I\rightarrow \mathrm{N}}(\Sigma) \tag{3}$$

The degree of nematic order is here symbolized by Σ. Quantity u expresses the strength of favorable interactions between neighboring molecules. The second term of Equation (3) gives the entropy loss accompanying the transition. The free energy has to be minimized with respect to order parameter Σ.

8.5 A QUICK GLANCE TO NEMATIC-ISOTROPIC TRANSITION

The transition from isotropic to nematic phase is characterized by formation of molecular orientation with respect to director $\vec{n}$. We may characterize it by order parameter Σ (Figure 17).

$$\Sigma=\frac{1}{2}\left\langle 3\cos^2\theta-1\right\rangle \tag{4}$$

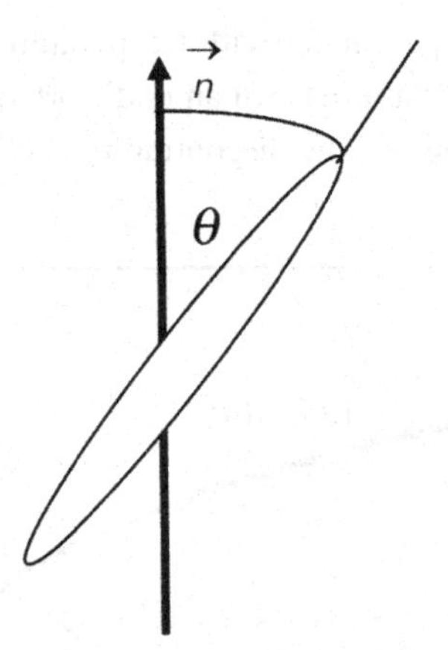

FIGURE 17 To definition of order parameter with respect to director $\vec{n}$ in nematic phase.

In an isotropic liquid all directions occur with the same probability. Therefore, order parameter $\Sigma = 0$. The complete order is characterized by only one angle, $\theta = 0$, and as a consequence $\Sigma = 1$. The average Equation (4) is calculated in the appendix.

Mayer and Saupe discussed Equation (3) quantitatively (Mayer and Saupe, 1958). It turns out that isotropic → nematic phase transition is of first order. The order parameter changes according to theory at T_{NI} from $\Sigma = 0$ to $\Sigma = 0.43$ (Mayer and Saupe, 1958). Experimentally, changes were found from $\Sigma = 0$ to $\Sigma_{exp} \approx 0.2$ ... 0.3 (Abe et al., 2005; Croucher and Patterson, 1980). We sketch briefly the phenomenological approach by Landau-de Gennes (de Gennes, 1974). Accordingly, the difference in free energy density between nematic and isotropic phase is developed in powers of order parameter Σ. It follows:

$$\frac{\Delta F}{k_B T} = a\left(T - T^*\right)\Sigma^2 + b\Sigma^3 + c\Sigma^4 \tag{5}$$

The first coefficient comprises the typical temperature dependence for a first order transition close to the transition point whereas variation of the other coefficients with temperature is neglected. The first order transition is characterized by:

$$\frac{\partial F}{\partial \Sigma} = 0 \text{ (equilibrium)} \qquad \frac{\partial^2 F}{\partial \Sigma^2} > 0 \text{ (phase stability)} \tag{6}$$

The variation of order parameter with temperature is shown in Figure 18. It gives the equilibrium order parameter of an undistorted nematic for a given set of Landau-de Gennes coefficients. The discontinuity is clearly visible.

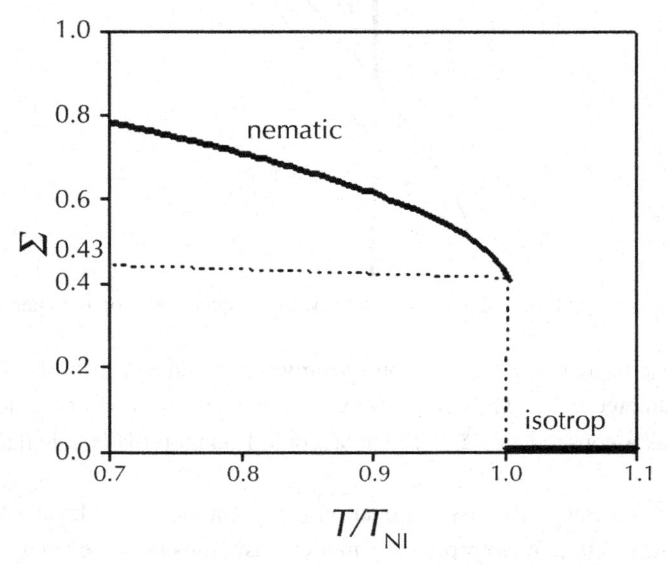

FIGURE 18 Order parameter *versus* temperature according to Equation (5).

8.6 POLYMER BLENDS WITH LIQUID CRYSTALLINE POLYMERS

We will discuss nematic ordering in polymer systems and we start with solutions of rigid rods as the simplest system in which isotropic-nematic transition occurs. Solutions of a flexible polymer and a nematic low molecular liquid crystal display at low polymer content, when cooled down from the isotropic phase, segregation into a nematic and an isotropic phase. At higher polymer content, the solution decays first in two isotropic phases, one rich in polymer, and the other poor in polymer. Further cooling leads to separation of the latter in a nematic phase very poor in polymer and the isotropic phase rich in polymer. This is sketched schematically in Figure 19. Phase behavior of the indicated type was observed in EBBA (*p*-ethoxy benzylidene-*bis*-4-*n*-butylaniline) mixed with polystyrene (Ballauf, 1986; Lee et al., 1994) and with poly(ethylene oxide) (Kronberg et al., 1978).

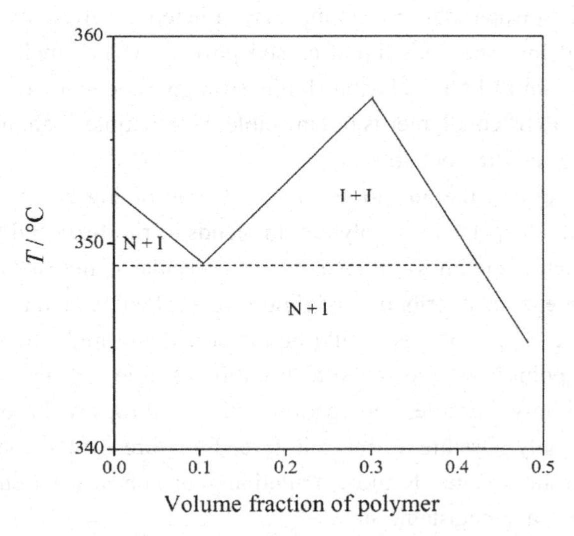

FIGURE 19 Schematic phase diagram of a nematogen-polymer solution.

The blending of flexible polymers with LC polymers is governed by competition of orientation ordering, resulting from LC polymer, and the driving force towards maximal entropy gain of the system by mixing. This competition leads to complex phase structures. In order to accommodate properly orientation of the LC moiety in blends with flexible polymer chains, the nematic ordered chain has to adopt adequate conformations. Therefore, very specific phase segregation will be preferably found. We restrict discussion only of a few simple examples.

Many studies selected LC copolyesters as one constituent. We will briefly consider a LC copolyester consisting of three kinds of monomers: terephthalic chloride, phenylhydroquinone, and naphthalene diol, with mole fractions 0.5, 0.44, and 0.06, respectively (Khan et al., 1995). Experiments revealed that the melting temperature is around 300 °C and the clearing temperature at approximately 410 °C. One also observed that degradation of the polymer destroys reversibility of the transitions.

The study of LC copolyesters started with the polyester of poly(ethylene terephthalate) (PET) and parahydroxy benzoic acid (PHB) by Jackson and Kuhfuss, 1974 (Jackson Jr. and Kuhfuss, 1996). It turned out that the copolyesters are heterogeneous consisting of flexible (ET-rich) and rigid (HB-rich) blocks. Thus, the copolyester itself, with a composition of 0.6 mole fraction of HB, segregates into a nematic and isotropic phase. One may say, a heterogeneous "blend" develops with phases of low and high liquid crystal polymer (LCP) melting point. The high melting point of LCP inclusions limits strongly processing of the copolyesters. This PET/PHB copolymer is not miscible with flexible chain molecules like poly(styrene) or poly(carbonate).

The blending had the aim to generate self-reinforcing blends with fibrillar morphology of a dispersed LC polymer. In blends of rigid-rod and flexible chain polymers, which sometimes are also called molecular composites, one tries to achieve homogeneous distribution of rigid rods in flexible chain polymers on a molecular scale. These blends should be materials with high strength and high modulus. The point is the molecular distribution of rigid rods in a flexible chain matrix is inherently unstable. Segregation of the rigid rods will proceed in those mixtures. It is only possible to freeze in forced mixtures of the constituents that constitute metastable states. Rapid coagulation of one phase solutions may lead to forced mixtures in a metastable state.

APPENDIX

Calculation of Average Equation (4) for the Isotropic Phase

The average over the orientation distribution function reads

$$\Sigma = \frac{1}{2}\int_0^{\pi/2} \left(3\cos^2\theta - 1\right) w(\theta) \sin\theta \, d\theta$$

For isotropy, we have $w(\theta) = (4\pi)^{-1}$, which leads to $\Sigma = 0$ since

$$\left\langle \cos^2\theta \right\rangle = \int_0^{\pi/2} d\theta \sin\theta \cos^2\theta \overset{\sin\theta = t}{=} \int_0^1 dt t\sqrt{1-t^2} = -\frac{1}{3}\left(1-t^2\right)^{3/2}\Big|_0^1 = \frac{1}{3}$$

which yields the above result for an isotropic liquid.

KEYWORDS

- **Liquid crystalline order**
- **Nematic-isotropic transition**
- **Nematics**
- **Nematics-polymer solutions**
- **Smectics**

REFERENCES

1. Abe, A., Furuya, H., Zhou, Z., Hiejima, T., and Kobayashi, Y. *Adv. Poly. Sci.*, **181**, 121 (2005).
2. Ballauf, M. *Molec. Cryst. Liq. Cryst.*, **4**, 15 (1986).
3. Croucher, M. D., and Patterson, D. *J. Solution Chem.*, **9**, 771 (1980).
4. de Gennes, P. G. *The Physics of Liquid Crystals*; Clarendon Press: Oxford, 1974.
5. Hiejima, T., Seki, K., and Kobayashi, K., Abe, A. *J. Macromol. Sci.*, **B42**, 431 (2003).
6. Jackson Jr., W. J., and Kuhfuss, H. F. *Polym. Chem.*, **34**, 3031 (1996).
7. Kelker, H., and Scheurle, B. *Angew. Chem. Int. Edn.*, **8**, 884 (1969).
8. Khan, N., Bashir, Z., and Price, D. M. *J. Appl. Polym. Sci.*, **58**, 1509 (1995).
9. Kosa, T., Sukhomlinova, L., Su, L. L., Taheri, B., White, T. J., and Bunning, T. J. *Nature*, **485**, 347 (2012).
10. Kronberg, B., Bassignana, I., and Patterson, D. *J. Chem. Phys.*, **82**, 1714 (1978).
11. Lee, S., Oertli, A. G., Gannon, M. A., Lin, A. J., Pearson, D. S., Schmidt, H. W., and Frederickson, G. H. *Macromolecules*, **27**, 3955 (1994).
12. Mayer, W. and Saupe, A. *Zeitschrift fuer Naturforschung*, **13a**, 564 (1958).

CHAPTER 9

SURFACE TENSION OF POLYMER BLENDS AND RANDOM COPOLYMERS

HANS-WERNER KAMMER

CONTENTS

9.1 INTRODUCTION

The surface properties of polymers are important in technology of plastics, coatings, textiles, films, and adhesives through their role in processes of wetting, adsorption, and adhesion. We will discuss only surface tensions of polymer melts that can be measured directly by reversible deformation or can be inferred from drop shapes. Those inferred from contact angles of liquids on solid polymers ("critical surface tension of wetting") are excluded, as their relations to surface tensions are uncertain.

Most of the polymers are very viscous even at high temperatures. Moreover, they exhibit non-Newtonian flowing behavior. Considerable care is required to ensure precise measurements. We will discuss here surface tensions of polymer blends and of random copolymers; especially, we will focus on its variation with composition.

9.1.1 THERMODYNAMICS

Surface tension γ is the two-dimensional analog to pressure:

$$\mathrm{d}f = -P\mathrm{d}v \qquad \mathrm{d}f = \gamma \mathrm{d}a \ (v = const) \tag{1}$$

In equilibrium, free energy and area are minimal, hence:

$$\gamma > 0 \tag{2}$$

Gibbs equation gives the dependence of γ on temperature and bulk composition:

$$a\mathrm{d}\gamma + s^{\mathrm{S}}\mathrm{d}T + \sum_{\mathrm{i}} N_{\mathrm{i}}^{\mathrm{S}}\mathrm{d}\mu_{\mathrm{i}} = 0 \tag{3}$$

All $\mathrm{d}\mu_{\mathrm{i}} = 0$ when no substance exchange takes place, then we have:

$$\frac{\mathrm{d}\gamma}{\mathrm{d}T} = -\frac{s^{\mathrm{S}}}{a} < 0 \tag{4}$$

9.1.2 EXPERIMENTAL METHODS FOR DETERMINATION OF SURFACE TENSION

Only methods based on analysis of drop profiles are suitable for surface (and interfacial) tension measurements. These include:

- Pendant drop method.
- Sessile drop method.

These methods are independent of liquid/solid contact angle, but require accurate knowledge of liquid density.

Pendant Drop Method

The profile of a pendant drop (Figure 1) is balanced by hydrodynamic and surface equilibrium. The surface tension is given by:

$$\gamma = \frac{1}{H} g D_e^2 \Delta\rho \tag{5}$$

The symbols in Equation (5) have the following meaning: g - gravitational constant, $\Delta\rho$ – difference in densities between drop and surrounding, D_e – equatorial diameter of the drop, H – correction factor determined by shape factor $S = D_s/D_e$ (D_s as given in Figure 1). It is important to assure that the drop attains this equilibrium shape. In our measurements it took around 20 min (Pefferkorn et al., 2010). The values of H as a function of S have been tabulated (Hansen and Rodsrun, 1991; Fleischer et al., 1994; Kamal et al., 1997; Demarquette and Kamal, 1998; Arashiro and Demarquette, 1999; Song and Springer, 1996; Gamarbi et al., 1998).

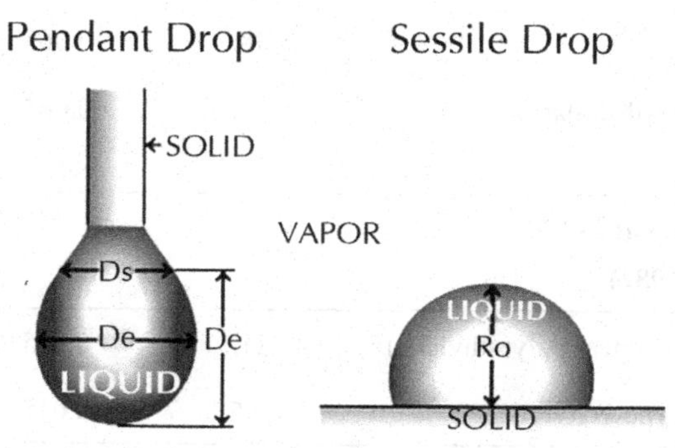

FIGURE 1 Schematic representation of pendant drop and sessile drop method.

SESSILE DROP METHOD

It is also based on drop profile and therefore, similar to pendant drop. Dimensions usually measured are the distance from equator to vertex, R_o, and the equatorial radius $D_e/2$. Constancy of D_e and D_e/R_o are the criteria for equilibrium. The surface tension is given by:

$$\gamma = KR_o^2 g\Delta\rho \tag{6}$$

The correction factor K is a function of D_e/R_o and has been tabulated (Hansen and Rodsrun, 1991; Fleischer et al., 1994; Kamal et al., 1997; Demarquette and Kamal, 1998; Arashiro and Demarquette, 1999; Song and Springer, 1996; Gamarbi et al., 1998). Table 1 gives some results.

TABLE 1 Surface tension and temperature coefficient

Low molecular at 273 K	γ/ **mN m^{-1}**	$-(d\gamma/dT)$/**mN m^{-1} K^{-1}**
Water	72.8	0.14
Mercury	476	0.2
Benzene	28.9	0.1
Polymers at 313 K (Wu,1974; Wu, 1982)		
Poly(tetrafluoro ethylene)	13.7	0.065
Nylon 66	38.7	0.065
Poly(ethylene terephthalate)	36.8	0.065
at 543 K	28.3	
Interfacial tension at 313 K (Wu,1974; Wu, 1982)		
Poly(vinyl acetate)/linear poly(ethylene)	11.3	0.015
Poly(chloroprene)/poly(styrene)	0.5	0.015

9.2 COMPARISON OF COPOLYMER AND POLYMER BLEND

Rastogi and St.Pierre (Rastogi and St.Pierre, 1969) compared random copolymers of ethylene oxide and propylene oxide with miscible blends of poly(ethylene oxide) (PEO) and poly(propylene oxide) (PPO). In the original paper, they claimed linear variation of surface tension of the random copolymer with composition:

$$\gamma_{\text{random copolymer}} = X_{PEO}\gamma_{PEO} + X_{PPO}\gamma_{PPO} \tag{7}$$

The Equation (7) means, there is no surface activity of the constituent with the lower surface energy. This statement is obviously not precise as Figure 2 shows. The difference in surface tensions is quite remarkable: 43.5 and 31.1 mN/m for PEO and PPO, respectively. We recognize considerable surface excess of the component with the lower surface energy. Although, negative deviation from additive is for the random copolymer less than in blends of PEO and PPO owing to conformational restrictions in the copolymer. These restrictions hamper preferential accumulation of the low-energy segments in the surface (see also Surface Tension of Polymer Blends).

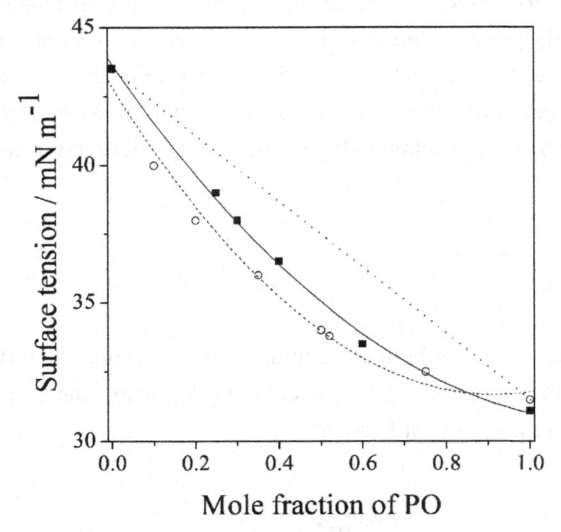

FIGURE 2 Surface tension of random poly(ethylene oxide-*co*-propylene oxide) (■) and of blends of PEO and PPO (o) at 85 °C; graph after data of Rastogi and St.Pierre (1969).

9.3 LANGMUIR ADSORPTION ISOTHERM

We discuss variation of surface tension of blends and random copolymers with composition in terms of a generalized Langmuir adsorption isotherm. The partition coefficient in surface and bulk might be expressed by equilibrium of occupancy of the surface region by segments of types 1 and 2 as follows:

$$s_1 + s_2^S \leftrightarrow s_1^S + \frac{1}{N} s_2 \tag{8}$$

where $N > 1$. Here, s_i refers to segments of type i in bulk and s_i^S refers to segments of type i in the surface region. Quantity $N > 1$, reflects self-association of component 2 in bulk. According to Equation (8), the partition coefficient k reads expressed in mass fractions and with $\lambda \equiv 1/N$

$$k = \frac{W_1^S / W_1}{W_2^S / W_2^\lambda} \tag{9}$$

In Equation (9) perfect mixing of the segments in the surface region is supposed, and nearly perfect mixing in the bulk phase. The representation of bulk concentration of component 2 by W_2^λ is justified as long as $1/N$ is close to unity. From Equation (9), it directly follows the relationship between surface and bulk concentration which represents a slightly generalized Langmuir adsorption isotherm

$$W_1^S = \frac{kW_1}{kW_1 + W_2^\lambda} \tag{10}$$

For $k > 1$ and $\lambda < 1$, but both close to unity, one sees that the ratio $W_1^S/W_1 < 1$ in the range where component 1 is in excess. Analogously, one finds for the ratio $W_2^S/W_2^\lambda < 1$. More precisely, it follows:

$$\frac{W_2^S}{W_2^\lambda} < \frac{W_1^S}{W_1} < 1 \tag{11}$$

These inequalities imply that both constituents are depleted in the surface region with respect to bulk concentration, but component 1 is less than component 2. Moreover, the ratio W_1^S/W_1 displays a minimum at:

$$W_2^{\min} = \left(\frac{k}{\lambda}\right)^{\frac{1}{\lambda-1}} \tag{12}$$

In equilibrium, the surface tension γ of a mixture is related to the ratio of surface and bulk concentration, thus it follows:

$$\frac{W_1^S}{W_1} = \exp\left(\frac{\gamma - \gamma_1}{\gamma^+}\right) \tag{13}$$

With $\gamma^+ \equiv k_B T/a$ being a reference surface tension, where k_B is the Boltzmann constant and a is the surface area which is covered by an average segment. After inserting Equation (10) in Equation (13), it follows for the surface tension:

$$\gamma = \gamma_2 - \gamma^+ \ln\left(kW_1 + W_2^\lambda\right) \text{ with } \gamma_2 - \gamma_1 = \gamma^+ \ln k \tag{14}$$

or

$$\gamma = \gamma_1 + \gamma^+ \ln\left(\frac{k}{kW_1 + W_2^\lambda}\right) \tag{14a}$$

The function $\gamma\,(W_2)$ displays a minimum at $W_2^{\min}$ given by Equation (12), and the corresponding surface tension is:

$$\gamma^{\min} = \gamma_1 - \gamma^+ \ln\left(\frac{W_2^{\min}}{\lambda} + W_1^{\min}\right) \tag{15}$$

From Equation (15) it is easily verified that $\gamma^{\min} < \gamma_1$ for $\lambda < 1$. The interesting point here is that the minimum in surface tension is caused by a reduction of entropy in bulk due to self-association of component 2.

Let us discuss Equations (14) and (14a) in limiting situations.

Ia) $k = 1 + \varepsilon$ with $\varepsilon << 1$ and $\lambda = 1$

We get

$$\frac{W_1^S}{W_1} = 1 + \varepsilon W_2 \text{ and } \frac{W_2^S}{W_2} = 1 - \varepsilon W_1 \tag{16}$$

For $\varepsilon > 0$, component 2 is depleted in surface region as compared to bulk and component 1 is enriched. Hence, it is $\gamma_1 \leq \gamma < \gamma_2$. For $\varepsilon < 0$, we get the opposite behaviour.

Ib) $k = 1 + \varepsilon$ with $\varepsilon << 1$ and $\lambda < 1$

We get again a similar expression as Equation (16), but the right-hand sides have to be subdivided by $(W_1 + W_2^\lambda) > 1$ and in the second equation W_2 on the left-hand side is substituted by $W_2^\lambda > W_2$

$$\frac{W_1^S}{W_1} = \frac{1 + \varepsilon W_2}{W_1 + W_2^\lambda} \text{ and } \frac{W_2^S}{W_2^\lambda} = \frac{1 - \varepsilon W_1}{W_1 + W_2^\lambda} \tag{16a}$$

This may lead, especially for small W_2, to depletion of both constituents in the surface region, but component 2 is more depleted than component 1 as indicated by Equation (11).

IIa) $k = \varepsilon$ with $\varepsilon << 1$ and $\lambda = 1$

$$\frac{W_1^S}{W_1} = \frac{\varepsilon}{W_2} \text{ and } \frac{W_2^S}{W_2} = \frac{1}{W_2}\left(1 - \varepsilon \frac{W_1}{W_2}\right) \tag{17}$$

One recognizes that component 2 has a dramatic surface excess whereas component 1 is depleted. Surface tensions are related oppositely to each other as compared to I), $\gamma_2 < \gamma < \gamma_1$.

IIb) $k = \varepsilon$ with $\varepsilon << 1$ and $\lambda < 1$

We get instead of Equation (17)

$$\frac{W_1^S}{W_1} = \frac{\varepsilon}{W_2^\lambda} \text{ and } \frac{W_2^S}{W_2^\lambda} = \frac{1}{W_2^\lambda}\left(1 - \varepsilon \frac{W_1}{W_2^\lambda}\right) \tag{17a}$$

Since $W_2^{\lambda} > W_2$, effects are reduced as compared to IIa).

9.4 SURFACE TENSION OF POLYMER BLENDS

9.4.1 PEO(1)/PPO(2)

In view of Langmuir approach, we observe excess of propylene oxide (PO) in the surface for both random copolymer and blend (Figure 2). This corresponds to Equation (17) of II). As one expects, depletion of PEO in the surface is more pronounced for the blend. With $\gamma_{PPO(2)} = 31.5$ mN/m and $\gamma_{PEO(1)} = 43.5$ mN/m at 25 °C, we get for

copolymer $\lambda = 1$ $k = 0.25$ $\gamma^{+} = 8.7$ mN/m $a^2 \approx 0.5$ nm^2

blend $\lambda = 1$ $k = 0.08$ $\gamma^{+} = 4.75$ mN/m $a^2 \approx 0.8$ nm^2

We observe that the average areas available for the segments in the copolymer and the blend are different. The area is restricted in the copolymer as compared to the blend or segment density in the surface is higher in the copolymer.

9.4.2 PS(1)/PVME(2)

The miscible blend of poly(styrene) (PS) and poly(vinyl methyl ether) (PVME). The phase diagram is shown in Figure 18 (see Chapter 5, Transitions in Polymers). We recognize lower critical solution temperature (LCST) behavior above 152 °C. A plot of interaction parameter times temperature, χT, *versus* temperature, indicates clearly the LCST (Figure 19). We get also energy and excess entropy contribution to parameter χ from this plot: $\Delta U/R = -43$ K, $S^E/R = -0.1$ (Equation (23)).

Figure 3 shows variation of surface tension with blend composition (Bhatia et al., 1988; Cowie et al., 1993a; Cowie et al., 1993b). In agreement with II), we observe huge surface excess of PVME ($\gamma = 21.9$ mN/m) as compared to PS ($\gamma = 29.7$ mN/m). We also recognize that the area available for an average segment in the surface region is quite large, $a = 4$ nm^2. Surface fraction of PVME, W^{S}_{PVME}, plotted in Figure 4, shows that PVME wets almost completely the surface of the blend. We have here preferential surface adsorption of PVME.

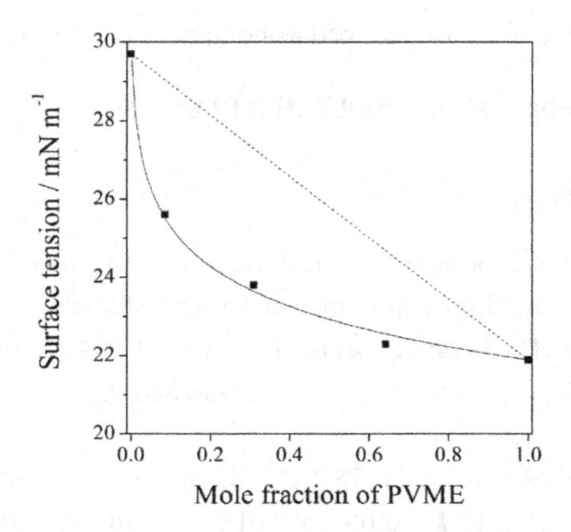

FIGURE 3 Surface tension of PS/PVME blend as a function of composition at 150 °C; graph after data of Cowie et al. (1993a); the solid curve was calculated after Equation (14) with $k = 0.005$, $\gamma^+ = 1.5$ mN/m, $\lambda = 1$.

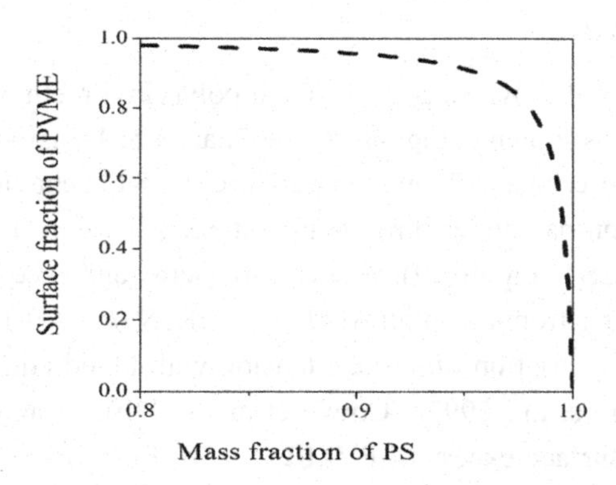

FIGURE 4 Surface fraction of PVME in blends with PS at 150 °C; the curve was calculated by use of Equation (10) with the parameters given in caption to Figure 3.

9.4.3 PS/PP

In principle such complete wetting, as just discussed for PS/PVME blends, could also occur in immiscible blends when spreading of the melt with the lower surface tension is possible. The immiscible blend of polystyrene (PS) and polypropylene (PP) does not show this wetting behavior (Figure 5). We observe:

$$\gamma = X_{PS}\gamma_{PS} + X_{PP}\gamma_{PP} \tag{18}$$

which means $k = 1$ or there is no difference in composition between surface and bulk. They display the same heterogeneity.

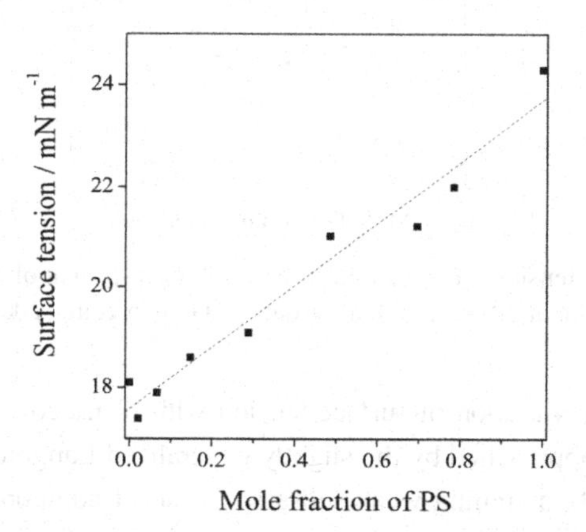

FIGURE 5 Surface tension of PS/PP blends at 230 °C; graph after data of Funke et al. (2001).

9.4.4 PMA(1)/PEO(2)

For a mixture of low molecular constituents, *n*-propanol and *n*-nonane, it was observed not only negative deviation of surface tension from linear variation with composition, but also a minimum in surface tension (Gaman et al., 2005). Figure 6 shows variation of surface tension with composition of the mixture. Similar behaviour was found for the miscible blend of poly(methyl acrylate) (PMA) and poly(ethylene oxide) (PEO) (Pefferkorn et al., 2010). As far as the author is aware, this is first time that an extreme value for surface tension was found in a miscible polymer blend. The polymers used had number average molecular masses of $M_n \approx 5000$ g/mol. Blends of PMA and PEO of those low molecular masses are completely miscible (Pedemonte and Burgisi, 1994). The Flory-Huggins interaction parameter was determined from *PVT* data to $\chi \approx -0.03$ at 120 °C (Pefferkorn et al., 2010).

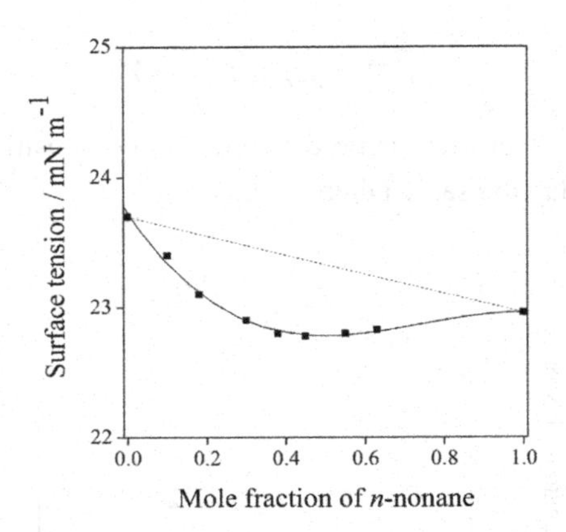

FIGURE 6 Surface tension of the mixture comprising *n*-propanol and *n*-nonane as a function of composition at 20 °C; graph after data of Gaman et al. (2005).

In the following, variation of surface tension with blend composition as shown in Figure 7 will be approached by the slightly generalized Langmuir adsorption isotherm, Equation (10), assuming weak self-association of component 2 in the bulk. Here, PEO is considered to be component 2. This implies variation of surface tension with composition according to Equation (14) with a minimum given by Equation (15).

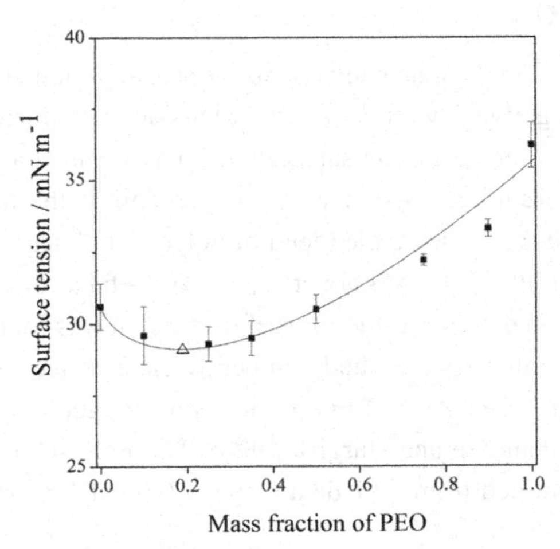

FIGURE 7 Surface tension of PMA/PEO blends as a function of composition at 120 °C; the solid curve was calculated after Equation (14) with the parameters given in Equation (19); the open triangle indicates the minimum of surface tension after Equation (15).

We analyze it according to the limiting situation Ib). This leads to Equation (11), telling us that both components are depleted in the surface region as compared to bulk in the range of low PEO content in the blend, but PEO more than PMA. This generates the minimum in surface tension. We fit experimental results to Equation (14). The best regression curve follows with:

$$k = 1.03 \quad \lambda = 0.955 \quad \gamma^{+} = 169.2 \text{ mN/m} \tag{19}$$

The corresponding curve is depicted in Figure 7. With these parameters, the minimum follows from Equations (12) and (15). It appears at:

$$W_{\text{PEO}}^{\text{min}} = 0.19 \quad \text{and} \quad \gamma^{\text{min}} = 29.1 \text{ mN/m} \tag{20}$$

Results are in good agreement with experimental findings. The reference surface tension γ^{+} corresponds to a surface coverage by an average segment of $a \approx 3$ Å².

The temperature coefficient of surface tension is plotted *versus* composition in Figure 8. One recognizes that the blend exhibits the temperature coefficient of PMA as long as PMA is in excess. At lower content of PMA, the coefficient steeply ascends to the temperature coefficient of PEO. This indicates that surface entropy of the blend is ruled by PMA in the range of low content of PEO.

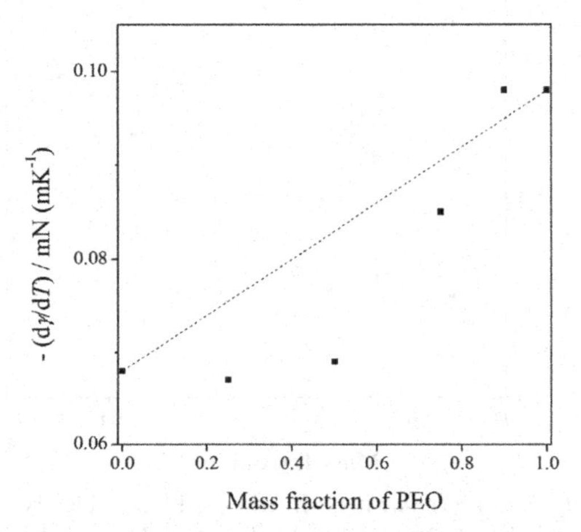

FIGURE 8 Temperature coefficient of surface tension as a function of blend composition; the dotted curve gives linear variation of temperature coefficient with composition.

For a wide variety of organic liquids, it was shown by Sanchez (Sanchez, 1983) that surface tension, isothermal compressibility and density form an invariant with $A_o = (2.78 \pm 0.13) \cdot 10^{-8}$ (J m² kg²)$^{1/2}$.

$$\gamma\left(\frac{\kappa}{\rho}\right)^{1/2} = A_o \tag{21}$$

Even random copolymers of ethylene and vinyl alcohol as well as of styrene and acrylonitrile follow Equation (21); see next paragraph (Kammer and Kressler, 2010). In reference Pefferkorn et al., 2010, *PVT* data of the blend were determined that allowed for calculation of the left-hand side of Equation (21). Results for the blend under discussion are presented in Figure 9 as a function of blend composition. The data obey Equation (21) in fair approximation as long as PMA is in excess. But, the calculated value of $A_o \approx 2.2 \cdot 10^{-8}$ (J m² kg²)$^{1/2}$ is considerably lower than the value found by Sanchez. When PEO is in excess, quantity $\gamma(\kappa/\rho)^{1/2}$ is no longer constant, but becomes a function of composition and eventually approaches $A_o = 2.78 \cdot 10^{-8}$ (J m² kg²)$^{1/2}$.

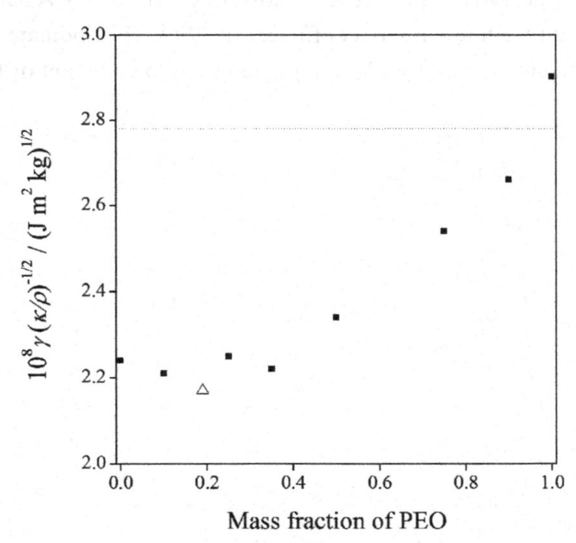

FIGURE 9 Quantity $\gamma(\kappa/\rho)^{1/2}$ *versus* mass fraction of PEO in blends of PEO and PMA. All data refer to 120 °C. The dashed line marks A_o as found by Sanchez (1983). The open square corresponds to minimum of surface tension after Equation (15).

9.5 SURFACE TENSION OF RANDOM COPOLYMERS

The composition dependence of surface tension of a random copolymer was already discussed in Figure 2 of Paragraph 9.2. This copolymer showed negative deviation from linearity with respect to composition.

The random copolymer poly(ethylene-*co*-vinyl alcohol) (EVOH) varies linearly with copolymer composition although there is a remarkable difference in surface tension of the parent polymers:

$$\gamma_{\text{lin PE}} = 22.8 \text{ mN/m} \qquad \gamma_{\text{PVA}}(\text{extrapolated}) = 43 \text{ mN/m}.$$

However, conformational restrictions prohibit enrichment of the low-energy constituent in the surface region. The variation of surface tension with copolymer composition is shown in Figure 10. The composition dependence is given by (Kammer and Kressler, 2010)

$$\gamma = \gamma_{\text{PVA}} - 20.3\, X_{\text{ethy}} \tag{22}$$

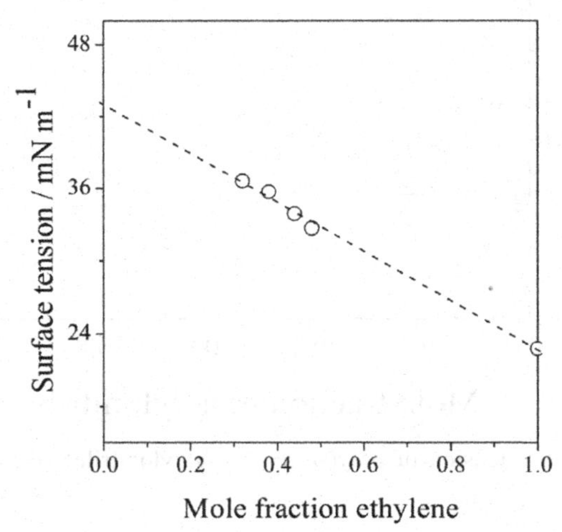

FIGURE 10 Surface tension of poly(ethylene-*co*-vinyl alcohol) as a function of composition at 220 °C.

Different behaviour shows poly(styrene-*co*-acrylonitrile) (SAN) (Funke et al., 2007). Although the parent polymers, PS and poly(acrylonitrile) (PAN) are immiscible, the random copolymer behaves similar to a miscible blend with negative deviation from additive that is surface excess of the constituent with the lower surface tension.

$$\gamma_{PS} = 26.3 \text{ mN/m and } \gamma_{PAN} \text{ (extrapolated)} \approx 46 \text{ mN/m at } 220\ °C$$

Figure 11 shows the result. In terms of Langmuir approximation Equation (14), experimental results are fitted best by:

$$k = 0.075, \gamma^{+} = 7.5 \text{ mN/m}, \lambda = 1, a = 0.9 \text{ nm}^2 \quad (23)$$

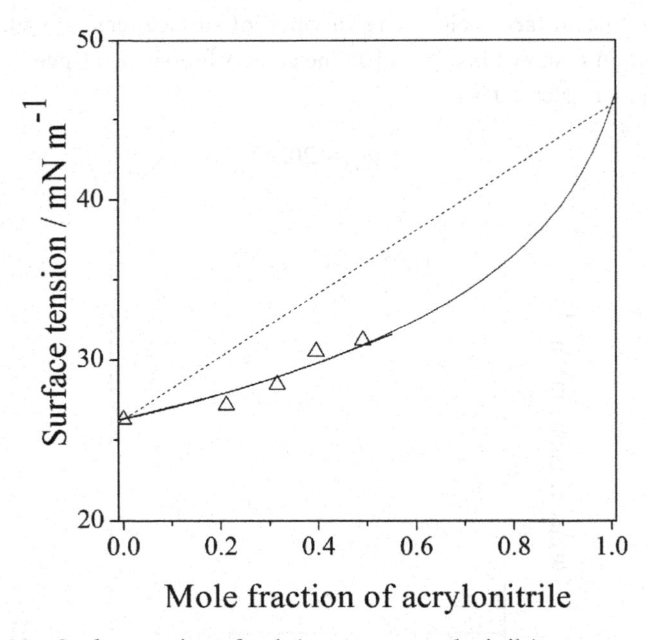

FIGURE 11 Surface tension of poly(styrene-*co*-acrylonitrile) *versus* composition at 220 °C.

We note a small area only is available for the segments in the surface region. Figure 12 compares the surface enrichment of styrene in the copolymer SAN with the depletion of polystyrene in the surface region when blended with PVME.

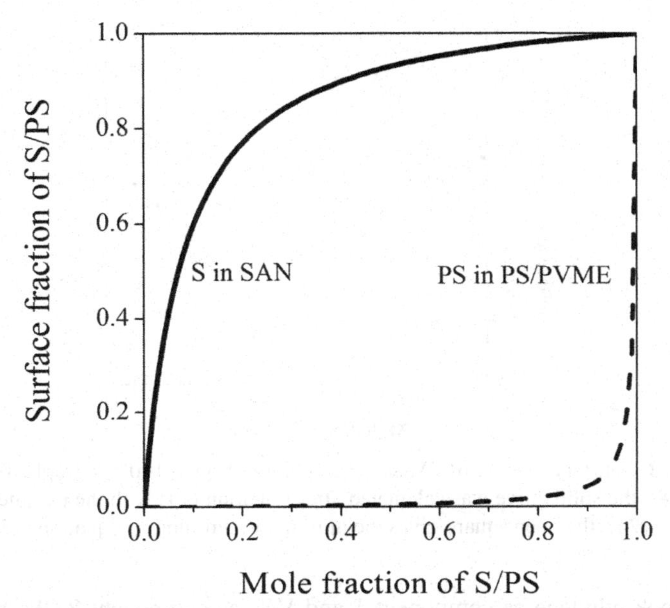

FIGURE 12 Surface fraction of S in SAN and PS in the blend of PS and PVME at 220 °C; curves were calculated after Equation (10) with the parameters given in Equation (23).

Very special behavior is displayed by the random copolymer poly(ethylene-*co*-vinyl acetate) (EVAc) (Hata, 1968) due to the fact that the parent polymers have very similar surface tension:

$$\gamma_{PE} = 27.6 \text{ mN/m}, \qquad \gamma_{PVAc} = 27.2 \text{ mN/m at } 160\ °C$$

We observe negative deviation from additive with a minimum in surface tension (Figure 13). Experimental data are fitted best to Equation (14) with the parameters:

$$k = 0.96,\ \gamma^{+} = 9.8 \text{ mN/m},\ \lambda = 0.65,\ a = 0.6 \text{ nm}^2 \qquad (24)$$

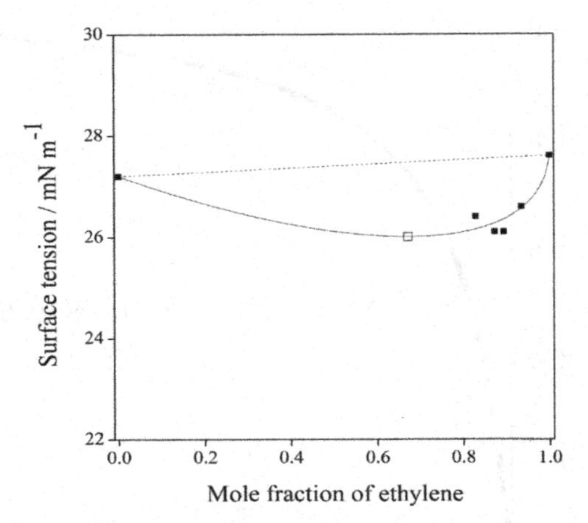

FIGURE 13 Surface tension of EVAc *versus* composition at 160 °C; graph after data by Hata (1968); the solid curve was calculated after Equation (14) with the parameters given in Equation (24); the open square gives the minimum according to Equation (15).

We took ethylene as component 1 and VAc as component 2. We recognize considerable self-association of VAc. The area per segment occupied in the surface is quite small. The minimum of surface tension occurs after Equation (15) at:

$$X_{ethyl}^{min} = 0.672 \qquad \gamma^{min} = 26.0 \text{ mN/m} \tag{25}$$

Figure 14 shows that ethylene units are somewhat depleted in the surface region.

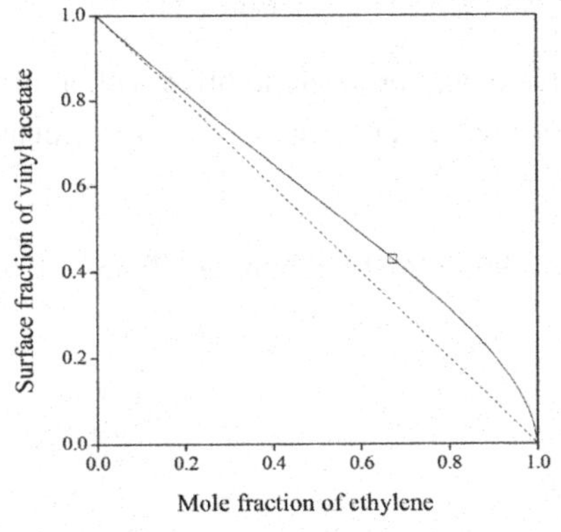

FIGURE 14 Surface fraction of VAc calculated after Equation (11) with parameters Equation (24); the open square indicates the composition at minimum of surface tension.

We tested if the random copolymers under discussion obey relationship Equation (21) by plotting surface tension *versus* $(\rho/\kappa)^{1/2}$ at temperatures as mentioned before. The result is shown in Figure 15. Data for EVOH and SAN follow fairly well Equation (21) with $A_o = (2.680 \pm 0.015)\ 10^{-8}\ (J\ m^2/kg)^{1/2}$ whereas EVAc copolymers deviate by approximately 10 % from Equation (21).

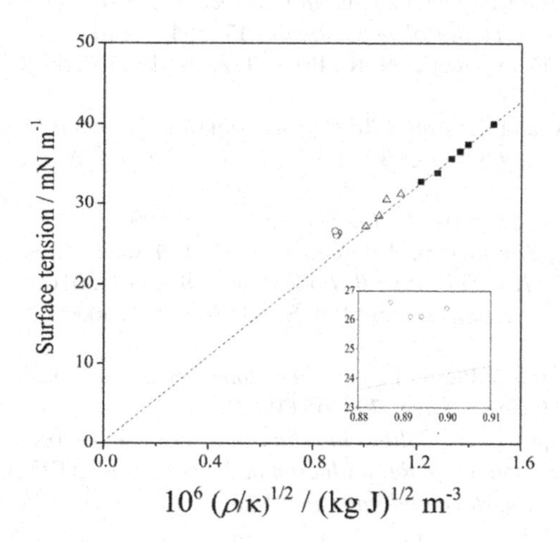

FIGURE 15 Equation (21) for random copolymers: ■ – EVOH, Δ – SAN, and o – EVAc.

KEYWORDS

- **Langmuir adsorption isotherm**
- **Pendant drop method**
- **Sanchez approximation**
- **Surface tension of polymer blends**
- **Surface tension of random copolymers**

REFERENCES

1. Arashiro, F. Y. and Demarquette, N. R. *Mat. Res.* Sao Paulo **2**(1) (1999).
2. Bhatia, Q. S., Pan, D. H., and Koberstein, J. T. *Macromolecules*, **21,** 2166 (1988).
3. Cowie, J. M. G., Devlin, B. G., and McEwen, I. J. *Polymer*, **34**, 4130 (1993a).
4. Cowie, J. M. G., Devlin, B. G., and McEwen, I. J. *Macromolecules*, **26**, 5628 (1993b).
5. Demarquette, N. R. and Kamal, M. R. *J. Appl. Polym. Sci.*, **70**, 75 (1998).
6. Fleischer, C. A., Morale, A. R., and Koberstein, J. T. *Macromolecules*, **27**, 379 (1994).

7. Funke, Z., Hotami, Y., Ouigizawa, T., Kressler, J., and Kammer, H. W. *Europ. Polym. J.*, **43**, 2371 (2007).
8. Funke, Z., Schwinger, C., Adhikari, R., and Kressler, J. *Macromol. Mater. Eng.*, **286**, 744 (2001).
9. Gaman, A. I., Napari, I., Winkler, P. M., Vehkamäki, H., Wagner, P. E., Strey, R., Viisanen, Y., and Kulmala, M. *J. Chem. Phys.*, **123**, 244502 (2005).
10. Gamarbi, H., Demarquette, N. R., and Kamal, M. R. *Int. Polym. Proc.*, **2**, 183 (1998).
11. Hansen, F. K. and Rodsrun, G. *Colloid Interface Sci.*, **141**, 1 (1991).
12. Hata, T. *Kobunshi (High Polymers, Japan)* **17**, 594 (1968)..
13. Kamal, M. R., Demarquette, N. R., Price, T. A., and Lai-Fork, R. A. *J. Polym. Eng. Sci.*, **37**, 813 (1997).
14. Kammer, H. W. and Kressler, J. In *Polymer Physics: From Suspensions to Nanocomposites and Beyond*, Chapter 8; Utracki, L. A., Jamieson, A. M., Eds.; Wiley: New York, 2010.
15. Pedemonte, E. and Burgisi, G. *Polymer*, **35**, 3719 (1994).
16. Pefferkorn, D., Sommer, S., Kyerematend, S. O., Funke, Z., Kammer, H. W., and Kressler, J. *J. Polym. Sci.: Part B: Poly. Phys.*, **48**, 1893 (2010).
17. *Polymer Interface and Adhesion*; Wu, S. Ed.; Marcel Dekker : New York and Basel, 1982.
18. Rastogi, A. K. and St.Pierre, L. E. *J. Coll. Interface Sci.*, **31**, 168 (1969).
19. Sanchez, I. C. *J. Chem. Phys.*, **79**, 405 (1983).
20. Song, B. and Springer, J. *Colloid Interface Sci.*, **184**, 64 (1996).
21. Wu, S. *J. Macromol. Sci. – Revs. Macromol. Chem.* C10, 1 (1974).
22. Wu, S. *Polymer Interface and Adhesion*; Marcel Dekker: New York, 1982."

CHAPTER 10

MACROMOLECULES IN THE CONDENSED STATE

HANS-WERNER KAMMER

CONTENTS

10.1 MELTS

We restrict here to amorphous situations in polymers since crystallization of polymers is extensively discussed in chapter 6 (see Chapter 6, Crystallization and Melting of Polymers).

PRELIMINARIES

As discussed in chapter 3 (see Chapter 3, Macromolecules in Solution), macromolecules may exist in the solid state, the molten state, and in solutions, but not in the gaseous state. This is so because the boiling temperature of a polymer melt is proportional to the degree of polymerization N:

$$T_{\text{boiling}} \sim N$$

This temperature is for $N >> 1$ much higher than the temperature of thermal decomposition of the chains.

$$T_{\text{boiling}} >> T_{\text{decomposition}}$$

Hence, polymer molecules do not exist in the gaseous state.

Polymer chains in a Θ-solvent behave perfectly with a size (Equations 36 and 37 chapter 3)

$$R_{\text{o}} \sim N^{1/2} \tag{1}$$

Dilute polymer chains in a good solvent are swollen.

$$R_{\text{F}} \sim N^{3/5} \tag{2}$$

What happens with the size of a chain molecule if we add to a solution more and more macromolecules generating a bulk molten state? Qualitatively, as Equation (2) indicates, in a good solvent, a force is acting on the chain and pointing outward. This force is responsible for swelling of the chain. If we have a densely packed melt of chains an exactly opposite force becomes operative that acts on the neighbor chains and keeps the total density constant. As a result, chains in a melt experience no force and remain ideal. The size is given by Equation (1).

Suppose the volume of a segment is a^3, then the number concentration of monomers in a perfect coil follows with Equation (1)

$$c_{\text{coil}} = \frac{N}{v_{\text{coil}}} = \frac{1}{a^3 N^{1/2}} \tag{3}$$

This concentration is very small compared to the concentration of monomers in a densely packed melt:

$$c_{\text{melt}} = \frac{1}{a^3} \tag{4}$$

Establishing of concentration Equation (4) needs interpenetration of chains in the melt.

ENTANGLEMENT VERSUS SEGREGATION

The previous discussion suggests that chains overlap in the melt. Even then, however, individual chains preserve the shape of a perfect coil in three-dimensional melts. There was a dispute among polymer scientists about two opposite views on polymer melts as sketched (Figure 1).

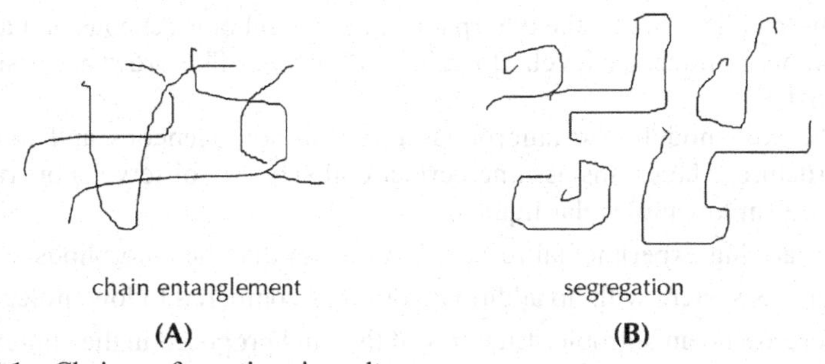

FIGURE 1 Chain conformations in melt.

In Figure 1(A), chains interpenetrate strongly whereas chains are segregated in Figure 1(B). Neutron scattering experiments showed that chains are ideal in a melt. Hence, interpenetration of chains is necessary in a three-dimensional melt to establish a concentration in the order of Equation (4). Figure 1(A) is adequate.

In a two-dimensional melt (approximated in a surface film), the concentration inside a coil is:

$$c_{\text{coil}}^{(2)} \cong \frac{N}{R_o^2} \cong \frac{1}{a^2}$$

In the two-dimensional case, the concentration inside a coil is comparable with the total concentration in the melt. Figure 1(B) seems to be more adequate in that case.

In conclusion, in three dimensions, the high flexibility of chain molecules ensures interpenetration and entanglement of coils and concomitantly, a high density of segment packing (small free volume) so that a polymer melt, as a low molecular liquid, has insignificant volume compressibility.

DISORDER VERSUS ORDER IN THE AMORPHOUS STATE

The physical structure of melts can be characterized by:

- Chain conformation studied by small-angle neutron scattering,
- Local order characterized by pair distribution functions (derived from electron scattering traces),
- Morphology studied by light or small-angle X-ray scattering.

Basically, two different models have been proposed to describe the equilibrium situation in melts and to interpret the experimental results:

- The bundle model—the amorphous phase is inhomogeneous and anisotropic on a molecular level. Domains with crystal-like order are assumed to exist.
- The coil model—the amorphous phase is homogeneous and random in structure. The chains assume perfect coil size. We observe short-range order as in low molecular liquids.

Summarizing experimental results, it turns out that the amorphous phase is a homogeneous system with no additional order as compared to low molecular liquids. There are no anisotropic structures of the kind proposed in the context of the bundle model. The chains assume random coil conformation. The near-range order, determined by chain segments belonging to different interpenetrating chains, cannot be described by alignment of neighboring chains or parallel packing of chains. It turns out that the coil model reflects reality to a good approximation.

The theoretical and experimental instrument for discussing pair correlations in dense systems is the so-called radial distribution function, $g(r)$. This function is provided by scattering experiments. What we are measuring in scattering experiments are particle correlations or density or concentration fluctuations. These correlations are given by expressions of the type:

$$\overline{X^2} - \overline{X}^2 \tag{5}$$

where the bars symbolize averages of quantity X. All correlations are ignored when $\overline{X^2} = \overline{X}^2$. The intensity of scattered radiation is monitored as a function of scattering vector q.

$$I = I(q) \quad \text{with} \quad q = \frac{4\pi}{\lambda} \sin(\Theta / 2) \tag{6}$$

λ—wavelength of radiation and Θ—scattering angle. Particle number fluctuations, Equation (5) are given by $I = I(q \to 0)$.

$$I(0) = \frac{\overline{N^2} - \overline{N}^2}{\overline{N}} \tag{7}$$

The relationship between scattering intensity $I(0)$ and pair correlation function $g(r)$ reads:

$$I(0) = 1 + 4\pi\rho \int \mathrm{d}r r^2 \left[g(r) - 1 \right] \tag{8}$$

with $\rho = N/v$—particle number density.

It is also possible to express the particle fluctuations by thermodynamic quantities:

$$\frac{\overline{N^2} - \overline{N}^2}{\overline{N}} = k_B T \rho \kappa_T \tag{9}$$

k_B—Boltzmann's constant and κ_T—isothermal compressibility.

Equation (19) is derived in Appendix. Combination of Equation (7) and Equation (9) displays a relationship between scattering results and thermodynamic quantities.

$$I(0) = k_B T \rho_T \tag{10}$$

Equation (10) might be seen as an equation-of-state for polymer melts.

When scattering solely arises from particle number or density fluctuations in the melt, experimental scattering results should be in agreement with Equation

(10). This would support the coil model. Scattering experiments yield similar results for polymers and low molecular liquids and show agreement with Equation (10). This is in favor of the coil model.

The X-ray scattering is only sensitive to density fluctuations whereas scattering of polarized light (so-called H_v-component) reacts to anisotropy fluctuations as well. It has been observed that the H_v-component does not show any angular dependence for amorphous polymers in equilibrium. It demonstrates that there are no anisotropic domains resulting from alignment or order of chains.

PAIR CORRELATIONS IN MELTS AND RANDOM PHASE APPROXIMATION (RPA)

We want to discuss density correlations caused by fluctuations in a dense system of interacting polymer chains. This is done in terms of the RPA.

We are discussing a system consisting of many chains each of them comprises N segments. The concentration of k-type segments might be Φ_k. We define now the fluctuation of Φ_k by:

$$\delta\Phi_k(r) = \Phi_k(r) - \Phi_{ko} \tag{11}$$

where $\Phi_k(r)$ is the local concentration at r and Φ_{ko} the average constant concentration of k-type segments. Fluctuations of type Equation (11) display a Gaussian distribution in equilibrium systems. All averages have to be seen as averages over normal (Gaussian) distributions. We abbreviate here the averages by <...> (instead by the bar as in Equation (5)). For concentration fluctuations of segments of types k and l, we have analogously to Equation (5) with definition Equation (11).

$$\langle \Phi_k(r)\Phi_l(r')\rangle - \Phi_{ko}\,\Phi_{lo} = \langle(\Phi_k(r) - \Phi_{ko})\,(\Phi_l(r') - \Phi_{lo})\rangle = \langle \delta\Phi_k(r)\,\delta\Phi_l(r')\rangle \tag{12}$$

The correlation function introduced with the last term of Equation (12) gives the average correlation between fluctuations at r and r'. This average is equal to the so-called response function or structure factor S:

$$\left\langle \delta\Phi_k(r)\,\delta\Phi_l(r')\right\rangle = S_{kl}\left(|r - r'|\right) \tag{13}$$

where the dependence on just the difference between r and r' is due to the translational invariance of the system.

The Fourier transform of S_{kl} reads:

$$S_{kl}(q) = \int d\,|r - r'|\, S_{kl}(|r - r'|)\ \exp(-iq\,|r - r'|) \tag{14}$$

For non-interacting perfect chains, the Fourier transform of Equation (13) is just given by:

$$S_{kl}^{o}(q) = <\exp[-i\boldsymbol{q}(\boldsymbol{r}_k - \boldsymbol{r}_l)]>_{perfect} \tag{15}$$

The average in Equation (15) is taken over all conformations of the ideal chain. The response function for segments k and l of non-interacting chains becomes when we denote the length of a segment by a:

$$S_{kl}^{o}(q) = \exp\left[-|k-l|\frac{q^2a^2}{6}\right] \tag{16}$$

When we summarize the scattering amplitudes Equation (16) on all monomers, we get Debye's scattering function $g_D(q)$:

$$g_D(q) = \frac{1}{N}\sum_{k,l}^{N} S_{kl}^{o}(q) = \frac{2N}{\Lambda}\left[1 - \frac{1}{\Lambda}\left(1 - e^{-\Lambda}\right)\right] \tag{17}$$

with:

$$\Lambda \equiv N\frac{q^2a^2}{6} = (qR_o)^2$$

(Compare the discussion in chapter 3). We abbreviate here the radius of gyration of the perfect chain by R_o (Equation 6, chapter 3) (see Chapter 3, Macromolecules in Solution). The limiting behavior of Equation (17) is given by:

(A) $\Lambda << 1$ or $qR_o << 1$ $\qquad g_D(q) = N\left(1 - \frac{\Lambda}{3}\right)$

(B) $\Lambda >> 1$ or $qR_o >> 1$ $\qquad g_D(q) = \frac{2N}{\Lambda}$

In A), the wavelength of fluctuation λ, is much larger than the coil size R_o, $\lambda >> R_o$ and g_D approaches N, $g_D(0) = N$.

In B), the wavelength of fluctuations is much smaller than the coil size, $\lambda << R_o$ and $g_D \rightarrow 0$.

With Equations (13) and (16), we have the tools for discussing the structure function of interacting systems in a self-consistent field approach RPA.

After Equation (13), the Fourier transform $\psi_k(q)$ of concentration fluctuations $\delta\Phi_k(r)$ is given by:

$$\psi_k(q) = -\sum_l S_{kl}(q)\frac{U_l(q)}{k_B T} \tag{18}$$

Equation (18) describes the response of the system, reflected in density fluctuations of monomers k, which is induced by potentials U_l that act on all monomers. Calculation of the structure function is now carried out in a self-consistent way. We approximate Equation (18) by:

$$\psi_k(q) = -\sum_l S^o_{kl}(q)\frac{U_l(q)+V(q)}{k_B T} \tag{19}$$

It means, we are looking for the response of ideal chains experiencing not only external potentials U_l but also a self-consistent potential V. Due to the chemical identity of all monomers, the self-consistent potential, generated by the surrounding chains in a dense chain system, is equal for all N monomers of a chain. The self-consistent potential can be easily calculated using the incompressibility condition of polymer melts. The sum of all density fluctuations has to be zero

$$\sum_k \psi_k(q) = 0 \tag{20}$$

Using Equation (20) with Equation (19), we have for the self-consistent potential V:

$$V = -\frac{1}{N g_D(q)}\sum_l S^o_l(q)U_l(q) \tag{21}$$

where:

$$S_l^o(q) = \sum_k S_{kl}^o(q)$$

Insertion of Equation (21) in Equation (19) and comparison with Equation (18) yields the self-consistent approximation of the structure factor for a dense system of interacting chains. We arrive at:

$$S_{kl}(q) = S_{kl}^o(q) - \frac{S_k^o(q)S_l^o(q)}{Ng_D(q)} \tag{22}$$

For $q \to 0$, we have $S_{kl}{}^o = 1$ and $S_k{}^o = S_l^o = g_D = N$. Hence, Equation (22) vanishes.

$$S_{kl}(0) = 0 \quad \text{for all } k, l \tag{23}$$

Equation (23) implies that also the scattering intensity vanishes since:

$$I(q) \propto \sum_{k,l} S_{kl}(q) \tag{24}$$

Equation (23) indicates that fluctuations with wavelength $\lambda \to \infty$ vanish, as it should be since the density cannot fluctuate at a large scale in equilibrium. For smaller wavelengths or larger values of q, one may observe density fluctuations although the total density is constant.

10.2 GLASS TRANSITION AND THE GLASSY STATE

10.2.1 BASICS

This chapter adds some supplements to part glass transition (refer to Chapter 7). When crystallization is suppressed in a melt, since the rate of crystallization is smaller than the rate of cooling, the melt solidifies to a glass. Phenomenological, the transition to the glassy state is characterized by changes of certain physical properties during isobaric cooling of a liquid. The temperature dependence of quantities as:

- Coefficient of isothermal expansion.

- Specific heat.
- Modulus of elasticity.
- Diffusion coefficient.

It is used to describe the glass transition. We illustrate the glass transition by an isobaric volume-temperature diagram as given in Figure 11 in chapter 5 $V = V(P,T)$ symbolizes the specific volume, $V = v/m$. The diagram reveals:

- The volume does not display a jump at the glass transition,
- The glass transition depends on cooling rate, the slower the cooling rate the smaller the difference between V_{glass} and $V_{crystal}$ at $T < T_g$, and $T = const$,
- The coefficient of expansion is not very different in glassy and in crystalline state.

The X-ray diffraction shows that glasses display the same short-range structure as liquids. The long-range order of crystals does not occur. This indicates that glasses are frozen-in liquids. Moreover, the transition temperature depends on the cooling rate. It means the glassy state is a non-equilibrium situation accompanied by relaxation processes with extremely low rates. The solidification process might be seen as competition of two rate processes governing two characteristic times:

- Relaxation time of segments $\tau(T) \propto \exp(\Delta E/k_B T)$ which means increasing τ with descending T

Characteristic time t imposed by the cooling rate $\dot{T}$, $t = \Delta T/\dot{T}$

We repeat here the statements about glass transition in terms of these two times, as given chapter 5 in glass transition:

- $\tau << t$ *melt*, segments can follow immediately any temperature change $\Rightarrow$ equilibrium
- $\tau \sim t$ *glass* transition
- $\tau >> t$ *glass*, segments cannot follow the temperature change $\Rightarrow$ non-equilibrium

The accompanying non-equilibrium phenomena are discussed in chapter 5.

10.2.2 GLASS TRANSITION TEMPERATURE FOR MISCIBLE BLENDS

For immiscible polymers forming a blend, one observes just the two glass transition temperatures of the constituents. In the case of miscible polymers, however, one observes one monotonously changing glass transition temperature between the two T_gs of the pure components. We illustrate the situation again in the V–T plane, Figure 2.

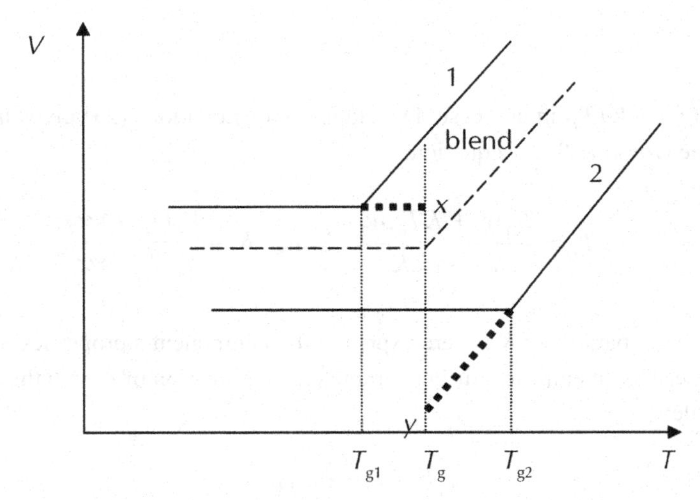

FIGURE 2 The V–T diagram illustrating T_g for polymer blends – x and y, extrapolations of the glassy and liquid state, respectively, for the corresponding reference systems.

We express specific volume of the blend by those of the constituents as follows:

$$V^{\mathrm{l}}\left(T_{\mathrm{g}}\right)=V_{1}^{\mathrm{l}}\left(T_{\mathrm{g}}\right)w_{1}+V_{2}^{\mathrm{l,h}}\left(T_{\mathrm{g}}\right)w_{2}+\Delta V^{\mathrm{l}}\left(T_{\mathrm{g}}\right) \tag{25}$$

$$V^{\mathrm{g}}\left(T_{\mathrm{g}}\right)=V_{1}^{\mathrm{g,h}}\left(T_{\mathrm{g}}\right)w_{1}+V_{2}^{\mathrm{g}}\left(T_{\mathrm{g}}\right)w_{2}+\Delta V^{\mathrm{g}}\left(T_{\mathrm{g}}\right)$$

The ΔV^{l} and ΔV^{g} at $T = T_{\mathrm{g}}$ denote deviations from additivity. We note, $V_2^{\mathrm{l,h}}$ (*h* for hypothetical; *y* according to the Figure) might be seen as the (specific) volume of component 2 after infinitesimally slow cooling from $T_{2\mathrm{g}}$ to T_{g} and analogously $V_1^{\mathrm{g,h}}$ that of component 1 after infinitely rapid heating from T_{g1} to T_{g}. Development of Equations (25) around T_{g1} and T_{g2}, respectively, shows that we refer in any case to the same reference system:

$$V_1^{\mathrm{l}}(T_{\mathrm{g1}})w_1 + V_2^{\mathrm{l}}(T_{\mathrm{g2}})w_2 \quad \text{or} \quad V_1^{\mathrm{g}}(T_{\mathrm{g1}})w_1 + V_2^{\mathrm{g}}(T_{\mathrm{g2}})w_2.$$

Hence,

$$\Delta V^{\mathrm{l}} = \Delta V^{\mathrm{g}}$$

since $V^{\mathrm{l}}(T_{\mathrm{g}}) = V^{\mathrm{g}}(T_{\mathrm{g}})$ in any case. Development of Equations (25) gives immediately. The *Gordon–Taylor* equation:

$$T_{\mathrm{g}} = \frac{T_{\mathrm{g1}} w_1 + K T_{\mathrm{g2}} w_2}{w_1 + w_2 K} \quad \text{with} \quad K = \frac{V_2\left(T_{\mathrm{g2}}\right)\Delta\alpha_2}{V_1\left(T_{\mathrm{g1}}\right)\Delta\alpha_1} \tag{26}$$

Although, parameter K is here expressed by component's properties, it is frequently seen as an empirical fitting parameter. Combination of K with the *Simha–Boyer* rule:

$$T_{\mathrm{g}}\,\Delta\alpha = const$$

leads to $K = K^{+}\, T_{\mathrm{g1}}/T_{\mathrm{g2}}$ which turns with $K^{+} = 1$ the *Gordon–Taylor* equation in the *Fox* equation:

$$\frac{1}{T_{\mathrm{g}}} = \frac{w_1}{T_{\mathrm{g1}}} + \frac{w_2}{T_{\mathrm{g2}}} \tag{27}$$

10.2.3 MECHANICAL GLASS TRANSITION

A polymer becomes not only glassy below T_{g} but also at higher constant temperature and sufficiently high rate of external force action. Therefore, one distinguishes:

- Structural glass transition at $T = T_{\mathrm{g}}$
- Mechanical glass transition when relaxation time of segments, τ, equal time of external force action, t: $\tau \approx t$

If the frequency of force action is in the order of 1 ... 10 Hz, mechanical and structural glass transitions are in quite good agreement.

Experimental results are given by thermomechanical curves that allow determining temperature ranges for existence of:

- Glassy,
- Rubber-like, and
- Viscous-flow state.

Each of these physical states of a polymer is of significance for its processing and use. One may conclude:

- Glassy state—stress causes small deformations only,
- High-elasticity region—stress causes large elastic strains that do not depend strongly on temperature,
- Melt—stress causes irreversible deformations (onset of flow).

10.3 THE STATE OF HIGH ELASTICITY

10.3.1 INTRODUCTION

The high elasticity or rubber elasticity is a specific state of matter displayed by polymers: Large reversible deformation is caused by small tensions. For instance, natural rubber can be stretched reversibly 10–15 times of its original length. Polymeric materials exhibiting high elasticity at room temperature are called elastomers or rubbers.

High elasticity behavior is different to that one what is usually observed in reversible (or elastic) deformations of solids (e.g. metals). Stress subjected to a specimen causes deformation. If the stress is removed and the specimen returns to its original shape, the deformation is called elastic or reversible. Hooke's law is the base law of deformation of a perfect elastic body (Chapter 7, Equation 15):

$$\text{deformation (strain)} \sim \text{stress}$$

The proportionality factor is called modulus of elasticity.

One may distinguish two classes of elastic behavior displayed by different materials:

- *Energy Elasticity*: Specimens have a high resistance against deformation or large stresses cause merely small deformations. This is usually detected in low molecular crystalline systems; values of the modulus of elasticity, E, are high.
- *Entropy Elasticity*: Small stresses effect large (reversible) deformations. This is displayed by gases and polymers, values of E are low.

In case 1, atoms are regularly arranged on a lattice and exert strong interactions to each other, which resist to changes of their arrangement. Hence, externally imposed stress has to overcome intermolecular interactions. Elasticity is energetic in nature and high values of E result. Moreover, if T increases, vibrations of atoms, forming the crystal, get more intense (and weaken interactions). As a result:

$$E\downarrow \text{ with } T\uparrow$$

In case 2, resistance of a gas against deformation is determined by thermal motion of the molecules. If the space for motion is restricted, as it occurs under stress, the entropy of the system alters. Therefore, elasticity is entropic in nature and low values of E are observed. Thermal agitation of molecules increases with T. As a result:

$$E\uparrow \text{ with } T\uparrow.$$

The same behavior as for gases is also observed for polymers. In that case, elasticity is related to stretching of the chains, which is accompanied by the change of entropy.

E modules at room temperature:

Crystals	$10^{13} - 10^{14}$ Pa
Vulcanized rubber	$10^{8} - 10^{9}$ Pa

In a stress-strain curve of a polymer, one may distinguish three regions:

1. Hooke's region (small), linear dependence of stress on strain with small E.
2. Small changes in stress cause large deformations (high elasticity). Note proportionality between stress σ and strain ε does not exist or E modulus is not constant but changes with σ.
3. Irreversible deformations develop to great extent ("flow").

10.3.2 THERMODYNAMICS OF HIGH-ELASTICITY

Elastomers consist of a more or less regular network formed by chemically linked chain molecules ("vulcanization"). The network points represent permanent links. The properties of the material depend strongly on the network density. For high densities of network points, the polymer turns out to be rigid whereas for low densities, it behaves like a plastic. Elastomers are characterized by medium densities of network points.

Under deformation, sections of chains between the network points are stretched. In a good approximation, elastomers do not change their volume under strain ($\Delta v = 0$). Hence, the work of deformation (stretching) of a specimen is due only to the action of the applied force, f. According to 1st and 2nd Law of Thermodynamics, it might be separated in an energetic and entropic contribution:

$$f\,\mathrm{d}l = \mathrm{d}u - T\mathrm{d}s$$

where dl is the change of length under the action of force f.

At $T,v = const$, it follows:

$$f = \left(\frac{\partial u}{\partial l}\right)_{T,v} - T\left(\frac{\partial s}{\partial l}\right)_{T,v} \tag{28}$$

Ideal elastomers are defined by constancy of internal energy u during deformation $\left(\frac{\partial u}{\partial l}\right)_{T,v} = 0$. After Equation (28), it follows:

$$f = -T\left(\frac{\partial s}{\partial l}\right)_{T,v} \tag{29}$$

In an ideal elastomer, stress is solely related to change in entropy with deformation. This entropy change results from stretching of the coiled molecules (between junction points) and the structure becomes more oriented. These effects are accompanied by a decrease in entropy.

In contrast, in the small strain limit of a perfect crystal, when the crystalline structure does not alter significantly, we have:

$$f = \left(\frac{\partial u}{\partial l}\right)_{T,v} \tag{30}$$

Equations (29) and (30) demonstrate explicitly the meaning of entropy and energy elasticity, respectively.

In slightly more generalized terms, we get for the entropy and energy changes with deformation:

$$\left(\frac{\partial s}{\partial l}\right)_{T,v} = -\left(\frac{\partial f}{\partial T}\right)_{l,v} \qquad \left(\frac{\partial u}{\partial l}\right)_{T,v} = f - T\left(\frac{\partial f}{\partial T}\right)_{l,v} \tag{31}$$

If one determines experimentally the stress and its temperature dependence at various elongations (under equilibrium conditions), we get from Equation (31) the contributions of internal energy ($\partial u/\partial l$) and entropy ($\partial s/\partial l$) to the stress of a real elastomer. Experiments show: At small elongations, when the entropy changes insignificantly, the contribution of the internal energy is dominant. At very small elongations (< 10 %), the entropy term is even positive. In a broad range of elongations, the magnitude of the acting stress is dominantly determined by the change in entropy. At elongations exceeding 300 %, crystallization processes may take place (stress-induced crystallization) which lead to evolution of heat. Therefore, eventually, the magnitude of $(\partial u/\partial l)_T$ increases.

10.3.3 STATISTICAL THEORY OF RUBBER ELASTICITY

The entropy change of an elongated chain ($R_o \rightarrow R$) is (compare Equation (28) chapter 3):

$$\frac{\Delta S}{k_B} = -\frac{3}{2}\frac{(R^2 - R_o^2)}{R_o^2}$$

If the chains are elongated in the same way by factors α_i in directions of space, one gets for n chains:

$$\Delta S = -\frac{1}{2}k_B n(\alpha_x^2 + \alpha_y^2 + \alpha_z^2 - 3) \tag{32}$$

There is no volume change in a perfect rubber at elongation, $\Delta v = 0$. Therefore, we have:

$XYZ = \alpha_X \alpha_Y \alpha_Z XYZ$ what yields $\alpha_X = \alpha$ and $\alpha_Y = \alpha_Z = \frac{1}{\sqrt{\alpha}}$

where $\alpha = \frac{l}{l_o}$. It follows for Equation (32):

$$\Delta S = -\frac{1}{2}k_B n(\alpha^2 + \frac{2}{\alpha} - 3) \tag{33}$$

Equation (33) reflects pure entropy elasticity for an ideal rubber. The first equation of Equation (31) recast to $f = -\frac{T}{l_o}\left(\frac{\partial s}{\partial \alpha}\right)_{T,v}$ yields with Equation (33):

$$f = \frac{k_B T n}{l_o}(\alpha - \frac{1}{\alpha^2}) \tag{34}$$

Introducing density ρ, the average molar mass between cross-linking points M_e, and $f/q = \sigma$, Equation (34) becomes:

$$\sigma = \frac{RT\rho}{M_e}(\alpha - \frac{1}{\alpha^2}) \tag{35}$$

The ratio ρ/M_e may be seen as the network density of the rubber. With $E = (\partial\sigma/\partial\alpha)_{\alpha=1}$, it is:

$$E = \frac{3RT\rho}{M_e} \tag{36}$$

which shows that:

$$E\uparrow \text{ if } T\uparrow$$

in agreement with experimental results.

Equation (35) implies the following approximations:

- Deformation occurs without change of volume ($\Delta v = 0$),
- Elasticity of the sample is coined by deformation of a single chain (single-chain approximation).

10.3.4 ELASTICITY OF REAL RUBBERS

Despite of the approximations, the statistical theory is of fundamental significance for understanding of the molecular mechanisms causing rubber-like elasticity. It serves as a starting point for generalizations that agree more precisely with experiments. One generalization is the Mooney-Rivlin equation. After Equation (35), we have:

$$\frac{\sigma}{\alpha - \alpha^{-2}} = C_1$$

as starting point. The extension given by Mooney-Rivlin leads to:

$$\frac{\sigma}{\alpha - \alpha^{-2}} = C_1 + C_2 \alpha^{-1}$$

where C_1 and C_2 are constants. Constant C_1 is proportional to the network density while C_2 takes into account deviations of the real network from the ideal network structure.

We list a few points that distinguish real rubbers from ideal ones:

- The $\Delta v \neq 0$. The distances between chains in the network change during deformation. This induces alteration of the interaction energy between chains. Hence, deformation is not only accompanied by entropy but also by energy changes. This becomes especially important at large deformations where crystallization may occur (stress-induced crystallization). When crystallization takes place (above around 300 % elongation), it involves a tremendous decrease of internal energy. Hence, the nature of the deformation mechanism changes, it becomes gradually more energetic in nature.
- One has to take into account the non-equilibrium situation of high-elastic deformations, that is its relaxation behavior (approach towards equilibrium in the course of time). This is so because deformation is never purely elastic. Stretching of chains (in the network) is accompanied by slippage of chains or creeping of the sample. As a result, deformation is a superposition of elastic (reversible) deformation and flow (irreversible deformation).
- Deformation does not concern solely deformation of single chains (trapped in a network) but also deformation of supermolecular structures.

APPENDIX

Brief derivation of Equation (9):

$$\overline{N^2} - \overline{N}^2 = \frac{k_B T}{\left(\frac{\partial \mu}{\partial N}\right)_{T,v}}$$

and

$$\left(\frac{\partial \mu}{\partial N}\right)_{T,v} = -\left(\frac{\partial \mu}{\partial P}\right)_{T,N} \frac{\left(\partial v / \partial N\right)_{P,T}}{\left(\partial v / \partial P\right)_{N,T}} = \frac{1}{\rho \kappa_T N} \Rightarrow \tag{9}$$

KEYWORDS

- **Entropy elasticity**
- **Glassy state of blends**
- **Gordon-Taylor equation**
- **Mooney-Rivlin equation**
- **Random phase approximation**

CHAPTER 11

MOLECULAR CHARACTERIZATION OF SYNTHETIC POLYMERS BY MEANS OF LIQUID CHROMATOGRAPHY

DUŠAN BEREK

CONTENTS

11.1 MOLECULAR CHARACTERIZATION OF SYNTHETIC POLYMERS

Polymers belong to the group of macromolecular substances. They are comprised of one or several kinds of repeating units called *mers* interconnected with chemical bonds and ordered according to certain statistics. Some macromolecular substances are built from the low-molecular moieties, mutual arrangement of which can be hardly described by any statistics (for example humin substances). The latter should not be designated "polymers". In other words, all polymers are macromolecular substances but not all macromolecular substances are polymers. In this chapter, we will deal almost exclusively with synthetic and man-made polymers. The important exceptions represent polysaccharides, especially cellulose and its derivatives. Cellulose is the most abundant organic polymer on the earth. The behavior of polysaccharides in many aspects resembles that of synthetic polymers. Polysaccharides are often chemically modified to adjust both their solubility and utility properties. Chemical modification of polysaccharides results in the specific class of natural/synthetic polymers.

Present chapter is devoted to the methods applied to determination of basic molecular characteristics of synthetic polymers with help of methods of (high-performance) liquid chromatography (HPLC). In other words, this chapter is focused on the methods designed to molecular characterization of synthetic polymers, not to their preparative or process fractionation. The general term *polymer HPLC* is employed in this chapter to designate entire group of liquid chromatography methods under discussion. The similarities and differences of polymer HPLC and HPLC of low-molecular substances are highlighted.

It is attempted to present easily comprehensible plain text with minimum equations. Number of figures is limited just to exemplify the backgrounds of the written information. The recommendations of general literature sources like the reviews and books are preferred over citations of specific publications. On the other hand, a tribute is paid to the scientists who stood at the cradle of methods discussed in this chapter and their inceptive, introductory works are quoted. Representation of applications of particular procedures is restricted to a few typical examples. The aim is mainly to illustrate the principles of methods. Numerous cross references among sections are introduced to facilitate readability of chapter.

11.2 CLASSIFICATION OF POLYMERS

In the next paragraphs, the introductory information on general behavior of polymers is outlined. The aim is to enable easier understanding of the backgrounds of determination of their molecular characteristics, which is the main task of present chapter.

11.2.1 GENERAL CONSIDERATIONS

Physical state of (high)polymers is liquid or solid.Considering molar mass of macromolecules one recognizes *oligomers* and *(high)polymers*. Oligomers possess molar mass up to few thousand $g.mol^{-1}$, while molar mass of synthetic polymers may reach many millions $g.mol^{-1}$. This is why it is useful to express molar mass of polymers in $kg.mol^{-1}$. For the sake of simplicity, oligomeric substances are included into the general term "polymers" except for the situation when specific properties of oligomers are discussed. To distinguish macromolecules according to their molar mass, the term *"(high)polymer"* will be employed in the latter case alongside with the term *"oligomer"*. High molar mass of macromolecules prevents their evaporation. Many oligomers are viscous liquids but most (high)polymers are solid at ambient temperature. In the highly-elastic state, segments of macromolecules behave as a liquid, while whole polymer behaves as a solid substance.

Energetic (enthalpic) interactions within bulk polymers largely affect their end-use properties. Conformation of polymer chains is affected by *intramolecular interactions.* The overall strength of *intermolecular interactions* among particular macromolecular chains in solid state, especially in the case of crystalline polymers and highly oriented macromolecules may be well comparable with the strength of chemical bonds. Both polymer-polymer and polymer-solvent enthalpic interactions significantly affect properties of solutions of macromolecules, which are as a rule employed for determination of their molecular characteristics.

Conformation of macromolecules in solid state and in solution exhibits both many contingencies and restrictions. *Coiled conformation* of macromolecules is most appropriate for majority of methods employed for molecular characterization of polymers. Statistical coils of macromolecules in equilibrium exhibit large *conformational entropy*. Any external intervention that leads to a change in the conformation of macromolecules has to surmount considerable resistance, which is connected with the loss of overall conformational entropy.

Effects of molar mass on numerous physical properties of macromolecules - let us call them "*O*" - such as heat capacity, specific volume, thermal expansion coefficient, refractive index, and so on. can be expressed with Equation (1).

$$\text{(property } O \text{ of a polymer at particular molar mass)} = (\text{property } O \text{ at } M_\infty) - K_s/M_n \quad (1)$$

where M_∞ is infinitive molar mass, K_s is a constant specific for both the given polymer and the property *"O"*, and M_n is number average molar mass. Evidently, there does exist a "critical molar mass" above which the property *"O"* remains practically molar mass independent that is the change of property *"O"* with molar mass is small enough to be neglected. This fact is important for unbiased determination of molecular characteristics of polymers, for example for monitoring polymer concentration and/or properties of macromolecules leaving the liquid chromatography column that is for detection of polymer in the column effluent.

The polymeric substances can be classified according to different qualities. The most appropriate classification parameters include molecular characteristics of polymers, such as above presented molar mass, and further chemical structure, as well as physical architecture of macromolecules:

- ***Homopolymers***: Homopolymers are formed by one single kind of low-molecular species, called *monomers.* Homopolymers may exhibit various physical architectures of macromolecules: linear, cyclic, branched, comb-like, star-shaped, ladder-like, and so on.
- ***Functional Macromolecules***: Functional macromolecules bear chemical groups, which differ in composition from the main polymer chain. The functional groups lend to polymer chains some specific properties, for example certain reactivity. This property is especially important for oligomers.
- ***Copolymers and Terpolymers***: *Copolymers and terpolymers* are constituted from two or three different monomeric units, respectively. Copolymers are classified according to the arrangement of the monomeric units into
- ***Statistical, Random and Alternating Copolymers***: In statistical copolymers, the monomeric units are arranged irregularly. If the arrangement of monomeric units follows the random statistics one speaks about random copolymers. The term "random copolymer" is often used improperly, instead of the term "statistical copolymer". Short blocks, sequences of identical monomeric units that exhibit different length are present in the statistical copolymers. This property is designated *blockiness* of statistical copolymers. Monomeric units regularly take turns in *alternating copolymers*
- ***di-, tri-, and multi-block Copolymers, or Segmented Copolymers***: In a diblock copolymer, a chain with certain chemical composition (a homopolymer or a copolymer chain) is linked by the chemical bond to the chain

with different chemical structure or architecture. Statistical copolymers and block copolymers may exhibit the same kinds of molecular architecture as homopolymers, the most common being linear macromolecules.
- ***Graft Copolymers***: In which the main chain, the macromolecular backbone supports side-chains that exhibit distinct chemical structure. If the side-chains possess the same chemical structure as the main chain, we deal with a *comb-like* polymer. A block copolymer, one part of which is formed by a linear chain and another part is a grafted chain is denoted *bottlebrush polymer*.

In the *star-like macromolecules,* three and more chains originate in the central locus whereas all chains can have the same chemical structure (for example homopolymers or statistical copolymers). Alternatively, each arm has a distinct composition – *miktoarm star copolymers*.

Mixtures of either chemically identical or dissimilar macromolecules are *polymer mixtures* and *polymer blends*, respectively. Polymer blends can be molecularly homogeneous or, more commonly, heterogeneous. Distinct phases are present in the latter polymer blends. The often employed term "polymer alloys" is not recommended by the International Union of Pure and Applied Chemistry, IUPAC.

A special group of modern polymeric materials is represented by *composites with polymer matrix*, also simply called *polymer composites*. The latter contain certain particulate or fibrous *fillers*. Molecular characteristics of polymer matrix play important role in the optimum properties of composites.

11.2.2 MOLECULAR CHARACTERISTICS OF SYNTHETIC POLYMERS AND THEIR DISPERSITY

Synthetic polymeric materials possess enormous variability of utility properties, which depend:
- On the molecular characteristics of their building species, macromolecules
- On the mutual arrangement of macromolecules in the solid state (also significantly affected by the processing method)
- On both, the nature and the amount of admixtures. The latter can be either low- molecular substances such as stabilizers, antioxidants, colorants, and so on, or other macromolecules, or solid particles, and/or fibers in the above mentioned polymer composites.

The already quoted primary or basic molecular characteristics of synthetic polymers are:

- *Molar mass* with units [g/mol] or [kg/mol], (NOT *molecular mass*! - because we deal with the moles of substance and NOT with single molecules). The unitless term *molecular weight* is also accepted by IUPAC. High molar mass is the general, fundamental property of polymers.
- *Chemical structure (composition)* of polymer chains (NOT in terms of overall content of C, H, N, O in the whole polymer but considering actual chemical structure of monomeric units or distinct chains, and so on.). This term also includes functional groups present in macromolecules: their nature and arrangement, as well as their number per macromolecule. As indicated in Section 11.2 above, functional groups to a large extent affect properties of oligomer molecules since their relative role is much more important than in the (high)polymers. Chemical reactions of functional oligomers enable creation of the new kinds of important - often crosslinked macromolecular systems. Molecules of oligomers can carry double bonds, which are able to enter polymerization reactions. In this case one speaks about *macromonomers.*
- *Physical (molecular) architecture*, also called *molecular topology.* The main groups of molecular architectures of polymer chains were listed in the previous paragraphs: linear versus branched versus cyclic macromolecules, stereoregular macromolecules (atactic, syndiotactic, isotactic, heterotactic polymers - for example poly(methyl methacrylate)s), *cis-trans-* macromolecules - for example poly(isoprene), head-to-head versus head-to-tail macromolecules; poly(catenane)s, optically active macromolecules (*D, L* and racemate *R* - for example poly(lactide)s). Important parameter of molecular architecture is the abovementioned blockiness of statistical copolymers with otherwise identical overall chemical composition that is the length of sequences formed with the same monomeric unit. It is hardly relevant to speak about molecular characteristics of crosslinked macromolecules.
- *Charges* from the point of view of their nature, number, as well as of their position and arrangement in a macromolecule. Charged macromolecules represent rather important constituent of many practical synthetic and natural polymeric materials. Considering determination of molar mass, chemical structure and architecture of polymers in solution, presence of charges in macromolecules is as a rule a complicating factor. With only few exemptions, measurements of molecular characteristics of polymers are performed in their electroneutral state. The dissociation of macromolecules in solution is usually suppressed by addition of the low-molecular counter-ions. Some macromolecules, for example poly(ethylene oxide)s may, however preferentially accommodate low-molecular ions in the domain of their chains. Consequently, the originally non-charged macromolecules behave similar

to polyelectrolytes. They are designated *pseudo-polyelectrolytes* and their molecular characterization in solution may be disturbed.

In this chapter, peculiarities of the charged macromolecules will be dealt only marginally – especially to demonstrate adverse impact of charges on molecular characterization of polymers with help of liquid chromatography.

Chemical structure and physical architecture of macromolecules may exhibit large interconnected variability, which is reflected in the *secondary molecular characteristics.* Differences in the secondary molecular characteristics may even appear within a group of polymers possessing the same overall chemical structure or architecture. For example, some block copolymers may possess the same overall chemical structure but *stereoregularity* of their chains can be different. It is evident that the actual properties of macromolecules may be extremely intricated if the effects of two or even all three basic molecular characteristics are combined.

Molecular characteristics of synthetic polymers are never uniform. They always exhibit certain *dispersity*. "Dispersity" is a new term, coined by IUPAC, which should substitute the former term: *distribution.* In fact all synthetic polymers represent multicomponent mixtures of macromolecules, which differ in one or several molecular characteristics. With a rather few exceptions such as for example some *polymer mixtures* and *polymer blends,* dispersities in molecular characteristics of common polymers are continuous in nature. For example, the molar masses of macromolecules that form particular members of typical homologous series usually differ only in the molar mass of a single monomeric unit. The resulting total molar mass of polymers ranges from a minimum, to a maximum value, while the latter may be several times higher than the minimum value. Therefore, the molecular characteristics are described with their *average values* or with the *dispersity functions*. Consequently, we have:

- mean or average molar mass, and molar mass dispersity
- mean or average chemical composition (chemical structure), and chemical-composition dispersity
- mean or average molecular architecture and dispersity in molecular architecture.

Similarly, one can speak about mean sequence-length and its dispersity in statistical copolymers, about mean tacticity and its dispersity in the stereoregular polymers, and so on.

Dispersity of molecular characteristics of polymeric substances can be expressed in different ways. One can *count* or *weight* macromolecules of the same kind to express their relative concentration in the system – and correspondingly

- their number or weight fraction, and to calculate the *number-* or *weight-average* of particular molecular characteristic.

11.2.3 MOLAR MASS DISPERSITY (DISTRIBUTION)

The molar mass dispersity is the best known of all dispersities in molecular characteristics of synthetic polymers and it is most frequently determined. Assessment of dispersities in chemical structure and in physical architecture is much more demanding. Let us stress again that molar mass dispersity fairly affects numerous utility properties of industrial polymers. The *width of molar mass dispersity* is usually expressed with *the ratio of particular molar mass averages*. This is habitually sufficient for a sound estimate of suitability of particular polymer for most particular applications. *Molar mass dispersity function* quantitatively reflects amount of macromolecules with certain molar mass present in the sample. It can be represented in the integral or in the differential form. The latter possesses a more informative nature and it is more frequently used.

Considering molar mass of a polymer, amount of macromolecules with given molar mass M_i can be expressed by their number N_i or by their mass M_iN_i. Hence the molar mass dispersity function is a relation between the number fraction p_i or the mass (weight) fraction q_i and logarithm of molar mass M_i of particular macromolecules. It holds for the number fraction p_i

$$p_i = N_i / \Sigma N_i \tag{2}$$

and for the weight fraction

$$q_i = N_i M_i / \Sigma N_i M_i \tag{3}$$

The "counting" of macromolecules gives number average molar mass, M_n, while the weighing of macromolecules leads to weight (mass) average molar mass, M_w.

$$M_n = \Sigma p_i M_i \tag{4}$$

and

$$M_w = \Sigma q_i M_i \tag{5}$$

The ratio M_w/M_n is called *polydispersity* (the previously used term was *polymolecularity*) of a polymer and it gives important information on the width of its molar mass dispersity. Less frequently, the viscosity average molar mass, M_v, and the artificial "z" and "z+1" averages of molar mass, M_z and M_{z+1} are considered. M_v is defined

$$M_v = \{\Sigma q_i M_i^a\}^{1/a} \tag{6}$$

where *a* is the exponent from the Kuhn-Mark-Houwink-Sakurada equation (compare section 11.2.4, Polymer Solutions). "z" and "z+1" averages are important for the very broad molar mass dispersed polymers such as polyolefins.

Unfortunately, average values of molecular characteristics of polymers say little about the *type* and *shape* of the molar mass dispersity function. For example the M_w/M_n values may be identical for both the broad unimodal and the multimodal distributed polymers.

Schematic representation of a dispersity function of molar mass is depicted in Figure 1.

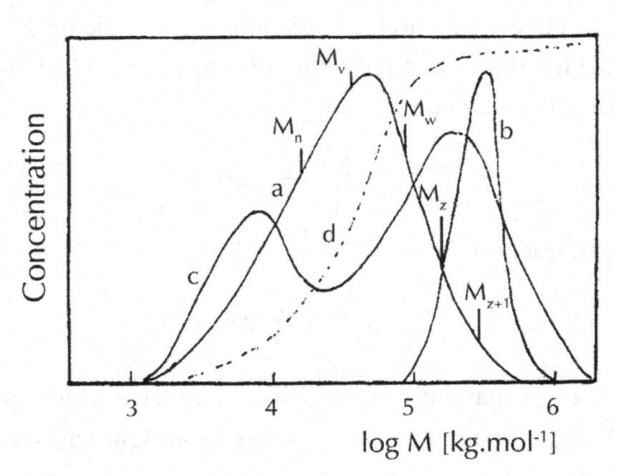

FIGURE 1 Typical courses of molar mass dispersity functions. a, b, c—differential, and d- integral representation: a- broad molar mass dispersity polymer; b-narrow molar mass dispersity polymer; c- polymer with bimodal dispersity.

Distinct shapes of molar mass dispersity functions exist for synthetic polymers of different origin: *log normal, Schultz-Zimm,* and so on. Graphical representations of quantitative dispersity functions are less commonly employed than simple ratios M_w/M_n or M_z/M_w.

Often, size exclusion chromatograms (SEC) (compare section 11.7, Size Exclusion Chromatography) of polymers under study are expressed as differential representations of molar mass dispersity. The chromatographic retention volumes are directly transformed into the molar masses. This approach renders useful immediate information about tendencies of molar mass evolution in the course of building or decomposition polyreactions but the absolute values of molar mass can be only rarely extracted from it. As a rule, polystyrene calibrations are applied for molar mass calculation so that one deals with *the polystyrene equivalent molar masses*, not with the absolute values. The resulting "dispersity (distribution) functions" may be heavily skewed because the "linear part" of the calibration dependence for the polymer under study may exhibit well different slope compared with the polystyrene calibration, which was employed for the transformation of retention volumes into molar masses.

The important feature of many synthetic polymers is the simultaneous presence of dispersities in two and even in several molecular characteristics. Polymers exhibiting multiple dispersities - while it must be accentuated again that molar mass dispersity is always present(!) - are called *complex polymers.* The representative complex polymers are different kinds of copolymers and functional polymers. A mixture of a complex polymer with another macromolecular substance is denoted *complex polymer system.* Typical complex polymers systems are many polymer blends , as well as the block copolymers, which contain their parent homopolymers. Molecular characterization of constituents of complex polymers and complex polymer system constitutes a particular challenge. Different approaches to address this task will be presented in this chapter.

Detailed discussion of molecular characteristics of polymers, about they average values and dispersities can be found in numerous monographs and reviews. Their list is attached as the Recommended Reading at the end of this chapter. For the present purpose, it is important to repeat that all synthetic polymers and also polysaccharides are polydisperse in their nature. Only Mother Nature is able to produce strictly uniform macromolecules, such as concerning both their molar mass and overall chemical composition – numerous proteins.

The terms "monodisperse" or "monodispersed" polymers often can be found in literature. IUPAC however recommend to substitute the latter designations with the term *"uniform"*.

It may be useful to define difference between analysis and characterization of polymers. Of course, the general term *analysis* can be used in numerous different aspects. However, in polymer science and technology, the term *analysis* is mainly employed in connection with the determination of overall chemical composition of a polymer in terms of chemical elements, as well as with the assignment of amount of either low-molecular or solid admixtures. On the contrary, the term *characterization* is reserved to determination of properties typical for macromolecular substances, such as for example just their molecular characteristics.

Thorough determination of molecular characteristics of synthetic polymers is a must in the research laboratories of companies that produce, process, and apply polymeric materials.

Surprisingly little attention is paid in many research laboratories

- to molecular characteristics of polymers employed in their studies. The batch-to-batch repeatability of the large-scale production of most synthetic polymers is limited and molecular characteristics of resulting industrial polymers may vary on the day-to-day basis over a quite sizeable range. In many companies, individual batches of polymers are mixed together to keep the utility properties of resulting polymeric product as constant as possible. If the exact molecular characteristics of a polymer employed in the actual research are not known the generalization of results obtained may be rather problematic or even impossible.
- to reliably characterize on molecular level the products of polymer syntheses and transformations. Under such conditions, some generalized optimistic conclusions may be imprecise or even wrong. Usually, only size exclusion chromatography is employed – and macromolecular admixtures, for example the unwanted products of polyreactions, are ignored. SEC measurements are often done superficially, the resulting chromatograms are not well handled and the data obtained are misinterpreted or even purposedly misused (compare Section 11.7.5, Advantages, Limitations and Drawbacks of SEC). The task of this chapter is not only to give basic information on the liquid chromatography methods available for molecular characterization of synthetic polymers but also to remind their limitations, shortages, drawbacks and pitfalls.

It is anticipated that the target group of readers of this book comprises novices in the field. However, it is hoped that also many practicing chromatographers can uncover information that are seldom available in the narrowly focused research papers.

This chapter is based on the contribution of the author to the practical, well attended Short course at Universiti Teknologi MARA in Malaysia organized by Assoc. Prof. Dr. Chin Han Chan in 2009, who should be thanked and praised for this activity.

11.2.4 POLYMER SOLUTIONS

The following basic, simplified information on polymer solutions are aimed at facilitation of understanding the discussion of backgrounds of polymer HPLC.

It is evident that only the molecularly dispersed systems, *polymer solutions* are of practical interest for characterization of macromolecules with help of liquid chromatography. Any particulate of fibrous material dispersed in polymer solution represents serious danger for the chromatographic column. Besides extraneous solid microparticles, the otherwise apparently homogeneous solutions of polymer samples may contain macromolecular associates, aggregates or micelles, which belong to sample proper and often perturb measurements. Similar to solid particles, the large supermolecular structures can obstruct the column inlet or the bed of column packing. If a heterogeneous "solution" is injected into a liquid chromatograph, the flow resistance and consequently the inlet pressure of the column rapidly increases. Both the microparticles and the above supermolecular structures often remain unnoticed by the naked eye. Therefore, it is recommended to carefully filter the unknown, apparently homogeneous sample solutions before their introduction into the column. Unfortunately, some soft supermolecular structures, such as the microgels can penetrate even through the very fine filters and clog the column end-piece or the bed of column packing because they may adhere to the surface of inlet filter and packing particles. In the favorable case, the supermolecular structures that were not retained by the sample filter or by the column inlet/outlet frits, and which do not adhere to the bed of packing would elute in the interstitial volume of column (compare sections 11.6.1.3 and 11.7.3.1). They can disturb molecular characterization of polymers, for example, determination of their molar mass with help of SEC because they are monitored by most detectors and the resulting peak(s) may interfere with the sample peaks.

Macromolecules in solution may assume various shapes: for example statistical coils, worm-like, rods or globules. The coiled structure is most suitable for molecular characterization by liquid chromatography. As indicated in section 11.2.1, polymer coils in solution possess large conformational entropy and any forced change of the coil dimension is accompanied with the entropy alteration.

Polymers are dissolved in their *solvents* and precipitated by their *nonsolvents*. The general term *solvent* is, however often used for any liquid so that a nonsolvent for given polymer is also a sort of "solvent". As a rule, polymers dissolve in liquids that exhibit certain similarity with macromolecules, for example, a likeness in polarity. Polarities of polymers and solvents are characterized by solubility parameters, δ, values of which can be found for example in Polymer Handbook (compare the Recommended Reading). Crystalline polymers are often soluble only above their melting point. Typical examples are polyolefins. Some solvents are able to disrupt crystalline structure of polymers and dissolve them at ambient temperature (compare sections 11.6.2 and 11.7.3.2). The practically important example is hexafluoro isopropanol, HFIP, which dissolves several polar crystalline polymers, for example aromatic polyesters. Already presence of 2% of HFIP enables to keep the latter polymers molecularly dispersed in the low polarity liquid, chloroform.

The dissolution of polymers in a solvent is usually a slow process. It is preceded by swelling, which is controlled by the diffusion rate of solvent molecules into the bulk polymer. To accelerate dissolution, it is useful to stir the system, to work with airy precipitates of polymer sample instead of bulk particles and to increase temperature. Crystallized polymers usually dissolve after melting of their crystallites at elevated temperature. After temperature has been decreased, some of them (for example polyolefins) precipitate, while other ones (for example poly(ethylene oxide)s) may remain molecularly dissolved.

The necessary condition for dissolution of a given polymer in a given solvent is the drop in Gibbs free energy (free enthalpy, Gibbs function) ΔG of the process, which is defined

$$\Delta G = \Delta H - T\Delta S \tag{7}$$

where ΔH and ΔS are changes of the overall enthalpy and entropy of system due to dissolution, respectively, and T is temperature. The gain in entropy that results from mixing of macromolecules with solvent molecules is much lower than in case of blending unlike low-molecular substances. Therefore, change of enthalpy is often decisive for the spontaneous dissolution of macromolecules. Value of ΔH depends on interactions between solvent molecules and segments of polymer chains. From the thermodynamic point of view, a solvent for given polymer can be *good* or *poor*. Segments of macromolecules exhibit rather strong attractive interaction with the molecules of a good solvent, while interaction of polymer segments with the poor solvent molecules is weak or even slightly repulsive. The otherwise minor contribution of dissolution entropy to ΔG becomes crucial in the latter case.

The important parameter of dissolved macromolecules is their radius of gyration R_G:

$$\overline{R}_G^2 = \frac{1}{Z}\sum_1^z r_i^2 \quad (8)$$

where r_i is distance of the ith-segment of the macromolecule from its center of gravity and Z is number of segments.

Compared with their size in the amorphous state, macromolecules are expanded in good solvents whereas they tend to shrink in poor solvents. In other words, size of macromolecules of given polymer with given molar mass in solution depends on thermodynamic quality of solvent. It is characterized by the expansion coefficient α, which is defined:

$$\alpha^2 = R_G/R_G^\theta \quad (9)$$

where R_G is radius of gyration of particular macromolecules in the given solvent and R_G^θ is their radius of gyration in the athermic, "*theta*" solvent, or

$$\alpha^3 = [\eta] / [\eta]^\theta \quad (10)$$

Here $[\eta]$ and $[\eta]^\theta$ are *limiting viscosity numbers* (*intrinsic viscosities*) of polymer in the given and in the theta solvents, respectively.

The coil dimensions in the theta solvent are designated "unperturbed".

$\Delta H < 0$ good solvents $\alpha > 1$

$\Delta H = 0$ athermic, theta solvents $\alpha = 1$

$\Delta H > 0$ poor solvents $\alpha < 1$

Unperturbed coil dimensions in solution are equal with those of macromolecules in amorphous solid state.

Interactions in polymer solutions are classified as:

- short range interactions (mutual interactions of neighbor segments in a macromolecule)
- long range interactions (mutual interactions between remote segments)
- intermolecular interactions

- interactions among polymer segments and molecules of solvent.

Under theta conditions, segment – segment interactions are equal to segment – solvent interactions.

From the point of view of polymer HPLC, size of macromolecules in solution and polymer-solvent interactions are of major importance. The behavior of macromolecules in solution can be expressed by the virial expansion

$$U/c = RT(A_1 + A_2c + A_3c^2...) \tag{11}$$

where U stands for the colligative property of polymer solution such as osmotic pressure (in membrane osmometry), or T (in vapor pressure osmometry), or light scattering parameters. R is gas constant, T is temperature, and A_i are virial coefficients: $A_1 = 1/M$ with M for molar mass; A_2 denotes interaction polymer - solvent. A_3 and the higher members of virial expansion can be usually neglected.

The change of Gibbs function for polymer solution is expressed with the famous Flory-Huggins equation:

$$\Delta G_m = RT\,(n_1 ln\varphi_1 + n_2 ln\varphi_2 + n_1\varphi_2\chi_{12}) \tag{12}$$

where 1 and 2 stand for solvent and polymer, respectively, n is number of moles, φ is volume fraction and χ_{12} is polymer-solvent interaction parameter, which describes thermodynamic quality of solvent for given polymer. Polymer – solvent interaction parameter χ_{12} is expressed as

$$\chi_{12} = \frac{V_1(\delta_1 - \delta_2)^2}{RT} \tag{13}$$

where V_1 is molar volume of solvent, and δ_1 and δ_2 stand for solubility parameters of solvent and polymer, respectively.

For polymer HPLC, important parameter is viscosity of polymer solution. It increases with the molar mass of macromolecules and also with the improving thermodynamic quality of solvent that is with the size of macromolecules in solution. High viscosity of polymer solutions prevents execution of liquid chromatography measurements at elevated concentration, usually above 1% or 10 mg.mL^{-1} for polymers with intermediate molar mass. Oligomers can be treated at even tentime higher concentration. On the other hand, extreme solution viscosity of ultra-high molar mass polymers with M >1,000 $kg.mol^{-1}$ may bring about important experimental problems. Dependence of viscosity of polymer solution on molar mass M and on the polymer-solvent interaction describes Kuhn-Mark-Houwink-Sakurada viscosity law

$$[\eta] = K_{[\eta]} M^a \quad (14)$$

where $[\eta]$ is limiting viscosity number, and $K_{[\eta]}$ and a are constants for given polymer – solvent pair. Equation (14) holds for linear macromolecules with molar mass above about 10 $kg.mol^{-1}$. The exponent a renders information on the thermodynamic quality of solvent toward polymer:

$a > 0.65$—refers to a *good solvent* that exhibits intense attractive interactions with polymer segments

$0.5 < a < 0.65$—stands for a *poor solvent* with the weak interactions with macromolecules

$a = 0.5$—marks a *theta (athermal) solvent,* in which the solvent – segment interactions are equal to the segment – segment interactions

at $a < 0.5$ the segment – segment interactions exceed the solvent - segment interactions. Under such conditions, the system is instable and polymer coils tend to collapse. This leads to the association or aggregation of macromolecules and eventually to the onset of phase separation, during which the liquid microphases and later the macrophases are formed in polymer solution. Usually but not always, both phases contain macromolecules. The solid particles of a precipitate appear when the solvent quality further decreases.

The product $[\eta].M$ is designated *hydrodynamic volume of macromolecules,* V_h and forms a basis for important *Benoit's universal calibration* in SEC (see section 11.7.3.1, Columns for Size Exclusion Chromatography and Their Calibration). Coiled, chemically different macromolecules in different solvents that exhibit equal V_hs elute from the given SEC column within the same retention volume - provided the pore structure of the column packing remains unaffected by the eluent change and the enthalpic interactions with the column packing are negligible.

Interaction parameter χ_{12}, second virial coefficient A_2 and exponent a in the viscosity law characterize *thermodynamic quality* of the solvent for given polymer. The most common in use is the exponent $a,$ as it is needed to express the hydrodynamic volume of macromolecules and also because its values are known for many different polymer-solvent systems (see "Polymer Handbook" in the Recommended Reading).

The solubility of macromolecules as a rule improves with the rising temperature. Solvent – polymer mixtures usually exhibit *the upper consolute temperature* or *upper critical solution temperature*, UCST, with a maximum on the plot of system concentration versus temperature. Above the critical solution temperature, polymer is fully soluble at any concentration. For practical work, the systems with UCST below ambient temperature are welcome. There are, however numerous polymer – solvent systems, in which the solvent quality decreases with increasing temperature. The plot of system concentration versus temperature exhibits a minimum. The phenomenon is called *lower consolute temperature* or *lower critical solution temperature*, LCST. Polymer is only partially soluble or even insoluble above lower critical solution temperature. This unexpected behavior can be explained by the dominating effect of entropy in case of the stiff polymer chains or by the strong solvent – solvent interactions. The possible adverse effect of rising temperature on polymer solubility must be kept in mind when working with low solubility polymers and with multicomponent mobile phases. It may lead to the unforeseen results especially in the polymer HPLC techniques that combine exclusion and interaction retention mechanisms, in *coupled methods of polymer HPLC* (see section 11.8, Coupled Methods of Polymer HPLC).

Application of coupled methods of polymer HPLC necessitates fine control of interaction of the separated macromolecules with the column packing. The easiest approach to do so is to compound two or several solvents of distinct polarities. In that event, one speaks about the *mixed solvents (two-* or *multi-component solvents).* In most cases, polymer chains interact preferably with one of the mixed solvent components. Its concentration increases in the domain of polymer coils compared with both the original solvent mixture and the bulk mixed solvent among macromolecules. This phenomenon is called *preferential solvation.* In the chromatographic column, macromolecules are separated from their bulk solvent, which gives rise to the *system peaks* or *solvent peaks* on chromatograms monitored by the non-specific detectors such as differential refractometers (see sections 11.6.1.4 and 11.7.3.2). The system peaks generated by preferential solvation appear at the end of size exclusion chromatograms, at high retention volume. System peaks usually accompany macromolecules or even occur before polymer peaks in the enthalpy driven methods of polymer HPLC (see section 11.8.2, Enthalpy Driven Polymer HPLC). Preferential solvation inevitably arises also in most solutions of low-molecular substances dissolved in the multicomponent solvents but its chromatographic manifestation is much less important than in case of polymer HPLC, especially because the samples are as a rule eluted much later than the bulk solvent zones and the most commonly employed photometric detectors do not markedly respond to eluent composition changes.

Another specific phenomenon in polymer solutions is the effect of *cosolvency* when a mixture of two nonsolvents forms a solvent or even a thermodynamically very good solvent for polymer under study. Much less common is the occurence of *co-nonsolvency*, when a mixture of two solvents for given polymer constitutes its precipitant. Both cosolvency and co-nonsolvency result from solvent – solvent interactions and from the entropy changes of polymer coils induced with their interaction with solvent molecules. It is advisable to consider these unexpected phenomena when working with mixed mobile phases.

11.3 LIST OF METHODS FOR DETERMINATION OF MOLECULAR CHARACTERISTICS OF POLYMERS

11.3.1 MEAN (AVERAGE) VALUES OF POLYMER MOLAR MASSES

Classical Methods:

- Ebuliometry
- Kryometry
- Vapor pressure osmometry – for oligomers
- Membrane osmometry – for (high)polymers
- Ultracentrifugation
- Static and dynamic light scattering
- Viscometry - for (high)polymers

The latter methods were almost fully substituted with size exclusion chromatography, though especially light scattering and viscometry are often used for detection in SEC (see section 11.7.3.2, Mobile Phases and Detectors for Size Exclusion Chromatography).

Modern Methods:

- Liquid chromatography – especially size exclusion chromatography
- Mass spectrometry, MS (matrix assisted laser desorption ionization MS, MALDI MS, and electrospray MS)
- Field-flow fractionation, FFF.

11.3.2 MEAN (AVERAGE) VALUES OF CHEMICAL STRUCTURE AND PHYSICAL ARCHITECTURE OF POLYMERS

The area is dominated by nuclear magnetic resonance, NMR. Valuable information can be obtained also by means of infrared spectroscopy and pyrolytic gas chromatography. However, the quantification of data obtained by the latter method is often difficult.

Important data on physical architecture of crystallable polymers can be obtained from temperature rising elution fractionation, TREF and crystallization fractionation, CRYSTAF.

11.3.3 MOLAR MASS DISPERSITY (DISTRIBUTION)

Classical methods:

- Solubility based methods of fractionation (see below),
- Ultracentrifugation
- Diffusion rate measurements (incl. thermo-diffusion)
- Membrane permeability.

All latter methods are experimentally demanding and often slow. Ultracentrifugation requires sophisticated and expensive instrumentation and membrane based methods suffer from the limited availability of membranes with the narrow pore dispersity. This is why these methods were practically fully replaced by SEC.

Modern methods:

- Mass spectrometry (matrix assisted laser desorption ionization mass spectrometry, MALDI MS, electrospray ionization mass spectrometry, ESI MS)
- Field-flow fractionation, FFF
- Liquid chromatography – especially size exclusion chromatography.

11.3.4 CHEMICAL STRUCTURE DISPERSITY

Coupled and two-dimensional liquid chromatography with detectors that monitor composition of polymers in effluent (see sections 11.8 and 11.9).

11.3.5 PHYSICAL ARCHITECTURE DISPERSITY

Methods based on differences in crystallizability of macromolecules - temperature rising elution fractionation, TREF and crystallization analysis fractionation, CRYSTAF.

Coupled and two-dimensional methods of liquid chromatography with detectors that monitor physical architecture of eluted macromolecules.

11.3.6 SIMULTANEOUS DETERMINATION OF DISPERSITY IN MOLAR MASS AND IN CHEMICAL STRUCTURE OR IN MOLECULAR ARCHITECTURE

The area is dominated with *coupled* and *two-dimensional polymer HPLC techniques.*

11.4 SEPARATION METHODS FOR SYNTHETIC POLYMERS

In order to determine dispersity of their molecular characteristics, polymers must be separated, fractionated. There are only few exceptions from this rule. For example, concerning molar mass dispersity, valuable information can be acquired from the rheological measurements.

The separation methods employ differences in polymer properties that depend on the molecular characteristic under interest. The attributes of macromolecules are presented in this section, which are suitable to form a basis for separation.

As indicated, molar mass both average and dispersity belong to the most important molecular characteristic of synthetic polymers. Molar mass of macromolecules affects their solubility and size in solution and to some extent also their interactivity. For both the synthesists and the technologists, also chemical structure and physical architecture of macromolecules are highly important because they either confirm successful synthesis or markedly affect the end-use properties of polymers. Similar to molar mass, the latter molecular characteristics impact the interactivity of macromolecules and partially also their size in solution. Development of methods for assessment of dispersities in chemical structure and physical architecture of macromolecules is still only in a rather initial stage.

Following qualities of polymers are employed in their separation according to molar mass:

- Solubility of macromolecules which as a rule decreases with the rising molar mass. This is at least qualitatively understandable if one considers reduction of entropy of mixing. The most common solubility based fractionation of macromolecules, which directly employs phase separation of (diluted) polymer solutions is called *successive precipitation.* Phase separation of polymer solution is invoked by the worsening thermodynamic quality of solvent for macromolecules. It can be attained by the adding of a nonsolvent to the solution of polymer or by the changing of temperature. During the successive precipitation, the initially homogeneous solution disintegrates into a diluted phase (inappropriately called the "sol") and a concentrated ("the gel") phase. Macromolecules are partitioned among the phases according to their molar mass and the highest molar mass species are concentrated in the concentrated phase. It is necessary to know that polymer concentration in the diluted phase may be zero and in this case, no fractionation would take place. Alternatively, polymers can be separated with help of *successive dissolution,* for example by applying the well-known classical continuous procedure, Baker-Williams column fractionation. In the latter case, the dissolution power of solvent is improved in a controlled way. In the course of the successive dissolution, thin polymer film deposited on the surface of inert (glass) beads is exposed to solvent with the continuously or step-wise increasing thermodynamic quality. Macromolecules with the lowest molar masses become soluble as first. The effect of temperature on polymer solubility was discussed in the previous paragraph. The fractionation methods based on the successive precipitation or dissolution were popular till the advent of size exclusion chromatography. They are still applied for selected preparative or production fractionations.

The latter solubility based methods do not directly belong to chromatography, however, they can be offline combined with polymer HPLC. Moreover, the solubility effects are directly employed in some coupled chromatographic methods (compare section 11.8.4, Eluent Gradient Polymer HPLC), while some of them employ the *tendencies* to phase separation rather than the complete precipitation processes (see section 11.8.6, Liquid Chromatography under Limiting Conditions of Enthalpic Conditions).

- Size of macromolecules in solution, from the point of view of their general behavior in flowing liquids that is from the aspect of *hydrodynamic processes.* Laminar flow of liquid in a capillary assumes the parabolic profile. The molecules of liquid near the walls are transported more slowly than those in the centre of capillary. Larger macromolecules cannot enter the slowest streamlines and, on average, they proceed along the capillary faster than smaller macromolecules. Schematic representation of this situation is presented in Figure 2.

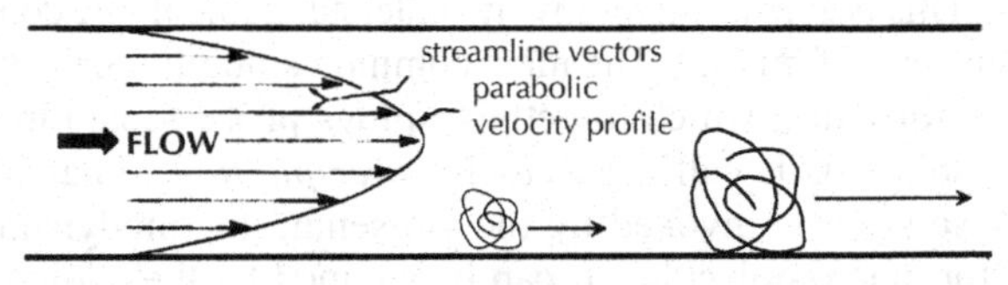

FIGURE 2 Flow patterns of a liquid in a capillary and their effect on progressing rate of macromolecules of different size. For detailed explanation, see the text.

In the first approximation, the column packed with small particles can be considered an array of capillaries. Hydrodynamic processes augment the exclusion based separation (see section 11.7.2, Retention Mechanisms and Accompanying Processes in SEC) and form the ground of a liquid chromatography-like method called *hydrodynamic chromatography*, HDC. HDC found application in separation of very large macromolecules, particles and dispersions. Its separation selectivity is rather low and it was partially substituted by the group of methods termed *field flow fractionation* (see section 11.3.3, Molar Mass Dispersity).

- Size of macromolecules in solution, based on their selective permeation through the porous particles of the column packing and resultant exclusion processes, which form the ground of size exclusion chromatography and most field flow fractionation methods. This section mainly deals with the SEC, (see section 11.7, Size Exclusion Chromatography).
- Enthalpic interactivity of macromolecules in the liquid chromatography systems. The interactions of macromolecules with the mobile phase and with the column packing are the most important parameters of this group. The interactions of macromolecules with eluent affect retention of macromolecules in different ways, for example indirectly, *via* their size (compare this section ad. 2 and 3). There exists strong interrelation between size of macromolecules and their solubility in eluent (see section 11.2.4, Polymer Solutions). Enthalpic interactions between column packing and separated macromolecules directly affect retention processes. The possible effect of packing-polymer interactions on the size and even on the solubility of macromolecules remains so far unexplored. Controlled combination of exclusion and interactions of macromolecules within column is the keystone of coupled methods of polymer HPLC, which will be analyzed in section 11.8.

11.5 CHROMATOGRAPHY OF SYNTHETIC POLYMERS

11.5.1 GENERAL CONSIDERATIONS AND BASIC TERMS

Gas chromatography, GC, evidently cannot be directly employed for separation and molecular characterization of macromolecules because they cannot be evaporated. On the other hand, GC can provide useful information on presence and quantity of volatile low molecular admixtures in polymeric materials. There were published numerous attempts aimed at separation and quantification of products of polymer pyrolysis by means of gas chromatography. Unfortunately, the manifold of processes taking place in the course of pyrolysis, difficulties of the exact control of the kinetics of process and its singularity for each distinct polymer makes the usability of gas chromatography in molecular characterization of polymer rather limited. GC was also applied to detection of the column effluent in liquid chromatography, including polymer HPLC (compare section 11.6.1.4).

Liquid chromatography, polymer HPLC – the subject of this chapter - is at present the most important tool for direct determination of mean values (averages) and dispersities of molecular characteristics including entire dispersity functions of macromolecular substances.

The instrument for execution of separation is called *liquid chromatograph* (see section 11.6, Instrumentation and Materials for Polymer HPLC). Its hearth is the *column* a straight tube, which is filled with the material on which the separation proper takes place, the *column packing*. The latter term is often substituted with the designation *stationary phase*, which can be an *adsorbent*, a chemically attached *bonded phase* or just timely immobilized *stagnant (quasi-stationary) phase* on the surface of a *carrier*. The sample constituents are transported along the column by means of a liquid denoted *eluent* or *mobile phase*. The liquid that leaves the liquid chromatography column is denoted *effluent*. The sample elutes from the chromatographic column in form of a *peak*, which exhibits one or several maxima. The average value of molecular characteristic is usually assessed from the peak position, that is from the peak retention volume, while dispersity of molecular characteristic reflects the peak shape, usually its width. Due to the slow diffusion rate of macromolecules within the column packing, as well as due to the parasitic mixing and diffusion phenomena both within column and extra column part of a chromatograph, chromatographic peaks are additionally broadened. The resulting phenomenon is termed *band broadening,* and it reduces precision and accuracy of chromatographic results (see also section 11.7.5). The extent of peak broadening due to separation process is called *separation efficiency*. The peak broadening can be barely determined with polymer samples in exact manner because the latter always also exhibit certain dispersity. This is why in polymer HPLC, the observed peak width reflects not only separation efficiency of particular chromatographic process but also the dispersity of sample applied for measurement. Determination of band broadening of the chromatographic instrument with help of low-molecular marker unfortunately gives only an information about the chromatographic system but hardly about the chromatographic process.

Separation of macromolecules in polymer HPLC is based on difference in their size or enthalpic interactivity. In the former case, the entropy of exclusion processes controls separation and the enthalpic interaction should be negligible. The corresponding method of *exclusion polymer HPLC* is designated SEC (see section 11.7). Since recently, the original term *gel permeation chromatography,* GPC has been employed more and more frequently to denote exclusion based separation method of lipophilic macromolecules and term *gel filtration chromatography,* GFC is frequently utilized to identify exclusion-based separation method of hydrophilic macromolecules. The reason may be the abbreviation "SEC", which also means "Security and Exchange Commission" in USA. In the interaction-based polymer HPLC methods, enthalpy drives polymer retention. However, changes of conformational entropy of macromolecules may strongly affect retention of macromolecules (compare section 11.5.2.3, Enthalpic Processes – Interaction in Polymer HPLC). The enthalpy/entropy based separation methods are often simply called *interaction polymer HPLC*. The term *coupled methods of polymer HPLC* will be employed in present chapter to stress that entropy and enthalpy are combined, coupled within the same column (see section 11.8 below).

Size exclusion chromatography is undoubtedly the most important method of polymer HPLC. Therefore, it is inappropriate to speak about HPLC *and* SEC. The reason for above uncertainty in terminology may be the fact that the exclusion based separations are rare in the case of low-molecular substances except mainly the situation when macromolecules are to be (pre)separated, or just removed from the complex sample. The terms *HPLC of low-molecular substances* and *polymer HPLC* are employed in present chapter as general designations to differentiate the two groups of liquid chromatography methods.

Separation of distinct macromolecules in polymer HPLC results from their different *retention* within column. *Retention mechanism of analyte* is the general term recommended by IUPAC. It denotes the mutual difference of elution rate of distinct macromolecules. Similar to HPLC of low-molecular substances, elution rate of macromolecules and molecules of mobile phase differ also in polymer HPLC. As a rule, in HPLC of low-molecular substances, the separated species elute with lower velocity than molecules of their original solvent, they are retained, decelerated. With few exceptions, elution rate of macromolecules is the same or slower than elution rate of eluent molecules also in various the coupled methods of polymer HPLC.

In the case of SEC, however, polymers elute faster than molecules of solvent so that eventually the sample macromolecules outrun molecules of their original solvent. Consequently, the authentic meaning of the term *retention* is somewhat vague in the case of SEC. Still, it became quite customary and therefore will be employed within this chapter. Correspondingly, the term *retention volume* will be used to represent volume of mobile phase needed to elute particular macromolecules from the column.

Retention volume of a chromatographic peak is the volume of mobile phase at the peak apex. It corresponds with the volume of the mobile phase, which elutes the maximum concentration of macromolecules. The term *elution volume* is also often employed in literature with the same meaning as retention volume. In contrast to HPLC of low-molecular substances, the value of retention volume is beside the concentration of analyte in the column effluent the most important direct, quantitative outcome of the chromatographic experiment. In modern polymer HPLC, the actual retention volumes are deduced from the time scale.

Injected sample concentration, c_i and volume, v_i are important experimental parameters of polymer HPLC, especially of SEC. Their effects on results are discussed in Section 11.7.5.

Temperature of experiment is less important in the exclusion polymer HPLC, SEC unless sample solubility is not an issue. On the other hand, interaction and coupled methods of polymer HPLC methods necessitate rather strict control of temperature.

11.5.2 PROCESSES IN COLUMNS FOR POLYMER HPLC – RETENTION MECHANISMS

11.5.2.1 GENERAL CONSIDERATIONS

In order to design the appropriate liquid chromatography separation system, it is necessary to understand on molecular level some basic principles and tendencies of the processes taking place in the chromatographic column. Above processes result in differences in retention of sample constituents to allow their mutual separation. Extent of retention of macromolecules within column reflects the volume of mobile phase needed for their elution, their abovementioned *retention volume*, V_R. For the sake of simplicity, let us consider constant overall experimental conditions that is the eluent flow rate, temperature and pressure drop. The latter two parameters are dictated not only by the inherent hydrodynamic resistance of column that is influenced by the eluent viscosity, size and shape of packing particles but also by the sample viscosity, which may be rather high in polymer HPLC. Further, only one variable molecular characteristic of separated macromolecules will be considered namely their size in solution, which depends on their molar mass and on the solvent (eluent) thermodynamic quality. All other molecular characteristics are kept constant and/or the role of their variation is considered negligible. This is the case of the linear homopolymers with no *side chains* and with equal *stereoregularity*, that is with uniform spatial orientation of their *side groups*. Polymer sample is dissolved in eluent. Certainly, this all is only an approximation – a model system.

One can write for retention volume V_R of macromolecules

$$V_R = V_0 + KV_S \qquad (15)$$

where V_0 is the interstitial volume of the column filled with particulate packing, K is the distribution constant defined as a ratio of solute concentration in the stationary and the mobile phase, c_s/c_m, respectively and V_s is volume of the stationary phase within column. Depending on the particular method of polymer HPLC, V_s is defined in different ways. It is the total *volume of pores*, V_p, in a porous packing but it also can be related to the total *surface* of the column packing (mostly to the surface situated within the pores) or to the *effective volume of bonded phase*. The volume of pores is relatively well defined for the non-compressible packings with the identified pore walls. This is assumed to be the case of most packings applied in polymer HPLC. Both packing pore size and volume play the decisive role in separations based solely on the solute exclusion (see section 11.7). The exclusion processes, however must also be considered in most *coupled methods of polymer HPLC* that employ combination of entropy and enthalpy based processes (see section 11.8). In coupled methods of polymer HPLC, the *surface of packings* or the effective *volume of bonded phase* (see section 11.6.1.3.3, Chemical Structure of Column Packings) are to be taken into account. Some theoretical approaches also regard *surface exclusion*. The total volume of liquid within column, V_m depends on both the interstitial volume, V_0 and the pore volume, V_p – that is $V_m = V_0 + V_p$. V_m can be estimated from the retention volume of either the deuterated analogue of eluent or the non retained low molar mass substance. This value is denoted void volume in HPLC of low-molecular substances. The problems connected with precise determination of the void volume are discussed in numerous HPLC papers and textbooks.

Thermodynamics provides general relation between retention volume and distribution constant of any chromatographic technique. It can be written

$$V_R \sim K\frac{V_s}{V_m} = \exp\left(-\frac{\Delta G}{RT}\right)\frac{V_s}{V_m} \tag{16}$$

where ΔG is the change of Gibbs function due to transfer of solute between the mobile and the stationary phase, R is the gas constant and T temperature. Considering Equation 16, one can write.

$$\ln V_R \sim \frac{-\Delta H}{RT} + \frac{\Delta S}{R} + \ln\frac{V_s}{V_m} \tag{17}$$

where ΔH and ΔS are changes of enthalpy and entropy associated with transfer of solute between eluent and stationary phase, respectively. For a given liquid chromatography system the ratio V_s/V_m is nearly constant.

Formally, one can speak about enthalpic and entropic contribution to retention volume. Division of liquid chromatographic processes into enthalpic and entropic ones is very useful though it alone says little about the phenomena within column on molecular level. Relative contribution of entropy and enthalpy to retention volume can be estimated with help of the van't Hoff plot that is the dependence of V_R on $1/T$. However, assessment of their absolute values and their prediction is hardly possible from the properties of chromatographic system such as porous structure and interactive ability of the column packing, solvent strength and thermodynamic quality of eluent or from the molecular characteristics of separated macromolecules. In fact, numerous theories attempted to derive the quantitative relation between the distribution constant K and the molecular characteristic of polymer samples but so far it was not succeeded. The previously mentioned uncertainty of V_s, as well as the mutual dependence of entropy and enthalpy of chromatographic processes belong to the reasons, why the methods of polymer HPLC are considered non absolute.

In summary, the quantitative description and prediction of entropic and enthalpic processes taking place within the polymer HPLC column was so far not achieved. Still, their contribution to retention volume can be graphically visualized and qualitatively explained with help of the plot *log M vs* V_R. In contrast to convention of physical chemistry, according to the accepted habit of chromatography, the independent variable, logarithm of molar mass M is plotted on the axis of ordinates.

Let us first consider the situation when macromolecules do not exhibit any attractive or repulsive interaction with the *porous* column packing surface except for the effects caused by the imperviousness of the pore walls. This corresponds to $\Delta H=0$ in Equation (17) and the sample retention volume is controlled exclusively by the entropy of process. The fictitious retention volume of eluent molecules corresponds to total volume of liquid within column, V_m because the small molecules of the mobile phase freely permeate all pores of the column packing. The same retention volume - with certain approximation – exhibit molecules of the monomer that forms the chain of polymer under study. Consequently, V_m is the maximum retention volume at $K = 1$ available in absence of enthalpic interactions.

Large molecules of polymers are partially or fully excluded from the packing pores and therefore they elute faster than the eluent molecules. Naturally, the macromolecules outrun also their original solvent. The acceleration of polymer relative to eluent molecules rises with increasing size of macromolecules because the volume fraction of accessible pores decreases correspondingly. As a result, both the distribution constant K and the retention volume V_R of polymer molecules defined by Equation (16) decrease with increasing size of macromolecules in solution – that is with their rising molar mass. The retention volume reaches its minimum value for the macromolecules, which are fully excluded from the packing pores. The corresponding molar mass is called the *excluded molar mass, M_{ex}*. For molar masses higher than M_{ex}, the acceleration of macromolecules compared to eluent molecules becomes constant. Under otherwise identical experimental conditions, value of M_{ex} reflects the size of the largest pores within the column packing. The minimum value of V_R corresponds to $K = 0$ in Equation (15) that is $V_R = V_0$. V_0 is is the *interstitial volume* of the column. The *interstitial volume of column*, V_0 for a given column size depends on the packing bed geometry that is on the arrangement of packing particles. For a well packed column, V_0 does not depend on size of (nearly) uniform spherical packing particles. Whereas the *exclusion processes* play a rather inferior role in HPLC of low-molecular substances, they form the basis for the most important procedure of polymer HPLC, SEC.

Let us now consider the case when the polymer molecules exhibit certain *attractive* interaction with the column packing. Such interaction decelerates elution of polymer species and increases observed retention volumes. If the extent of enthalpic interaction is small, retention volumes of macromolecules increase but the exclusion processes still dominate (see Figure 3 curves b and c).

Curve b reflects the effect of interacting end-groups, which is the situation typical for oligomers. The role of end-groups decreases with the increasing molar mass of macromolecules. When the enthalpic interaction between the macromolecules and the column packing is large enough its effect may overrun the impact of exclusion. As a result the retention volumes increase with rising polymer molar mass (see Figure 3, curve e). In this case, K in Equation (15) is larger than unity.

Due to the attractive interaction of monomer with the column packing, also its retention volume exceeds V_m. An important case of the coupled exclusion – interaction processes is in Figure 3 represented by curve d. It mirrors the full mutual compensation of exclusion and interaction, which leads to the independence of retention volume of polymer molar mass. This situation is important for characterization of complex polymers. It allows elimination or at least suppression of the molar mass effect of macromolecules so that their unbiased separation according to another molecular characteristic can be performed. The experimental conditions leading to this specific elution behavior are called *critical conditions of enthalpic interactions*. They will be more in detail discussed in section 11.8.3.

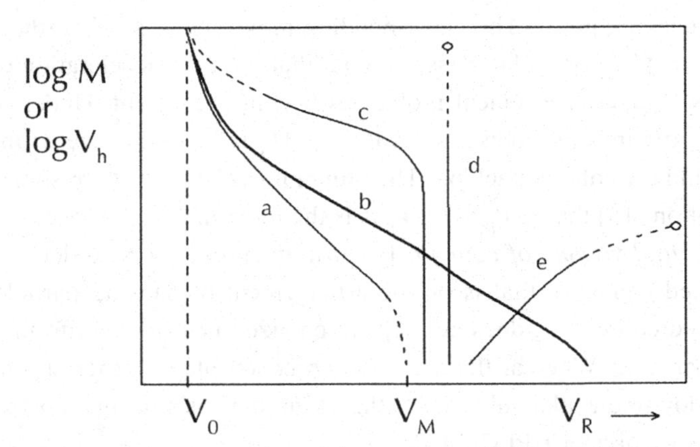

FIGURE 3 Schematic representation of depencdence of polymer retention volume on size, molar mass of macromolecules. For detailed explanation, see the text..

Enthalpic interactions in the polymer HPLC columns can lead to the situation when a part or even entire polymer sample stays fully retained within the column packing. These instances are in Figure 3 marked with the broken lines and with the open circles, respectively. The reduced overall sample recovery may badly affect accuracy of results of separation (compare sections 11.8.3 and 11.8.4). The eluted part of sample may be non-representative because the largest macromolecules exhibit increased tendency to be fully retained. Polymer fully retained within column can be released only by changing the experimental conditions – for example by increasing the eluent strength. In many practical polymer HPLC systems, the complete release of previously fully retained macromolecules from the column may represent a difficult task.

To summarize, the processes taking place in the column of polymer HPLC cause either acceleration or deceleration of separated macromolecules in relation to elution rate of mobile phase molecules. Only in case of critical conditions the elution rate of macromolecules and molecules of mobile phase is equal.

11.5.2.2 Entropic Processes – Exclusion in Polymer HPLC

In his pioneering works, Cassassa (1967 and 1976) has shown that the extent of pore exclusion of macromolecules is controlled by the changes of their (conformational) entropy. The principle is in a simplified form explained in Figure 4 (a–c). A zone of polymer solution with a non-zero concentration travels along a column packed with porous particles. Initially, the concentration of macromolecules within pores is zero (see Figure 4(a)). Inequality of solute concentration between the bulk solution outside pore ($c > 0$) and within pore ($c = 0$) "pulls" macromolecules into the pores. There is a tendency to equalize the chemical potentials in both volumes. In the course of pore permeation, the macromolecules shrink and therefore they lose part of their conformational entropy. When the penalty in entropy outbalances the pulling force, macromolecules discontinue their pore permeation – in other words the macromolecules cannot enter the pores, which would be large enough to accommodate their completely compressed size. Next the polymer concentration outside the pores becomes zero and the macromolecules are pulled out of the pores to continue their elution.

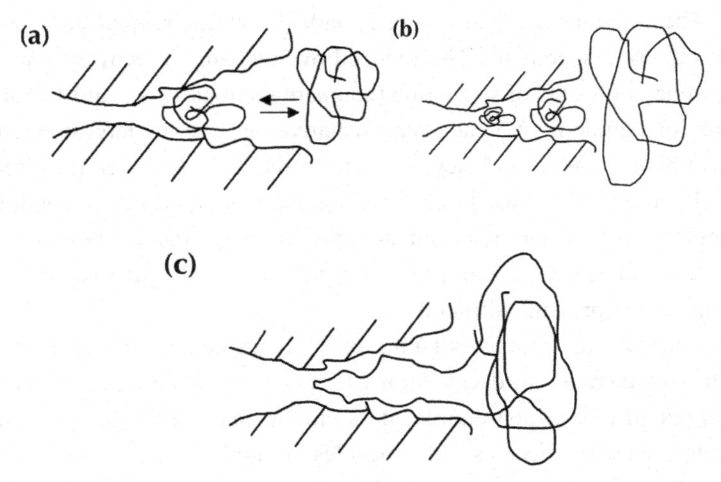

FIGURE 4 Schematic representation of exclusion processes of macromolecules (a, b, and c).For detailed explanation, see the text.

If the polymer solution contains macromolecules of distinct sizes, these will permeate different pore volume. The smallest species find the largest accessible pore volume. Consequently, they are longer accommodated within the column packing pores and their retention volume is higher than that of larger species (see Figure 4(a)). Smallest retention volume $V_R = V_0$ (K in Equation (15) equals zero) exhibit macromolecules, which are too large to enter any pore (see Figure 3(a)). This is the principle of separation based on size exclusion of (macro)molecules – the keystone of SEC.

It is to be noted, that the long segments of the very large macromolecules may partially permeate the packing pores (see Figure 4(c)). A frequent occurence of such phenomenon is anticipated in the case of ultra-high molar masses over 1,000 and, especially, over 5,000 kg.mol^{-1}. This kind of partial permeation likely contributes - together with the shearing of macromolecules on the outer surface of column packing particles – to the mechanical degradation of ultra-high molar mass polymers in the SEC columns (see section 11.7.5).

Surprisingly enough, the above actions are very fast and the exclusion based separation of macromolecules can be considered an equilibrium process. Numerous precise measurements of retention volumes of macromolecules under conditions of their partial pore permeation were performed and practically no effect of the eluent flow rate has been revealed.

In summary, the size exclusion of macromolecules in absence of any energetic interactions with the column packing is an *entropy controlled process.* It is based on the distribution of solute molecules between two *volumes* of eluent – that is between the interstitial volume and the pore volume within column. According to the terminology of physical chemistry, such process is to be designated the *partition* and because the present one is controlled by entropy, it should be termed the *entropic partition.*

Skvortsov and Gorbunov showed that it holds for the slit-like pores and for the (gaussian-type) coiled macromolecules

$$K \sim 1 - 2\pi^{-0.5}\frac{\overline{R}}{\overline{d}} \text{ for } \overline{R} < \overline{d} \tag{18}$$

and

$$K \sim 8\pi^{-2}\exp\left[-\left(\frac{\pi\overline{R}^{2}}{2\overline{d}}\right)^{2}\right] \text{ for } \overline{R} > \overline{d} \tag{19}$$

Where $\overline{R}$ is the effective mean radius of macromolecules dissolved in the eluent and $\overline{d}$ is the effective mean pore radius (1986).

Equations (18) and (19) are in agreement with the experimental observations that in SEC does exist a distinctive relation between effective sizes of column packing pores and polymer coils.

As demonstrated by numerous experiments, temperature does not much influence the exclusion processes (compare Equation (17)) in eluents, which are thermodynamically good solvents for polymers. In the latter case, dependence of molecular size, more exactly of *hydrodynamic volume of macromolecules* (see sections 11.2.4 and 11.7.3.1) on temperature is not pronounced. The situation is different in poor solvents (see section 11.2.4), in which the hydrodynamic volume of macromolecules extensively responds to temperature changes.

Eluent flow patterns contribute to the retention/acceleration of macromolecules within the polymer HPLC columns (see section 11.4, Separation Methods for Synthetic Polymers). The flow rate of liquid is lower near the walls of the capillary than in its middle part (see Figure 2). Therefore, the average velocity of the larger macromolecules is higher compared to the smaller species. As mentioned, this phenomenon can be employed in separation of very large macromolecules and particles. The corresponding method is denoted *hydrodynamic chromatography* (see section 11.4). SEC can be hardly used for particles because the Brownian motion is unable to release them from the packing pores.

11.5.2.3 ENTHALPIC PROCESSES – INTERACTION IN POLYMER HPLC

Various kinds of enthalpic processes, energetic interactions can be present in the column for polymer HPLC. Actually every constituent of this multicomponent system can interact with all other partners. The situation is very complicated and this is why the folowing discussion is restricted to the binary interactions, while the solvent-solvent interactions are not considered. Following enthalpic processes and their consequences for retention of separated species are regarded.

- interactions of macromolecules *with the column packing*, which are controlled by the strength of mobile phase, as well as by both temperature and pressure - that bring about *adsorption, enthalpic partition*:
- interactions *of macromolecules with the mobile phase:* thermodynamic quality of mobile phase toward macromolecules induces shrinking or expansion of polymer coils, as well as their enthalpic partition, and phase separation
- intermolecular *interactions among entire macromolecules* in samples that lead to association and aggregation phenomena
- intramolecular *interactions among segments* of the same macromolecule that evoke coil expansion or coil collapse.

As notified, the pecularities are only briefly mentioned of *ionic interactions* among segments of macromolecules that induce extensive coil expansions, as well as between macromolecules and packing that may lead to ion-exchange, ion-inclusion, and ion-exclusion processes.

It is to be stressed that the conformational barriers do not allow the full unfolding of polymer coils. Therefore, only a fraction of segments present in a polymer coil can simultaneously interact with the column packing. In other words, behavior of a macromolecule within the HPLC column cannot be obtained by the simple summing the interactions of monomeric units.

From the point of view of polymer HPLC the interactions of individual macromolecules with both column packing and eluent are most important and association or aggregation of polymers within chromatographic system should be avoided.

In practice, the operators deliberately adjust experimental conditions to control enthalpic interactions keeping the pore exclusion (nearly) constant. The resulting retention volumes are evaluated in order to optimize the separation of given sample. Some general rules do exist in this respect to facilitate the system adjustment procedures. They will be discussed in the following parts of this practically oriented chapter.

In conclusion, large differences exist in the behavior of small molecules and large (chain) macromolecules in the chromatographic systems. These mainly result from substantial role of conformational entropy of macromolecules and are augmented by distinctions in the viscosity, flow patterns, as well as in the mobility (diffusibility) of solutes of different sizes. It is necessary to consider these differences in order to devise the appropriate chromatographic system for efficient HPLC separation of particular polymer samples.

11.5.2.3.1 Adsorption of Macromolecules

Adsorption is a phenomenon frequently encountered in nature. It is caused by attractive interactions between the adsorbing species and the substrate. Adsorption is a typical *enthalpy driven process*. According to the accepted terminology, adsorption from a solution is the distribution of solute molecules between the *volume of solution* and the *adsorbent surface.* The adsorbent surface can be formed by a solid or by a liquid. Respectively, one speaks about the *surface adsorption* and the *interface adsorption.* The extent of adsorption depends on the affinity between the solute molecules and the surface – as well as on the nature of solvent molecules. The solute-surface affinity is dictated by the nature and concentration of the interacting groups on both adsorbent (column packing) and adsorbate (separated macromolecules). Generally, polar-polar interactions are much stronger than the interactions due to non polar London forces. Therefore the adsorption of polymers of the medium to high polarity on the polar surfaces is as a rule more intensive than the adsorption of the nonpolar macromolecules on the nonpolar surfaces. Adsorption of macromolecules on solid surfaces or liquid interfaces is always accompanied by large changes of conformational entropy of polymer coils. The extent of solute adsorption strongly depends on the magnitude of interactions between the adsorbent and the solvent molecules, on the *solvent strength* ε^0 (compare mobile phases for polymer HPLC). The solvent strength concept was originally introduced by Snyder (1967) in the description of adsorption based retention mechanism in liquid chromatography of low-molecular substances. Sample molecules compete with the solvent molecules for the surface or interface of sorbent. The *strong solvents* intensively interact with the surface of adsorbent and suppress adsorption of sample molecules, while the interaction of the *weak solvents* with the adsorbent surface is rather mild so that adsorption of sample molecules is less hindered. The Snyder's concept can be also applied to comprehension of adsorption of macromolecules (compare section 11.6.2). If the solvent is very strong, it fully prevents adsorption of polymer molecules on the given surface, it is a desorption promoting liquid, a *desorli.* This is the situation welcomed in size exclusion chromatography, where the strong mobile phases and the non-adsorptive column packings are to be used. The very weak solvent promotes extensive adsorption of macromolecules on the adsorbent, it is an adsorption promoting liquid, an *adsorli.* The latter extreme case is unwanted in the methods of polymer HPLC because polymer samples may stay irreversibly retained within the column flushed with an adsorli mobile phase. It may be even difficult to quantitatively release the macromolecules captured within porous column packing by subsequent application of a desorli (see section 11.8). The solvent strength of mobile phases can be adjusted by mixing appropriate liquids. The solvent strength of a mixture of liquids is usually considered additive so that the solvent strength of a mixture of liquids A and B, ε^0_{AB} can be roughly estimated from the relation

$$\varepsilon^0_{AB} = \phi_A . \varepsilon^0_A + \phi_B . \varepsilon^0_B \quad (20)$$

where ε^0_A and ε^0_B are the solvent strengths of eluent components and ϕ_A and ϕ_B their volume fractions, respectively. The difference between thus guessed and the real ε^0_{AB} values rapidly increases with the difference in polarity of the liquids A and B. In practice, the trial-and-error approach is widely employed in polymer HPLC to adjust the appropriate solvent strength of a mixed eluent.

Adsorption of macromolecules on the solid surface is schematically depicted in Figure 5(a). If the solvent is relatively strong, the *trains* of macromolecules attached to adsorbent surface are short, while the free ends and loops are large. Under otherwise identical conditions – the same adsorbent, polymer and temperature – the trains of adsorbed polymer chain are longer when the solvent is weaker.

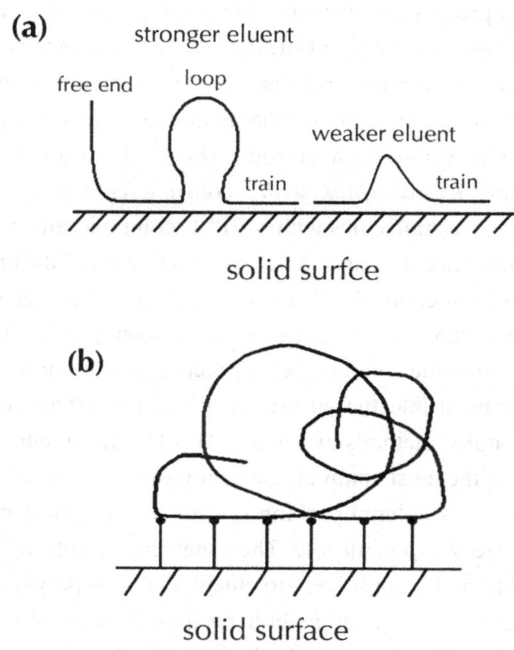

FIGURE 5 Schematic representation of macromolecules adsorbed (a) on the solid surface and (b) on the interface between liquid and quasi-liquid phases. For detailed explanation, see the text.

The typical example of the interfacial adsorption of macromolecules represents their interaction with the groups chemically attached to the surface of carrier, with the *bonded phase* (see section 11.6.1.3.3). The bonded phase can be considered a quasi liquid carrying either polar (for example –CN, –OH, $-NH_2$) or nonpolar ($-CH_3$) groups (Figure 5(b)). The polar interacting groups are screened off the solid surface by appropriately long (usually *n*-propyl) groups, *spacers*.

It is to be stressed again that the polymer coils markedly change their conformation during the adsorption process and simultaneously they lose large part of their conformational entropy.

The exact role of the solvent *thermodynamic quality* in the polymer adsorption is not well understood though numerous experiments indicate that it plays by far less important role than the solvent strength. Therefore, polymer adsorption in the chromatographic columns is in practice almost exclusively controlled by adjustment of eluent strength.

The adsorption process can be well affected by *temperature.* Rising temperature as a rule decreases extent of adsorption of small molecules. This is not the general case for macromolecules because their conformational entropy strongly depends on temperature. The gain in enthalpy due to adsorption may be exceeded by the large loss of conformational entropy. This could result in the infrequently observed *rise of polymer adsorption with increasing temperature.*

Adsorption of macromolecules within HPLC columns can be affected also by *pressure,* which changes extent and sometimes even sense of the preferential sorption of the mixed eluent components on the column packing (see section 11.6.2).

With only few exceptions, the extent of adsorption and resulting retention of macromolecules within the column packing increases with molar mass.

It can be concluded that the adsorption processes extensively affect retention volumes in coupled methods of polymer HPLC. Eluent nature (composition) and temperature are the most common tools employed in control of adsorption in the particular polymer – column packing system. The thermodynamic quality of eluent likely plays less important role. The statement, which can be found in the literature: "…addition of a nonsolvent to eluent increases polymer adsorption…" is misleading. The nonsolvent can be either a desorli or an adsorli for the given polymer so that a nonsolvent present in the solvent mixture can correspondingly either decrease or increase the extent of adsorption of macromolecules. Considerations on the role of conformational entropy of macromolecules in the adsorption processes may help explain some unexpected results in coupled methods of polymer HPLC.

11.5.2.3.2 Enthalpic Partition (Absorption) of Macromolecules

Another enthalpy driven process in polymer HPLC is the distribution of solute molecules between the *volume* of the mobile phase and the *volume* of the chromatographic stationary phase. The accepted terminology designates the processes of this kind *absorption* or the *partition,* which is driven by enthalpy and therefore it is the *enthalpic partition.* To prevent confusion between processes of adsorption and absorption, the term enthalpic partition is preferred in present chapter - also as the opposite to entropic partition, exclusion. Surprisingly, in many works the processes of *adsorption* and *enthalpic partition* are confused though the principal differences in their qualities are evident.

The most important stable, "non-bleeding" liquid chromatographic *stationary phases* with the defined *volume* are those *chemically bonded* to the surface of a solid carrier (compare section 11.6.1.3.3). The stationary phases, which are only physically anchored to the surface of the solid carrier are rather instable. The commonly available chemically homogeneous bonded phases with a large enough volume are mainly the *"monomeric"* or *"polymeric" alkyl bonded phases* on silica gel. They are extensively employed in HPLC of low-molecular substances. Numerous producers supply –C-4, –C-8, –C-14, and –C-18 "monomeric" bonded phases, in which the alkyl groups form a sort of "brush". –C-22, and –C-30 monomeric alkyl bonded phases are also on the market. The "polymeric" alkyl bonded phases are usually formed by the multiple layers of alkyl groups (mostly C-18), which are mutually crosslinked by means of the –Si–O–Si– bridges. Monomeric alkyl bonded phases are preferred in polymer HPLC. Synthetic and natural macromolecules immobilized on the surface or within the pores of the solid particles constitute a not yet fulfilled promise for their applications in polymer HPLC. Numerous distinct polymers were deposited on the silica gel surface and subsequently crosslinked. Unfortunately, they are habitually inappropriate for polymer HPLC because of their low permeability for macromolecules. Promising are the composite materials, in which the pore volume is filled with a homogeneously crosslinked sparse polymer network. The general problem with most bonded phases is a relatively slow mass transfer and consequent broadening of chromatographic zones. Naturally, this shortage is more pronounced with the polymer samples than with the low-molecular analytes.

A stationary phase capable of the enthalpic partition of macromolecules can also originate from the preferential sorption of one mixed eluent component on the bare surface of column packing. In this case a dynamic "stagnant phase" is formed because of the exchange processes between mobile phase and the stagnant phase. The composition of eluent layer, which is adjacent to the packing surface differs from the overall composition of both stagnant and mobile phases. Specific stationary phases with the non-negligible volume can also be formed by the solvated polymer chains, which protrude over walls of otherwise solid particles of the heterogeneously crosslinked porous organic particles (see section 11.6.1.3.3).

The role of mobile phase in the adsorption and enthalpic partition processes is substantially different. As explained, solvent quality toward macromolecules only marginally affects their adsorption on a given surface. Solvent strength controls the adsorption based retention of polymer species in the particular column packing – polymer system. On the contrary, solvent strength toward silica gel practically does not affect enthalpic partition of low-polarity polymers between the eluent and the (alkyl) bonded phase – provided adsorption of macromolecules on the residual silanols (compare Figure 6) is suppressed. The driving force of enthalpic partition is the solvent quality. The (solvated) stationary phase must be a little better solvent for separated macromolecules than is the mobile phase. The solvated alkyl groups however represent a rather poor "solvent" for most polymers. This means that eluent must be even poorer to push macromolecules into such stationary phase. The situation is schematically depicted in Figure 6.

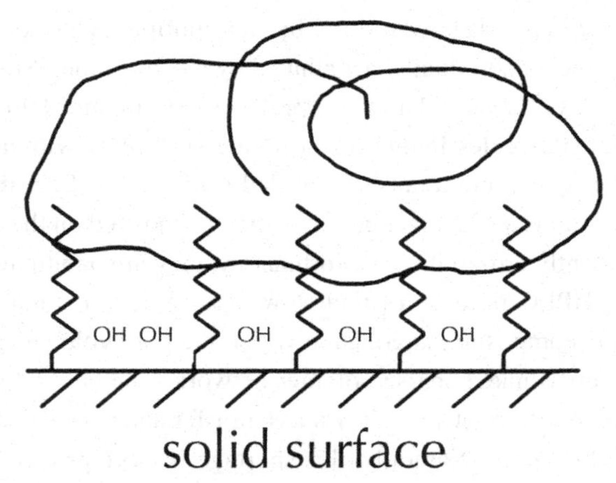

FIGURE 6 Schematic representation of enthalpic partition of a macromolecule in favor of the bonded phase. For detailed explanation, see the text.

The quantitative prediction of extent of enthalpic partition is difficult because composition of the bulk eluent differs from eluent situated within the stationary phase, such as for example that formed with the alkyl bonded groups. The bonded groups are preferentially solvated by one of the mixed eluent components. The magnitude of preferential solvation of the bonded groups by an eluent component may be very large. Surprisingly enough, the nonpolar alkyl groups bonded to the silica surface can be well solvated with the molecules of polar solvents such as methanol. Typical examples are mixed eluents methanol – water widely employed in HPLC of low-molecular substances. Due to preferential solvation of the bonded groups the driving force of the enthalpic partition caused by repulsion of macromolecules by the mobile phase may be either reduced or augmented.

For example, if the bonded alkyl groups are preferentially solvated with a good solvent for separated macromolecules, the difference in the dissolving power between the mobile phase that contains large concentration of a nonsolvent and the bonded phase may increase rather dramatically. As a result, even the polymers of medium polarity such as poly(methyl methacrylate) can be partitioned from certain eluents in favor of alkyl-bonded phases. Some liquids, for example dimethylformamide, intensively promote partition of macromolecules in favor of alkyl-bonded phases even though they themselves are incompatible with the alkanes, which form the bonded phase. The enthalpic partition usually does not take place over the entire length of the bonded alkyl groups (compare scheme in Figure 6). Therefore, the application of –C-18 alkyl-bonded groups may turn inappropriate in some coupled methods of polymer HPLC. The shorter bonded alkyl-groups may be sufficient. The stretched –C-18 groups namely occupy about half of the pore volume of silica gel with 10 nm (100Å) pores and excessively decrease its effective pore diameter.

The important feature of all available silica gel alkyl-bonded phases is presence of rest- silanols on the carrier surface. In spite of employment of the sophisticated, strictly proprietary *end-capping procedures*, which are aimed at elimination of free silanols, about 50% of starting -SiOH groups remain unreacted in the best commercial –C18 bonded phases. Unexpectedly, these silanols are accessible for polar macromolecules that somehow reptate around alkyl groups. As a result, polar macromolecules can became retained with adsorption on the alkyl bonded silica column packings. Simultaneous effect of two independent enthalpic retention mechanisms is usually undesired in coupled methods of polymer HPLC because of difficulties with both control of polymer elution and interpretation of results. This is why the possible adsorption of medium- to high-polarity macromolecules should be prevented when working with alkyl-bonded column packings. As strong as possible (polar) solvent should be employed as one of the eluent components. It is useful to remember the above facts when selecting the (mixed) mobile phase and the column packing for separation of complex polymer systems that contain macromolecules of well distinct polarities.

To summarize, the enthalpic partition processes in the columns for polymer HPLC substantially differ from the adsorption processes. Enthalpic partition on the alkyl-bonded phases can be employed for separation of not only nonpolar macromolecules but also of the low- to medium- polarity polymers. The extent of the enthalpic partition and consequently also resulting polymer retention is controlled primarily by the thermodynamic quality of eluent toward separated species and by the extent of the bonded phase solvation.

11.5.2.3.3 Phase Separation of Macromolecules

As shown in section 11.2.4, the solubility of polymers is generally much lower than the solubility of low-molecular substances. When the solvent quality is deteriorated by adding a nonsolvent or by a change of temperature, polymer coils shrink, they begin to associate/aggregate and, eventually the phase separation occurs. In the initial stage of phase separation, the micro-droplets of two immiscible liquids are usually formed. Existence of the micro-droplets in the HPLC column is undesired. It should be repeated again that not always the decrease of temperature but sometimes also its increase can cause phase separation of originally homogeneous polymer solution. Other important, frequently unexpected features of polymer solubility to be considered in coupled methods of polymer HPLC are the *cosolvency* and *co-nonsolvency* phenomena cited in section 11.2.4.

Phase separation of macromolecules is generally a slow process. This sometimes allows application of a nonsolvent mobile phase for polymer under study for the fast separations, which must be concluded before the onset of phase separation. Generally, solubility of polymers markedly depends on all their molecular characteristics. This enables very efficient separations of macromolecules employing phase separation and re-dissolution processes in special procedures of coupled methods of polymer HPLC (see section 11.8.4). On the other hand, due to the complexity of the phase separation phenomena the resulting retention volumes may simultaneously depend on several molecular characteristics of separated macromolecules. This complicates interpretation of the separation results. Both precipitation and re-dissolution of most polymers may be even affected by presence of seemingly inactive surface of the column packing. Therefore, careful control of the phase separation phenomena is recommended in coupled methods of polymer HPLC.

11.5.3 SEPARATION ARRANGEMENTS IN POLYMER HPLC

The liquid chromatography separations can be performed in different ways:

- ***Static, Batch Procedures***: Which are mainly employed for preparative or process fractionations. Sample is mixed with the separation medium, for example with the adsorbent, occasionally stirred and after time needed to attain equilibrium, the supernatant is removed and subject to further processing. The sample constituent, which was retained in the separation medium is released with help of a displacer. The entire process is slow and labor-intensive. It is hardly suitable for analytical characterization purposes

- ***Dynamic, Continuous Procedures***: In which not only the mobile phase but also the sample is continuously introduced into the chromatographic column, which is an annulus or simulated moving bed. The latter disposition is most suitable for preparative fractionation, while only simple, two- to three-component polymer blends can be efficiently fractionated in an annulus
- ***Dynamic, Interrupted Procedures:*** Which are most widely used for analytical purposes. They can assume *planar* or *column* configuration. The latter one dominates modern polymer HPLC. The sample is introduced in the form of a defined volume into the straight tube filled with a *packing*, which is designated *the column*. The column is flushed with the *mobile phase* and the *column effluent* is continuously monitored with a *detector*. Primarily, the sample concentration is determined in the column effluent. Often also selected molecular characteristic of polymer in the column effluent are assessed, usually its molar mass or chemical structure. As a rule, volume of sample introduced into the column represents only a tiny fraction of the column volume. The exceptions represent the methods of liquid chromatography under limiting conditions of enthalpic interactions (see section 11.8.6).
- ***Differential Procedures***: In which the mobile phase is the solution of a reference polymer and the polymer under study is dissolved and injected in the neat eluent. Small differences between molecular characteristics of macromolecules in the mobile phase and in the injected solution are registered. The high viscosity of mobile phase causes large flow resistance and low diffusion rate of macromolecules. As a result, the detector response is often extensively distorted. Owing to the above shortages the applicability of this approach is well limited.
- ***Vacancy Procedures***: In which the mobile phase is a solution of polymer to be characterized and the neat eluent is injected into column as a "sample". Similar to differential procedures, high viscosity of the mobile phase causes important flow resistance and also decelerates diffusion of macromolecules so that the observed *vacancy peaks* are significantly broadened. Large sample consumption further prevents application of these procedures for characterization of research materials.

11.6 INSTRUMENTATION AND MATERIALS FOR POLYMER HPLC

11.6.1 APPARATUS FOR POLYMER HPLC

Basic features of instruments for polymer HPLC are briefly outlined in this chapter. As a rule, the operator can only a little affect its construction. The differences between apparatus for polymer HPLC and for HPLC of low-molecular substances are described more in detail in section 11.6.1.3, with emphasis on the column packings. To avoid advertisement, the names of instrument suppliers are not presented. A general practical hint for operators in polymer HPLC: Instrument performance quoted by producers is often too optimistic.

The basic scheme of a liquid chromatograph is depicted in Figure 7.

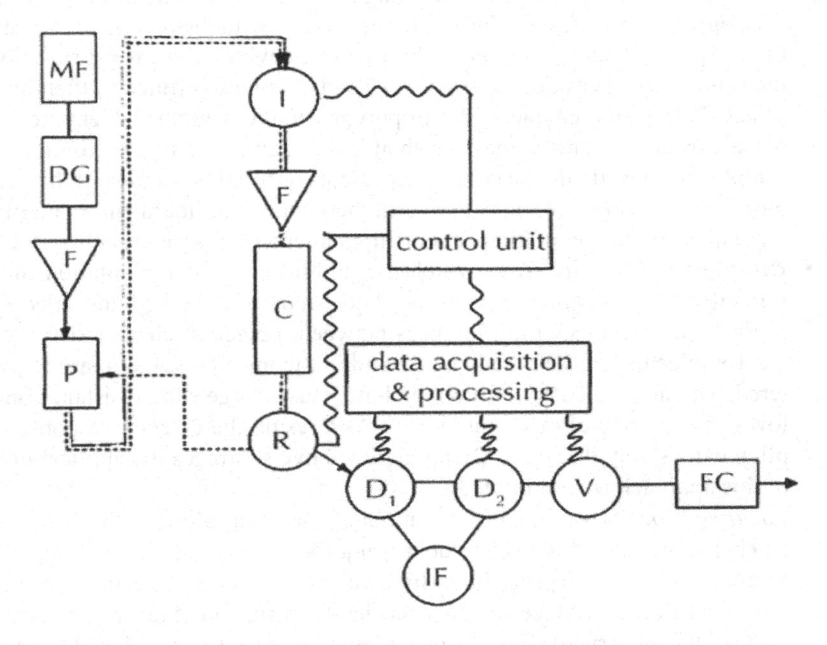

MF - eluent container, DG - degasser, F - filters, P - pumping system, I - sample injector, C - column(s), R - recycling valve, D - detector(s), V - volumeter, FC - fraction collector

FIGURE 7 Scheme of apparatus for polymer HPLC.

11.6.1.1 MOBILE PHASE HANDLING

The eluent container MF is a vessel made of a strong and chemically resistant material. It should be well closed and provided with a drying cap to prevent absorption of humidity and to minimize the (preferential) evaporation of eluent (components).

The degasser DG removes dissolved air components, mainly oxygen from the mobile phase. The classical heated degassers were substituted with small, simple and reliable vacuum systems.

The filters F remove microparticulate matter from eluent, including both that contained in sample and released from the moving part of system such as pumping systems and sample injector. It is useful to filtrate the sample with exchangeable filter mounted on the sample syringe. Fixed filters are made of inoxidable steel while the syringe filters embody polymer membrane usually with pores as small as 0.2 μm that are compatible either with aqueous (cellulose made) or organic (fabricated of poly(tetrafuoroethylene)) sample solvent.

11.6.1.2 PUMPS, SAMPLE INJECTORS AND RECYCLING VALVES

The pumping systems P must provide pulseless, highly precise and accurate, very constant but resetable flow rate of mobile phase at pressures up to about 40 MPa (~400 bar). Flow rate resetability is important and often overlooked attribute of the well-working pumping systems. The pump can slightly change eluent delivery after each restart. This may cause important errors, especially in SEC. It is advisable to check frequently the actual flow rate, for example with a precise burette and stopwatch. The ultra high-pressure pumping systems that generate pressure above 100 MPa are so far not common in polymer HPLC. Constant flow rate allows expression the retention volumes on the time scale. The pumping systems must exhibit low wear and high corrosion resistance against mobile phases employed.

The latter requirement may be crucial when working with highly aggressive, for example acidic mobile phases. The pneumatic pumps that produce constant pressure, not flow rate are with advantage applied for packing the columns.The positive displacement (single piston) are applied only rarely. The non-negligible compressibility of liquids complicates their use. Reciprocating pumps dominate on the marked. They are equipped with two pistons, the opposed movement of which is controlled with specially designed cam to (practically) remove pulsations.

More precise is the electronic control of piston movement, especially when the pistons are actuated with two otherwise independent motors. In the latter cases, the additional pulse dampeners that were common with classical pumps are less important. Often, a long capillary mounted between pump and sample injector is satisfactory. Its volume is to be considered in the eluent gradient polymer HPLC procedures (see section 11.8.4). Special pumping systems commonly employed in gradient HPLC of low-molecular compounds are frequently also used in eluent gradient polymer HPLC.

Sample injectors I allow applying sample solution into the column without flow interruption. A scheme of the most common injecting valve is depicted in Figure 8.

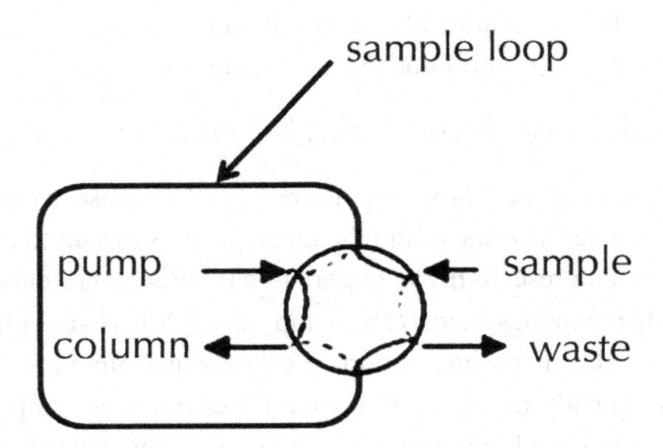

FIGURE 8 Six-port two-way sample injecting valve equipped with the capillary loop of appropriate volume, usually 20 to 100 μL. The sample loop is filed with sample solution from a syringe. By the valve operation, the content of sample loop is displaced into column.

Instead of the hand-driven injection valves autosamplers are often under use. They facilitate changing sample volume and enable automation of analyses.

The recycling valve R is in its construction similar to injecting valve, however it has only four ports. It allows returning selected fraction of column effluent *via* the pump back to the column C. Recycling procedure enables increasing selectivity of separation but its application is rather limited because the front fraction easily catches the rearmost one.

11.6.1.3 COLUMNS

The column C is the hearth of a liquid chromatograph. The columns for polymer HPLC well differ from those for HPLC of low-molecular substances. Therefore, increased attention is paid to the column issues in this chapter, bearing in mind that the beginners that may cope with the introduction of polymer HPLC methods in their laboratories.

11.6.1.3.1 General Considerations

Two parameters of the column for polymer HPLC are widely considered. *Column efficiency* reflects broadening of the chromatographic bands in the course of the chromatographic process. It can be judged from the width of peak of appropriate low-molecular marker. From the viewpoint of polymer HPLC, such test is important primarily for unveiling erroneous behavior of a column, such as presence of larger packing heterogeneities. Second characteristic, the separation *selectivity* is mainly important for the SEC columns (see section 11.7).

Straight tubings from stainless steel with polished inner walls are the common column material. Their diameter ranges from less than one mm (capillary columns) to several cm (preparative columns). The standard columns for analytical polymer HPLC have diameter of four to ten mm and length of (100 to 300) mm. Dimensions of columns for analytical SEC are (7.5-10 x 250-300) mm. Columns for the fast SEC are short and wide, with the diameters as large as 50 mm.

Regular packing of small particles into the column is a demanding task. However, it can be performed quite successfully under laboratory conditions with help of a pneumatic pump, which transports slurry of the packing material into the column under high and constant pressure. Creation of irregularities and even voids in the packing bed is to be avoided. As stated above, the quality of packing, the column efficiency is controlled by the peak shape of an appropriate low-molecular marker.

Optimized, repeatable fabrication of high-quality columns for polymer HPLC is as a rule a top secret, proprietary process of the column producers. Service-life of columns for polymer HPLC is generally much longer than in case of columns for HPLC of low-molecular substances. Substantially higher is also price of the former. Therefore, it is advisable to consider carefully future applications of columns before their purchase. Recommendation of cautious handling with the columns is another important hint for the beginners. Mechanical shocks and sudden changes of mobile phases with substantially different polarities should be avoided, especially in case of organic polymer based column packings. If possible, always the same eluent for particular SEC column should be applied. When it is necessary to alter mobile phase, the resistance of the column packing to the new mobile phase should be learned from the producer's information and the eluent change is to be done in the small steps. It is useful to clean the column after prolonged use by flushing with mobile phase at a slightly elevated temperature or, very cautiously, with an appropriate displacer, which is a medium polarity solvent in the case of polystyrene/divinylbenzene SEC column packings. Silica gel column packings that are usually employed in the coupled methods of polymer HPLC are in general much more robust than the packings based on organic polymers.

Since recently, monoliths of silica gel or organic polymers became popular in HPLC of low-molecular substances. They exhibit reduced flow resistance and high mechanical strength. Their internal structure forms the solid matrix and two kinds of pores. Macropores or channels allow fast transport of mobile phase, while the separation processes take place in the smaller mesopores. So far, however, most commercially available monoliths possess too small effective volume of the mesopores to compete with particulate column packings in polymer HPLC, especially in SEC.

11.6.1.3.2 Physical Structure of Column Packings

With only few exceptions, the column packings for polymer HPLC are *totally porous* materials. They are designated: gels, sorbents, carriers and the like. A carrier supports stationary phase, usually chemically bonded organic groups to its surface. Modern column packings in polymer HPLC are solid to withstand elevated pressure up to tens of megapascal. Ultra-high pressure separations are since recently considered also in polymer HPLC. In such case, column packing must resist pressure over 100 MPa. The classical soft, homogeneously crosslinked gels are mechanically instable and can be applied only for low pressure and low performance separations. Particles of practically all modern column packings assume spherical shape. The latter is created by dispersing the starting material within an immiscible liquid. Internal porous structure of the column packing particles is formed with help of different methods, most commonly by adding a chemically inert diluent to the charge, see below. Schematically depicted internal structure of column packings is shown in Figure 9.

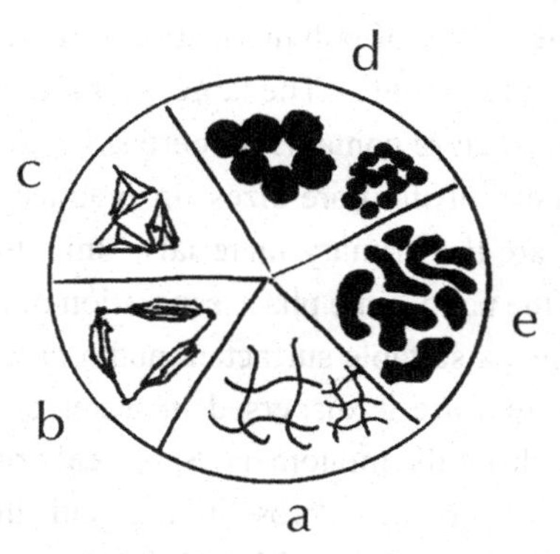

FIGURE 9 Schematic representation of internal physical structure of column packings for polymer HPLC. For explanation, see the text. Note that the outer shape of packing particles is spherical.

The internal physical structure of column packing particles is responsible for their both porous structure and mechanical properties.

Texture 11.9a represents *homogeneously crosslinked soft, swelled polymer gel.* Molar mass of polymer chains between crosslinks determines the pore size, and also the pressure stability of particles, which hardly exceeds 0.3 MPa. Diffusion of macromolecules within particles and control of their interactivity is undeniable but the low pressure stability of homogeneously crosslinked gels prevents their use in polymer HPLC. Column packings of this kind, typically crosslinked dextran gels were employed in the first stage of polymer separation with help of liquid chromatography. The attempts to create *composite column packings*, in which the pores of solid, pressure resistant, for example inorganic matrix are filled with homogeneously crosslinked network of organic macromolecules were so far not successful.

Structure 11.9b depicts arrays of polymer crystallites. Typical representative are agarose materials. The mechanical stability of such semi-soft gels can reach about 1 MPa.

Scheme 11.9c represents array of crystals, in which the pores are formed by vacancies among crystals. Typical representative is porous alumina. Diffusion rate of macromolecules within structure of this material is likely hindered and separation efficiency is low.

Structure 11.9d is a cluster of nonporous nano-spheres called globules or nodules. It is often designated *globular structure*. This is the most common texture of the organic column packings for polymer HPLC. Globules incur by the phase separation process in the course of polymerization and crosslinking process. The pores are free spaces among globules. The larger globules the larger are the pores. As a rule, the globular structure contains numerous irregularities. This results in the rather broad dispersity of the pore sizes. In practice, appropriate monomer and crosslinking agent are the primary materials. Initiator of polymerization, a solvent, which induces the nano-scale phase separation of macromolecules generated during polyreactions, a suitable surfactant and a protective macromolecular colloid are added. The mixture is dispersed in water or in a nonpolar organic solvent in the case of hydrophilic monomers. Spherical droplets of above mixture are formed, in which polymerization, crosslinking, and simultaneous nano-phase separation takes place. Porous silica gel incurs by the controlled polycondensation of sylicic acid dispersed in organic solvent. Alternatively, the dispersed silica sol is destabilized to create solid array of globules.

Image 11.9e depicts to the "sponge-like" structure of the column packing. It may exhibit very narrow pore size dispersity and good mechanical stability even at high pore volume. Typical representatives are porous glasses and wide-pore silica gels. Porous glass is prepared from the common borosilicate glass, which undergoes *slow phase separation* at elevated temperatures around 1000 °C, during which SiO_2 separates from B_2O_3. Before the phase separation is completed, the material is cooled down, crushed, and the irregular particles are size separated. Afterwards, B_2O_3 is leached by acid to generate pores. Wide pore silica gels are prepared from the narrow pore materials by their *hydrothermal treatment*, which denotes the impact of water steam at high temperature and pressure or by the attack of organic basic substances at elevated temperature. Sponge-like organic porous systems result from the process, which is in principle analogous to that employed in preparation of globular systems, however the nano-phase separation is interrupted before its completion. The operation is rather difficult to control.

As indicated, globular (see Figure 9(d)) and partially also sponge-like (see Figure 9(e)) structures of the column packings are most commonly employed in polymer HPLC. It is necessary to stress that both latter kinds of column packings are prepared under conditions that include two independent phase separations, one microscopic and another one on the nano-scale.

In polymer HPLC, particles with different pore sizes and large pore volume are needed. However, the extensive increase in pore volume - similar to rise in the pore size – brings about reduction of mechanical stability of the column packing. Materials with various effective pore size averages are commercially available. Following terminology of physical chemistry, pores under two nm are called *micropores*, pores between two and about 10 nm are designated *mesopores,* between about 10 and 100 nm are *macropores* and above 100 nm *gigapores*. Mass transfer into and from the pores, which are smaller than about 2 nm is generally constrained. The modern column packings are synthesized keeping in mind this limitation. Pore size averages of the column packings, which are designated for SEC range from about 4 to 400 nm. They exhibit distinct size dispersity. Selection of effective pore size depends on the purpose of use, whether oligomers or high polymers are to be characterized. Evidently, for separation of larger macromolecules in the exclusion mode, the greater pores are needed. In SEC, very popular are the column packings that contain pores of broad but controlled dispersity so that they can be employed for separation over a wide range of macromolecular sizes. Advantageously, such column packings can produce linear dependences of retention volume versus logarithm of molar mass (see section 11.7.3.1, Columns for Size Exclusion Chromatography and Their Calibration) and the particular columns are called *universal, linear SEC columns*. Evidently, SEC columns that separate in the wide range of molar masses exhibit reduced separation selectivity. The effective pore sizes of the column packings for the liquid chromatography procedures that combine exclusion (entropy) and interaction (enthalpy) based retention mechanisms are generally smaller than in the packings for SEC, usually 10 and up to 100 nm (compare section 11.8). For the unclear reasons, pore volume of most available column packings drops with their decreasing pore size. As a result, columns packed with particles that exhibit pores of narrow size and large pore volume are hardly accessible.

Particle sizes of the column packings for polymer HPLC range from three to 20 μm. Larger particles are applied for separation of ultra-high molar mass polymers, which are easily shear-degraded on the surface of small-sized particles. Particles in the region of 5 or 10 μm are suitable for conventional polymer HPLC. Particles with the sizes down to about 2 μm are infrequently applied for some fast SEC separations at very high pressure. However, experimental work with such small particles is difficult. High quality pumping system is needed and polymer solutions must be carefully filtered.

To keep the band broadening process as low as possible, particles must exhibit narrow size-dispersity. Sophisticated polymerization reactors or advanced procedures for sizing, size-fractionation of particles are employed by the column packing producers to reach this goal.

At present, only few researchers prepare their own column packings. It is to be stressed again that future employment of the column should be kept in mind before purchase. For the scouting and preliminary experiments, columns packed with 20 μm are often sufficient. As to column efficiency, methods of coupled polymer HPLC are less demanding (see section 11.8). In many cases, 10 μm particles serve sufficiently.

11.6.1.3.3 Chemical Structure of the Column Packings

In the course of development of polymer HPLC, column packing matrices with various chemical compositions were tested. The following list presents only the most important column packings, which have found wide practical application. Often, the column producers keep secret the actual chemical structure of their column packings. The composition of most often applied packings is presented in italics.

Organic column packings -

polysaccharides, poly(styrene-co-divinyl-benzene)s, poly*(divinylbenzene)s, poly(acrylate)s, poly(methacrylate)s, poly(urethanes),* ...

Inorganic column packings -

oxides of metals – *silica*, zirconia, titania, alumina.

Carbon –

(amorphous – glassy carbon), graphite

Combined inorganic-organic column packings materials -

In which surface of mechanically strong (inorganic) matrices such as silica gel is chemically modified with the grafted, chemically bonded organic groups. The most important are *bonded alkyl groups* from C1 – to C30, while the series is dominated with octadecyl, C18 groups, further amino-, nitril-, glyceryl-, phenyl- and few other bonded groups (compare section 11.5.2.3.2, Enthalpic Partitions of Macromolecules). The surface of pores can also be coated with crosslinked organic polymers. Volume of pores can be filled with homogeneous networks of organic polymers such as poly(styrene-*co*-divinylbenzene), crosslinked poly(siloxane)s, poly(acrylate)s, poly(methacrylate)s and the like.

It is to be accentuated again that the enthalpic interactivity of the column packing with separated macromolecules should be minimized in the instance of SEC while it should be well controllable in the case of interaction polymer HPLC. From the latter points of view, not only the overall composition of both packing matrix and surface but also presence of impurities plays important role. In the case if silica gels appearance of metal (ions) is crucial. The classical silica gels of "A" type contained large concentration of latter impurities. At present, almost exclusively the high-purity silica gels of "B" type are on the market. As result of the irreversible arresting of sample parts or, less commonly, due to the chemical reaction with eluent, the surface properties of column packings for polymer HPLC can be altered in the course of their employment. This is one of the reasons why the columns for polymer HPLC should be periodically tested, recalibrated and if necessary properly cleaned – or just exchanged. A warning signal of the column deterioration offer unexpected shifts of retention volumes and peak broadening of appropriate markers.

At present, the by far most popular column packings for SEC are polystyrene/divinylbenzene copolymers, while coupled methods of polymer HPLC are dominated with bare or alkyl-bonded silica gels.

11.6.1.4 DETECTORS

The main task of analytical polymer HPLC is to afford information on molecular characteristics of macromolecules under study. This is attained with help of separation processes within column. The column effluent contains macromolecules with changing both concentration and molecular characteristics. Online monitoring concentration, composition or molecular characteristics such as molar mass of macromolecules in the column effluent is the basic objective of devices designated *detectors*. The detectors belong together with columns and pumping systems to the most important parts of instruments for polymer HPLC. The relevant online polymer HPLC detectors are briefly introduced in this part. The off-line procedures of detection in polymer HPLC are presented only marginally. The in-situ detection applied in thin layer chromatography and rather infrequently also within the separation columns for polymer HPLC is not treated.

The essential information supplied by detectors is the concentration of macromolecules in the column effluent. For comprehensive molecular characterization of polymers, additional data on eluted macromolecules are also important such as their molar mass, composition or architecture.

Concentration of macromolecules in the column effluent can be determined either indirectly or directly. In the former case, the detector monitors certain property of the whole column effluent, from which the sample concentration is derived. Such devices are also called non-specific or universal detectors. The typical representatives are differential refractometers and densitometers. On the contrary, the properties of separated macromolecules are directly tracked in the latter case. The major members of this detector class are photometers and detectors that monitor light scattering from the "semi-solid" aggregates of macromolecules.

The site of detection proper, usually the measuring cell, must be very small. Otherwise, the sample zones that leave the column are extensively broadened in the course of detection itself. In the case of the flow-through cells, their usual volume is just 10 μL.

Following parameters of detectors for polymer HPLC are to be considered in the first range:

- sensitivity
- linearity of response toward sample concentration (called linear dynamic response of detector)
- dynamic range of detector – area of measurable concentration for the given analyte noise and drift of detector response either inherent or caused by the external reasons such as changes of temperature.

Four different constructions of *differential refractometers* are in use - deflexion (Snellius type), interferometric, reflection (Fresnel type), and Christiansen type, while the former two types dominate. Refractometers can detect any polymer provided its refractive index sufficiently differs from that of mobile phase. Practically all commercial SEC instruments are equipped with the differential refractometers. The sensitivity of differential refractometers is generally rather low and the concentration of sample injected into polymer HPLC column must be relatively high. Too elevated sample concentration in the case of high polymers however largely brings about numerous problems (see sections 11.7.3.2 and 11.7.5). On the other hand, the allowed concentration of oligomers can be much higher. Sample capacity of most polymer HPLC techniques, especially those of SEC, is rather limited. Peaks are broadened when concentration of injected sample increases. At a certain threshold concentration, the column becomes overloaded and completely loses its separation ability. SEC retention volumes as a rule increase with the sample concentration and when larger amounts of samples are injected, the chromatograms are to be corrected in order to obtain exact molar mass values (compare section 11.7.5). Similarly, the sample volume is limited in most polymer HPLC techniques.

The flow-through *densitometers* monitor changes in density of column effluent. Their sensitivity is relatively low and therefore they are mainly applied for oligomers at higher concentrations.

The flow-through *light scattering detectors* measure intensity of light scattered by macromolecules in the column effluent. Devices that monitor intensity of scattered light at a single angle or simultaneously at several angles are available. Molar mass and size of macromolecules of homopolymers in single eluents can be obtained with high precision by means of flow-through light scattering detectors. This is why these devices are often called *absolute detectors*. Information obtained by the flow-through light scattering detector also enable characterization of polymer branching. Other "absolute" detectors are *flow-through viscometers*. The designs of the flow-through membrane *osmometers* were published by several authors but the commercial instruments are so far not available. It is likely that the main problem – similar to static osmometry – lies in the unavailability of reliable membranes.

Photometers that work in the *ultraviolet* domain of spectra and devices monitoring *fluorescence* of eluted macromolecules are much more sensitive than differential refractometers provided separated macromolecules bear appropriate chromophores. This is however an exception rather than rule. Absorption of *infrared light* renders valuable information not only about concentration of eluted macromolecules but also about their chemical composition and in some cases also about their physical architecture. The IR transparency of most solvents is poor and this restrains applicability range of this kind of detectors for the direct monitoring the column effluent. IR flow-through detectors are mainly employed in SEC of polyolefins. On the other hand, appropriate *interfaces* can be applied for monitoring IR absorption of separated macromolecules. Column effluent is continuously deposited on the rotating germanium disc, which is after solvent evaporation off-line processed in the IR photometer. The problem is the uniformity of polymer film created and in order to ensure appropriate correction, the measurements are done at two different wavelengths.

Devices that continuously monitor *chemiluminiscence* of the column effluent so far have not found noticeable application in polymer HPLC.

It is often useful to monitor *simultaneously* not only concentration and molar mass but also composition of eluting macromolecules. Detectors that for example combine a differential refractometer with the ultraviolet photometer, and the flow-through light scattering monitor or the flow-through viscometer are comercially available.

Another group of detectors with sample transfer employs *pyrolysis of macromolecules*. Column effluent is deposited continuously on the appropriate moving transporter such as a wire, chain, or net. Next, the solvent is evaporated, sample is pyrolyzed and polymer concentration is assessed from the amount of carbon dioxide formed. The problems with cleaning of sample transporter to prevent base line noise and drift prevented broad use of detectors of this kind. The attempts to apply similar principle for monitoring polymer composition by engagement of complete gas chromatography of the fractions so far remained only on the level of laboratory experiments. Detection of such kind would provide valuable information on the sample composition, for example for statistical copolymers. There are, however, so far unsolved problems with the dependence of composition of pyrolytic products on presence of the neighboring units in copolymers.

Particular class of detectors that are frequently employed in coupled methods of polymer HPLC (see section 11.8), are the devices that monitor intensity of light scattered by the stream of (semi) solid polymer particles generated after evaporation of mobile phase from the droplets of column effluent. Intensity of light scattered by polymer particles is measured under 90° angle. It depends on mass of polymer particles in the measuring cell volume. The devices are denoted *evaporative light scattering detectors*, ELSD. ELSD are able to monitor polymer concentration in many single and mixed mobile phases. This is a necessary condition for successful detection of column effluent for practically all coupled methods of polymer HPLC, which employ mixed mobile phases in both isocratic and gradient arrangements. The sensitivity of ELSD is up to 100 times higher than that of differential refractometers even if the difference of refractice indexes of polymer and eluent is favorable. Sensitivity of detectors employing the nebulization principle can be further augmented by application of corona or the mist-chamber principle. Though ELSD "does not see" eluent, its response to some extent depends on both eluent and polymer nature and even on polymer molar mass. ELSD also exibit only limited linearity of response.

Very valuable data on the eluted macromolecules can provide *nuclear magnetic resonance*. In the online arrangement, the operator must however cope with the rather low sensitivity of the flow-through NMR instruments and with the necessity to suppress the background solvent signals. The latter problems are often solved by the off-line procedures. Selected fractions of column effluent are collected. If necessary, repeated or preparative separations are applied. Solvent that formed original mobile phase is evaporated, new solvent is employed, and the resulting solution is off-line measured in the NMR instrument. The entire process is material-, labor- and time-intensive. Similar technique is also frequently employed for a combination of polymer HPLC with the matrix *assisted laser desorption-ionization mass spectrometry* and with the *electrospray mass spectrometry*. In contrast to NMR, sample consumption for mass spectrometry is substantially lower. On the other hand, the information value of NMR for the complex polymers is very high and in some instances hardly replaceable.

11.6.1.5 VOLUMETERS, FRACTION COLLECTORS

In the past, various kinds of the volumetric devices were in use for monitoring volume of column effluent. These were for example siphons, drop-counters and thermopulse devices. All of them exhibited drawbacks, especially a dependence of the determined volume on eluent nature, flow rate, temperature and on presence of sample in the column effluent. Therefore, their precision was rather limited. Thanks to the recent significant improvements in efficiency of the pumping systems, the volumeters are no more frequently employed. However, they can make a good service in checking the pump resetability (see section 11.6.1.2, Pumps, Sample Injectors and Recycling Valves).

The fraction collectors are needful for procuring enough material for the subsequent off-line analyses, for example for NMR – by means of repeated separations.

11.6.1.6 SOFTWARE

Generally, software for polymer HPLC is a “black box”, which hardly allows any intervention from the operator. The handling of injecting valves, fraction collectors, electronic interfaces and other control units is usually integrated in the software employed for data acquisition and processing.

11.6.2 MOBILE PHASES FOR POLYMER HPLC

In section 11.2.4 the role was discussed of liquids in dissolving and precipitation of macromolecules, as well as in affecting properties of polymer solutions. It was stated that depending on the *thermodynamic quality of solvent*, the same macromolecule can assume various conformations and consequently also different sizes in solution. Similar thoughts are to be applied also considering role of solvents in liquid chromatography systems, in which macromolecules necessarily interact with eluent molecules. However, the third component has to be taken into account in polymer HPLC, namely the column packing and its interactions not only with the macromolecules of sample but also with the eluent molecules.

As explained in section 11.5.2.3.1, in HPLC of low-molecular substances with bare silica gel, the extent of eluent interactions with the column packing surface is denoted *solvent strength,* ε^0. Principle of solvent strength approach can also be adopted in polymer HPLC. Retention of macromolecules is suppressed with a *strong mobile phase,* which intensively interacts with the column packing. Very strong mobile phases may completely prevent enthalpic interaction of macromolecules with the column packing surface so that the sample elutes in the ideal exclusion mode (see Figures 3(a) and 13(a)). In connection with the adsorption retention mechanism, such solvents are denoted *desorlis*. Vice versa, the *weak mobile phases* do not obstruct interactions of macromolecules with the column packing and allow their more or less extensive retention. *Very weak* mobile phases, which interact with the column packing surface with a very low intensity may even permit full retention of sample molecules within column. Considering adsorption retention mechanism, they are called *adsorlis*.

The chromatographic strength of solvents related to bare silica gel is closely tied with their polarity. In the first approximation, the higher the solvent polarity the higher its strength. The low polarity eluents combined with the polar bare silica gel originally dominated the area of HPLC of low-moleculars substances. They are denoted *normal phases* to distinguish them from the presently more frequent polar mobile phases and the non-polar column packings that are designated *reverse(d) phases*. Numerous ε^0 values of different solvents toward bare, unmodified silica gel are tabulated.

The typical unitless ε^0 values are as follows: Cyclohexane 0.03; carbon tetrachloride 0.14; toluene 0.22; chloroform 0.31; dichloromethane 0.32; tetrahydrofuran 0.35; acetonitrile 0.50; *N,N*-dimethylformamide high; water >0.73. However, silica gels of diverse origin exhibit both different concentration and topology of the interactive silanol groups and, consequently rather unlike surface properties. Therefore literature values of ε^0 suggest tendencies rather than the universally valid extent of solvent - column packing interaction. Still, they are useful for the first orientation when selecting eluent components to control retention of analytes in coupled polymer HPLC with bare silica based column packings (see section 11.8).

Considering the action of adsorlis in the polymer – adsorbent systems, the situation differs from that observed with low-molecular substances and can be depicted by *the high affinity isotherms* (see Figure 10). In the case of adsorption of low-molecular compounds, their concentration in the supernatant is only exceptionally zero. On the contrary, up to a certain threshold concentration, which is often called *the saturation onset*, actually all molecules of many polymeric adsorbates may be adsorbed so that their concentration in the supernatant is negligible.

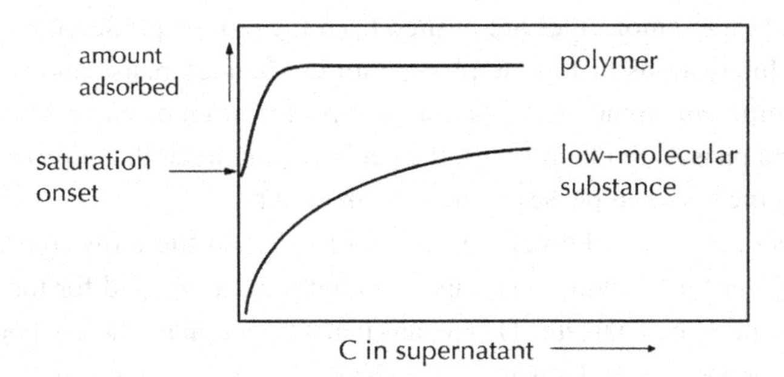

FIGURE 10 Schematic representation of model adsorption isotherms for low-molecular and polymer substances. For detailed explanation, see the text.

The occurence of the complete adsorption of macromolecules forms a base of the *full adsorption-desorption liquid chromatography-like separation method*, FAD (Berek and Nguyen, 1998). A very weak mobile phase is employed, which acts as an adsorli for n-1 sample constituents. They are completely retained within the appropriate *full adsorption-desorption column* packed with nonporous particles. The unretained sample constituent is directly forwarded to an online SEC column for its molecular characterization. In the next stage, eluent strength is stepwise increased and sample constituents are successively one-by-one desorbed and forwarded into the SEC column. Successful separation of up to six distinct polymers with help of FAD method was reported. In the FAD method, solvents are chosen so that they always act as a desorli just for one sample constituent.

To recapitulate, the *eluent strength* decides about adsorption and desorption of given macromolecules onto/from the given solid surface.

The behavior of many low-polarity *bonded phases* and the column packings based on *crosslinked organic polymers* (see section 11.5.2.3.2, Enthalpic Partition of Macromolecules) differs from that of polar column packings such as bare silica gel and the classical solvent strength concept should be re-evaluated. This is especially important for the alkyl bonded phases. In this case, both the surface and the interface adsorption of polymer species (see section 11.5.2.3.1, Adsorption of Macromolecules) play less important role and macromolecules are mainly retained by the *enthalpic partition* (absorption) (see section 11.5.2.3.2). As explained, in order to ensure this kind of retention of macromolecules, the mobile phase must be their poorer solvent than the *solvated bonded phase.* Only in that event, macromolecules are pushed from the mobile phase into the stationary phase. Interactions of mobile phase with the bonded phase and (especially) with the sample macromolecules largely control retention of polymers within the alkyl bonded phases. In other words, the decisive parameter that governs polymer retention in the reversed phases is the solvent quality.

For practical reason, however it is useful to retain the term *eluent strength* also in polymer HPLC with the nonpolar column packings and for the enthalpic partition retention mechanism. This means that a thermodynamically good mobile phase is termed *strong* and a poor mobile phase is denominated *weak.*

The bonded phases carrying polar groups such as $-NH_2$, $-OH$, $-CN$, and so on, which are detached from the packing surface by a *spacer* (usually the *n*-propyl groups) assume intermediate position between the polar solid surfaces and the nonpolar bonded phases though the effect of the polar end groups usually strongly prevails.

Thermodynamic quality of mobile phase may be so poor that macromolecules tend to precipitate within column. The occurence of phase separation is generally undesired in the isocratic coupled methods of polymer HPLC (see section 11.8). Controlled phase separation of macromolecules within column represents a specific feature of certain procedures of polymer HPLC, which are as usual connected with application of the eluent gradient.

In conclusion, two distinct properties of mobile phases are to be distinguished in polymer HPLC namely their *strength* toward the column packing and their *thermodynamic quality* toward both the column packing and the separated macromolecules.

The most important requirements imposed on solvents employed as mobile phases in polymer HPLC are:

- solubility of sample
- chemical inertness toward sample, column packing and also toward material of instrument
- compatibility with the column packing, its wetting
- adjustability of eluting strength – polarity or thermodynamic quality
- miscibility with other potential eluent components
- detector criteria (refractive index, UV or IR transparency, and so on)
- environmental aspects, including toxicity
- viscosity
- availability and price.

Multicomponent eluents are very frequently applied in coupled methods of polymer HPLC with the aim to control interaction between sample and column packing (see section 11.8) They are employed less commonly in SEC, except when they should ensure sample solubility and its stability in solution. Inevitably, mixed eluents cause problems resulting from the phenomena of preferential solvation of sample molecules (see sections 11.2.4 and 11.6.1.4), preferential sorption on the column packing, and preferential evaporation from the eluent container. The latter occurences often give rise to the base line perturbances including the appearance of system peaks. Depending on the particular case, the mixed mobile phases may either enhance or impair the sample detectability.

As mentioned, mobile phases must help control (prevent or enhance) the enthalpic interactions between sample and column packing and suppress the dissociation of macromolecules.

Tetrahydrofuran (THF) is by far the most popular eluent for polymer HPLC and especially for SEC. It possesses high dissolving power for many low- to medium-polarity polymers and it is well compatible with numerous column packings, especially with polystyrene/divinylbenzene gels. Further, THF exhibits high UV transparency, low viscosity and toxicity, as well as fair price. However, tetrahydrofuran easily oxidizes to form explosive peroxides (precaution at distillation!) and it creates charge – transfer complexes with oxygen, which show the UV light absorption. THF is highly hygroscopic and absorbs water from air. The boiling point of THF/water azeotrope, which contains about 5 wt.% of water is only 3 °C below the boiling point of pure THF. This complicates purification of THF with help of distillation. Both *di- and trichlorobenzene* are used with polymers that are soluble only at elevated temperature, especially with polyolefins. *Dimethyl-*

formide is applied in SEC of high-polarity polymers. It dissolves some inorganic salts such as lithium chloride, lithium bromide or potassium rhodanide, and the corresponding solutions are employed as eluents to prevent dissociation of macromolecules that could form polyelectrolytes. *Hexafluoroisopropanol* and *trifluoroethanol* dissolve numerous crystalline polymers such as poly(terephthalate)s. Their price is, however very high. The solvation power of hexafluoroisopropanol is enormous and already its low concentration in a mixed solvent prevents precipitation of some, concerning their solubility, problematic polymers. *Water* and *salt solutions* incl. *buffers* are irreplaceable mobile phases in exclusion chromatography of most biopolymers and water-soluble synthetic polymers.

Similar to SEC, the most commonly employed solvents in the coupled methods of polymer HPLC are *tetrahydrofuran, toluene, dimethylformamide, dichloromethane, chloroform, and cyclohexane. Hexafluoroisopropanol* may help solve problems with the low-solubility polymers.

HPLC grade solvents are quite popular in the HPLC of low-molecular substances. UV absorbing impurities are carefully removed from these solvents, which, on the other hand may contain rather high concentration of water. Naturally, presence of humidity in the original solvent hardly plays any role in the reverse-phase HPLC of low-molecular substances where mobile phases as a rule contain high concentration of water. However, already traces of humidity in the eluent may significantly affect its chromatographic strength in case of polymer HPLC.

11.6.3 MACROMOLECULAR STANDARDS FOR POLYMER HPLC

As stated in section 11.5, methods of polymer HPLC are not absolute. This also holds for the most important one of them, SEC. Therefore, the instruments have to be calibrated or molecular characteristics of analyzed macromolecules that leave the separation column are to be monitored with an independent method (compare sections 11.6.1.4 and 11.7.3.2). Polymers with the known molecular characteristics serve for the column calibration. They are called – not always entirely correctly – macromolecular standards. The sets of polymers with different known molar masses and as a rule with low molar mass dispersity are available from several producers. They have been previously characterized with other physicochemical method(s) or just by the independent SEC measurements. In some cases, molecular characteristics of above standards are derived or even only estimated taking into account method of their preparation. As a result, quality of macromolecular standards provided by some suppliers can remarkably fluctuate and this may constitute a weakpoint of any method of polymer HPLC. The most important application of narrow molar mass dispersity macromolecular standards lies in the construction of SEC calibration dependence *log M versus* V_R (see section 11.7.3.1). There were also attempts to calibrate the SEC columns with help of broad molar mass dispersity poplymers but this is less reliable. The most common and well credible SEC calibration standards are linear polystyrenes, PS, which are prepared by the anionic polymerization. As indicated in section 11.7, according to IUPAC, the molar mass values determined by means of SEC based on PS calibration standards are to be designated "polystyrene equivalent molar masses". Other common SEC calibrants are poly(methyl methacrylate)s, which are important for eluents that do not dissolve polystyrenes, such as hexafluoroisopropanol, further poly(ethylene oxide)s, poly(vinyl acetate)s, polyolefins, dextrans, pullulans, some proteins and few others. The situation is much more complicated with complex polymers such as copolymers. For example, block copolymers often contain their parent homopolymers (see sections 11.8.3, 11.8.6 and 11.9). The latter are hardly detectable by SEC, which is often applied for copolymer characterization by the suppliers (compare Figure 16). Therefore, it is hardly appropriate to consider them standards. Molecules of statistical copolymers of the same both molar mass and overall chemical composition may well differ in their blockiness and therefore their coils may assume distinct size in solution. In the case of complex polymers and complex polymer systems, the researchers often seek support in other characterization methods such as nuclear magnetic resonance, matrix assisted desorption ionization mass spectrometry and like.

11.7 SIZE EXCLUSION CHROMATOGRAPHY

11.7.1 GENERAL CONSIDERATIONS

Various names are used for the method that separates macromolecules according to their size in solution. The term *size exclusion chromatography* is recommended by IUPAC but the original term *gel permeation chromatography*, reserved for the size separation of synthetic polymers in organic solvents is returning to professional literature, while the designation *gel filtration chromatography* is mainly employed for the size separation of biopolymers in aqueous eluents (see section 11.5, Chromatography of Synthetic Polymers). In order to make a difference of

SEC from the *coupled methods of polymer HPLC* also the term *exclusion polymer HPLC* is sometimes applied.

The pioneers of size exclusion chromatography are considered *Porath and Flodin* (1959), who as first published successful separation of proteins on polysaccharide gels. *Moore* (1964) independently employed the exclusion principle to separation of polystyrenes on porous polystyrene beads. *Benoit* (1966) is regarded the "father" of universal calibration and *Casassa* (1967) formulated the idea of entropy control of exclusion based separation of macromolecules. Numerous other researchers contributed to the method development, especially in its first stages.

SEC is presently the most important method for separation and molecular characterization of synthetic polymers. The method enjoys enormous popularity and most institutions involved in research, production, testing and application of synthetic polymers are equipped at least with a simple SEC instrument. Size exclusion chromatograms are often directly transformed into the molar mass dispersity functions (compare section 11.3.3, Molar Mass Dispersity). Often, the molar mass data presented are not absolute, because polystyrene or other polymer "standards" distinct from polymer under study have been employed for the column calibration (see sections 11.6.3 and 11.7.3.1). Still, the data equivalent to the polymer applied to the column calibration, more or less precisely represent the tendencies of molar mass evolution in the course of building-up or decomposition polyreactions.

Numerous textbooks and review articles were devoted to this marvelous method (see the Recommended Readings at the end of this chapter). SEC, however exhibits several weak-points (see section 11.7.5), which are often ignored by good many users. This chapter attempts to present a realistic picture of SEC indicating not only its benefits but also its shortcomings, which deserve further research that would lead to method amelioration.

11.7.2 RETENTION MECHANISMS AND ACCOMPANYING PROCESSES IN SEC

The basic retention mechanisms of SEC were presented in section 11.5.2.2. These mainly include the steric exclusion of macromolecules from the pores of the column packing *controlled* by entropy and *supported* by the flow patterns. The separation procedures that employ exclusively the latter retention mechanisms are often denoted the *ideal SEC*. In other words, the complete absence of enthalpic interactions in the system ($\Delta H = 0$ in Equation (17)) is a prerequisite of ideal SEC. Several undesired processes, however may influence behavior of macromolecules in the SEC columns. These affect both the position and the width of chromatographic peaks and may lead to erroneous results. Moreover, the additional peaks may appear on the chromatograms. One deals with the *real SEC*. The most important contributors to the change-over from the ideal to the real SEC are the following:

1. The secondary retention mechanisms, which result from the enthalpic interactions within the SEC column and cause adsorption, enthalpic partition, phase separation, incompatibility and ionic effects. As shown in section 11.5.2.3, both *adsorption* and *enthalpic partition* decelerate the elution of polymer species and increase the observed retention volumes of analytes. The processes that take place in the first stages of the *phase separation* - that is the aggregation and association of macromolecules - increase effective size of separated species and decrease their retention volumes. The *incompatibility* of polymers is a consequence of the repulsive interactions between the macromolecules of different nature. The incompatibility of synthetic polymers is a general phenomenon of large technological importance: great majority of the mixtures of unlike synthetic macromolecules undergo phase separation. The incompatibility between the column packing and the polymer sample decreases the effective volume of eluent in the packing pores, which is accessible for separated macromolecules and thus it reduces their retention volumes. The incompatibility may play a non-negligible role in the SEC systems, in which the injected polymer solutions are rather concentrated and the macromolecules that form the column packing substantially differ in polarity from the separated species. Ion interactions between the sample molecules and the column packing may either increase *(ion inclusion)* or decrease *(ion exclusion)* retention volumes. Most above enthalpic interactions respond to temperature changes.
2. The *side processes* bring about the changes of sizes of macromolecules in the course of their transport along the SEC column. The size variations may result from the de-coiling of macromolecules due to the flow, as well as from the degradation of polymer species by the mechanical shearing, or by the oxidization – or just by interaction with the solvent

molecules (for example degradation of dissolved poly(hydroxy butyrate). The shear degradation of macromolecules (see section 11.5.2.2, Entropic Processes – Exclusion in Polymer HPLC) can be reduced when applying low eluent flow rates and large (up to 20 μm) packing particles (see section 11.6.1.3.2, Physical Structure of Column Packings). The opposite processes are the aggregation and association of macromolecules within column, especially at the onset of the phase separation (compare ad 1). The continuous measurements of changes in protein size due to association within the column packing were attempted already in the first stage of the SEC development. Interesting results were obtained by the simultaneous monitoring of light scattering and concentration evolution of the zones of macromolecules in-situ that is directly within the column.

The side processes may bring about either an increase or a decrease in the determined retention volumes. Often, the presence of side processes in the SEC system is signalized by the increased dependence of polymer retention volumes on the operational parameters such as the sample concentration, flow rate, and temperature.

3. The *parasitic processes* include the *axial* and *longitudinal diffusion* in the interstitial volume, as well as the local flow irregularities due to the *mixing,* and the *viscosity effects.* Most parasitic processes taking place in the interstitial volume of the well-packed SEC columns do not play important role - except for the viscosity effects, which may cause extensive broadening of chromatographic peaks and appearance of multiple peaks. The injected polymer solutions usually exhibit much higher viscosity than the mobile phase. This difference produces sample zone perturbances in the form of "fingers" and the process is sometimes denoted *viscous fingering.*
4. The *osmotic effects* play an important role mainly at the high injected polymer concentrations. They may selectively affect the retention volumes of smaller macromolecules contained in the polymer sample. The osmotic effects within porous column packings form the basis of interesting preparative liquid chromatographic method denoted the *high osmotic pressure liquid chromatography.*
5. The phenomenon of *secondary exclusion* originates from the higher diffusion rate of the smaller polymer species compared with the larger ones. The smaller macromolecules may timely occupy the packing pores,

which were otherwise accessible for the larger polymer species. As a consequence, the elution rate of large macromolecules is further accelerated.

6. The *concentration effects.* This term describes variations in V_R of the polymer samples due to changes of injected concentration, c_i. The retention volumes as a rule *increase* with the *rising* c_i in the area of practically applicable concentrations (compare section 11.7.5). This is mainly due to the phenomenon of *crowding* when the concentration of macromolecules is high enough so that they touch each other in solution - and shrink due to their mutual repulsive interactions. The concentration effects in SEC are expressed by the slopes κ of the mostly linear dependences of V_R on c_i. The κ values rise with the molar mass of samples – up to the exclusion limit of the SEC column, M_{ex}. The concentration effects for the oligomers and also for the pore-excluded (high)polymer species are usually small or even negligible (compare Figure 15). The κ values also depend on the thermodynamic quality of eluent. Concentration effects may slightly contribute to the reduction of the band broadening effects in SEC because retention volumes of species with the higher molar masses are more reduced than those of the lower molar masses.
   Retention volumes also may rapidly *increase* with the *decreasing* c_i in the area of very low injected polymer concentrations, below the *critical concentration,* c^* - that is when the crowding of macromolecules in solution starts to disappear.
7. The *preferential interactions* of macromolecules with the mixed eluent components. The backgrounds of the preferential solvation of polymer chains in two-component solvents were elucidated in section 11.2.4. In SEC, the consequences of preferential solvation are visible by the non-specific detectors such as differential refractometers, which monitor the *system-peaks* on chromatograms (see sections 11.6.1.4 and 11.7.3.2). Evidently, the system-peaks must be well separated from the sample peaks to avoid their mutual interference and consequent impairment of results. To do so, the SEC column system must contain also narrow-pore packing material. It is well comprehensible that the extraneous peaks may also appear with the eluents that contain special additives, for example stabilizers or (inorganic) salts added to the mobile phase in order to help dissolve the sample, to prevent its degradation or to suppress the undesired ion interactions.

8. The *SEC band-broadening,* which affects the calculated data of the molar mass averages and dispersity, as well of the long-chain branching – as these are directly assessed from the peak width. Especially affected are the M_n (see section 11.3.1) values. The extra-column band-broadening is caused by the diffusion and the flow irregularities in the sample injectors, connecting capillaries, column end-pieces and detectors. Still, the most important source of the band-broadening in SEC is the slow mass transfer within the pores of the column packing. Consequently, the shape of SEC peaks reflects not only molar mass dispersity of sample but also all undesired contributions to the band width. The extent of the inherent band-broadening usually depends on the molar mass of sample. The strictly uniform (incorrectly called "monodisperse") synthetic homopolymers are not available. Therefore, it is difficult to determine the net band-broadening effect and to introduce efficient corrections. The band-broadening phenomenon was substantial with the classical SEC columns that employed large, 30 to 70 μm particles of packing. Several methods were proposed to quantitatively express the SEC band-broadening and to introduce corresponding corrections. The well known is the famous *Tung's method of reversed flow,* which allows elimination of sample polydispersity by reversing elution of macromolecules from the mid of column back to the injector. The band-broadening correction methods were largely abandoned when the SEC column technology was improved and the small column packing particles 10, 5 and even 3 μm in diameter were commercialized. Thanks to the decreased particle size and the improved pore shape of the SEC column packings, the extent of band-broadening was strongly reduced. Still, the ameliorated column technology did not completely eliminate errors in the molar mass values caused by the band-broadening phenomenon. At present, the re-introduction of the band-broadening corrections is attempted in order to improve exactness of SEC measurements. Some companies build simple procedures of band- broadening correction into their software designed for processing the SEC chromatograms. It is hoped that a universal, efficient and user-friendly band-broadening correction software will become available – possibly allowing also compensation of the concentration effects.

To summarize, the most important processes in the SEC column can be divided into:

Basic separation mechanisms – size exclusion from pores of column packing - including also partial exclusion and partial penetration, and from all surfaces present in column – entropy controlled processes

Auxiliary separation mechanisms – that depend on the size of macromolecules - and support the SEC separation: (restricted) diffusion, hydrodynamic processes.

Thermodynamic *secondary separation mechanisms* – enthalpic interactions, which can either support or suppress separation process. They affect retention volumes and if not under control they well complicate data interpretation.

Side separation mechanisms – which are induced by the changes in sizes of separated macromolecules before or in the course of the separation process.

Parasitic processes – which include mixing, both radial and axial diffusion, viscosity effects, broadening and deformation of sample zone, as well as the secondary exclusion.

Combined occurences – for example concentration effects that result from crowding plus viscosity plus secondary exclusion and that cause increase of retention volumes with injected polymer concentration

It is evident that the processes in the SEC columns are intrinsically complex. Complicated are also SEC systems themselves due to *dynamics of macromolecules in solution* which result in permanent changes of their size, shape and orientation because of *intricated pore shapes* and *not well defined pore sizes*

As a consequence, SEC has no comprehesive theory and there does not exist any direct quantitative correlation between pore size (dispersity) of the column packing determined for example by mercury porometry and size (dispersity) of macromolecules, their radii of gyration determined by light scattering of static systems or by viscometry.

This situation also results in the uncertainty of characterization of structure of porous bodies with help of SEC designated "inverse SEC". In other words, SEC is a relative, non-absolute method. Each column (or column system) must be calibrated, or molar mass of species leaving column must be monitored by special "absolute" detectors (light scattering measuring devices, viscometers, mass spectrometers, and so on (see sections 11.6.1.4 and 11.7.3.2).

11.7.3 INSTRUMENTATION AND MATERIALS FOR SEC

11.7.3.1 COLUMNS FOR SIZE EXCLUSION CHROMATOGRAPHY AND THEIR CALIBRATION

General information on the columns for polymer HPLC were presented in section 11.6.1.3. In the following, the specific requirements laid on the SEC columns will be elucidated.

Pore sizes of columns for SEC must match the sizes of dissolved macromolecules. The column also must reliably sort the analyzed macromolecules with lowest molar mass from the molecules of initial solvent and the low-molecular substances present in sample. The latter substances - that is various polymer additives, as well as humidity and gases, mainly oxygen dissolved in sample solution - are recognized with the non-specific detector such as differential refractometer, and the resulting peaks could inferfere with the peaks of sample.

The same problems bring about the system-peaks, which are common to mixed mobile phases. The latter result from the preferential solvation of dissolved macromolecules (see section 11.2.4), from the preferential evaporation of one eluent component from the sample solution, and also from the displacement of the mobile phase component, which is preferentially sorbed on the column packing surface. To stress again, the enthalpic interactions between macromolecules and column packings are to be carefully suppressed (compare section 11.5.2.3). Unfortunately, this is often not the case in the routine laboratory practice.

The column packings, which are inherently well interactive for the polar macromolecules - such as bare silica gel - are only rarely applicable for SEC separations. To avoid enthalpic interactions within the SEC columns, polarity of macromolecules, mobile phase and column packing should be similar. The sense of latter requirement is evident from the scheme in Figure 11, called by its deviser "magic triangle" (Kilz, 2004): The shape of a triangle, which is formed by the lines connecting the polarity of the column packing, the mobile phase and the separated macromolecules should be as symmetrical as possible. To achieve this goal, not only polarity of eluent but also that of column packing should be adjusted to the polarity of polymer characterized.

In the common practice the latter qualification is met only rarely and most SEC separations in organic eluents are done with the low-polarity polystyrene/divinylbenzene column packings. Application of polar mobile phases such as dimethylformamide, DMF with the low-polarity polystyrene/divinyl benzene column packing brings about effect of enthalpic partition (see section 11.5.2.3.2) and causes the excessive rise of retention volumes of low polarity polymers such as polystyrene calibration standards.

This is why the polystyrene/divinylbenzene columns employed with DMF or solutions of salts in DMF as eluents are habitually calibrated with poly(methyl methacrylate), PMMA "standards". Evidently, the resulting values are to be designated *PMMA equivalent molar masses.*

Only few companies supply SEC columns devised for separation of polar polymers in polar eluents. SEC of basic polymers still represents a challenge because interaction of macromolecules with the available column packings is rather strong. On the other hand, hydrophilic SEC column packings are well available. Many acidic polymers can be reliably separated also with help of otherwise well interactive silica gel packings.

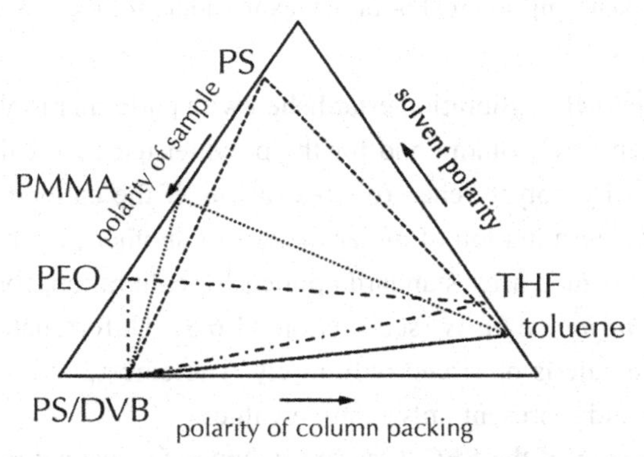

FIGURE 11 "MAGIC TRIANGLE"—the role of polarity of eluent, column packing, and analyzed polymer in SEC. For detailed explanation, see the text.

SEC columns are calibrated with a series of narrow dispersity polymer "standards" with known molar masses (see section 11.6.3, Macromolecular Standards for Polymer HPLC). They are injected either one-by-one or in the packs that contain polymers with sufficiently different molar masses. Retention volumes in the peak apex are located and the dependence *log M versus* V_R is constructed. The schematic representation of the procedure and resulting *SEC calibration dependence* is depicted in Figure 12 (Figure 3(a)).

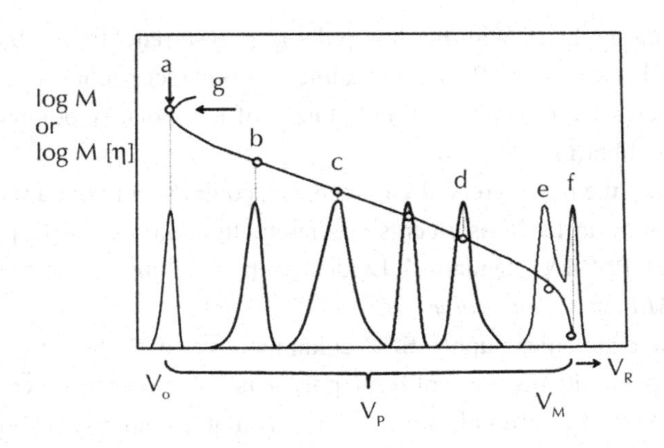

FIGURE 12 Schematic representation of construction of the SEC calibration dependence. M is the peak molar mass (most abundant in the sample), [η] is the limiting viscosity number (intrinsic viscosity) of polymer. $[\eta] = K_{[\eta]} M^a$ (for linear, coiled macromolecules) where $K_{[\eta]}$ and a are constants from viscosity law (see section 11.2.4) for given polymer and given solvent at given temperature. For further explanation, see the text.

The *log M versus V_R* calibration established with particular mobile phase holds only for the given SEC column and for the polymer used for calibration. As indicated, it is usually polystyrene. To stress again, if the calibration dependence is employed for determination of molar masses of another polymer, the obtained values should be named after standard applied for calibration, for example *polystyrene equivalent molar mass* (see section 11.6.3). Unfortunately, this IUPAC coined important rule is observed only rarely. The corresponding molar masses are not absolute and represent only relative values.

Direct processing of the SEC retention volumes for distinct polymers enables *Benoit's universal calibration* (see section 11.7.1) (Benoit et al., 1966). It applies the concept of *hydrodynamic volume*, V_h as the general-purpose parameter that governs retention of macromolecules in ideal SEC. Hydrodynamic volume of a macromolecule is defined as the product of its molar mass M and limiting viscosity number (intrinsic viscosity) $[\eta]$ (see section 11.2.4), in given eluent (compare Figure 12). The universal calibration dependence *log M[η] versus V_R* also can be employed for transfer of data among various eluents -provided geometry of both the bed and the pores of packing remain unchanged, and the enthalpic interactions within the corresponding separation systems are negligible or at least similarly small.

K and *a* values are tabulated for numerous polymer - solvent systems (compare section 11.2.4). Unfortunately, the scatter of the literature data is quite large of K and a. Moreover, the viscosity law is consistently valid only above $M \sim 10$ kg/mol (10,000 g/mol). For macromolecules with lower molar mass, the K and a values may change rather abruptly, or even the conventional form of the viscosity law is no more valid. If the molar masses of samples are situated in vicinity of the above molar mass area, the universal calibration dependence should be employed with precaution. This is the important limitation of the universal calibration concept for oligomers. The constants K and a determined with linear polymers are not valid for branched macromolecules.

Polar interactivity of the SEC columns may cause large errors in the determined molar mass values. The effect is to some extent unexpected for the low-polarity polystyrene/divinylbenzene column packings. Still, these materials may contain polar groups on their surface, which were introduced during production of the column packing or generated in the course of the column use. The latter may attract polar groups situated within macromolecules. The resulting effects are vizualized in Figure 13. The situation can be improved by employing relatively strong, medium polarity mobile phase. Often – but not always – tetrahydrofuran fulfills this requirement. The attractive interactions between column packing and separated macromolecules can be often suppressed by small amount of acidic or basic additives to mobile phase.

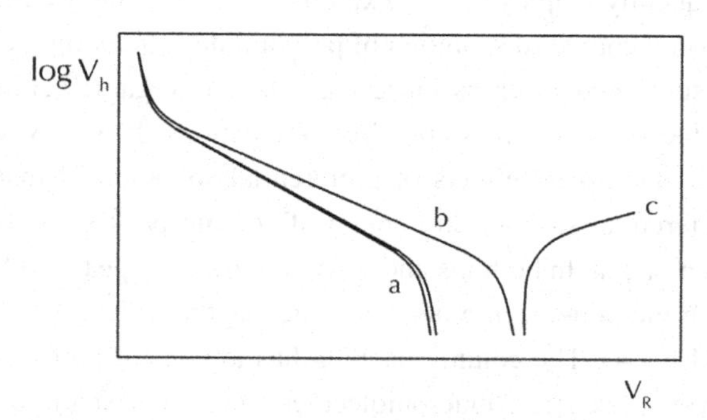

FIGURE 13 Schematic representation of enthalpic interactivity of P(S-co-DVB) SEC columns from different producers: Universal calibration dependences for a polymer of medium polarity in the low-polarity mobile phase or for a polar polymer in the medium polarity eluent are considered. (a) low interactivity columns, (b) medium interactivity column, (c) high interactivity column, in which enthalpic retention mechanism overruns exclusion retention mechanism.

It can be stated that the situation with SEC separation of polar polymers employing PS/DVB column is ambiguous. Polystyrene calibration standards are retained within polar eluents such as dimethylformamide due to their enthalpic partition, and polar macromolecules may be substantially decelerated within low polarity mobile phases due to their adsorption. It is evident that the column packings and mobile phases with matched polarity are to be applied for characterization of polar polymers.

11.7.3.2 MOBILE PHASES AND DETECTORS FOR SIZE EXCLUSION CHROMATOGRAPHY

General issues of mobile phases and detectors for polymer HPLC were in detail discussed in sections 11.6.2 and 11.6.1.4, respectively. In this section, only peculiarities of mobile phases and detectors for SEC are highlighted.

The eluents for SEC have not only to well dissolve separated macromolecules but also to efficiently suppress their interactions with the column packing (see section 11.7.3.1), and to allow polymer detection. The most popular eluent for SEC is tetrahydrofuran, properties of which were expressly described in section 11.6.2, including the danger of presence of both humidity and peroxides. For polyolefins, which are insoluble at ambient temperature, chlorobenzenes are often utilized as eluents. In order to attain sample solubility, and to suppress enthalpic interactions between packing and macromolecules, two- and multi-component mobile phases are frequently employed. For example, small amount of expensive hexafluoro isopropanol enables dissolution of poly(amide)s and aromatic poly(esters). Highly polar substances such as inorganic salts and organic acids or bases can be added to eluents to suppress enthalpic interactions. It is however, necessary to consider the mentioned effects of preferential solvation of macromolecules, as well as preferential sorption in domain of column packing and resulting displacement phenomena. In both instances, system peaks appear on chromatograms monitored with the non-specific detectors such as the differential refractometers (see section 11.6.1.4). The column packing has to contain pores, which are narrow enough to well separate macromolecules from the system peaks, otherwise the results are necessarily biased. Irrespectively of the system peaks presence, the mixed mobile phases also complicate application of some other detection modes, especially monitoring of the flow-through light scattering. The thermodynamic quality of eluent toward polymer a little affects selectivity of separation: the better quality the higher selectivity. On the other hand, the concentration dependence of retention volume and the viscous fingering effects increase with the rising thermodynamic quality of eluent toward separated macromolecules.

The basic information supplied by the SEC detectors is concentration of macromolecules contained in the column effluent. Additional data on eluted macromolecules are welcome such as their molar mass, composition or architecture, however a knowledge of polymer concentration in the column effluent is usually inevitable also for assessment of the later characteristics. The most important devices that measure the SEC effluent concentration are *differential refractometers*, followed with *photometers* - for macromolecules that contain appropriate chromophores, and *evaporative light scattering detectors*. It is necessary to bear in mind that the concentration response of most detectors depends on nature, mainly on chemical structure of separated macromolecules. In some cases, detector response also depends on molar mass of macromolecules. This is right important in the case of oligomers, in which end goups, generally functional groups, play especially notable role. Molar mass of eluting macromolecules can be assessed with the *flow-through light scattering detectors* and with the *flow-through viscometers*. Real, absolute molar mass values are obtained with help of former detectors for homopolymers in single eluents. However, application of above detectors for copolymers and for mixed mobile phases usually provides only molar mass estimates. *Infrared detectors*, especially in connection with appropriate *interfaces*, provide valuable data on the chemical structure of eluted macromolecules. However, as shown in section 11.8, irrespective of detectors employed, SEC alone is only exceptionally suitable for molecular characterization of complex polymers with a dispersity not only in their molar mass but also in their chemical structure.

The combined detectors, which simultaneously monitor not only concentration and molar mass but also chemical composition of eluting macromolecules became quite popular in modern SEC. Most of combined detectors include a differential refractometer, an ultraviolet photometer, a flow-through light scattering monitor and sometimes also a flow-through viscometer.

Monitoring of the *nuclear magnetic resonance* of the SEC column effluent possesses enormous potential but its common employment is hampered with the low sensitivity of instruments, their rather large measuring cells and their high price. For certain time, it was anticipated that *matrix assisted laser desorption-ionization mass spectrometry,* MALDI MS and *the electrospray mass spectrometry* could completely substitute SEC. Soon, the limitations of MALDI MS concerning suitable matrices, chemical structure of detectable polymers and the molar mass range were revealed. Still, mass spectrometries are irreplaceable in the area of oligomer characterization and they can render very valuable information when employed as the additional detection systems in characterization of macromolecules previously separated by means of SEC or other liquid chromatographic technique.

It is to be stressed again that the measuring cell of the detector must be very small. Otherwise, the sample zones that leave the column are extensively broadened. Together with the applicability, sensitivity – and price - this detector parameter also should be considered. The volume of the common flow-through measuring cells is just 10 μL.

In their choice of instruments, the operators are almost entirely dependent on the producers of detectors. There is no univeral apparatus for polymer HPLC and no universal detector for SEC. It is important to plan future use of the instrument before its purchase.

11.7.4 APPLICATIONS OF SIZE EXCLUSION CHROMATOGRAPHY

The main task of SEC is determination of molar mass averages and dispersities. As it was repeatedly emphasized, the obtained values are strictly exact only for linear homopolymers because in this case the retention volume of macromolecules can be directly and quantitatively related to their molar mass. Retention volumes of complex polymers only reflect relative values of molar mass (see section 11.8). However, even the estimates of molar masses are often important for assessing tendencies of the course of various polyreactions. SEC also renders other valuable data on polymers such as their long-chain branching, limiting viscosity numbers (intrinsic viscosities), intermolecular interactions, and radii of gyration in solution, futher on their preferential solvation in mixed solvents, as well as on particulars of mutual association and aggregation of macromolecules, and on estimates of their diffusion rate in porous bodies. SEC can also afford notable data on conjugates of proteins with drugs. The detailed analysis of all SEC applications significantly exceeds the range of this chapter and the potential users of the method are referred to the specialized literature, see the Recommended Reading at the end of this chapter.

The procedure of molar mass values determination from a SEC chromatogram is demonstrated in Figure 14.

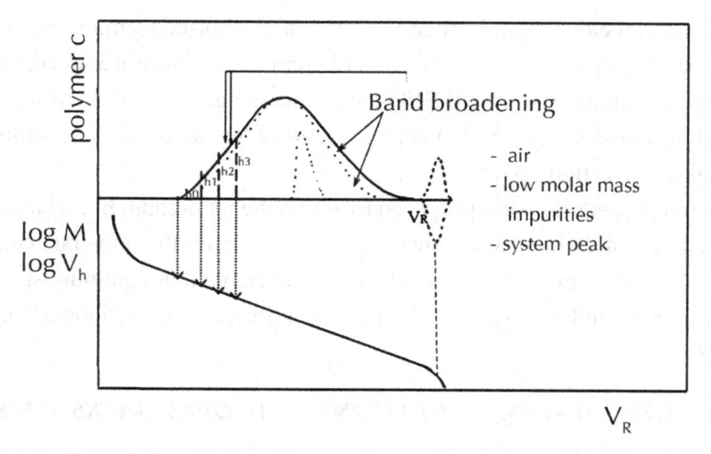

FIGURE 14 Schematic representation of processing of a SEC chromatogram, the calculation of molar mass averages. For detailed explanation, see the text.

Chromatogram is divided into the narrow slices. Surface or height of each slice *i* is measured. Concentration of polymer within the slice *i*, C_i is calculated from equation

$$C_i = A_i / \Sigma A_i \tag{21}$$

where A_i is the surface or height of particular slice *i*. Next, the molar mass of macromolecules within each slice is read from the calibration dependence and the molar mass averages are calculated from equations

$$\overline{M}_n = \frac{1}{\sum_1^k c_i / M_i} \tag{22}$$

and

$$\overline{M}_w = \sum_1^k c_i M_i \tag{23}$$

where M_i is the molar mass of macromolecules in the slice *i*.

The manual calculation is of course time- and labor-consuming. At present, it is done practically exclusively with help of computers. There are several software available for computer-assisted SEC data processing. Most of them include application of universal calibration dependence and few also comprise simple band broadening correction option.

Several procedures were proposed to assess the long-chain branching of macromolecules with help of SEC. They are based either on the universal calibration concept or on the employment of "absolute" detectors that continuously monitor light scattering and viscosity of the column effluent (see sections 11.6.1.4 and 11.7.3.2).

11.7.5 ADVANTAGES, LIMITATIONS AND DRAWBACKS OF SEC

It should be said over-and-over again that SEC is an ingenious method, which deeply affected polymer science and technology. The method is fast, simple, relatively cheap, and it renders high intra-laboratory repeatability of results so that it exhibits remarkable *precision* of measurements. Strange enough, this latter feature of SEC may also be considered a problem! Several times repeated measurements of particular sample by the same operator, with the same instrument and under equivalent experimental conditions - that provide almost identical molar mass values may lead to a notion about the high data *accuracy*. However, the round-robin tests organized under auspices of IUPAC have indicated rather low inter-laboratory reproducibility of common SEC measurements: Selected polymer samples were dispatched to several laboratories and the participants performed measurements without any strict protocol. The resulting data exhibited unexpectedly large scatter.

The best agreement in molar mass values was recorded in the laboratories of companies, while the largest irregularities were attained in the universities. Surprisingly better reproducibility of data that is a higher results accuracy was reached with "the difficult polymers", for example with polyamides and polyolefins, which were measured at elevated temperature. It is evident that experts that dared to measure the difficult polymers made much fewer errors than the beginners, usually students, that analyzed the well soluble, "easy" polymers of low polarity. It was concluded that the mediocre inter-laboratory reproducibility, which signalize inadequate accuracy of many SEC results may be caused by the approach that can be designated: "Switch-on, inject, switch-out – computer provides some data."

The above observations clearly demonstrate the necessity of standardization of both measurements and data processing. Special attention deserves appropriate baseline and peak limit setting. As a rule, SEC software responds to the abrupt changing course of the chromatographic peak, to larger base-line and peak shape variations and it may not recognize gradual convergence between the chromatographic peak and the base-line (tailing peak), which often indicates presence of a non-negligible amount of macromolecules with lower molar mass. Software may not identify minor peak irregularities and the overlap of sample peak with the system-peaks. The latter may cut-off substantial part of chromatogram with high retention volume (compare Figure 12). Therefore, the visual check of each chromatogram is highly recommended before drawing quantitative conclusions from the computer-produced SEC data.

It is also necessary to note that the assessed molar mass dispersity of a polymer sample can be significantly affected by the accidental or intentional erroneous baseline and peak-limits setting. A small deliberate upward shift of baseline brings about substantial “improvement” of calculated molar mass dispersity of the product of synthesis. Small bulges can be frequently seen on the published SEC chromatograms. Their presence is often ignored. They may indicate either bimodality in the molar mass dispersity of the sample or presence of an admixture with the hydrodynamic volume similar the sample. In some cases, such admixture may not at all belong to the polymer under study.

It must be stressed again that unwanted and uncontrolled enthalpic interactions that may appear in the SEC column affect retention volumes of samples. One speaks about *exclusion – interaction mixed retention mechanism*. Often marginalized are also the successive changes in column both pore size (effective diameter) and volume due to retained fractions of analyzed samples.

A threshold situation is the full occupation of the possible active sites on the column packing surface with adsorbed macromolecules. The resulting drop of pore volume can cause large decrease of retention volume of the nonadsorbed polymer and in effect bring about substantial increase of its apparent molar mass.

To exclude effect of the changing properties of the columns and also to rule out the inadequacies in the pump operation, the SEC systems are to be frequently recalibrated and the eluent flow rate controlled after each restarting the pump. The latter requirement may be imperative with the low-cost pumping systems, which often exhibit limited resetability. It is generally appropriate to perform the verification measurements on the daily basis utilizing a known, well characterized broad-dispersity test sample and to continue serial measurements only when the molar mass of such "calibrant" is validated.

A specific drawback of SEC is the considerable widening of chromatographic zones, called band-broadening, which was outlined in section 11.7.1.

Injected sample concentration c_i and volume v_i, are frequently ignored experimental parameters, which affect polymer retention volumes. The increase in c_i or v_i causes rise in polymer V_R (see section 11.7.1). The practical sample volume and concentration depends on the detectability of polymer concerned, for example on its refractive index increment in the case of refractive index detector.

The schematic display of influence of c_i on the polymer retention volume is depicted in Figure 15.

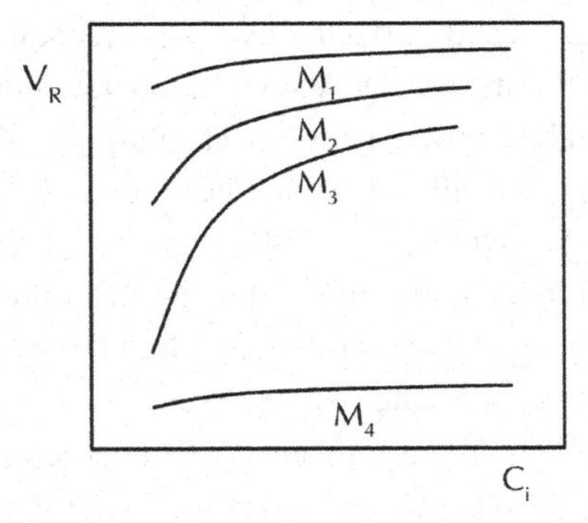

FIGURE 15 Schematic representation of effect of sample concentration on the corresponding retention volume in SEC. M is polymer molar mass, M_1 is low molar mass, macromolecules are eluted not far from V_M, $M_1 < M_2 < M_3$, M_4 denotes a high molar mass, macromolecules are eluted near V_0. For detailed explanation, see the text.

The practical hints that follow from above notes and Figure 15 can be summarized as follows:

1. v_i should be similar for the calibration standards and for the sample. Apparently, this advice is convenient to follow experimentally, however it may complicate treatment of wide and flat chromatograms of broad molar mass dispersity polymers. The effect of v_i on the peak apex position of narrow calibration standard differs from the influence of v_i on the retention volumes of fractions comprised in broad molar mass dispersity sample. Quantitative evaluation of the above difference is a complex matter

and the resulting improvement of results may be rather small. The adjustments of injected sample volume requires employment of an autosampler

2. A potentially similar situation occurs regarding the effect of c_i on V_R. In this case, however the difference between the narrow calibration standard and (broad) polymer sample is much more pronounced. Compared to the calibration standards, it is useful to increase the sample concentration two- to five-times. The distinction between concentration of sample and calibration standard is to be kept constant for entire series of measurements of similar samples.

Appropriate injected sample concentration and volume for a column with ~ 8 mm diameter can be roughly estimated from the empirical formula:

$$c_i\,[\text{mg.mL}^{-1}] = 10^{-3} \text{ x column length [mm]}/v_i\,[\text{mL}]$$

For a common SEC system with one column of (250 to 300) mm length and (7.5 to 10) mm diameter, typical v_i is (50 -100) μL and c_i equals 1 – 5 $mg.mL^{-1}$. c_i should be decreased as M rises over 100 $kg.mol^{-1}$, and especially over 1,000 $kg.mol^{-1}$. Correspondingly, v_i is to be increased. Generally, c_i is to be raised for broad molar mass distributed samples. The latter specifications are only approximate and the actual optimum experimental conditions depend mainly on the sample detectability. For example, in the case of refractometric detector, sample detectability is dictated by the refractive index increment, *dn/dc* for the given polymer in eluent. In any case, the c_i and v_i should be kept as low as possible. It is useful to start preliminary experiments with a higher sample concentration and to decrease it gradually down to the experimentally bearable limit, which is mainly dependent on the base line stability of chromatogram.

When c_i and/or v_i exceed certain limiting value, the SEC column becomes overloaded. The peaks are enormously broadened and the selectivity of separation is lost.

As known, SEC separates macromolecules according to their size in solution. Molecular sizes and the SEC retention volumes of complex polymers (copolymers, functional polymers) depend on all molecular characteristics of polymer. It is therefore evident that SEC cannot give quantitative data on molar masses of complex polymers that exhibit more than one dispersity (distribution) in ther molecular characteristics.

From its principle, SEC suffers from the limited separation selectivity because the retention volumes of samples are restricted by V_o and V_m values (Figure 3(a) and 12). This drawback prevents application of SEC to quantitative characterization of numerous complex polymer systems that contain macromolecules of distinct chemical structure or physical architecture possessing similar molecular sizes (compare Figure 16). In turn, low sample capacity and often insufficient sensitivity of detection makes it impossible to identify and characterize the minor components of complex polymer systems that are present in a matrix of a major constituent in the amount below about 10% - even if molar masses of minor and major constituent differ substantially. Still, SEC is applied in many laboratories for just the above- mentioned purposes. To solve the latter analytical challenges, coupled and two-dimensional methods of polymer HPLC are to be employed (see sections 11.8 and 11.9).

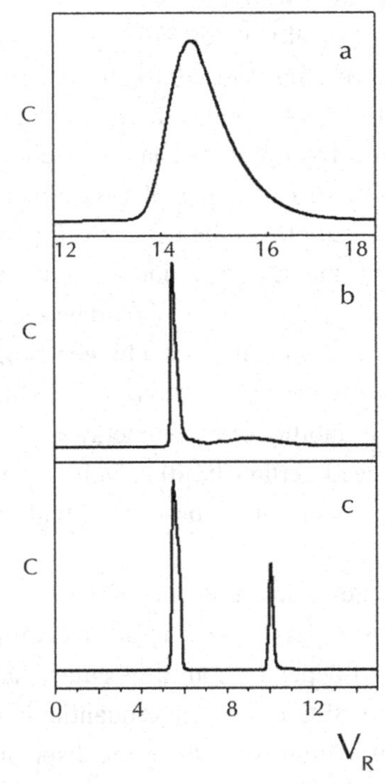

FIGURE 16 Real chromatograms of model 4:1 blends of polystyrene and PMMA. Molar masses of narrow molar mass dispersity polymers were 270 and 293 kg mol^{-1}, respectively. For detailed explanation, see the text.

11.8 COUPLED METHODS OF POLYMER HPLC

11.8.1 GENERAL CONSIDERATIONS

According to the terminology proposed by IUPAC, the term *hyphenation* is reserved to the combinations of different physical principles for analytical purposes, for example the simultaneous monitoring of refractive index and light absorption in the case of polymer HPLC. On the other hand, the term *coupling* denotes the combinations of distinct analytical procedures based on the same general physical principle. In the instance of polymer HPLC, the term coupling designates the combination of *different retention mechanisms* within the same HPLC column. When applying the porous column packing the entropy controlled retention mechanism, exclusion, is always present. The addition of any enthalpy driven retention mechanism leads to a *coupled method of polymer HPLC* (compare Equation (16)). Practically all *interaction* or *interactive* procedures of polymer HPLC are in fact the coupled methods. The general aims of the deliberate coupling of the retention mechanisms within the same polymer HPLC column are:

1. Either to substantially increase the separation selectivity according to certain molecular characteristic so that the effect of other characteristics can be – at least in the first approximation – neglected. For example, the effect of molar mass can be so much enhanced that the impact of other molecular characteristics of macromolecular sample on its retention volume can be ignored. or
2. To suppress the effect of certain molecular characteristic on sample retention volume so that the resulting chromatogram reflects mainly or even exclusively other molecular characteristic(s) of sample. In practice, it is usually attempted to partially or fully suppress the influence of polymer molar mass. In this instance, the coupling of LC retention mechanisms may allow assessment of chemical structure or physical architecture of a complex polymer irrespective of its molar mass average and dispersity. Under favorable conditions, also the constituents of a complex polymer system with similar molar masses can be discriminated and molar mass of one constituent determined. For example, in the case of a two-component polymer system, the molar mass effect can be suppressed selectively for one constituent so that it elutes in a completely different retention volume compared with the retention volume pertaining to SEC. In some cases, the

molar mass of the second constituent can be assessed from its simultaneous elution in the SEC mode.

A more general approach is represented by the *two-dimensional procedures of polymer HPLC* (see section 11.9), in which a coupled method of polymer HPLC is employed as an important (usually the first) separation step followed by SEC. This means that in the two-dimensional polymer HPLC a tandem of two distinct separation *systems* is needed. The three-dimensional polymer HPLC is still only in the incoming stage. The task of both coupled and two-(multi-) dimensional precedures of polymer HPLC is the comprehensive molecular characterization of complex polymers and complex polymer systems.

Practical liquid chromatography methods based on the coupling of entropic and ethalpic retention mechanisms in the same polymer HPLC column will be briefly elucidated in present section. The two-dimensional polymer HPLC will be discussed in Section 11.9.

11.8.2 ENTHALPY DRIVEN POLYMER HPLC

In this case, the enthalpic interactions within the HPLC system exceed the exclusion effects (see Figure 3(e)). The retention volumes of polymer species as a rule exponentially increase with their molar masses. The important limitation of the resulting procedures was presented in section 11.5.2.3. The retention of (high) polymers is usually so intense that the latter do not elute from the column any more. Therefore, the majority of enthalpy controlled HPLC procedures is applicable only to oligomers – up to molar mass of few thousands g.mol^{-1}. Still, the reduced sample recovery may affect results of separation even in case of oligomers. The selectivity of enthalpy driven HPLC separation is much higher than in the case of SEC but, naturally, the sequence of molar masses eluted from the column is reversed. If the effect of enthalpy is reduced, problems with sample recovery are mitigated – but at the same time the separation selectivity is reduced.

11.8.3 LIQUID CHROMATOGRAPHY UNDER CRITICAL CONDITIONS OF ENTHALPIC INTERACTIONS

The principle of *liquid chromatography under critical conditions of enthalpic interactions* (LC CC) was elucidated in section 11.5.2.3. Mutual compensation of the exclusion – entropy, and the interaction – enthalpy based retention of macromolecules (see Figure 3(d)) can be attained when the interactions that lead to either adsorption or enthalpic partition of sample are in the controlled way added to the exclusion retention mechanism. The resulting methods are called *LC at the critical adsorption point* (LC CAP) or *LC at the critical partition point* (LC CPP), respectively. The important feature of LC CC is that the elution is strictly isocratic, composition of mobile phase is constant and sample is dissolved and eluted in eluent. These requirements are sometimes neglected and this may lead to the unexpected results (see section 11.8.6). The procedures, in which the above conditions are not fullfiled should not be designated LC CC. The compensation of exclusion and adsorption also takes place in other coupled polymer HPLC procedures (compare sections 11.8.4 and 11.8.6) for which the general term *LC at the point of exclusion – adsorption transition* (LC PEAT) was proposed. It is anticipated that not only adsorption and enthalpic partition but also other kinds of enthalpic interactions, for example the ion interactions between the column packing and the macromolecules can be utilized for the exclusion – interaction compensation.

LC CC principle was disclosed by Belenkii and his coworkers (1976) in Skt. Petersburg. Other scientists from Skt. Petersburg, Skvortsov and Gorbunov (1986) presented theoretical explanation of the unexpected experimental results, when the retention volumes became independent of polymer molar masses. They introduced the concept of "chromatographic invisibility" for polymer chains eluted from porous packings irrespectively of their molar mass. In fact, the detector does "see" all macromolecules and the term "invisibility by column" seems to be more appropriate than the denomination "chromatographic invisibility". Gorbunov and Skvortsov (1995) also proposed several original applications of LC CC: molecular characterization of functional oligomers, and the assessment of molar masses of either a block in the block copolymer or a constituent of the polymer blend, which would co-elute with other constituents under the SEC conditions.

The above researchers are considered founders of LC CC.

Numerous LC CC applications were tested in Moscow, Russia, in Berlin and Darmstadt, Germany, and in Graz, Austria (for review see Pasch and Trathnigg, 1998). Presently, the LC CC method is rather widely employed in separation and molecular characterization of various complex polymers and of binary polymer systems.

In order to readily control enthalpic interactions within the liquid chromatography column, usually two-component mobile phases are utilized. One solvent suppresses and another one promotes enthapic interactions, either adsorption or enthalpic partition within the chromatographic systems. Mixing both solvents in appropriate proportion allows adjusting the critical conditions. If necessary - for example because of the solubility problems - multicomponent mobile phases can be applied. Enthalpic interactions between macromolecules and column packing employing a single mobile phase can be controlled by temperature. However, the applicable range of temperatures is usually insufficient. Still, temperature can be employed for the fine tuning of interactions.

Unfortunately, the ingenious and very useful idea of entropy – enthalpy compensation and the resulting method of LC CC also exhibits numerous shortages. These include:

1. The extremely high *sensitivity of critical conditions* to the eluent composition and possibly also to the continuous variations in the *column interactivity*, for example due to the irreversible retention of some sample components. The sensitivity of critical conditions rapidly increases with the molar mass of macromolecules concerned. It may be very difficult to repeatably prepare mixed eluents with exactly the same composition, as well as to prevent preferential evaporation from-, and absorption of humidity into- the eluent container. The columns should be frequently cleaned by flushing with an efficient displacer and then (laboriously) re-equilibrated with the original eluent. Often, the *log M versus* V_R dependence is not perfectly vertical and the exact critical conditions cannot be reached at all. Therefore, many measurements are done near the LC CC conditions, not at the proper *critical point* but in the *critical range.* This brings about additional uncertainty on accuracy of results.
2. The sensitivity of critical conditions to *temperature and pressure.* As it was shown in section 11.5.2.3 the adsorption and even more the enthalpic partition of macromolecules depends on temperature. Temperature together with pressure also affects the composition of the mixed eluent within both the sample polymer chains and the column packing. Though, as stated above, temperature impact on the interactions is usually not very large, it still can result in the undesirable departure from the critical conditions. It is important to remember that neither temperature nor pressure within column are strictly constant. They exhibit certain gradients. The friction between eluent molecules and column packing produces heat, which increases temperature within the packing. From principle, pressure markedly decreases along any HPLC column. As a result, the column interactivity may exhibit important positional distinctness. Fur-

ther uncertainty in temperature and pressure gradients within column is brought about by the accidental variation of experimental conditions. For example, the partially blocked column end-pieces may cause significant pressure changes in the course of the LC CC experiment. Unwillingly, one often works in the critical range rather than at the critical point.

3. The *excessive peak broadening.* Under critical conditions, the effect of molar mass dispersity of sample disappears. Therefore, the peak width should only reflect the mixing and diffusion occurences in the column end-pieces, connectors, detector, pores of packing and the interstitial column volume, as well as the processes taking place during the actual chromatographic process. The latter seems to be rather slow in the case of LC CC and this is likely the reason of the increased peak broadening, which rises with the rising polymer molar mass and with the decreasing packing pore size. The LC CC band broadening is demonstrated in Figure 16, in which the real chromatograms of a model polymer blend of polystyrene, PS and poly(methyl methacrylate), PMMA with similar both molar mass averages and dispersities are compared as obtained with SEC, LC CC and liquid chromatography under limiting conditions of desorption, LC LCD (see section 11.8.6). It is evident that selectivity of the linear, universal PS/DVB SEC columns is too low to even partially separate macromolecules with similar molar masses (Figure 16(a)). The resulting SEC peak is broadened due to overlapping of peaks of distinct polymers with slightly different molar masses. LC CC and LC LCD separations were performed with narrow pore columns from which the polymers were excluded. The LC CC well separates both homopolymers but the peak of critically eluted PMMA, which is "invisible for the column", is quite broad. Evidently, most PMMA was retained within the LC CC column. By contrast, the peak of LC LCD eluted PMMA is much narrower (see Figure 16(c), compare also section 11.8.6).

4. The *detection problems.* Due to preferential solvation of macromolecules in mixed solvent (see section 11.2.4), the composition of the bulk solvent among polymer chains differs from the composition of the original sample solvent, which is mobile phase. In LC CC, macromolecules elute together or in the vicinity of their bulk solvent. This prevents application of the non-specific detectors such as differential refractometers (see sections 11.6.1.4 and 11.7.3.2), which sensitively respond to changes in

the overall effluent composition. Even employment of photometric detectors with the non-absorbing eluent components and macromolecules bearing appropriate chromophore may be complicated by the phenomenon of preferential solvation. This is why the most commonly used LC CC detectors are evaporative light scattering devices, ELSD (see section 11.6.1.4). ELSD, however suffer from the limited linearity of response. Their response also depends on eluent composition and partially on the sample molar mass. The detection problem is especially serious in the case of LC CC of block copolymers because ELSD simultaneously monitors both coeluting blocks. As a result, the resulting concentration response of detector depends on the nature and the relative length of both blocks.

5. The *limited sample recovery* (compare Figure 16(b)). Part of LC CC eluted sample may remain trapped within the column. The irreversible sample retention increases not only with the increasing sample molar mass but also with the decreasing packing pore size. The retained macromolecules may extensively affect interactive properties of the column (compare also section 11.7.3.1). Unfortunately, the problem of reduced sample recovery in LC CC of (high)polymers is mostly ignored. In order to qualitatively check the sample recovery, the second injection valve with a large loop can be mounted between the pump and the sample injector. The loop of this additional valve is filled with a liquid efficiently promoting polymer elution that is with a potent displacer. Its injection unveils whether a portion of sample was retained within the column in the course of the LC CC experiment. Evidently, the detector must not respond to the displacer itself. Alternatively, the LC CC column is substituted with a long capillary and the peak surfaces obtained with the LC CC column and with the capillary are compared. For the quantitative evaluation of impact of limited sample recovery, a non-interactive SEC column is to be online attached to the LC CC column. Molar mass of polymer under study leaving the LC CC column is measured with help of the SEC column. The difference in molar mass averages and dispersities determined with help of SEC with and without presence of the LC CC column indicates imperfect sample recovery. The highest molar masses are often selectively retained within the LC CC column.

LC CC proved especially useful in the characterization of functional oligomers. Under critical conditions, the main chain of the oligomer is eluted irrespectively of its molar mass and the observed variation of retention volumes results from both nature and amount of the functional groups. In many cases, even topology of chemically identical functional groups can be assessed with help of LC CC. It is important to stress again that the LC CC sample recovery is much less problematic with the oligomers than with the (high)polymers. Also the LC CC separation and molecular characterization of binary polymer blends and block copolymers attracted rather wide attention. As has been indicated above, one constituent of the polymer blend under study is subject to critical retention and thus it is separated from another constituent, which is eluted in the SEC mode. If the pore size of the LC CC column allows, the molar mass of the unretained blend constituent can be calculated in the conventional way. Similarly, LC CC allows separation of the interactive parent homopolymer from the block copolymer that contains a non-interactive second block. The interactive homopolymer and one block of the block copolymer are made "invisible for the column". At the same time, the noninteractive block is eluted in the SEC mode so that its molar mass can be supposedly estimated without effect of the "invisible" block. However, the unretained parent homopolymer – if present – is coeluted with the block copolymer and it can well affect the results. Moreover, the detector usually responds to both blocks (compare above, ad. 4). A new set of LC CC experiments with a distinct both column packing and eluent that is by employing different retention mechanism is needed to separate second parent homopolymer from the block copolymer. This likely allows characterization of the second block of the block copolymer. However, the chains of the noninteractive part of the block copolymer "dilute" the interactive blocks and may bias its "critical" retention. Vice versa, the interactive block may affect hydrodynamic volume of the SEC eluted chains. The general shortages of LC CC, which were outlined above further complicate quantitative molecular characterization of the block copolymers. Nevertheless, LC CC can produce quite valuable information on the block copolymers, for example concerning presence and amount of parent homopolymers. However, section 11.8.6 demonstrates that the latter data can be obtained much easier with help of liquid chromatography under limiting conditions of enthalpic interactions.

The application of LC CC to the graft copolymers seems to be even more problematic. The densely grafted copolymers with the long side-chains often behave in solution similarly as the grafts alone. The reasonable results were reported only for the graft copolymers with the short or sparse side-chains.

The concept of entropy-enthalpy compensation that results in the critical conditions of enthalpic interactions and the molar mass independent sample retention turned out useful for the understanding of several other coupled methods of polymer HPLC. It is accepted that macromolecules tend to elute under critical conditions also when either strength or quality of mobile phase is changed within the HPLC system in the course of the HPLC experiment that is in the procedures that employ continuous or local gradient methods (see sections 11.8.4 and 11.8.6, respectively). Apart from shortages and limitations of LC CC, its concept belongs to the important breakthroughs in polymer HPLC.

The identification of critical conditions for the given polymer - column packing system necessitates a series of preliminary measurements. Either the eluent strength (in case of adsorption retention mechanism, see section 11.5.2.3.1) or the thermodynamic quality (in instance of enthalpic partition retention mechanism, compare section 11.5.2.3.2) is tentatively adjusted to attain the molar mass independent elution of polymer under study at constant temperature. The retention volumes are measured for at least two different molar masses in the given column using eluents of different compositions. The plots *retention volume versus eluent composition* for particular molar masses are constructed. Their intercept indicates the vicinity of the critical conditions. It is advised to successively *increase* the elution power of the mobile phase to prevent slow column equilibration. Further, it is to be emphasized again that samples must be always dissolved in particular eluent, because otherwise different elution patterns are obtained instead of LC CC - *viz.* liquid chromatography under limiting conditions of enthalpic interactions, (LC LC) (see section 11.8.6). A useful survey of published LC CC systems was compiled by Macko and Hunkeler (2003)

11.8.4 ELUENT GRADIENT POLYMER HPLC

Porath (1962) proposed to employ a gradient in the mobile phase composition for separation of macromolecules. The experimental basis for *eluent gradient polymer HPLC* (EG-LC) was laid in the thin layer liquid chromatographic arrangement independently in Kyoto by the team led by Inagaki (1968) and in Skt. Petersburg by the Belenkii's group (1969). Teramachi (1976) was the first to publish the successful polymer separations of polymers in the column arrangement applying adsorption retention mechanism. Glöckner (1992) pioneered the column

EG-LC polymer separations with the phase separation (precipitation – redissolution) retention mechanism. These authors can be considered founders of EG-LC of synthetic polymers.

Similar to other coupled methods of polymer HPLC, the choice of the column packing and the mobile phase components for EG-LC depends on the employed retention mechanism. Adsorption is preferred for polar polymers with polar column packings, usually bare silica or silica bonded with the polar groups. The *eluent strength* controls polymer retention (see section 11.5.2.3.1). The enthalpic partition is the retention mechanism of choice for the nonpolar polymers or polymers of low polarity applying the nonpolar or low polarity column packings, typically silica gel C18 bonded phases (see section 11.6.1.3.3). In this case, similar to the phase separation mechanism, mainly the *eluent quality* governs the extent of retention (see sections 11.5.2.3.2). It is to be reminded that even the low-polarity polymers such as polystyrene may adsorb on the surface of bare silica gel from the very weak, nonpolar mobile phases and vice versa, the polymers of medium polarity such as poly(methyl methacrylate) can be retained from their poor solvents (eluents) due to their enthalpic partition within the nonpolar alkyl-bonded phases.

In EG-LC, sample is injected into the porous column packing flushed with the mobile phase, which prevents elution of macromolecules. Depending on the retention mechanism employed, the eluent is an adsorli, a poor solvent or a nonsolvent. Subsequently, strength or quality of eluent is continuously or stepwise increased.

Macromolecules of different chemical structure or architecture successively commence eluting along the column and eventually leave the EG-LC column in distinct retention volumes. Sample solvent must be weak enough or poor enough to avoid premature start of elution called the sample break-through. The breakthrough appears when the sample solvent, which is different from eluent effciently prevents the full retention of macromolecules near the column inlet. In this case sample constituents jointly travel along the column within the zone of the original solvent without separation (compare section 11.8.6).

The rationale of EG LC is evident from Figure 17, in which application of the linear eluent gradient is depicted. Different gradient slopes and shapes can be employed to adjust appropriate selectivity of separation.

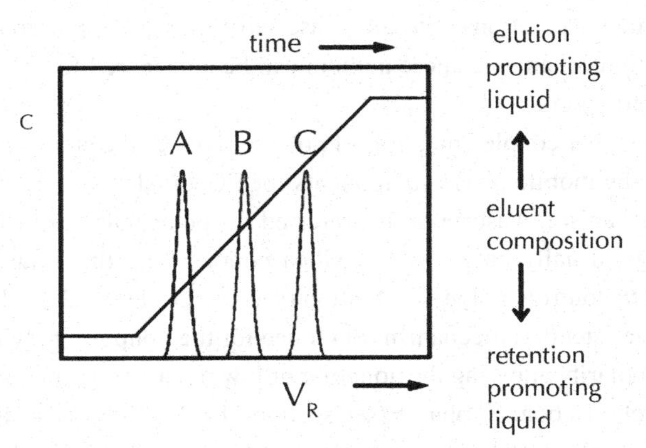

FIGURE 17 Schematic representation of EG LC principle. A fictitious three-component polymer blend A, B, and C is separated by means of the continuous linear eluent gradient. For further explanation, see the text.

Various semi-quantitative theories were developed to predict the course of polymer elution in EG-LC. For a simplified qualitative elucidation of the elution patterns, let us consider the adsorption retention mechanism. The injected sample is fully retained by adsorption near the column inlet. As the eluent strength is increased, the sample starts desorbing and eluting. It is supposed that the macromolecules with the lowest molar masses desorb first. Molecules of eluent permeate the packing pores and proceed along the column with a low velocity. The released macromolecules elute rapidly as they are partially of fully pore excluded. Eventually, macromolecules of particular adsorptivity reach the eluent composition, which is just weak enough to decelerate their continuing fast elution. It is likely the critical eluent composition. From this moment on, the polymer species of the same nature keep moving together, irrespectively of their molar mass. Macromolecules of a distinct adsorptivity that is of distinct nature – composition or architecture – are eluted from the column packing at another eluent composition, within a different retention volume. The "barrier effect" of the eluent gradient is responsible for the peak focusing, which is typical for many well-designed EG-LC systems based on adsorption and enthalpic retention mechanisms. In the case of the phase separation retention mechanism, the situation is often less favorable, mainly due to the slowness of the dissolution processes (see section 11.2.4). Moreover, the concept of critical conditions seems not to apply to the phase separation retention mechanism. Retention volumes of macromolecules depend on their molar mass so that the interpretation of chromatograms is difficult. On the other hand, selectivity of separation is very high and the method may provide valuable data when combined with appropriate detection option, for example with MALDI MS (see section 11.6.1.4).

EG-LC was successfully employed in many polymer separations, especially in those aimed at the characterization of polymer blends and statistical copolymers. The important shortage of numerous EG-LC procedures is the reduced sample recovery, which results from the full retention of a part of sample within the column packing. This is rather comprehensible because the deep pores of the column packing may stay equilibrated with the retention promoting mobile phase for a quite long time. Retained macromolecules may form the "*flower-like conformation*" (see section 11.5.2, Processes in Columns for Polymer HPLC – Retention Mechanisms) and the "crown of the flower" blocks pore entrance preventing accession of eluent with increased elution strength. Surprisingly, the retained macromolecules are released when the blank gradient is realized. As a result, the chromatograms with similar or even identical peak retention volumes are generated without injection of any sample – just by the repeated running of the eluent gradient. The amount of released sample is small but not negligible. This unexpected phenomenon needs further study. It can perturb serial analyses.

In conclusion, eluent gradient polymer HPLC is a useful tool for separation of complex polymer systems. It belongs to the important constituents of the two-dimensional polymer HPLC procedures (see section 11.9, Two-Dimensional Polymer HPLC).

11.8.5 TEMPERATURE GRADIENT POLYMER HPLC

Chang with coworkers (1996) in Pohang reported the high selectivity separations of some homopolymers according to their molar mass applying the *temperature gradient* within the polymer HPLC columns. This is considered the birth of the method, which is called *temperature gradient polymer HPLC* or *temperature gradient interaction chromatography* (TGIC). In TGIC, the column temperature is increased continuously or in the (small) steps. It is likely that - similar to the eluent gradient polymer HPLC - the fast elution of polymer species in the warmer eluent is decelerated by the slowly moving molecules of the cooler eluent. The barrier of cooler eluent is presumably at its critical conditions for given macromolecules – in this case at critical temperature. TGIC was used for the high selectivity analytical separations of homopolymers, and star-like polymers, as well as for the discrimination of polymer blend components including one parent homopolymer from the diblock copolymers. In the latter case, one component of the mixture eluted unretained in the exclusion mode and the other one was subject to the temperature controlled elution. TGIC was also applied for the preparative fractionation of polystyrene with the aim to create the extremely narrow molar mass dispersity fractions. These fractions were used in the SEC band broadening studies. Chang has likely managed to solve problems connected with the radial temperature gradients within TGIC columns. Still, the question of sample recovery may remain open. In any case, TGIC represents an interesting contribution to the coupled methods of polymer HPLC.

11.8.6 LIQUID CHROMATOGRAPHY UNDER LIMITING CONDITIONS OF ENTHALPIC INTERACTIONS

In *liquid chromatography under limiting conditions of enthalpic interactions* (LC LC) the combination of exclusion with enthalpic interactions is attained in an unconventional way (Berek, 2000). LC LC is an incoming, so far little known group of methods and this is why it will be described in more detail. In LC LC, the isocratic elution principle is combined with the local gradients. Similar to polymer HPLC methods discussed in above sections, porous column packing is applied also in LC LC. Macromolecules tend to elute rapidly from the column provided they are fully or at least partially excluded from the packing pores. The low-molecular substances permeate (practically) all pores of the column packing and therefore their transport rate is low. A low-molecular substance is selected, which promotes enthalpic interactions of macromolecules of certain nature within the LC LC column: adsorption, enthalpic partition or phase separation. With advantage, this interaction promotor is one of the eluent components. A sufficient excess of interaction promoting low-molecular substance, is transported along the column in front of sample. It decelerates elution of the interactive polymer sample constituent. In fact, the interaction boosting low-molecular substance creates for interactive macromolecules a *selective,* slowly progressing impermeable *barrier*. The barrier can be formed by the eluent itself while the sample is dissolved and injected in an elution promoting solvent: The interactive macromolecules cannot leave the zone of their original solvent – otherwise they are retained within column (compare the discscussion on the sample break-through in section 11.8.4). In the alternative approach, mobile phase promotes sample elution, whereas a narrow zone of retention promoting substance is injected in front of sample solution.

The latter option is well flexible and experimentally feasible. Moreover, a series of barriers with distinct compositions can be employed to separate multicomponent complex polymer systems. In any case, the interactive polymer is eluted slowly behind the barrier while the non-interactive macromolecules elute rapidly in the exclusion mode. In this way, the interacting macromolecules are quickly and efficiently separated from the non-interacting ones.

Employing the two above experimental arrangements and the three so far tested retention mechanisms (see section 11.5.2.3), six LC LC procedures were designed and successfully tested. They are called:

LC under limiting conditions of

- solubility, LC LCS
- adsorption, LC LCA
- (enthalpic) partition, LC LCP

if mobile phase creates the decelerating barrier. For the case, when the barrier is generated with a narrow zone of interaction promoting liquid, the following terms were proposed:

Liquid chromatography under limiting conditions of

- desorption, LC LCD
- (enthalpic) unpartition, LC LCU
- insolubility, LC LCI.

The basic review of LC LC procedures, as well as their initial applications was reviewed by Berek (2000).

The operation principle of LC LCD with a narrow barrier of adsorli is evident from Figure 18.

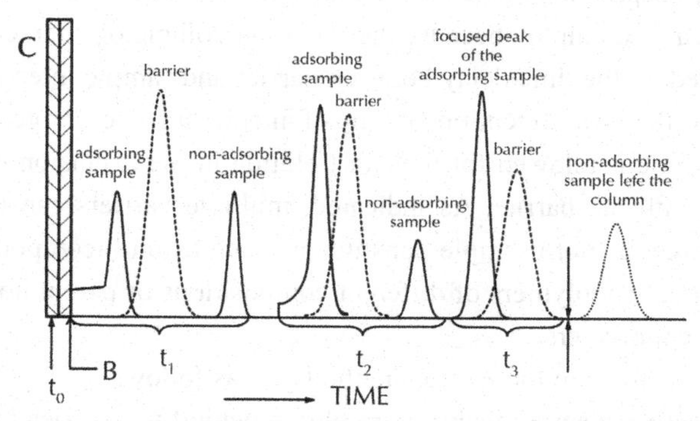

FIGURE 18 Schematic representation of liquid chromatography under limiting conditions of enthalpic interactions. The adsorption retention mechanism is exemplified. Narrow zone of retention promoting zone, barrier B is injected in front of sample. For further explanation, see the text.

The non-interactive macromolecules elute in the exclusion mode without any impact of barrier. The initially rectangular barrier is subject to typical chromatographic band broadening process. On the other hand, the zone of interactive macromolecules undergoes a sort of focusing. The initial volume of interactive sample can be large because it is successively decreased and the band broadening due to usual diffusion and mixing chromatographic processes are partially or even fully offset. It is supposed that interactive macromolecules accumulate on the edge of barrier near the particular critical conditions. With advantage, narrow pore column packings with as large pore volume as possible are employed. If the noninteractive macromolecules are pore excluded they also elute in a relatively narrov zone.

In practice, two-component solvent mixtures are employed as eluents and sample solvents in LC LC. One constituent of mixture supports elution of interactive polymer from the particular column, while another one induces its retention within column. To adjust polymer interactivity or to cope with the limited solubility of analyzed polymers, multicomponent solvents can be employed. Typical examples are mixtures of hexafluoropropanol with chloroform, which dissolve aromatic polyesters and some polyamides at ambient temperature. The sample solvents, eluents and barriers usually contain the same liquids but their composition is adjusted to fulfil their particular role: Sample solvent must dissolve all its constituents and barrier must efficiently decelerate interactive macromolecules. Eluent serves either as a barrier in LC LCS, LC LCA and LC LCP or it promotes unhindered sample elution in LC LCD, LC LCU and LC LCI.

The LC LC separation mode that employs narrow zone of retention promoting substance is easy to employ because the retention volume of interacting polymer can be adjusted by the time delay between barrier and sample injection. The longer time delay, the lower retention volume of interacting macromolecules because the extended time is allowed for their fast elution in the exclusion mode, before their impact with the barrier. As indicated, multiple barriers can be employed to separate three or more sample constituents such as multicomponent polymer blends, statistical copolymers of different compositions or parent homopolymers from diblock copolymers.

The basic features of the LC LC methods are as follows:

1. Interacting macromolecules are eluted behind the barrier independently of their molar mass. This is an important prerequisite for the undisturbed

separation of complex polymers and complex polymer systems according to other molecular characteristics, without molar mass interference.

2. As mentioned, LC LC as a rule produces narrow, focused polymer peaks because macromolecules accumulate on the edge of barrier. The resulting narrow peaks possess typical sharp front. It is supposed that here again the critical conditions of adsorption or enthalpic partition represent the *limits* of the barrier composition, which actually hinders the fast progression of polymer species. The onset of polymer precipitation locates barrier composition in the case of phase separation retention mechanism.
3. LC LC can be applied to the (ultra) high molar mass polymers. Contrary to the liquid chromatography under critical conditions, LC CC, the liquid chromatography under limiting conditions of enthalpic interactions apparently does not exhibit any upper limit concerning sample molar mass. On the other hand, if the difference is too small between the elution rate of the low-molecular substances that form the barrier and the oligomeric sample constituents, the LC LC separation selectivity may become insufficient. Consequently, LC CC and LC LC can be considered mutually complementary methods from the point of view of their applicability in the two-dimensional polymer HPLC of complex polymer systems (see section 11.9): by preference LC CC for oligomers, and LC LC for (high) polymers with molar mass above about 1 $kg.mol^{-1}$.
4. Availability of instrumentation. Common HPLC instruments can be utilized, preferably equipped with the column thermostat. In the case of the LC LC procedures, which apply the zone of the barrier liquid injected in front of sample, an additional injecting valve is to be installed between the sample injector and the LC LC column. This valve is equipped with a large-volume loop employed for the barrier injection. Application of the additional valve substantially simplifies experimental work. Alternatively, an autosampler is to be employed, which allows adjustments of injected volumes over a wide range. HPLC columns of different sizes are utilized. The small columns with volume of (1–2) mL serve for the scouting experiments aimed at identification of suitable experimental conditions, while the columns with the volume in the range of (10-15) mL are adequate for most analytical separations. The LC LCD, LC LCU and LC LCI methods can be easily expanded to the preparative scale. It should be stressed again that the narrow pore column packings with as

large pore volume as available are to be employed in LC LC. Depending on the polymers to be separated, the column packing is bare silica gel or silica gel bonded with polar groups for adsorption retention mechanism, while alkyl bonded silica gel or usual polystyrene/ divinylbenzene packings are appropriate for enthalpic partition based separations. The polarity of column packing plays less important role when the phase separation retention mechanism is applied. As elucidated, it can be quite large. The occurence of peak focusing allows injections of extremely large volumes of samples, which may easily reach 20% of the total volume of the LC LC column. The practical limit depends on the column size because the rear part of sample has to catch its front part before the latter leaves the column. Of course the peak width of the non-interactive sample constituents is to be considered. This rises with the injected sample volume. Therefore the optimization of sample volume is necessary. Sample detectability is an important parameter. Volume of barrier injected in front of sample is often ten times larger than the sample volume. Its actual appropriate size is determined by the introductory experiments. Too small barrier becomes permeable for the interactive macromolecules in the course of its elution, while too large barrier takes needlessly much column space and in effect it decreases selectivity of separation of multicomponent complex polymer systems.

5. The unprecedented robustness and experimental feasibility of most LC LC procedures. While variations in the eluent composition as small as 0.1 wt.% may produce large shifts in the LC CC retention volumes (see section 11.8.3), changes in eluent composition larger than 10 wt.% are allowed in many LC LC systems. Similarly, rather sizeable variations in temperature are often, though not always, tolerated. The decelaration of interactive macromolecules is primarily controlled by the composition of barrier, which is to be carefully optimized before application of method. The necessary experiments are, however fast and simple because just the same both sample and mobile phase is employed and only barrier composition is altered.
6. The sample capacity of the optimized LC LC procedures is very high also in term of injected concentration. High sample capacity is an important advantage considering identification and characterization of minor sample constituents in polymer blends, preparative polymer separations,

as well as two-dimensional polymer liquid chromatography. For example, LC LC allows discrimination and further direct characterization of minor macromolecular admixtures (< 1% and in favorable cases even < 0.1%) in polymer blends, including parent homopolymers present in the block copolymers.

7. The important advantage of LC LC procedures that utilize controlled volume barriers lies in the high sample recovery. This is rather comprehensible because the pores of the column packing are in a continuous equilibrium with the elution promoting mobile phase. The equilibrium is likely perturbed only within the pore orifices in the course of the brief contact of the column packing with the retention promoting barrier.
8. The detection problems of the LC LC procedures are similar to LC CC (see section 11.8.3). The zone of interacting macromolecules is partially overlapped with the barrier and this fact complicates quantitative sample monitoring. This is why the evaporative light scattering detectors with all their drawbacks (see section 11.6.1.4) are so far mostly employed in the LC LC measurements.
9. Presence of barrier can complicate further processing of fractions obtained. For example, sample is contaminated by the nonsolvent in LC LCI that applies the phase separation retention mechanism.

A unique application of LC LCD and LC LCU is the already mentioned one-step separation of both parent homopolymers from the diblock copolymers. Advantageously the system of column packing, eluent and barrier is selected, in which one block in the block copolymer and the corresponding homopolymer do not exhibit any retention induced by enthalpic interactions. The non-interactive blocks of the block copolymer act as a sort of "diluent" of the interacting blocks and reduce their interactivity (compare section 11.8.3). This is why the chains of block copolymer are less interactive than the chains of interactive homopolymer. A block copolymer that contains both parent homopolymers is in fact a three-component system. Therefore two distinct barriers that possess different decelerating ability are introduced into the column with appropriate time delay. The first barrier is more efficient. It hinders fast progression of the block copolymer. The less efficient barrier is introduced into the column later. It does not decelerate the block copolymer but it is still able to hinder the fast elution of the interacting homopolymer. The action of two barriers is evidenced in Figure 19.

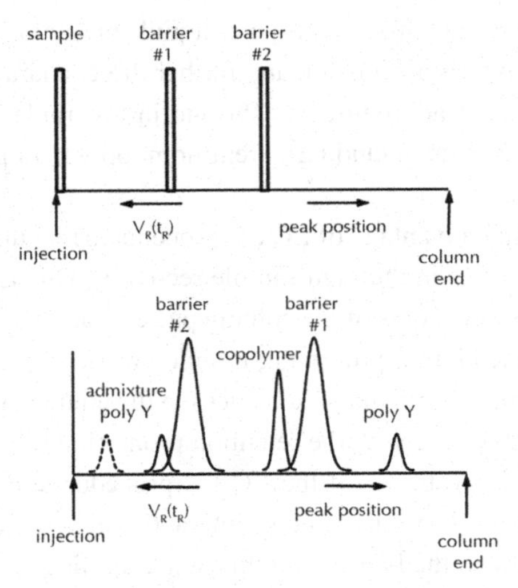

FIGURE 19 Schematic illustration of separation of a three-component polymer system (a block copolymer that contains both parent homopolymers) plus a low-molecular admixture (marker) with help of the LC LCD method. Two barriers of distinct efficiency are employed. For further explanation, see the text.

A real experimental example ot this highly powerful procedure is demonstrated in Figure 20. As evident, the method is capable to base line separate all three macromolecular constituents and the low-molecular admixture within less than 15 min.

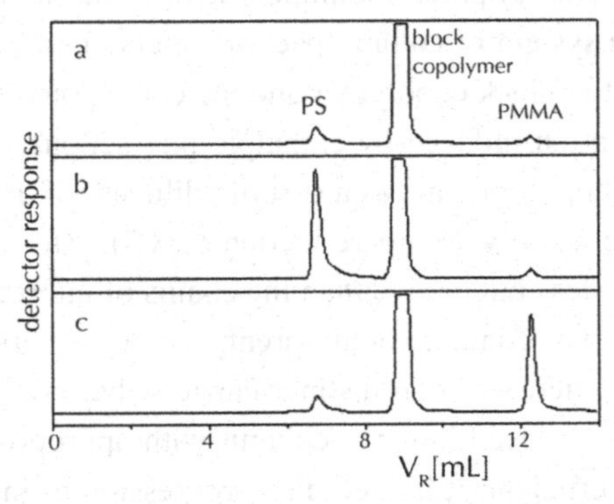

FIGURE 20 The LC LCD separation of polystyrene and poly(methyl methacrylate) homopolymers from the real sample of block copolymer PS-block-PMMA with help of LC LCD a) original chromatogram, b) sample spiked with PS, c) sample spiked with PMMA. The column packing was bare silica gel with pore size of 6 nm. Eluent was a mixture THF/toluene 70/30, barrier #1 was neat toluene, and barrier #2 was a mixture of THF with toluene 30/70. All mixed solvent compositions are in weight percent.

Parent homopolymers were successfully separated from various block copolymers, for example PMMA-*b*-poly(*n*-propyl methacrylate), PMMA-*b*-poly(*n*-butyl acrylate), PMMA-*b*-poly(*n*-pentyl methacrylate), poly(lactic acid)-*b*-poly(butylene terephthalate-*stat*-butylene adipate), PS-*b*-poly(methyl acrylate), poly(*N*-vinyl pyrollidone)-*b*-poly(caprolactone) and others.

LC LCD was also employed for separation of poly(methyl methacrylate), and poly(ethyl methacrylate) according to their stereoregularity, as well as for discrimination of low-solubilty polymers poly(ethylene terephthalate) from poly(butylene terephthalate) at ambient temperature.

It is anticipated that the LC LC procedures will find broad application in polymer science and technology, especially in the molecular characterization of outcomes of advanced polymer syntheses and in control of production of various industrial complex polymer systems.

11.8.7 ENTHALPY ASSISTED SEC

The principle of *enthalpy assisted SEC* (ENA SEC) is evident from Figures 3(b) and 3(c) (see section 11.5.2.1). In ENA SEC, the exclusion mechanism governs the order of elution that is the retention volumes of macromolecules decrease with their rising molar mass. The controlled presence the enthalpic interactions, however, significantly increases selectivity of separation according to molar mass.

The simpliest case of ENA SEC represents the separation of oligomers. The experimental conditions are selected so that only the end groups on the oligomer molecule are retained due to the enthalpic interactions. Small molecules are strongly retained and their V_R increase. As the molar mass of oligomer rises, the effect of end groups decreases. This leads to the SEC-like elution order with increased selectivity of separation (see Figure 3(b)).

A rather new approach to ENA SEC of high polymers has roots in the concept of the critical conditions (see section 11.8.3). Near the critical conditions - but on the exclusion side - retention volumes of macromolecules, which are excluded from the column packing pores rapidly decrease with their increasing molar mass (see Figure 3(c)). The SEC-like separation with a high selectivity can be accomplished. Both adsorption and enthalpic partition retention mechanisms were applied in ENA SEC of (high)polymers near the critical point. The important shortage of the ENA SEC methods is the strongly reduced recovery macromolecules, which rapidly accrues with the rising molar mass.

11.9 TWO-DIMENSIONAL POLYMER HPLC

In this section, the principles of the modern, increasingly popular two-dimensional polymer HPLC, 2D polymer HPLC or just 2D-LC are briefly elucidated. As explained in sections 11.7 and 11.8, direct comprehensive characterization of *complex polymers* is hardly possible by the SEC alone. SEC employs difference in size and consequently in extent of exclusion of macromolecules from the column packing pores. However, size of macromolecules in solution simultaneously depends on all molecular characteristics of sample. Coupled methods of polymer HPLC often help to at least partially solve the problem. On the other hand, to provide complete molecular characterization of *complex polymer systems*, which are mixtures of macromolecules of distinct nature and therefore exhibit multiple dispersities in their molecular characteristics, all sample constituents must be mutually discriminated and then independently characterized. Particular constituents of complex polymer systems can be fully separated with help of SEC only if they possess well different molar masses and if their relative concentrations in system are similar. As explained in section 11.7.1, limited both selectivity and sample capacity of SEC does not enable to assess minor macromolecular constituents (<1%, in fact often < 10%) in complex polymer systems even if their size in solution well differs from that of polymer matrix. Coupling different retention mechanisms within one single column may allow separation of several distinct constituents of complex polymer system but it only exceptionally enables their comprehensive molecular characterization. In order to tackle this task, the procedures of two-dimensional polymer liquid chromatography, 2D-LC are to be applied. 2D-LC employs distinct retention mechanisms in two independent chromatographic systems, between which certain degree of both complementarity and orthogonality is to be attained. This is the basic idea of the two-dimensional polymer HPLC. It is anticipated that also the three- and multi-dimensional polymer HPLC will be developed in future.

Several researchers attempted the two-dimensional separations of synthetic polymers. For example, Balke and Patel (1980) employed two different SEC or SEC – like procedures for separation of statistical copolymers. It is likely that the first practically applicable 2D-LC separations of macromolecules were done by Kilz, Schulz et al. (1995) who pioneered modern two-dimensional polymer HPLC.

Fundamentals of 2D-LC are evidenced in Figure 21.

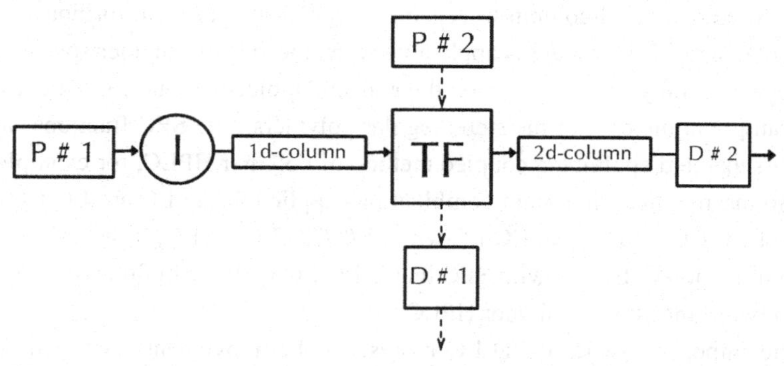

FIGURE 21 Schematic representation of 2D-LC principle. P#1 and P#2 are pumps, D#1 and D#2 are detectors. 1d- and 2d-columns employ distinct retention mechanisms. TF stands for procedures and devices that ensure transfer of fractions among columns. For detailed explanation, see the text.

Sample is separated in the first-dimension (1d) column and the fractions obtained are forwarded into the independent second-dimension (2d) column for further separation and characterization. Depending on the methods utilized for the particular separation dimensions, mobile phases in both columns can be either identical or unlike. P#2 delivers the same mobile phase as P#1 in the former case or it is not employed ar all. Results of separation in the 1d and the 2d columns are monitored with help of detectors D#1 and D#2, respectively. In the first step, the experimental conditions for the first-dimension separation are optimized. The result of separation is registered with detector D#1. Based on results of such scouting experiments, the optimized separation in the 1d column is repeated and fractions obtained are forwarded into the 2d column. As discussed below, there are several different options for practical implementation of such transfer.

The presently most popular approach to two-dimensional polymer HPLC avails partial or preferably full suppression of the molar mass effect in the 1d column so that the complex polymer or complex polymer system is separated mainly or even exclusively according to chemical structure or physical architecture of macromolecules occuring in sample. Appropriate coupled methods of polymer HPLC are to be applied to this purpose (compare section 11.8). In the 2d separation column – it is usually SEC - the fractions from the 1d column are further discriminated according to their molecular size. In other words, fractions obtained in the first-dimension column are separated in the self-existent second-dimension column, which applies distinct separation mechanism(s). Only exceptionally SEC can be used as the 1d column to separate complex polymers according to their molecular size. This concept is applicable when the size of polymer species does not depend or only little depends on their second molecular characteristic, as it is for example in the case of the stereoregular polymers. The SEC fractions can be further separated applying a coupled method of polymer HPLC, for example LC CC. So far, the most important combinations applied in 2D-LC are LC CC with SEC (or LC CC × SEC) and EG LC with SEC (EG LC × SEC). It is believed that the combination of LC LC with SEC that is LC LC x SEC will find wide application in two-dimensional polymer HPLC.

The important issue of 2D-LC represents the abovementioned transfer of column effluent between the 1d and the 2d columns, which can be done either off-line or online. In the off-line approach, the fractions from the 1d column are collected and successively re-injected into the 2d column. In this case, the unit TF is just a fraction collector. The macromolecules within particular fractions from the 1d column are immixed so that resulting overall separation selectivity may be challenged. Moreover, entire procedure is laborious and slow. Various approaches were elaborated for the online transfer of fractions from the 1d column into the 2d column. Often, the fractions from the 1d column are cut into small parts that are one-by-one gradually transported into the 2d SEC column for independent characterization. This is the method of choice if the first-dimension separation produces broader peaks, such as it does liquid chromatography under critical conditions of enthalpic interactions, LC CC (see section 11.8.3). The operation principle of such "chop-and-reinject" method is evident from Figure 22. In this case, the TF unit from Figure 21 is a switching valve.

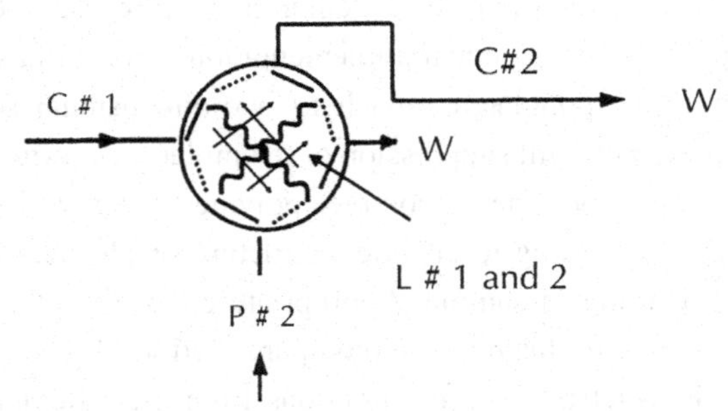

FIGURE 22 Scheme of the eight-port switching valve for successive online transfer of sample fractions between the 1d and 2d columns. The effluent from the 1d column (C#1) is transferred into the 2d column by means of the switching valve with two loops. P#2 is the second pump.

The switching valve that connects the 1d and the 2d column is equipped with two sample loops. While one of them is filled with the effluent from the 1d column, the content of other loop is displaced into the 2d column. As evident from Figure 22 the filling and emptying direction varies in one of the loops of switching valve. This somewhat decreases efficiency and selectivity of separation. The problem can be avoided by applying either one ten-port two-way switching valve or two six-port two-way valves. The operation of electrically or pneumatically driven switching valve(s) can be automated with help of appropriate software. The exacting requirement of such continuous chop-and-reinject online 2D-LC approach is that the elution rate in the 1d column must be slow, while the separation process in the 2d SEC column has to be very fast because it needs to be concluded before the next portion of sample from the 1d column arrives at the 2d column. Then only the entire effluent leaving the first-dimension column can be processed continuously in the second-dimension column without flow interruption. This requisite may bring about experimental problems, which can be solved with help of the stop-and-go elution method. The flow in the 1d column is periodically interrupted to allow time for the second-dimension separation. Alternatively, the fractions from the 1d column can be stored in a set of the capillaries or in the full-retention-elution columns (compare section 11.6.2), where they do await for their transfer into the second-dimension column. The full-retention-elution approach could even enable the re-concentration of fractions diluted in the course of the first-dimension separation. In any case, also above described 2D-LC analysis is well time-consuming. The typical graphical representation of the 2D-LC separation results, LC CC x SEC is a *contour plot* that depicts mutual dependence of sample composition and its molar mass (Figure 23).

In order to avoid the long-lasting analyses, only selected part of the 1d column effluent is forwarded into the 2d column. This approach is denoted the "heart-cut method", and the unit TF in Figure 21 is just a simple four-port two-way switching valve. The drawback of the heart-cut approach is that a part of sample is lost and (important) details on sample dispersity may get overlooked.

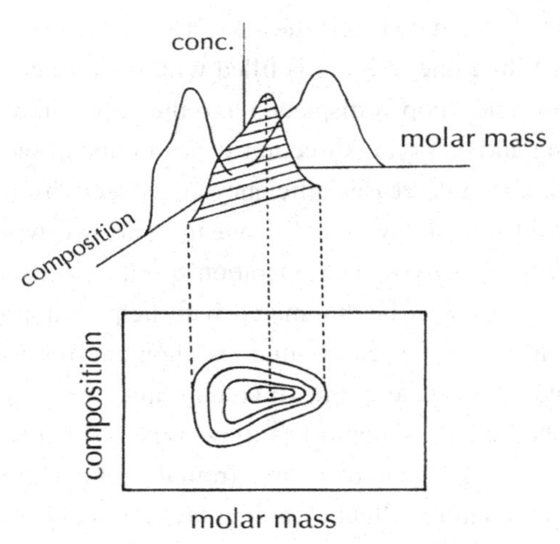

FIGURE 23 Representation of the contour plot: composition versus molar mass of a complex polymer. For detailed explanation, see the text.

A favorable situation occurs if the 1d column produces narrow, base-line separated fractions of macromolecules of distinct composition irrespectively of their molar mass. In that event, entire sample peaks from the 1d column can be directly forwarded into the 2d column. This is the case of many LC LC separations (compare section 11.8.6), and the resulting method is called "sequenced two-dimensional polymer HPLC", S2D-LC because entire sequences of the effluent leaving the LC LC 1d column are online forwarded into the 2d column, Berek, Siskova (2010). The scheme of the corresponding experimental assembly is depicted in Figure 24.

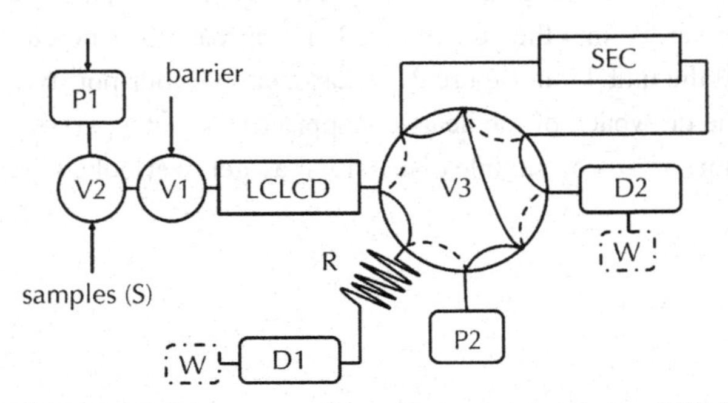

FIGURE 24 Scheme of the sequenced two-dimensional polymer HPLC. For further explanation, see the text.

The 1d column is LC LCD or LC LCU. The LC LCI approach is not suitable as the the sequences forwarded into the 2d column would contain barriers of nonsolvent. Evidently, the 1d (LC LC) column is to be interactive to enable employment of appropriate interaction retention mechanism. Eluent in the 2d column may differ from that applied in the 1d column. However, if the 2d SEC column is non-interactive, eluent in both columns can be identical. This further facilitates analyses.

The stop-and-go approach can be easily applied in the LC LC column because the chromatographic zones within porous column packing are under static conditions only very little broadened by diffusion. Alternatively, the first-dimension separation is repeated several times, once for each distinct sample constituent. Always one particular fraction that is one entire peak from the 1d column is forwarded into the 2d column and the non-used fractions go to waste. The concentration or volume of sample introduced into the 1d column is so adjusted that the polymer detection in the 2d column effluent is feasible. For the characterization of minor macromolecular constituents of complex polymer systems, large amount of original sample can be injected into LC LC column as its sample capacity is customarily very high. It has been demonstrated that the sequenced two-dimensional liquid chromatography LC LCD × SEC enables reliable determination of molar mass averages and dispersities of constituents of multicomponent complex polymer systems, including those of the parent homopolymers present in the block copolymers. The method is fast and repeatable. It allows identification and molecular characterization of minor admixtures in complex polymer systems at hitherto unattainable low concentrations.

Figure 25 demonstrates the importance of the 2D-LC for separation and molecular characterization of complex polymer system, the block copolymer polystyrene-block-poly(methyl methacrylate). It is evident that SEC alone could not disclose presence of homopolymers in the block copolymer. On the other hand, LC LCD has well separated both parent homopolymers from the block copolymer within a single step and the online SEC column afforded their molar mass averages and dispersities.

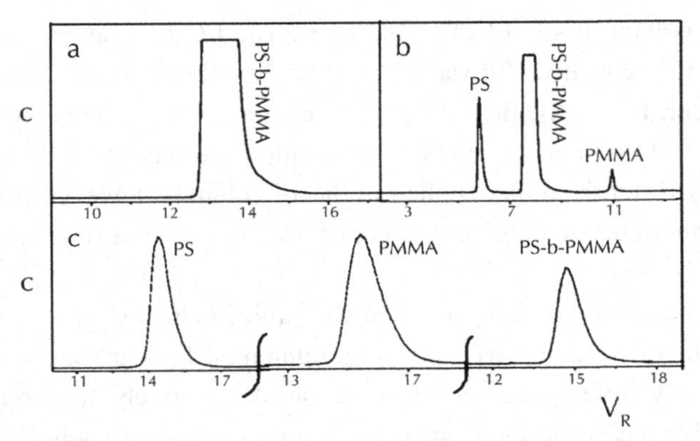

FIGURE 25 Comparison of SEC, LC LCD and 2D-LC real chromatograms of a typical commercial block copolymer PS-b-PMMA that contained parent homopolymers. For detailed explanation, see the text.

The SEC separations were done with a “universal” column (300 × 7.5) mm (compare section 11.7.3.1) packed with the non-interactive polystyrene/divinylbenzene gel. The LC LCD column (7.5 × 300) mm was packed with 10 nm bare silica gel. Eluent in both the 1d and the 2d columns was a mixture of THF and toluene 70/30 wt./wt. Barrier #1 was pure toluene and barrier #2 was a mixture THF/toluene 30/70 wt./wt. Time delays between the barrier injections were 3 and 5 min.

11.10 CONCLUSION

Liquid chromatographic methods were presented in this chapter, aimed at molecular characterization of synthetic polymers. To facilitate understanding of principles of particular experimental procedures the behavior of macromolecules in solution and various kinds of interactions in the liquid chromatography column were briefly elucidated. The practical backgrounds of particular methods were highlighted and the pitfalls lurking on their users were illuminated. This latter feature of the chapter is rather unconventional but, in the author´s opinion, this makes it useful not only for the beginners but also for those who seek reliable results rather than just embellishment of their publications or even deceptive arguments to support their notions.

Good luck in molecular characterization of synthetic polymers with help of liquid chromatography!

KEYWORDS

- **chemical composition**
- **complex polymer systems**
- **complex polymers**
- **enthalpic partition**
- **molar mass**
- **molecular architecture**
- **molecular characteristics of polymers**
- **phase adsorption**
- **phase separation**

ACKNOWLEDGMENT

The financial support of Slovak grant agencies VEGA (project 2-0001-12) and APVV (projects 0109-10 and 0125-11) is acknowledged.

REFERENCES

1. Balke, S. and Patel, R. D. *J. Polym. Sci. Polym. Lett. Ed.*, 18, 453 (1980).
2. Belenkii, B. G., Gankina, E. S., Tennikov, M. B., and Vilenchik, L. Z. *Dokl. Akad. Nauk SSSR*, 231 (1976).
3. Belenkii, B. G. and Gankina, E. S. *Dokl. Akad. Nauk SSSR*, 186, 573 (1969).
4. Benoit, H., Grubisic, Z., Rempp, P., and Decker, D. *J. Chim. Phys.*, 63, 1507 (1966).
5. Berek, D. *Prog. Polym. Sci.*, 25, 873 (2000).
6. Berek, D. and Nguyen, S. H. *Macromolecules*, 31, 8243 (1998).
7. Berek, D. and Siskova, A. *Macromolecules*, 43, 9627(2010).
8. Casassa, E. F. *Macromolecules*, 9, 182 (1976).
9. Casassa, E. F. *J. Polym. Sci. B*, 5, 773 (1969).
10. Glöckner, G. *Gradient HPLC of Copolymers and Chromatographic Cross-Fractionation*, Springer-Verlag, Berlin (1992).
11. Gorbunov, A. A. and Skvortsov, A. M. *Adv. Colloid Interface Sci.*, 62, 31 (1995).
12. Inagaki, H., Matsuda, H., and Kamiyama, F. *Macromolecules*, 1, 520 (1968).
13. Kilz, P. Methods and Columns for High-Speed Size Exclusion Chromatography, In Handbook of size exclusion chromatography and related techniques (chromatographic sciences series, vol. 91, C. S. Wu (Ed.), Marcel Dekker, Inc.,New York, p. 561 (2004).
14. Kilz, P., Krüger, R. P., Much, H., and Schulz, G. *Adv. Chem.*, 247, 223 (1995).
15. Lee, H. C. and Chang, T. *Macromolecules*, 29, 7294 (1996).
16. Lee, H. C. and Chang, T. *Polymer*, 37, 5747 (1996).

17. Macko, T. and Hunkeler, D. *Adv. Polym. Sci.*, 163, 161 (2003).
18. Moore, J. C. *J. Polym. Sci. A*, 2, 835 (1964).
19. Pasch, H. and Trathnigg, B. *HPLC of Polymers*, Springer, Berlin (1998).
20. Porath, J. *Nature*, London, 196, 47 (1962).
21. Porath, J. and Flodin, P. *Nature*, 183, 1657 (1959).
22. Skvortsov, A. M. and Gorbunov, A. A. *J. Chromatogr.*, 358, 77 (1986).
23. Snyder L. R. and Kirkland J. J. An Introduction to Modern Liquid Chromatography, 2nd ed., Wiley Interscience, New York (1979).
24. Snyder, L. R. *Principles of Adsorption Chromatography*, Arnold, London and Marcel Dekker, New York (1968).
25. Teramachi, S., Hasegawa, A., Shima, Y., Akatsuka, M., and Nakajima, M. *Macromolecules*, 12, 992 (1979).

RECOMMENDED READINGS

General handbook

Polymer Handbook; Brandrup, J.; Immergut, E.H.; A. Gruelke, E.A.; Abe, A.; Bloch, D.R.; Eds.; Wiley: New York, 1999

Books and chapters in books on liquid chromatography of polymers from 1991

Modern Methods of Polymer Characterization; Barth, H. G., Mays, J. W., Ed.; John Wiley&Sons, 1991.

Glöckner, G. Gradient HPLC of Copolymers and Chromatographic Cross-Fractionation; Springer: Berlin, 1992.

Handbook of SEC; Wu, C. S., Ed.; Marcel Dekker: New York, 1995.

Mori, S. HPLC application to polymer analysis. In Handbook of HPLC; Katz, E., Eksteen, R., Schoenmakers, P., Miller, N., Eds.; Marcel Dekker: New York, Basel, 1998.

Pasch, H.; Trathnigg, B. HPLC of Polymers; Springer: Berlin, 1998.

Mori, S.; Barth, H. G. Size Exclusion Chromatography; Springer: Berlin, New York, 1999.

Column Handbook for Size Exclusion Chromatography; Wu, C. S., Ed.; Academic Press: San Diego, London, 1999.

Encyclopedia of Chromatography; Cazes, J., Ed.; Marcel Dekker: New York, 2001.

Handbook of Size Exclusion Chromatography and Related Techniques (Chromatographic Sciences Series, 91); Wu, C. S., Ed.; Marcel Dekker, Inc.: New York, 2004.

Radke, W., Chromatography of polymers, in macromolecular engineering. In: Structure-

property correlation and characterization techniques, Matyjaszewski, K., Gnanou, Y., Leibler, L; Eds.; Wiley-vch: Berlin, 2007.

Striegel, A. M.; Yau, W. W.; Kirkland, J. J., Bly, D. D. Modern Size-Exclusion Liquid Chromatography, Practice of Gel Permeation and Gel Filtration Chromatography; John Wiley & Sons, 2009.

Berek, D., Polymer HPLC. In: Handbook of HPLC; Corradini, D., Phillips, T. M.; CRC Press, Taylor and Francis Group, Boca Raton: London, New York, 2010.

Selected reviews related to this chapter

Berek, D. Liquid chromatography of macromolecules at the point of exclusion-

adsorption transition. Principle, experimental procedures and queries concerning feasibility of method.Macromol.Symp.1996, 110, 33.

Kilz, P.; Pasch, H. Coupled Liquid Chromatographic Techniques in Molecular Characterization. In: Encyclopedia of analytical chemistry, 9, 7495, Meyers, R.A.; Ed.; Wiley: Chichester, 2000.

Pasch, H. Hyphenated techniques of liquid chromatography of polymers, Adv. Polym. Sci. 2000, 150, 1.

Berek, D. Coupled liquid chromatographic techniques for the separation of complex polymers, Prog. Polym. Sci. 2000, 25, 873.

Chang, T. Recent advances in liquid chromatography analysis of synthetic polymers. Adv. Polym. Sci. 2001, 163, 1.

Berek, D. Size exclusion chromatography – A blessing and a curse of science and technology of synthetic polymers.J. Sep. Sci. 2010, 33, 315.

Philipsen, H. J. A. Determination of chemical composition distribution of synthetic polymers, J. Chromatogr.A 2004, 1037, 329-350.

Berek, D. Two-dimensional liquid chromatography of synthetic polymers.Anal.Bioanal. Chem. 2010, 396, 421.

Gorbunov, A. A.; Skvortsov, A. M. Statistical properties of confined macromolecules, Adv.

Coll. Interface Sci. 1995, 62, 31-108.

Baumgaertel, A.; Altunas, E.; Schubert, U.S. Recent developments in the detailed

characterization of polymers by multidimensional chromatography. J. Chromatogr. A 2012, 1240, 1.

CHAPTER 12

IMPEDANCE SPECTROSCOPY: BASIC CONCEPTS AND APPLICATION FOR ELECTRICAL EVALUATION OF POLYMER ELECTROLYTES

TAN WINIE and ABDUL KARIEM AROF

CONTENTS

12.1 IMPEDANCE SPECTROSCOPY (IS)

A brief discussion on applying IS to investigate the electrical properties of polymer electrolytes has been dealt with in Chapter 7. This chapter, on the other hand, aims to introduce and provide the necessary background for beginners to use IS as a method of analysis. Thus, we begin the chapter with the definition of impedance, and then discuss the basic principles of IS followed by impedance data presentation and interpretation. Ideal and real impedance data will be presented, compared and discussed in order to enable the readers to grasp a clearer picture on the electrical properties and electrochemical processes in a polymer electrolyte system.

The IS is a technique in which a small sinusoidal or time varying potential or voltage of the order tens of millivolt (mV) is applied across a sample cell and the current through the cell is measured.The applied signal is small so that the current response of the sample to the sinusoidal voltage is linear or pseudo-linear. When the response is linear or pseudo-linear, the current through the sample will have the same frequency as the applied voltage, but shifted in phase.The complex ratio of the voltage to the current in an alternating current circuit $V(t)/I(t)$ is impedance and denoted as Z. Since the applied voltage and measured current changes with frequency, a spectrum of impedance as a function of frequency or angular frequency can be obtained.

12.2 IMPEDANCE: A COMPLEX QUANTITY

According to Ohm's law, if V is the voltage applied across a load and I is the current flowing through the load, then:

$$R = \frac{V}{I} \tag{1}$$

R is known as the resistance of the load.

Suppose the voltage V in the circuit of Figure 1 varies with time so that $V(t)$ is given by:

$$V(t) = V_o \exp(j\omega t) \tag{2}$$

where $j = \sqrt{-1}$. The resulting current response is then given by:

$$I(t) = I_o \exp(j\omega t - j\theta) \tag{3}$$

Here V_o and I_o are the voltage and current amplitudes, ω the angular frequency is equal to $2\pi f$, with f as frequency of the signal.The θ is the phase shift or simply phase between voltage and current. Figure 2 shows the phasor diagram of $V(t)$ and $I(t)$.

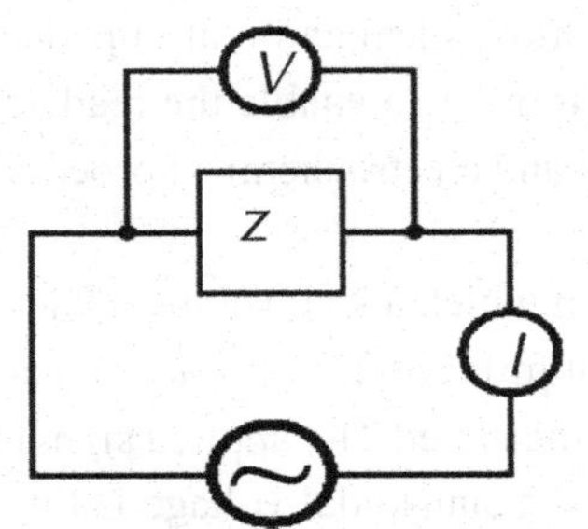

FIGURE 1 An alternating current (AC) supplying a voltage across a load Z driving a current I.

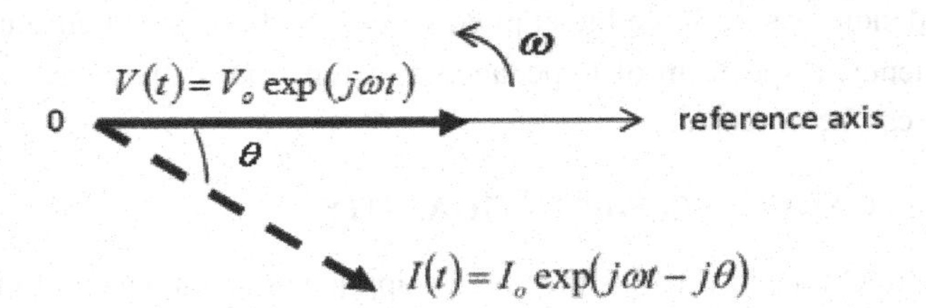

FIGURE 2 Phasor diagram of $V(t)$ and $I(t)$.

Impedance, Z is the complex ratio of the voltage to the current in an AC circuit and therefore can be written as:

$$Z = \frac{V(t)}{I(t)} = \frac{V_o \exp(j\omega t)}{I_o \exp(j\omega t - j\theta)} = Z_o \cos\theta + jZ_o \sin\theta \tag{4}$$

From Equation (4), $Z_o \cos\theta$ is the real component and $Z_o \sin\theta$ is the imaginary component. Impedance extends the concept of resistance to AC circuits, and comprises both magnitude and phase.

The relationship between real impedance *Z*’, imaginary impedance *Z*’ and phase shift θ is illustrated in Figure 3.

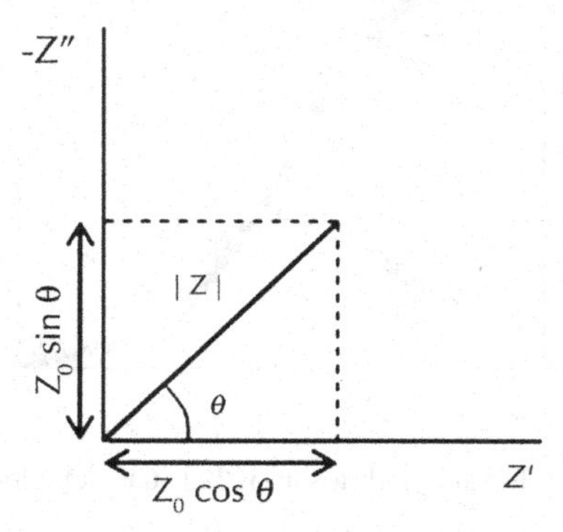

FIGURE 3 Magnitude and phase angle of impedance *Z*.

The magnitude of complex impedance, $|Z|$ is given in Chapter 7 as $|Z|^2 = \left(Z'^2 + Z''^2\right)$ and the magnitude of the phase angle θ between *Z*’ and *Z*’ is then given by Equation (5).

$$\theta = \tan^{-1}\left(\frac{Z''}{Z'}\right) \tag{5}$$

From Figure 3, when θ equals zero, $|Z| = Z_0 \cos\theta$ equals the real part of complex impedance and is actually the resistance, *R*. Hence, resistance can be thought of as impedance with zero phase angle.

12.3 REPRESENTING IMPEDANCE DATA: THE BODE AND NYQUIST PLOTS

The complex impedance is a function of frequency (see Equation (4)). Thus, both the real, *Z*’ and imaginary components of impedance, *Z*’ can be plotted against frequency as shown in Figure 4. The plots in Figure 4 are known as the Bode plots. Alternatively, one can plot a Bode plot by plotting log $|Z|$ or phase shift θ on the vertical axis and log f on the horizontal axis.

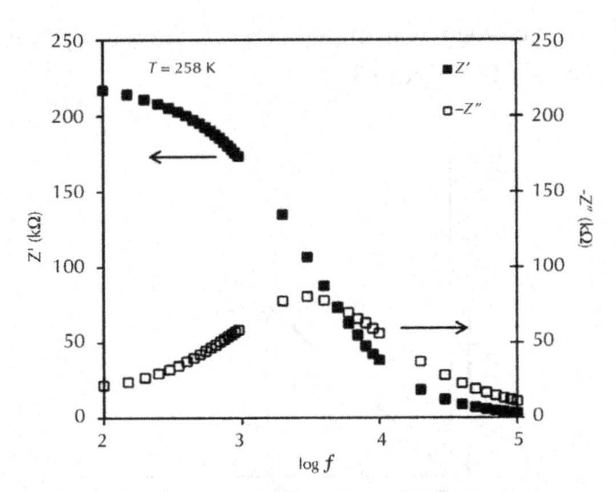

FIGURE 4 Bode plot for hexanoyl chitosan-30% $LiCF_3SO_3$ electrolyte system.

Both *Z*' and *Z*" can be combined in a single plot. A Nyquist plot is obtained by plotting *Z*' on the horizontal axis and *Z*" on the vertical axis. An example of a Nyquist plot is illustrated in Figure 5. As compared to a Bode plot, a Nyquist plot does not indicate the frequency response of a material directly. A Nyquist plot represents the electrical characteristic of a material. This electrical characteristic can be represented by an equivalent circuit that may consist of a resistor and capacitor, resistor in series with capacitor, resistor in parallel with capacitor, and so on.

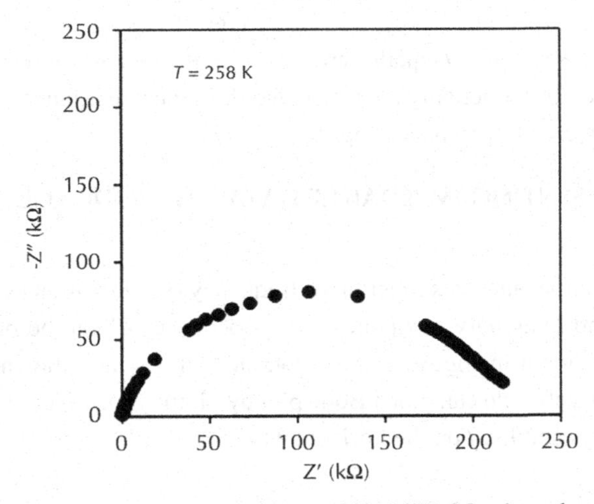

FIGURE 5 Nyquist plot for hexanoyl chitosan-30% $LiCF_3SO_3$ electrolyte system.

12.4 NYQUIST PLOTS OF SIMPLE CIRCUITS

The relationship between the real and imaginary impedance of circuit that consists of elements such as resistors and capacitors can be represented by a Nyquist plot. If the impedance of a circuit exhibits a Nyquist plot that is similar to that exhibited by a polymer electrolyte, then this circuit model can be used to represent the electrolyte and is known as the equivalent circuit for the electrolyte. The impedance data is commonly analyzed by fitting them to the impedance equation of the equivalent circuit. Now, we will discuss the selected circuit elements together with their corresponding Nyquist plots.

12.4.1 PURE RESISTOR, R

If the Nyquist plot shows a point on the Z' axis for all values of frequency (Figure 6), then the sample can be represented by a pure resistor. $Z' = R$ and $Z'' = 0$, the total impedance, $Z = R$.

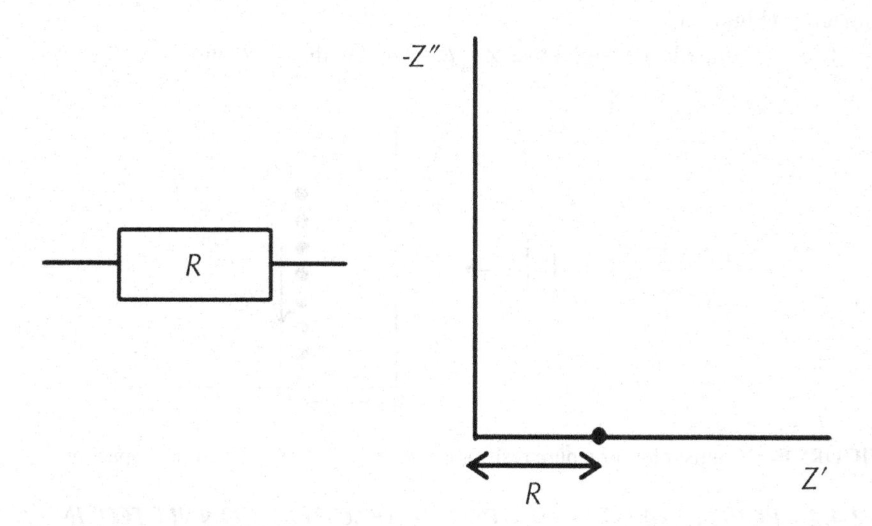

FIGURE 6 Nyquist plot for a pure resistor.

12.4.2 PURE CAPACITOR, C

If the Nyquist plot is a straight line lying along the Z'' axis, then the sample can be represented by a pure capacitor (Figure 7).The $Z' = 0$ and $Z'' = -1/\omega C$, thus, the total impedance, $Z = -j/\omega C$ and decreases with increasing ω.

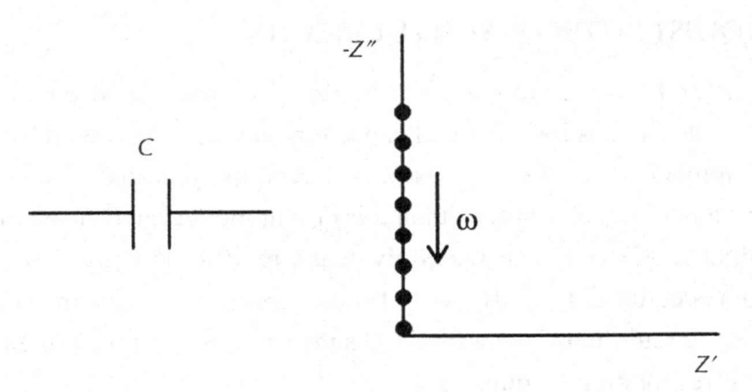

FIGURE 7 Nyquist plot for a pure capacitor.

12.4.3 PURE RESISTOR AND PURE CAPACITOR CONNECTED IN SERIES

If the Nyquist plot shows a straight line at $Z' = R$ and parallel to the Z'' axis, then the sample can be represented by a pure resistor connected in series with a pure capacitor (Figure 8).

The total impedance is given as $Z = R - j/\omega C$ with $Z' = R$ and $Z'' = -1/\omega C$.

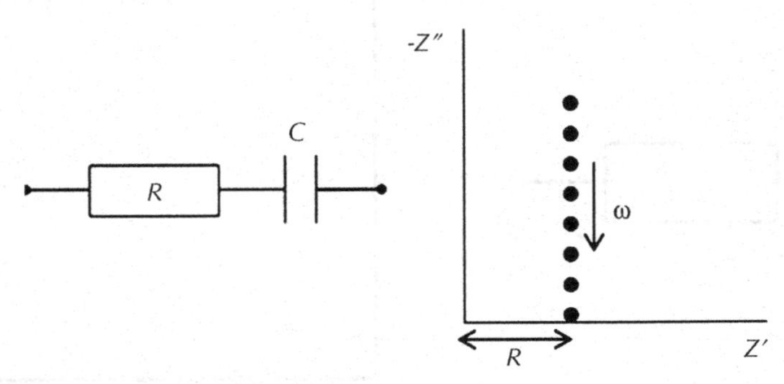

FIGURE 8 Nyquist plot for a pure resistor connected in series with a pure capacitor.

12.4.4 PURE RESISTOR AND PURE CAPACITOR CONNECTED IN PARALLEL: SINGLE RELAXATION TIME

The impedance of the equivalent circuit comprising a pure resistor and a pure capacitor in parallel connection is given by:

$$\frac{1}{Z} = \frac{1}{R} + j\omega C \tag{6}$$

which leads to:

$$Z = \frac{R}{1 + j\omega RC} = \frac{R}{1 + \omega^2 R^2 C^2} - j\frac{\omega R^2 C}{1 + \omega^2 R^2 C^2} \tag{7}$$

Here,

$$Z' = \frac{R}{1 + \omega^2 R^2 C^2} \tag{8}$$

and

$$Z'' = \frac{\omega R^2 C}{1 + \omega^2 R^2 C^2} \tag{9}$$

Eliminating ω in Equation (9) and Equation (8), it follows that:

$$\left(Z' - \frac{R}{2}\right)^2 + (Z'')^2 = \left(\frac{R}{2}\right)^2 \tag{10}$$

Equation (10) is the equation of a semicircle with radius $R/2$ and center at $(R/2,0)$ as depicted in Figure 9. Thus, this is why both the Z' and Z'' axes must have the same scales to see the semicircle.

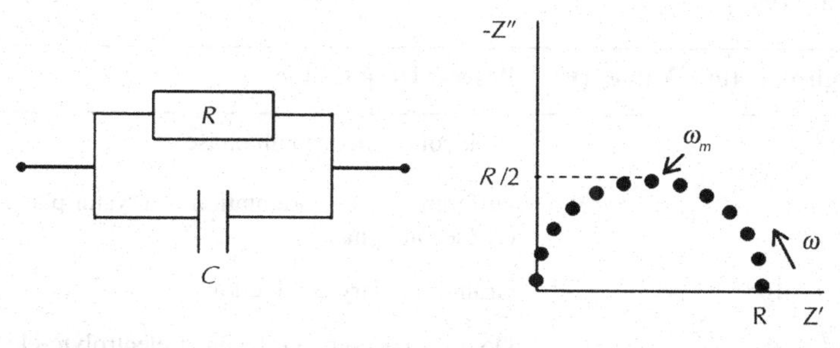

FIGURE 9 Nyquist plot for a pure resistor connected in parallel with a pure capacitor.

The angular frequency at the maximum point on the semicircle, ω_m is given by:

$$\frac{Z''}{Z'} = \omega_m RC = 1 \tag{11}$$

or

$$\omega_m = \frac{1}{RC} = \frac{1}{\tau} \tag{12}$$

τ is the time constant or relaxation time. Relaxation time is the result of the efforts carried out by ionic charge carriers in a material to align with the direction of the applied field.

The Nyquist plot in Figure 9 is a classical Debye response. Debye response is the dielectric response of an ideal, non-interacting population of dipoles to an alternating external electric field. It involves only a single relaxation time because the dipoles do not interact with each other or induce other dipoles surrounding it. They aligned themselves with the direction of external field within the same time. This relaxation model was introduced by Peter Debye in 1913.

From Equation (12), knowing ω_m and R, the capacitance, C can be evaluated. The C value varies with different conduction processes in a material (Table 1). The various conduction processes in a material is one of the reasons why its electrical characteristics cannot be represented by pure resistors and pure capacitors. Thus, a new circuit element known as constant phase element (CPE) is introduced (Bottelberghs and Broers, 1976).

TABLE 1 The variation of capacitance values with conduction processes (Sinclair, 1995; Irvine et al., 1990)

Approximate *C* Values (F)	**Process Responsible**
10^{-12}	Bulk conduction (main phase)
10^{-11}	Different phases, orientation of crystal planes, etc (second phase)
$10^{-11} - 10^{-8}$	Grain boundary conduction
$10^{-7} - 10^{-5}$	Double space-charge layer at electrolyte-electrode interface
10^{-4}	Electrochemical reactions

12.5 NYQUIST PLOTS OF MODEL SYSTEMS

Figure 10 shows a solid electrolyte with two non-blocking electrodes. For non-blocking electrodes, no accumulation of charge occurs at the electrode-electrolyte interface.The Nyquist plot is expected to show a semicircle. The equivalent circuit may take the form of a resistor connected in parallel with a capacitor. In the presence of R_s, the expected Nyquist plot together with its equivalent circuit is depicted in Figure 11. On the other hand, for blocking electrodes, charge accumulates at the electrolyte-electrode interfaces. This contributes to the double layer capacitances at the interfaces, C_{dl}. A vertical spike is thus expected to arise in the Nyquist plot due to the double layer capacitance (Figure 12).

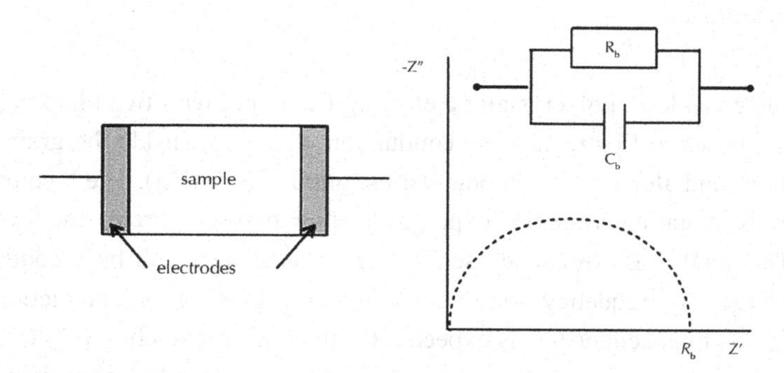

FIGURE 10 Expected Nyquist plot and equivalent circuit for a solid electrolyte with two non-blocking electrodes.

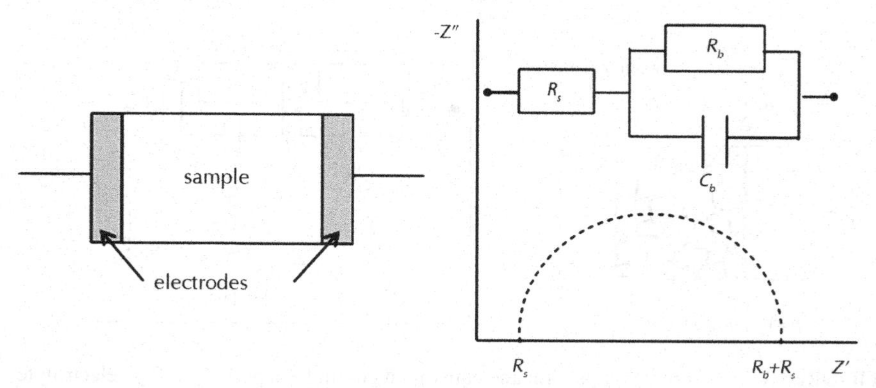

FIGURE 11 Expected Nyquist plot and equivalent circuit for a solid electrolyte with two non-blocking electrodes (semicircle shifted at a distance R_s from the origin).

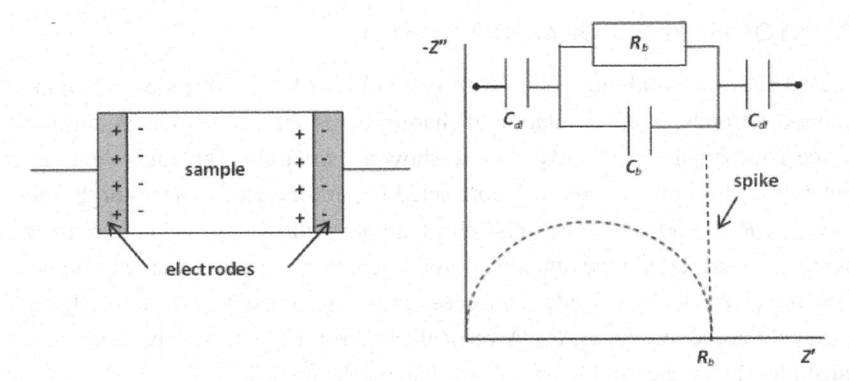

FIGURE 12 Expected Nyquist plot and equivalent circuit for a solid electrolyte with two blocking electrodes.

Let us consider a polycrystalline electrolyte sample with two blocking electrodes as shown in Figure 13. The conduction will occur inside the grain (bulk conduction) and along the grain boundaries (grain conduction). The Nyquist plot for the polycrystalline system is expected to show two semicircles and a vertical spike. The small high frequency semicircle is associated with bulk conduction and the large low frequency semicircle is associated with grain conduction. The larger value of capacitance, C is expected for the grain conduction ($C_g > C_b$) since the grain boundaries are usually thinner. C_g is the grain boundary capacitance and C_b is the bulk material capacitance.

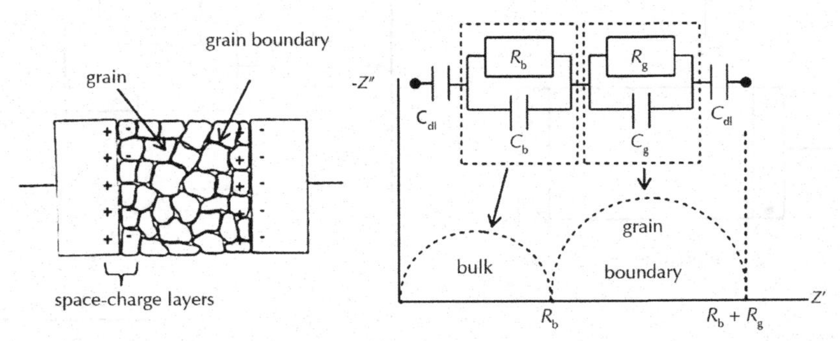

FIGURE 13 Expected Nyquist plot and equivalent circuit for a polycrystalline electrolyte with two blocking electrodes.

12.6 EQUIVALENT CIRCUITS FOR REAL SYSTEMS

12.6.1 CONSTANT PHASE ELEMENT (CPE)

The Nyquist plots of many real electrolyte systems deviate from the ideal Debye response, they tend to show distorted or depressed semicircles, tilted or curved spikes. The deviations can be fitted if a CPE replaces the capacitor in the equivalent circuit. A CPE is viewed as a leaky capacitor with impedance given by:

$$Z_{CPE} = \frac{Z_o}{(j\omega)^{\alpha}} = Z_o(j\omega)^{-\alpha} \tag{13}$$

where $0 \leq \alpha \leq 1$. When $\alpha = 0$, the impedance Z_{CPE} is frequency independent, $Z_o = R$ and CPE acts as a pure resistor. When $\alpha = 1$, $Z_{CPE} = Z_o/j\omega$, which implies that with $Z_o = 1/C$, the CPE acts as a pure capacitor with impedance $1/j\omega C$. When α is between 0 and 1, the CPE acts as intermediate between a resistor and a capacitor.

Equation (13) can be recast by using the de Moivre's theorem:

$$e^{j\theta} = \cos(\theta) + j\sin(\theta) \tag{14}$$

If $\theta = \pi$, Equation (14) becomes:

$$e^{j\pi} = \cos(\pi) + j\sin(\pi) = -1 \tag{15}$$

The term $j^{-\alpha}$ in Equation (13) can thus be written as:

$$j^{-\alpha} = (-1)^{-\alpha/2} = \left(e^{j\pi}\right)^{-\alpha/2} = e^{-j(\alpha\pi/2)} \tag{16}$$

From Equation (14), if $\theta = -\alpha\pi/2$, Equation (16) is then given as:

$$j^{-\alpha} = \cos\left(\frac{\alpha\pi}{2}\right) - j\sin\left(\frac{\alpha\pi}{2}\right) \tag{17}$$

Thus, substituting Equation (17) into Equation (13), the impedance of a CPE is read as:

$$Z_{CPE} = Z_o\left[\cos\left(\frac{\alpha\pi}{2}\right) - j\sin\left(\frac{\alpha\pi}{2}\right)\right]\omega^{-\alpha} \tag{18}$$

For a resistor and a CPE connected in parallel, The Z' and Z'' are given by [Badwal, 1988]:

$$Z' = \frac{R\left[1 + (\omega\tau_m)^{1-\alpha}\sin\left(\frac{\alpha\pi}{2}\right)\right]}{1 + 2(\omega\tau_m)^{1-\alpha}\sin\left(\frac{\alpha\pi}{2}\right) + (\omega\tau_m)^{2(1-\alpha)}} \tag{19}$$

and

$$Z'' = \frac{R(\omega\tau_m)^{1-\alpha}\cos\left(\frac{\alpha\pi}{2}\right)}{1 + 2(\omega\tau_m)^{1-\alpha}\sin\left(\frac{\alpha\pi}{2}\right) + (\omega\tau_m)^{2(1-\alpha)}} \tag{20}$$

where τ_m is the inverse of the peak frequency.

A pure resistor and a pure capacitor in parallel connection gives a perfect semicircle in a Nyquist plot. A resistor and a CPE in parallel connection, on the other hand, gives a depressed semicircle with its centre below the horizontal axis by an angle $\alpha\pi/2$ as shown in Figure 14.

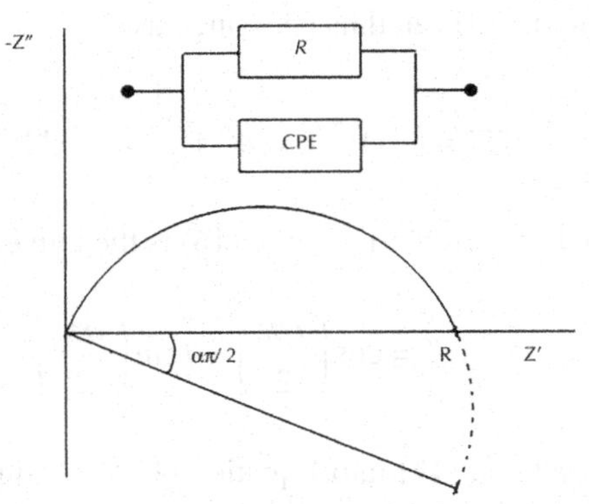

FIGURE 14 Depression of semicircle caused by replacing the pure capacitor with a CPE.

12.6.2 WARBURG ELEMENT

The fraction α has values between 0 and 1. When $\alpha = 0.5$, the CPE is called the Warburg element, W. The Warburg element is used to describe ionic diffusion (Macdonald, 1992) and the impedance is termed as Warburg impedance. In the case of intercalation electrodes, ionic species can diffuse at the interfaces, in the electrolyte or electrode and charge transfer can occur across the interfaces with resistance R_{ct}.

In a Nyquist plot, Warburg impedance is described by a straight line titled at 45° from the Z' axis as shown in Figure 15. Warburg impedance is frequency dependent. The value of Warburg impedance is low at high frequencies and high at low frequencies. This is because at high frequencies, due to the high periodic reversal of the applied field, the mobile ionic species do not have time to diffuse far. At low frequencies, due to the longer periodic reversal of the field, there is sufficient time for the ionic species to diffuse a greater distance.

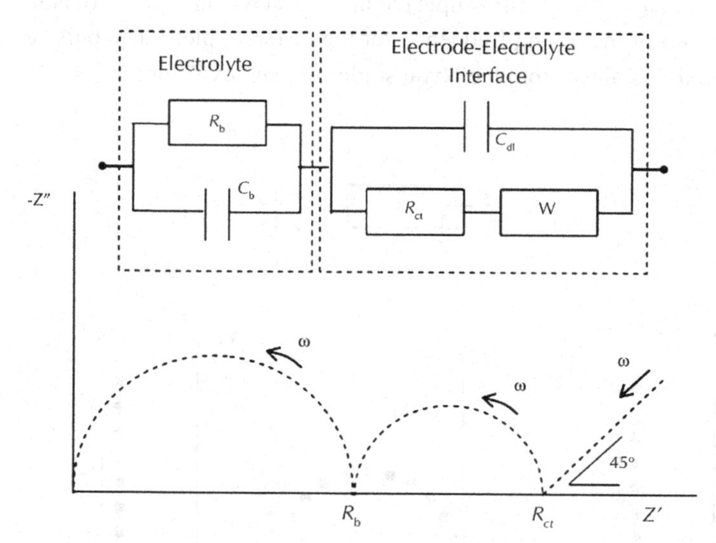

FIGURE 15 Nyquist plot and equivalent circuit for electrode-electrolyte interface.

12.7 IMPEDANCE RELATED FUNCTIONS: IMMITTANCES

There are other quantities related to impedance. These quantities are generally known as immittances and can be denoted by $I(\omega) = I' + jI''$. Table 2 lists the various immittance quantities such as complex admittance, Y, complex permittivity,

ε, and complex electrical modulus, M and their relationship to each other. Here, $\mu = j\omega C_o$ with $C_o = \varepsilon_o A/t$. Permittivity of free space, $\varepsilon_o = 8.854 \times 10^{-14}$ Fcm^{-1}, A is the sample-electrode contact area and t is the thickness of the sample.

TABLE 2 Relationships of immittance functions (Macdonald and Johnson, 2005)

	M	***Z***	***Y***	***ε***
M		μZ	μY^{-1}	ε^{-1}
Z	$\mu^{-1}M$		Y^{-1}	$\mu^{-1}\varepsilon^{-1}$
Y	μM^{-1}	Z^{-1}		$\mu\varepsilon$
ε	M^{-1}	$\mu^{-1}Z^{-1}$	$\mu^{-1}Y$	

The Z, Y, and M plots for simple circuits are shown in Figures 16 and 17. Note the direction of increasing frequency for the different plots. The bulk resistance, R_b is usually obtained from the Nyquist plot (Z" versus Z' plot).

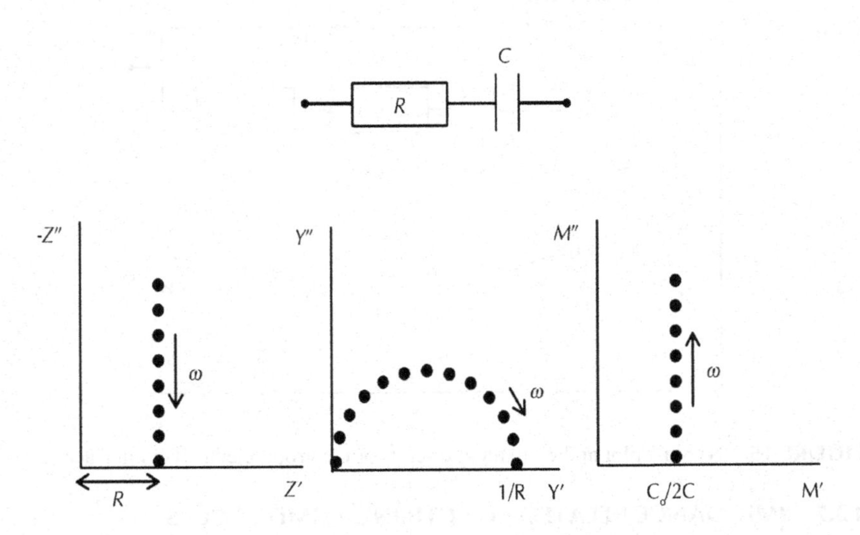

FIGURE 16 The Z, Y, and M plots for a pure resistor connected in series with a pure capacitor.

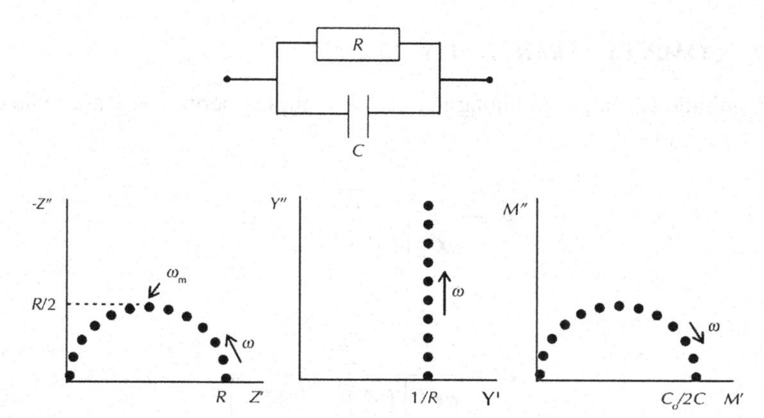

FIGURE 17 The Z, Y, and M plots for a pure resistor connected in parallel with a pure capacitor.

12.7.1 COMPLEX ADMITTANCE

From Table 2, the relationship between complex admittance and impedance is shown by:

$$Y(\omega) = \frac{1}{Z(\omega)}$$

$$= \frac{1}{Z' + jZ''}$$

$$= \frac{Z'}{(Z')^2 + (Z'')^2} - \frac{jZ''}{(Z')^2 + (Z'')^2} \quad (21)$$

with the real component, Y' and imaginary component, Y'' are given by Equation (22) and (23), respectively.

$$Y' = \frac{Z'}{(Z')^2 + (Z'')^2} \quad (22)$$

$$Y'' = \frac{Z''}{(Z')^2 + (Z'')^2} \quad (23)$$

12.7.2 COMPLEX PERMITTIVITY

The magnitude of real, ε' and imaginary, ε" of complex permittivity are related to Z' and Z'' as follows:

$$\varepsilon' = \frac{Z''}{\omega C_o\left[(Z')^2 + (Z'')^2\right]} \tag{24}$$

$$\varepsilon'' = \frac{Z'}{\omega C_o\left[(Z')^2 + (Z'')^2\right]} \tag{25}$$

where C_o and ω have their usual meanings.

12.7.3 COMPLEX ELECTRICAL MODULUS

The relationship between complex electrical modulus and impedance is given by:

$$\begin{aligned} M(\omega) &= \mu Z \\ &= -\omega C_o Z'' + j\omega C_o Z' \end{aligned} \tag{26}$$

Alternatively, complex electrical modulus can also represented by:

$$M(\omega) = \frac{1}{\varepsilon(\omega)} = \frac{\varepsilon'}{\left[(\varepsilon')^2 + (\varepsilon'')^2\right]} - \frac{j\varepsilon''}{\left[(\varepsilon')^2 + (\varepsilon'')^2\right]} \tag{27}$$

where the real, M' and the imaginary, M''components of the electrical modulus can be calculated from ε' and ε".

Each immittance highlights different features of an electrolyte system. The impedance, Z plots give prominence to resistive elements whereas the modulus, M plots give prominence to capacitances elements. The R_b values are usually extracted from the Z plots and the C values are extracted from the modulus peaks. Thus, for a complex electrolyte system, it is worthwhile to plot the impedance data in more than one immittance formalisms in order to extract all possible information.

12.8 EXTRACTING BULK RESISTANCE FROM THE NYQUIST PLOTS

The conduction property of a material could be identified by extracting the bulk resistance, R_b from the Nyquist plot. This is important in the study of polymer electrolytes. Figure 18 shows how to extract R_b from a Nyquist plot by graphical means. In order to do so, both the Z' and Z" axes must have the same scales. Extracting R_b from various forms of the Nyquist plots will be demonstrated in Figure 19. The accuracy of this graphical analysis will be discussed in the following section.

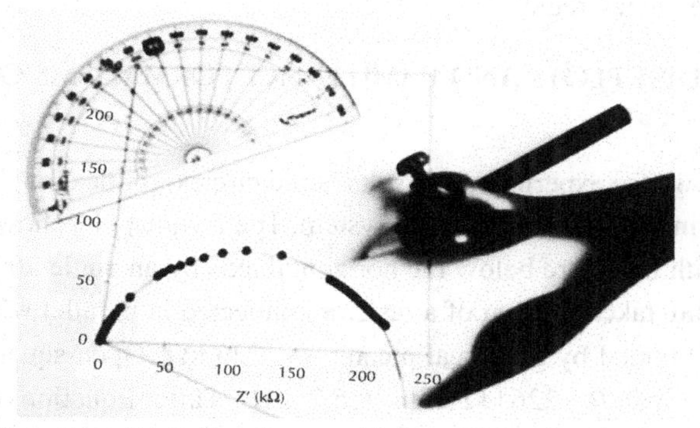

FIGURE 18 Extracting the bulk resistance, R_b from a Nyquist plot by graphical means (Note—Z' and Z" axes must have the same scale).

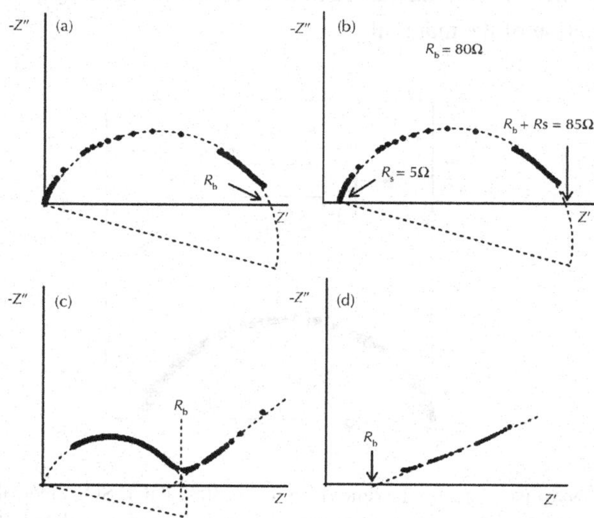

FIGURE 19 Extracting the bulk resistance, R_b from various forms of the Nyquist plots.

If a Nyquist plot takes the form of a semicircle (very often the semicircle is depressed in real systems) as shown in Figure 19(a), then the intercept of the semicircle on the Z' axis at the low frequency region gives the value of R_b. The plot in Figure 19(b) consists of a semicircle shifted at a distance R_s from the origin (0,0). The intercept of the semicircle on the Z' axis gives the value of $(R_b + R_s)$. Figure 19(c) shows a Nyquist plot with a semicircle at higher frequency followed by a spike at the low frequency end. R_b is obtained from the intersection of the semicircle and the spike. If the plot consists of only a steeply rising spike as depicted in Figure 19(d), the R_b is thus determined from the intercept on the Z' axis at the high frequency region.

12.9 NYQUIST PLOTS AND IMMITTANCE FORMALISMS OF A REAL SYSTEM

Figure 20 shows the experimental results (solid circles) for hexanoyl chitosan (M_w of ~2×10^5 g mol^{-1})-based electrolyte system. The Nyquist plot shows a depressed semicircle with its centre below the horizontal axis by an angle 20°. The equivalent circuit may take the form of a resistor connected in parallel with a CPE.The value of R_b obtained by graphical means was 226 kΩ. Open squares semicircle was calculated with $R = 226$ kΩ and $\tau_m = 5.3 \times 10^{-5}$ s after Equations (19) and (20). The intercept of the open squares semicircle on the Z' axis gives $R_b = 220$ kΩ. Thus the equivalent circuit that has been used to simulate the Nyquist plot gives a true representation of the material.

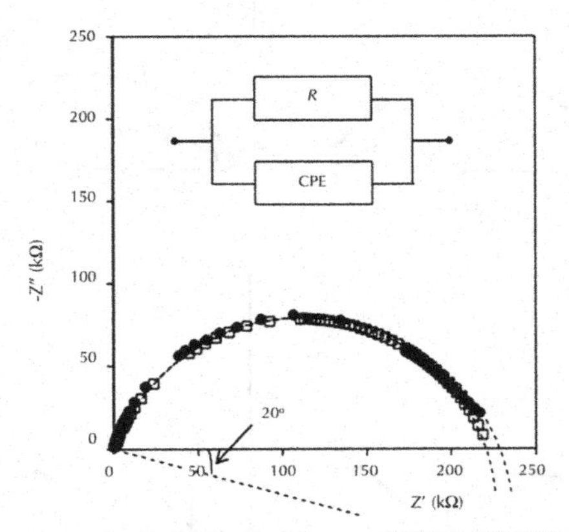

FIGURE 20 Nyquist plots for hexanoyl chitosan-30% $LiCF_3SO_3$ electrolyte system—solid circles experimental result, open squares semicircle was calculated after Equations (19) and (20).

If a capacitor replaces the CPE in the equivalent circuit (inset Figure 20), the Nyquist plot is expected to show a semicircle with center on the Z' axis. Substituting $R = 226$ kΩ and $C = 0.23$ nF into Equations (8) and (9), the simulation result shows a perfect semicircle center at (113 kΩ, 0 kΩ) as depicted in Figure 21.

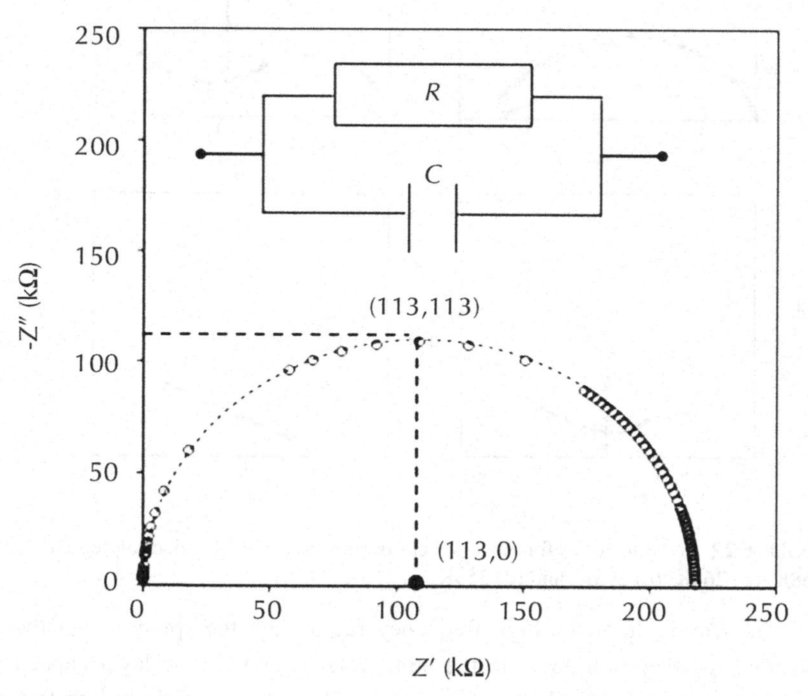

FIGURE 21 Nyquist plot for hexanoyl chitosan-30% $LiCF_3SO_3$ electrolyte system obtained after Equations (8) and (9).

Figure 22 shows the influence of temperature on the pattern of Nyquist plot. At 258K, the Nyquist plot takes the form of a depressed semicircle (Figure 22(a)). In contrast, the plot at temperature 268K consists of a depressed semicircle and a tilted spike. However, as the temperature rises, a semicircular spur is observed at the high frequency side followed by a spike at the low frequency end as shown in Figure 22(c). At higher temperatures (>300K), the absence of the high frequency semicircular portion is observed. The plot consists of only a steeply rising spike as depicted in Figure 22(d).

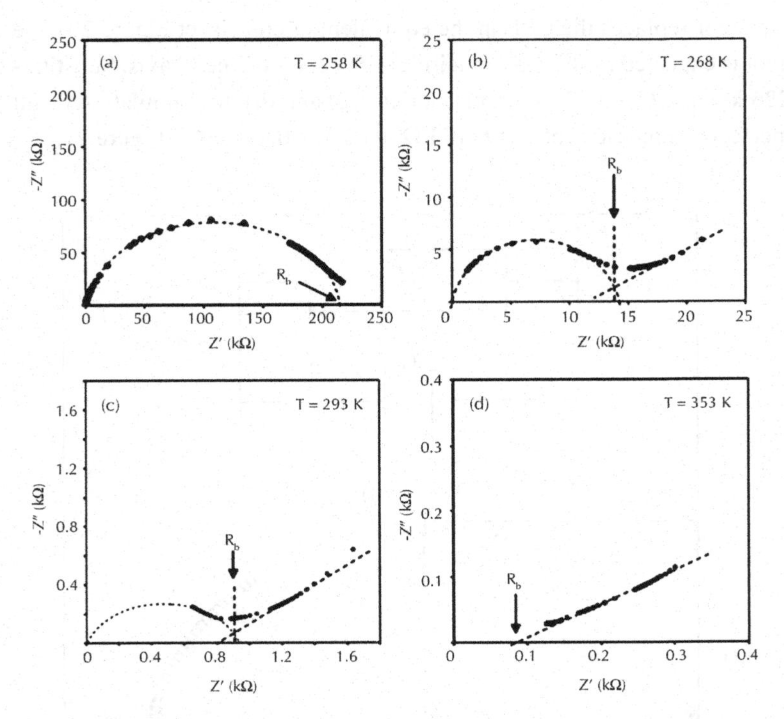

FIGURE 22 Nyquist plots for hexanoyl chitosan-30% $LiCF_3SO_3$ electrolyte system at (a) 258K, (b) 268K, (c) 293K, and (d) 353K.

The semicircle at the high frequency region and the spike at the low frequency region are attributed to the bulk material and double layer capacitance at the electrolyte-electrode interface, respectively. The semicircle was found to diminish with temperature increment. On the other hand, the spike becomes more prominent as the temperature rises. This may be due to the ions moving to the electrolyte-electrode interface as temperature increases.

It is clearly seen from Figure 22 that the value of R_b is strongly influenced by the experimental temperature, and is found to decrease with increase in temperature. The R_b value is related to the conductivity by the relation:

$$\sigma = \frac{t}{R_b A} \tag{28}$$

where t and A have their usual meanings.

The conductivities were calculated from Equation (28) and plotted against temperature as shown in Figure 23. Within the temperature range under investigation, the linear relationship with a regression value of 0.99 reveals that the temperature dependence of conductivity follows the Arrhenius law, suggesting that the conductivity is thermally assisted. The Arrhenius law can be expressed as:

$$\sigma = \sigma_o \exp\left(-\frac{E_a}{kT}\right) \tag{29}$$

where σ_o is the pre-exponential factor, E_a is the activation energy of ionic conduction, k is the Boltzmann constant, and T is temperature in Kelvin.

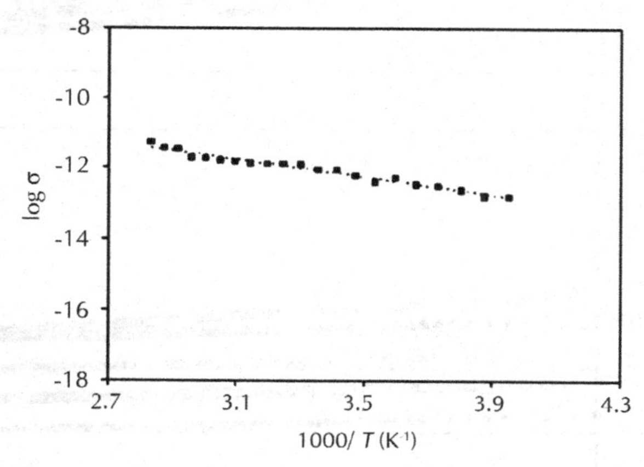

FIGURE 23 The temperature dependence of conductivity for hexanoyl chitosan-30% $LiCF_3SO_3$ electrolyte system.

Figure 24 shows the frequency dependence of dielectric constant, ε' at various temperatures. The dielectric constant is a measure of the capacitive or polarization nature of a material (Núñez et al., 2004). A dispersion with high value of dielectric constant is observed in the low frequency region. This is attributed to space charge polarization. In the polymer electrolyte system, under the influence of an applied field, ions tend to diffuse along the field appropriately. However, the ions are unable to cross the electrolyte-electrode interfaces due to the blocking electrodes used in the experiment. Thus, these ions accumulate at the interfaces and results in space charge polarization. The space charge polarization is more prominent at low frequencies. This is because at low frequencies, due to rather long periodic reversal of the applied field, there is sufficient time for more ions to accumulate at the interface.

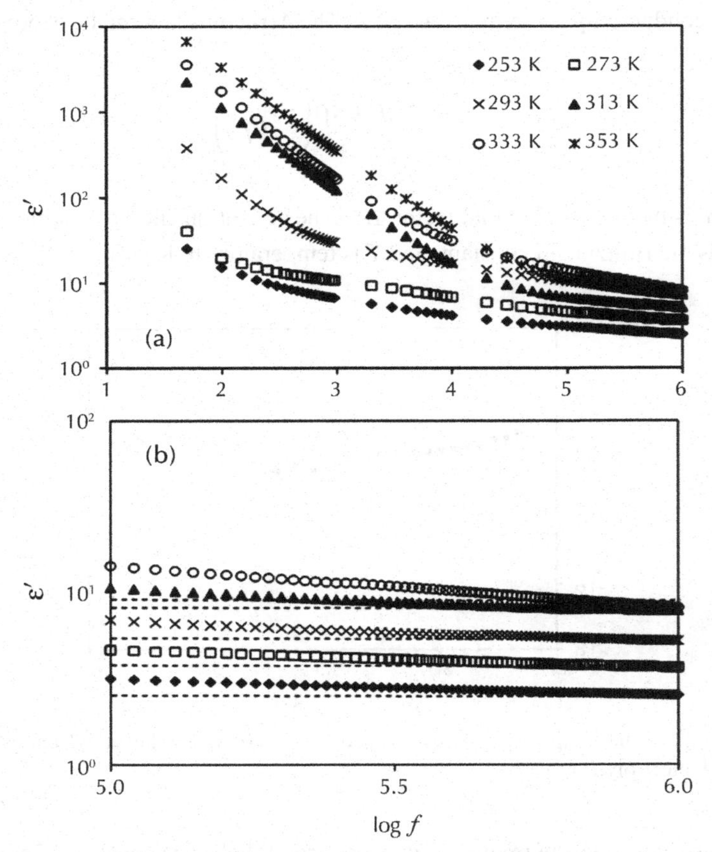

FIGURE 24 Frequency dependence of dielectric constant, ε' for hexanoyl chitosan-30% $LiCF_3SO_3$ electrolyte system at selected temperatures. (Dotted lines serve as guide to the eyes showing the values of dielectric constant are taken at the constant part of the log ε'-log f plots)

In the high frequency region, dielectric constant decreases with frequency and eventually becomes frequency independent as shown in Figure 24(b). The value of dielectric constant of the polymer electrolyte can be extracted and is tabulated in Table 3 for various temperatures. It is to be noted that the dielectric constant is taken at the constant part of the log ε'-log *f* plot. In case there is no constant region, dielectric constant is taken at the highest frequency available.

TABLE 3 The values of dielectric constant were extracted at the constant part of the log ε'-log *f* plots

Temperature (K)	Dielectric Constant
253	2.6
273	3.8
293	5.5
313	8.0
333	9.0

The dielectric constant is also a measure of the coulombic interaction between the anion and cation of a salt (Wintersgill and Fontanella, 1987). The higher degree of salt dissociation into free ions could be achieved with reduced anion-cation coulombic attraction. Thus, it can be deduced that the increase in dielectric constant implies the increase in the number of free ions (Winie and Arof, 2004). It can be observed from Table 3 that the value of dielectric constant increases with temperature indicating an increase in the number of free ions with temperature.

The frequency dependence of dielectric loss is depicted in Figure 25. Similar variation of dielectric loss is observed as in dielectric constant of Figure 24. The dielectric losses consist of conduction losses, dipole relaxation losses, and vibrational losses (Stevels, 1957). The conduction losses involve the migration of ions in the polymer network (bulk) under the influence of an applied field. As the ions migrate within the polymer matrix in the direction of the applied field, they give off part of the energy obtained from the field and transmit it to the network in the form of heat. The amount of heat lost per cycle is proportional to (σ/ω). The conduction, dipole relaxation and vibrational losses are proportional to (σ/ω). As the conductivity increases with temperature (see Figure 23), the conduction, dipole relaxation and vibrational losses increase causing a similar increase in the values of dielectric loss.

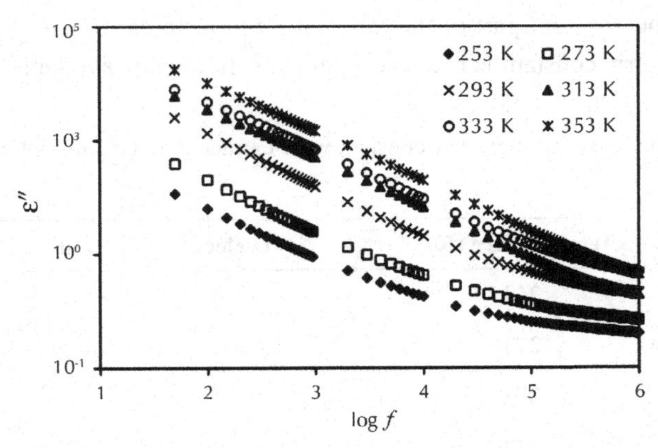

FIGURE 25 Frequency dependence of dielectric loss, ε" for hexanoyl chitosan-30% $LiCF_3SO_3$ at selected temperatures.

The effect of space charge polarization at low frequencies could be suppressed by plotting the data according to the electrical modulus formalism. This is manifested by *M*' and *M*" values approaching zero at low frequencies (Figures 26 and 27). *M*' shows an increase at high frequencies but well-defined dispersion peaks are not observed. On the other hand, plotting the *M*" with respect to frequency, results in the manifestation of dispersion peaks at 253K, 273K, and 293K. For higher temperatures, the dispersion peaks are expected to appear at frequencies beyond 1 MHz.The presence of dispersion peak in the *M*" versus log *f* plot is an indication that the material is an ionic conductor (Mellander and Albinsson, 1996).

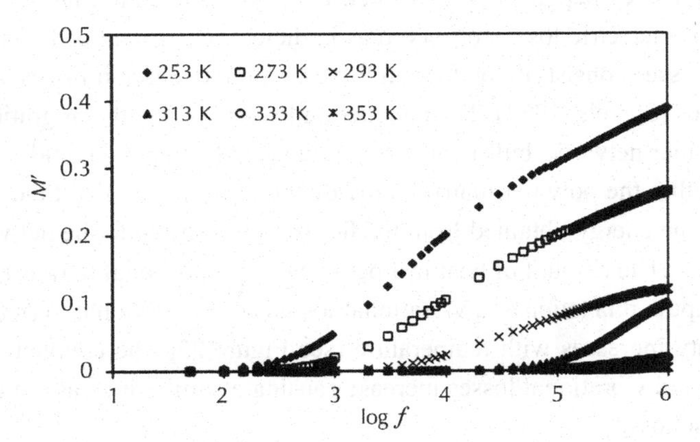

FIGURE 26 Frequency dependence of real part of electrical modulus, *M*' for hexanoyl chitosan-30% $LiCF_3SO_3$ electrolyte system at selected temperatures.

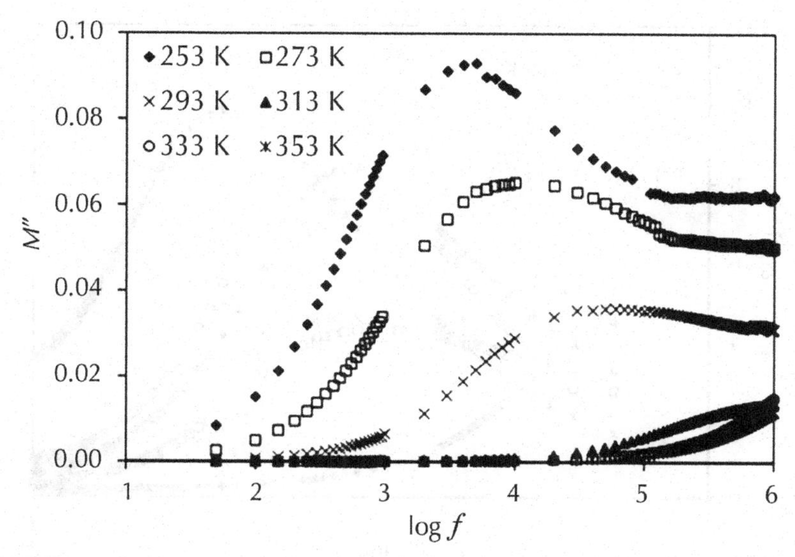

FIGURE 27 Frequency dependence of imaginary part of electrical modulus, *M*'' for hexanoyl chitosan-30% $LiCF_3SO_3$ electrolyte system at selected temperatures.

The loss tangent, tan δ is a ratio of energy loss to energy stored and can be calculated from:

$$\tan \delta = \frac{\varepsilon''}{\varepsilon'} \tag{30}$$

Figure 28 shows the variation of loss tangent as a function of frequency at various temperatures for the same sample. Relaxation times can be extracted from the loss tangent peak based on Equation (12) and are tabulated in Table 4. Relaxation time is found to vary with temperature. Within the frequency range under investigation, in this polymer electrolyte system, sources of dipoles may come from the salt that dissociate into cation-anion pairs, localized molecular polar groups and polar groups of hexanoyl chitosan. These dipoles may interact with each other or induce other dipoles surrounding it, causing a momentary delay in the alignment with the direction of the applied field. In other words, the charge and dipoles obey the change in the field direction at different times. This causes a distribution of relaxation times.

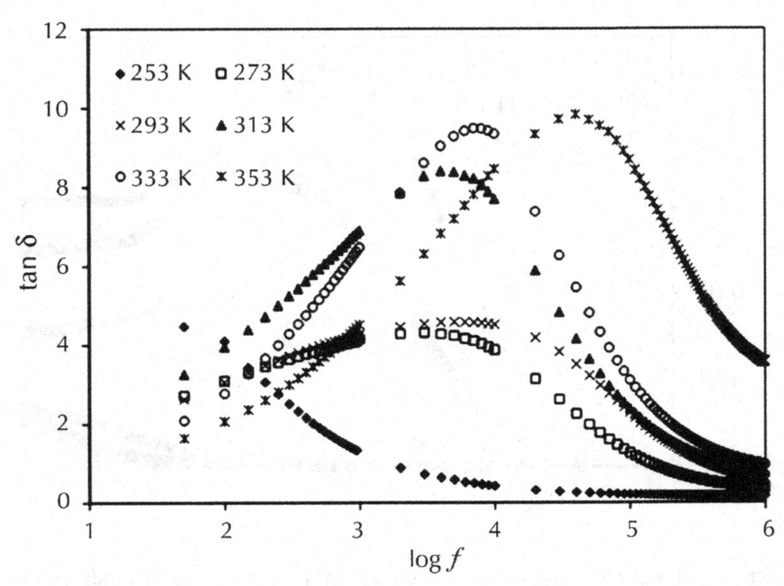

FIGURE 28 The variation of loss tangent with frequency for hexanoyl chitosan-30% $LiCF_3SO_3$ electrolyte system at selected temperatures.

TABLE 4 Variation of relaxation time with temperature for hexanoyl chitosan-30% $LiCF_3SO_3$ electrolyte system

Temperature, T (K)	Relaxation Time, τ (s)
273	5.30×10^{-5}
293	3.18×10^{-5}
313	3.98×10^{-5}
333	2.27×10^{-5}
353	3.98×10^{-6}

The asymmetric shape of the normalized plot of tan δ/ (tan $\delta)_{max}$ is a non-Debye behavior and can be interpreted as the consequence of the distribution of relaxation time. The normalized plot of tan δ/ (tan $\delta)_{max}$ shown in Figure 29 displaying a stretched exponential response which can be described by Kohlrausch–Williams–Watts (KWW) law (Macdonald, 2005):

$$\Phi(t) = \exp\left[-\left(\frac{t}{\tau_\sigma}\right)^\beta\right] \tag{31}$$

where τ_σ and β are the relaxation time and Kohlrausch exponent, respectively. The full width at half maximum (FWHM) of the tan δ/ (tan δ)$_{max}$ peaks are found to be broader than the Debye peak. FWHM of a Debye peak equals 1.14 decade (Dutta et al., 2002; Raistrick et al., 2005). The β values (β = 1.14/FWHM) are in the range of 0.35 to 0.56. The deviation of β value from unity implied the non-Debye relaxation (Dieterich and Maass, 2002).

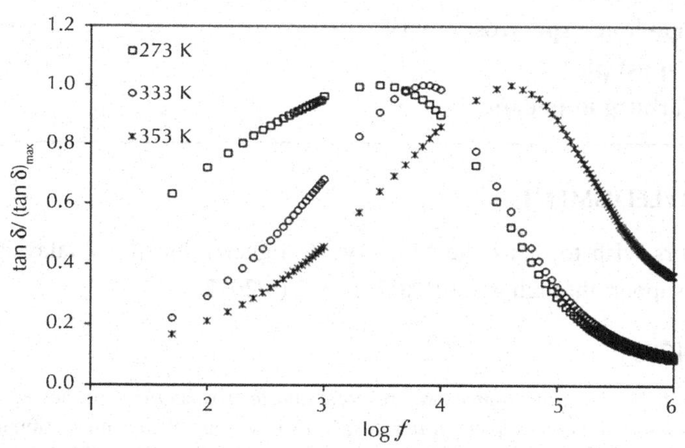

FIGURE 29 Normalized plot of loss tangent for hexanoyl chitosan-30% $LiCF_3SO_3$ electrolyte system at various temperatures.

12.10 CONCLUSIONS

An overview on the topic of IS, with emphasis on its application for electrical evaluation of polymer electrolytes is presented. This chapter begins with the definition of impedance and followed by presenting the impedance data in the Bode and Nyquist plots. Impedance data is commonly analyzed by fitting it to an equivalent circuit model. An equivalent circuit model consists of elements such as resistors and capacitors. The circuit elements together with their corresponding Nyquist plots are discussed. The Nyquist plots of many real systems deviate from the ideal Debye response. The deviations are explained in terms of Warburg and CPEs. The ionic conductivity is a function of bulk resistance, R_b, sample thickness and area. The extraction of R_b from various forms of the Nyquist plots is demonstrated.The immittances such as complex admittance, Y, complex permittivity, ε, and complex electrical modulus, M are related to impedance, Z. Each immittance highlights different features of the system. The loss tangent is a ratio of energy loss, ε" to energy stored, ε'. The relaxation time,τ can be extracted from the loss tangent peak. A plurality or distribution of relaxation times indicates the deviation from the ideal Debye response.

KEYWORDS

- **Constant phase element (CPE)**
- **Immittances**
- **Impedance spectroscopy (IS)**
- **Nyquist plot**
- **Warburg impedance**

ACKNOWLEDGMENT

The authors wish to thank the Ministry of Higher Education Malaysia for the financial support through grant RACE 16/6/2 (4/2012).

REFERENCES

1. Badwal, S. P. S. *Impedance spectroscopy and microstructural studies on materials for solid state electrochemical devices.* In Proceedings of the International Seminar Solid State Ionics Devices, B. V. R.Chowdari and Radhakrishna(Eds.), World Scientific: Singapore, pp. 165–189 (1988).
2. Bottelberghs, P. H. and Broers, H. H. J. *J. Electroanal.Chem.*, **67**, 155–167 (1976).
3. Dieterich, W. and Maass, P. *Chemical Physics*, **284**, 439–467 (2002).
4. Dutta, P., Biswas, S., and De, S.K. *Materials Research Bulletin*, **37**, 193–200 (2002).
5. Irvine, J. T. S., Sinclair, D. C., and West, A. R. *Adv. Mater.*, **2**, 132–138 (1990).
6. Macdonald, J. R. Characterization of the electrical response of high resistivity ionic and dielectric solid materials by immittance spectroscopy. *In Impedance spectroscopy theory, experiment, and applications*, E. Barsoukov and J. R. Macdonald (Eds.), John Wiley & Sons Inc.: New Jersey, pp. 264–281 (2005).
7. Macdonald, J. R. and Johnson, W. B. Fundamentals of impedance spectroscopy. *In Impedance spectroscopy theory, experiment, and applications*, E. Barsoukov and J. R. Macdonald (Eds.), John Wiley & Sons Inc.: New Jersey, pp. 1–26 (2005).
8. Macdonald, J.R. *Annals of Biomedical Engineering*, **20**, 289–305 (1992).

9. Mellander, B. E. and Albinsson, I. Electric and dielectric properties of polymer electrolytes. In Solid state ionics: new developments, B. V. R. Chowdari, M. A. K. L. Dissanayake, and M. A.Careem (Eds.), World Scientific: Singapore, pp. 83–96 (1996).
10. Núñez, L., Barreiro, S. G., Fernández, C. A. G., and Núñez, M. R. *Polymer*, **45**, 1167–1175 (2004).
11. Raistrick, I. D., Franceschetti, D. R., and Macdonald, J. R. *Theory. In Impedance spectroscopy theory, experiment, and applications*; E. Barsoukov and J. R.Macdonald (Eds.), John Wiley & Sons Inc.: New Jersey, pp. 27–128 (2005).
12. Sinclair, D. C. *Bol. Soc. Esp. Cerám. Vidrio*,**34**, 55–65 (1995).
13. Stevels, J. M. *Hantbuch der physic*, pp. 350–391 (1957).
14. Winie, T. and Arof, A. K. *Ionics*, **10**, 193–199 (2004).
15. Wintersgill, M. C. and Fontanella, J. J. Low-frequency dielectric properties of polyether electrolytes. *In Polymer electrolyte reviews II*, J. R.MacCallum and C. A.Vincent (Eds.), Elservier Applied Science: London, pp. 43–60 (1987).

Part 2: Advanced Polymeric Materials—Macro to Nanoscales

CHAPTER 13

PREPARATION OF CHITIN-BASED NANOFIBROUS AND COMPOSITE MATERIALS USING IONIC LIQUIDS

JUN-ICHI KADOKAWA

CONTENTS

13.1 INTRODUCTION

The polysaccharides are widely distributed in nature and have been regarded as structural materials and as suppliers of water and energy (Schuerch, 1986). In the many kinds of polysaccharides, cellulose and chitin are the most important biomass resources because they are the first and second most abundant natural polysaccharides on the earth, respectively (Klemm et al., 2005, Muzzarelli, 2011). Chitin is a structurally similar polysaccharide as cellulose, but that has acetamido groups at the C-2 position in place of hydroxy groups of the glucose residues in cellulose (Figure 1) (Kurita, 2006, Rinaudo, 2006, Pillai et al., 2009). Thus, the main-chain structure of chitin consists of β-(1→4)-linked 2-acetamido-2-deoxyl-D-glucopyranose (*N*-acetyl-D-glucosamine) residues. Despite its huge production in nature and easy accessibility, chitin still remains as an unutilized biomass resource primary because of its intractable bulk structure and insolubility in water and common organic solvents, and thus, only limited attention has been paid to chitin, principally from its biological properties.

Chitin is the second abundant natural polysaccharide after cellulose as aforementioned, there is major interest in conversion into various useful materials after proper dissolution in suitable solvents. Chitin occurs mainly in the exoskeletons of crustacean shells such as crab and shrimp shells in nature. Native chitin in crustacean shells is arranged as microfibrils embedded in a protein matrix (Raabe et al., 2006). The microfibril consists of nanofibers with 2–5 nm diameters. Moreover, three types of crystalline forms, that is, α-, β-, and γ-chiti ns, are known. The most abundant form is α-chitin (e.g., from crab and shrimp shells) (Minke and Blackwell, 1978), where the polymeric chains are aligned in an antiparallel fashion as same as cellulose II. This arrangement is favorable for the formation of strong intermolecular hydrogen bonding, leading to the most stable form in the three types of crystalline structures. In β-chitin (e.g., from squid pen), the polymeric chains are packed in a parallel arrangement (Gardner and Blackwell, 1975), that corresponds to cellulose I, resulting in weaker intermolecular forces. Accordingly, β-chitin is considered to be less stable than α-chitin. The γ-chitin may be a mixture of α- and β-forms.

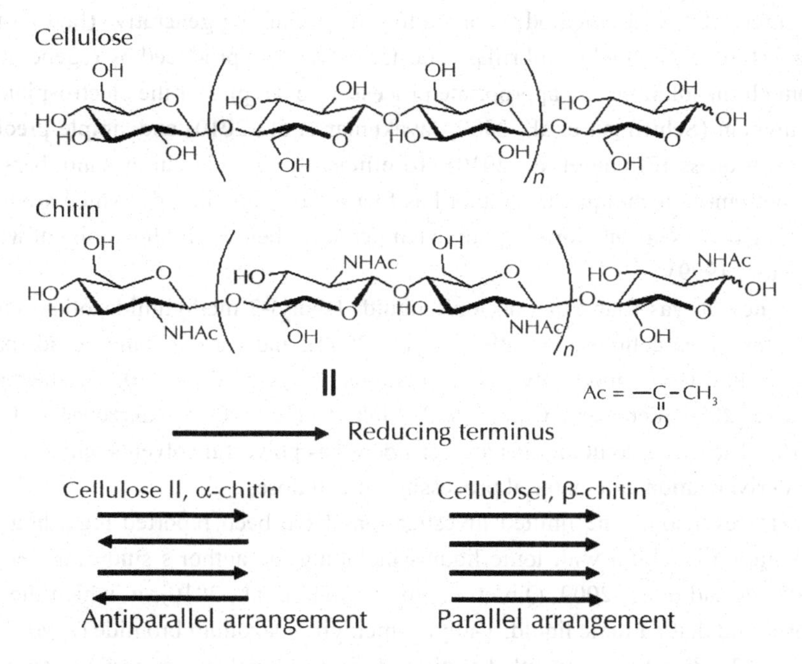

FIGURE 1 Chemical and crystalline structures of cellulose and chitin.

The preparation of nano-scaled polymeric assemblies such as nanofibers is one of the most useful methods to practically utilize polymeric materials as observed in the case of cellulose (Abe at al., 2007, Saito et al., 2006, Saito et al., 2007). For example, self-assembled fibrillar nanostructures from cellulose are promising materials for the practical applications in bio-related research fields such as tissue engineering (Isogai et al., 2011, Abdul Khalil et al., 2012). The efficient methods have also been developed for the preparation of chitin nanofibers. The conventional approaches to the production of chitin nanofibers are mainly performed upon top-down procedures that break down the starting bulk materials from chitin resources (Figure 2).

As the representative techniques, acid hydrolysis (Revol and Marchessault, 1993, Li et al., 1996, Li et al., 1997, Goodrich and Winter, 2007), 2,2,6,6-tetramethylpiperidine-1-oxyl radical (TEMPO)-mediated oxidation (Fan et al., 2008, Fan et al., 2009), and grinding technique (Ifuku et al., 2009, Ifuku et al., 2010) were previously performed to produce chitin nanofibers from the native chitin resources. The other method accorded to self-assembling generative (bottom-up) route (Figure 2), in which fibrillar nanostructures were produced by regeneration from chitin solutions *via* appropriate process, as examples of the electrospinning equipment (Schiffman et al., 2009, Jayakumar et al., 2010) and simple precipitation process (Zhong et al., 2010). To efficiently provide chitin nanofibers by the bottom-up technique, the author has focused on ionic liquids, which are low-melting point salts that form liquids at temperatures below a boiling point of water (Welton, 1999).

Since, it was found that an ionic liquid, 1-butyl-3-methylimidazolium chloride, dissolved cellulose (Swatloski et al., 2002), and various ionic liquids have been found to be good solvents for cellulose (Seoud et al., 2007, Liebert and Heinze, 2008, Feng and Chen, 2008, Pinkert et al., 2009, Zakrzewska et al., 2010). Therefore, ionic liquids are considered as powerful solvents and are used for derivatization and material processing of cellulose.

However, only the limited investigations have been reported regarding the dissolution of chitin with ionic liquids including the author's study (Wu et al., 2008, Prasad et al., 2009, Qin et al., 2010, Wang et al., 2010), in which the author found that an ionic liquid, 1-allyl-3-methylimidazolium bromide (AMIMBr) (Figure 3), dissolved or swelled chitin to form weak gel-like materials (ion gels) (Kadokawa, 2011, Kadokawa 2012).

In the following study, furthermore, it was reported that chitin nanofiber films were facilely obtained by regeneration from the gels using methanol, followed by filtration (Kadokawa et al., 2011). This chapter reviews the studies on the preparation of chitin nanofibers and the following fabrications of chitin-based nanofibrous and composite materials using the ionic liquid, AMIMBr.

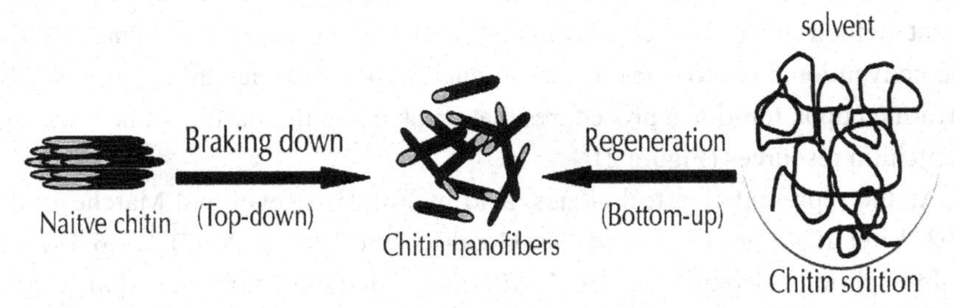

FIGURE 2 Top-down and bottom-up approaches for preparation of chitin nanofibers.

13.2 DISSOLUTION OF CHITIN WITH IONIC LIQUID

As aforementioned, the limited studies on dissolution of chitin with ionic liquids have been reported. For example, it was reported that 1-butyl- and 1-ethyl-3-methylimidazolium acetates dissolved chitin (Figure 3) (Wu et al., 2008, Qin et al., 2010). For the dissolution study of chitin with ionic liquids, the author noted the previous study reporting that the imidazolium-type ionic liquids having a bromide counter anion were the good solvents for synthesis of polyamides and polyimides (Vygodskii et al., 2002). The result inspired to use the same kind of ionic liquids for dissolution of chitin, because chitin has the –N-C=O groups as same as polyamides and polyimides. Thus, three ionic liquids consisting a bromide anion were prepared, which were AMIMBr, 1-methyl-3-propylimidazolium bromide, and 1-butyl-3-methylimidazolium bromide, and the dissolution experiments were conducted.

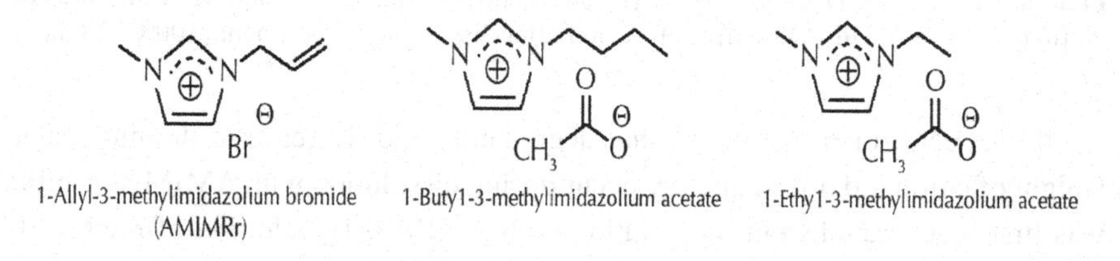

FIGURE 3 The representative ionic liquids that dissolve chitin.

A mixture of chitin with each ionic liquid was heated at 100°C and the heating process was simply followed by a charge coupled device (CCD) camera on glass plate with 200 times magnification scale. When AMIMBr was used for the experiment, the clear solutions of chitin were formed in the concentrations up to ~5% (w/w) (Figure 4(a) and (b)), whereas the other ionic liquids did not show ability to dissolve chitin even in 1% (w/w) concentration.

The disappearance of chitin powder in the solution of 5% (w/w) chitin with AMIMBr was confirmed further by the SEM measurement (Figure 4(c) and (d)). The SEM image of the mixture of chitin with AMIMBr after heating it at 100°C for 48 hr did not show any solid of chitin, unlike that observed in the SEM image before heating, suggesting that 5% (w/w) chitin was solvated with AMIMBr at least at the μm scale level.

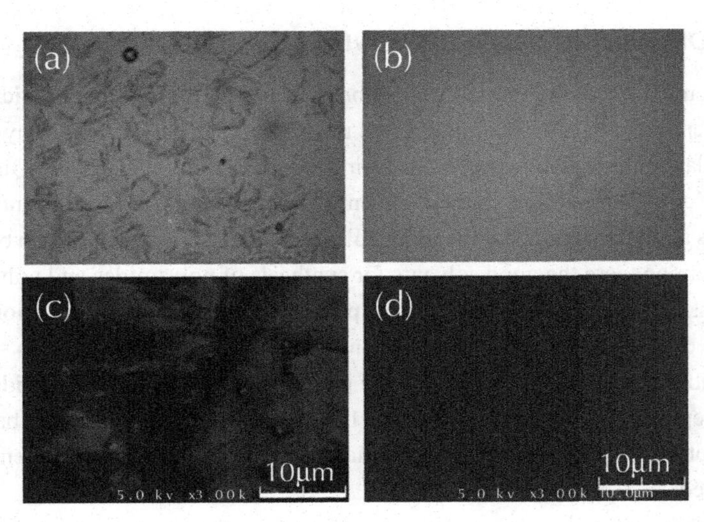

FIGURE 4 The CCD camera views (a and b) and SEM images (c and d) of mixtures of chitin (5% (w/w)) with AMIMBr before and after dissolution experiment (100°C, 48 hr).

It was confirmed as follows that degradation and decrease in the molecular weight of chitin did not frequently occur during dissolution with AMIMBr. Chitin was first regenerated from the solution with AMIMBr by addition of methanol, followed by filtration and dryness of the precipitate. The powder X-ray diffraction (XRD), thermal gravimetrical analysis (TGA), and IR results of the regenerated material were almost same as those of the original chitin. In addition, the hexanoyl derivative of the regenerated chitin, which was prepared by the reaction of the regenerated material with hexanoyl chloride (Kaifu et al., 1981), showed the exact same ^{1}H nuclear magnetic resonance (NMR) pattern as that of the hexanoylated sample from the original chitin. The GPC profiles of both the derivatives were not much different from each other.

When 7–12% amounts (w/w) of chitin were immersed in AMIMBr at room temperature for 24 hr, followed by heating at 100°C for 48 hr and cooling to room temperature, the mixtures gave gel-like materials (ion gels) with higher viscosity. Indeed, the obtained 7% (w/w) chitin with AMIMBr did not flow upon leaning a test tube, whereas the aforementioned 5% (w/w) chitin with AMIMBr started to flow upon leaning (Figure 5). However, the dynamic rheological measurements showed that both 5% (w/w) and 7% (w/w) chitins with AMIMBr behaved as the weak gels. Furthermore, the much lower yield stress of 5% (w/w) chitin with

AMIMBr than that of 7% (w/w) chitin with AMIMBr supported that 5% (w/w) chitin with AMIMBr can flow under gravitation.

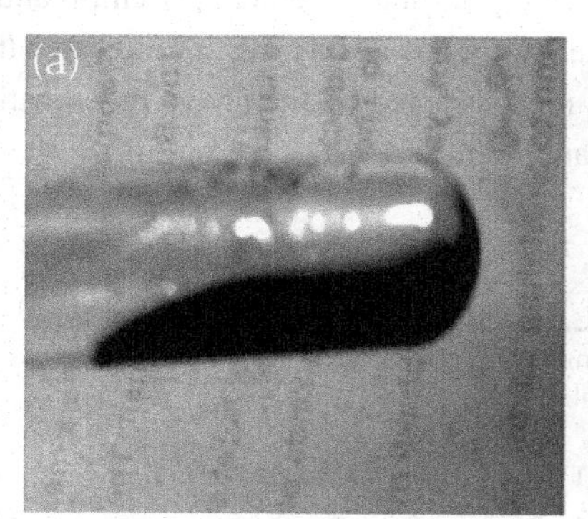

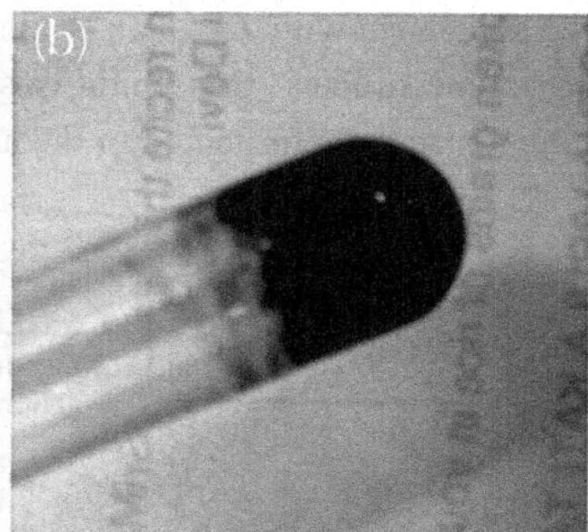

FIGURE 5 Photographs of 5% (w/w) (a) and 7% (w/w) (b) chitin with AMIMBr.

13.3 PREPARATION OF NANOFIBER FILMS FROM THE CHITIN GEL WITH IONIC LIQUID

The author has reported that chitin nanofiber films were facilely obtained by regeneration from the aforementioned ion gels with AMIMBr using methanol, followed by filtration (Figure 6) (Kadokawa et al., 2011). First, chitin was swollen with AMIMBr according to the procedure reported in the previous study as aforementioned to give chitin gels with AMIMBr (10–12% (w/w)). It was found that chitin dispersions were obtained when the gels were treated with methanol at room temperature for 24 hr to slowly regenerate chitin, followed by sonication. The resulting dispersion was diluted with methanol, which was subjected to the SEM measurement. The morphology of nanofibers with around 20–60 nm in width and several hundred nanometers in length was seen in the SEM image of the sample from the dispersion (Figure 7(a)), indicating the formation of the chitin nanofibers by the above gelation and regeneration procedures of chitin. When the dispersion was subjected to the filtration to isolate the regenerated chitin, the residue formed a film, which was further purified by Soxhlet extraction with methanol. The SEM image of the resulting film was also measured to confirm the nano-scaled morphology of chitin, which showed the pattern of highly entangled nanofibers (Figure 7(b)). Such entangled structure of the nanofibers probably contributed to formability of the film. The weight of the obtained chitin film indicated that AMIMBr used was mostly removed by the above regeneration and Soxhlet extraction procedures with methanol. The XRD pattern of the chitin nanofiber film mainly showed four diffraction peaks at around 9.5°, 19.5°, 20.9°, and 23.4°, which typically corresponded to crystalline structure of α-chitin and is in good agreement with that of an original chitin powder. The result indicated that the crystalline structure of α-chitin was reconstructed by the above regeneration procedure during the formation of the nanofibers.

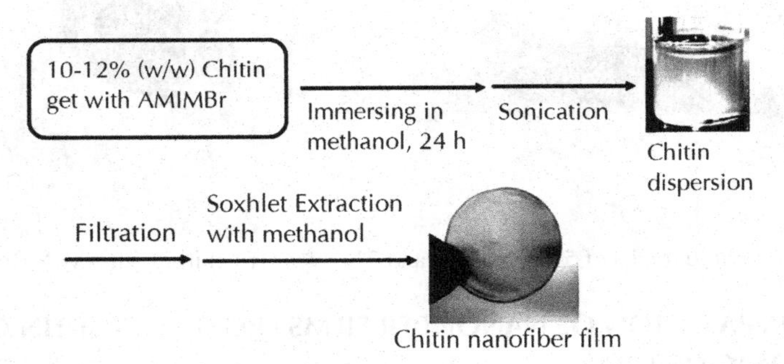

FIGURE 6 Procedures for the preparation of chitin dispersion and nano fiber film.

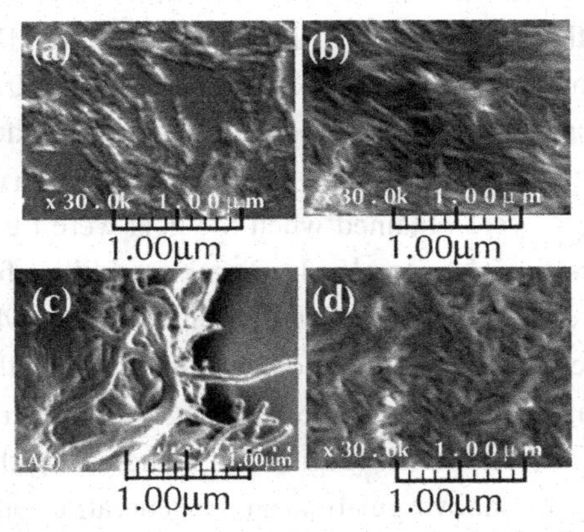

FIGURE 7 The SEM images of (a) chitin dispersion obtained using AMIMBr, (b) chitin nanofiber film, (c) regenerated chitin obtained using 1-butyl-3-methylimidazolium acetate, (d) and chitin nonofiber-PVA composite film.

An attempt was also made to prepare the chitin film using 1-butyl-3-methylimidazolium acetate, which was reported as another ionic liquid dissolving chitin as aforementioned (Wu et al., 2008), according to the same procedure using AMIMBr as described. However, the regenerated chitin was not dispersed in methanol and was precipitated as aggregates. Indeed, the isolated chitin by filtration of the resulting mixture did not form film. The SEM image of the regenerated chitin exhibited the morphology of large aggregates and did not obviously show the nano-scaled fiber morphology (Figure 7(c)). These results indicated no formation of the chitin nanofibers using 1-butyl-3-methylimidazolium acetate, suggesting that the chitin nanofibers were specifically formed using AMIMBr. The specificity of AMIMBr was probably owing to the difference in the dissolution states of chitin in the two ionic liquids because viscosity of the chitin solution in AMIMBr was much lower than that of the other solution.

13.4 PREPARATION OF CHITIN NANOFIBROUS COMPOSITE MATERIALS

The chitin has limited applications besides such traditional purposes. Therefore, considerable efforts have been still devoted to compatibilization of chitin with synthetic polymers to provide chitin-based new functional materials. As one of the possible applications of the present chitin nanofiber film, therefore, attempts were made to prepare the chitin nanofiber composite materials with synthetic polymers. Two kinds of approaches, that is, physical and chemical approaches have been considered to yield the polysaccharide-synthetic polymer composite materials (Figure 8). In former case, the polysaccharide and synthetic polymer chains construct material components by physical interaction in the composites, whereas the latter approach results in the formation covalent linkages between two polymer chains in the composites.

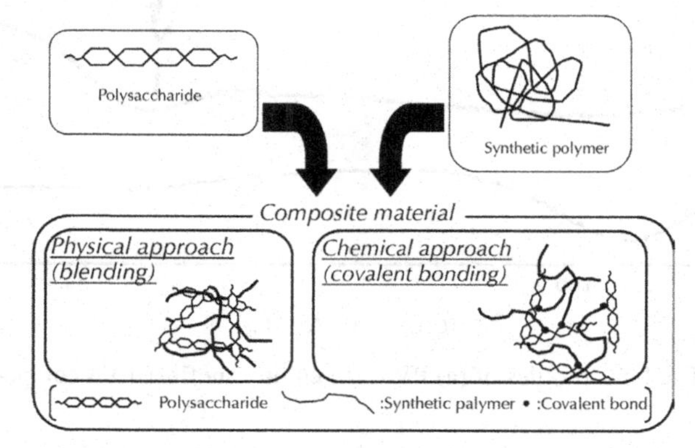

FIGURE 8 Polysaccharide-synthetic polymer composite materials by blending and chemical bonding.

By the physical approach, the chitin nanofiber poly(vinyl alcohol) (PVA) composite films were prepared (Kadokawa et al., 2011). First, the 10% (w/w) chitin gel with AMIMBr was prepared according to the aforementioned procedure and solution of the desired amount of PVA (DP = 4300 approximately) in hot water were added to the gel (the feed weight ratio of chitin to PVA = 1:0.30). Then, the regeneration, filtration, and Soxhlet extraction procedures gave the chitin nanofiber PVA composite material, which also became film form. It was determined that most of AMIMBr was removed, but a small amount still remained in the composite. The SEM image of the composite showed that the nanofiber-like morphology was maintained and PVA components probably filled in spaces among the fibers (Figure 3(d)), indicating relative immiscibility of chitin and PVA in the composite. Indeed, the DSC profile of the composite film exhibited an endothermic peak ascribed to a melting point of PVA (indicated with an arrow, Figure 9(b)). These results suggested some immiscibility of PVA with chitin in the composite. However, the melting point peak of PVA in the composite film was broadened, indicating that the crystallinity of PVA decreased in the composite film. The DSC result suggested that chitin and PVA might be partially miscible at the interfacial area between the two polymers in the composite film probably by the formation of hydrogen bonding between them or by the presence of a small amount of AMIMBr.

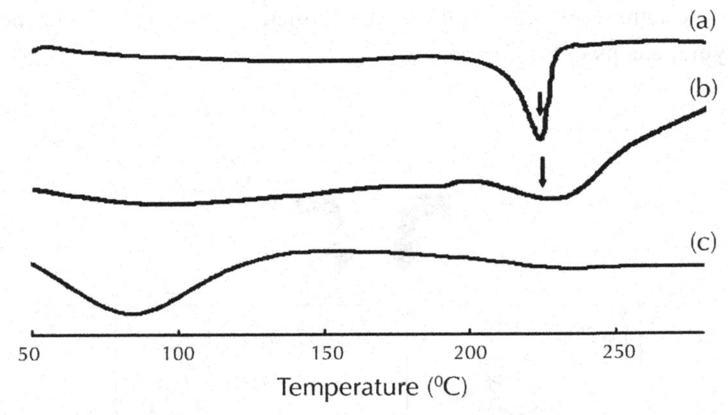

FIGURE 9 The DSC profiles of (a) PVA, (b) chitin nanofiber-PVA composite film, and (c) chitin.

The mechanical properties of the composites were evaluated by the stress–strain curves under tensile mode. Both the values of fracture stress and strain of the composite film were larger than those of the chitin nanofiber film (Figure 10). These results suggested the presence of PVA components in the composite film contributed to enhancement of the mechanical property.

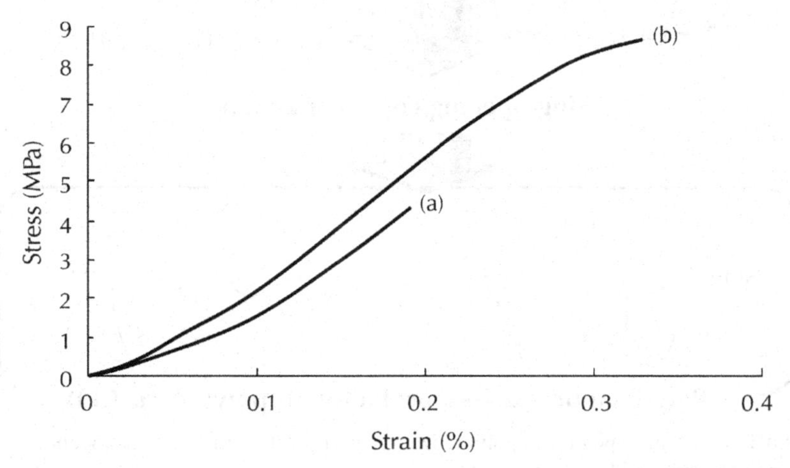

FIGURE 10 Stress-strain curves of (a) chitin nanofiber film and (b) chitin nanofiber PVA composite film.

Chitin-based eco-friendly composite materials have been prepared by the physical compatibilization with aliphatic biodegradable polyesters such as poly(L-lactide) (polyLA) and poly(ε-caprolactone) (polyCL) (Min et al., 2004, Li and Feng, 2005, Kim et al., 2009, Li et al., 2009). Because the biodegradable polyesters are efficiently synthesized by ring-opening polymerization of the corresponding cyclic ester monomers, such as L-lactide (LA) and ε-caprolactone (CL), initiated with hydroxy groups in the presence of Lewis acid catalysts such as tin(II) 2-ethylhexanoate (Figure 11) (Lou et al., 2003, Jérôme and Lecomte, 2008), on the other hand, the ring-opening graft polymerization technique using chitin as a multifunctional initiator with a number of hydroxy groups has also been employed as the chemical approach to provide chitin/biodegradable polyester composite materials connected by covalent linkages (Kim et al., 2002, Fujioka et al., 2004, Fujioka et al., 2005, Jayakumar and Tamura, 2008).

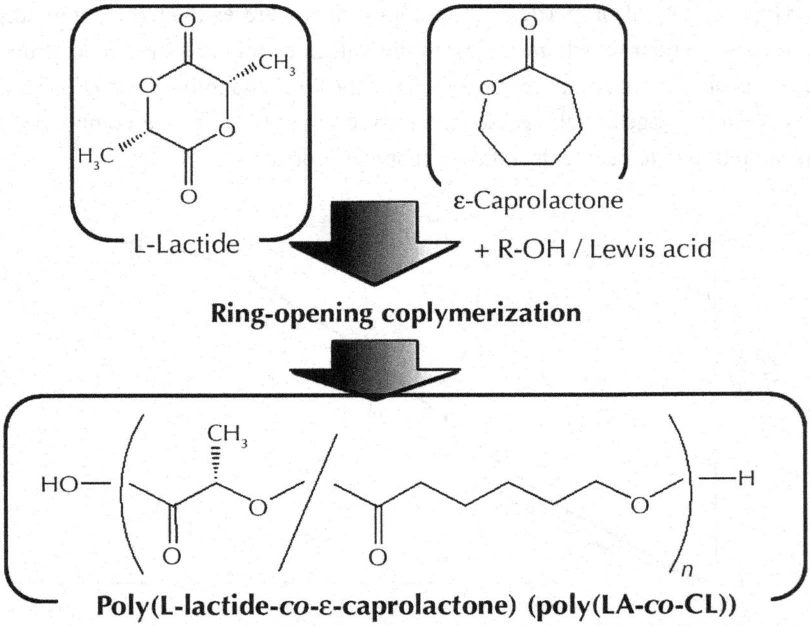

FIGURE 11 Ring-opening copolymerization of LA/CL initiated with alcohol in the presence of Lewis acid.

To incorporate the biodegradable polyesters on the aforementioned chitin nanofiber films, surface-initiated ring-opening graft copolymerization of LA/CL monomers initiated from hydroxy groups on the chitin nanofiber was attempted to produce chitin nanofiber-*graft*-poly(L-lactide-*co*-ε-caprolactone) (chitin nanofiber-*g*-poly(LA-*co*-CL)) films (Figure 12) (Setoguchi et al., 2012). To efficiently initiate the ring-opening graft copolymerization of LA/CL on surface of the nanofibers, spaces among the fibers were made by immersing the film in water for 10 s. The SEM image of the pre-treated film indicated that more spaces among the fibers were made by immersing in water (Figure 13(a) and (b)). However, the XRD result indicated the crystalline structure in the fibers was still remained even after the pre-treatment.

Then, the surface-initiated ring-opening graft copolymerization of LA/CL (feed molar ratio = 20:80) from the pre-treated film was carried out. First, the film was immersed in a solution of LA/CL in toluene. Then, the polymerization catalyst, tin(II) 2-ethylhexanoate (around 5 mol% for the monomers), was added and the mixture was heated at 80°C for 48 hr to take place the ring-opening copolymerization. After the resulting film was washed with chloroform and further subjected to Soxhlet extraction with chloroform, it was dried under reduced pressure to give chitin nanofiber-*g*-poly(LA-*co*-CL) film.

The IR spectrum of the resulting film exhibited carbonyl absorption at 1739 cm^{-1} due to the ester linkage of poly(LA-*co*-CL) (Figure 14), suggesting the presence of poly(LA-*co*-CL) in the product, which was explanatorily bound to the nanofibers by covalent linkage. The unbounded poly(LA/CL) was also produced by the ring-opening copolymerization initiated from the moisture present in the system. It, however, was removed during the purification procedures.

The amount (weight ratio to the original nanofiber film) and the LA/CL composition ratio of the grafted poly(LA-*co*-CL) on the film were evaluated by the weight difference of the films before and after the graft copolymerization and the ^{1}H NMR analysis of the alkaline-hydrolysis product to be 12.2% and 40/60, respectively. The LA/CL composition ratio in the polyester was higher than that in the feed (20/80) because of the higher reactivity of LA than CL in the copolymerization (Grijpma and Pennings, 1991).

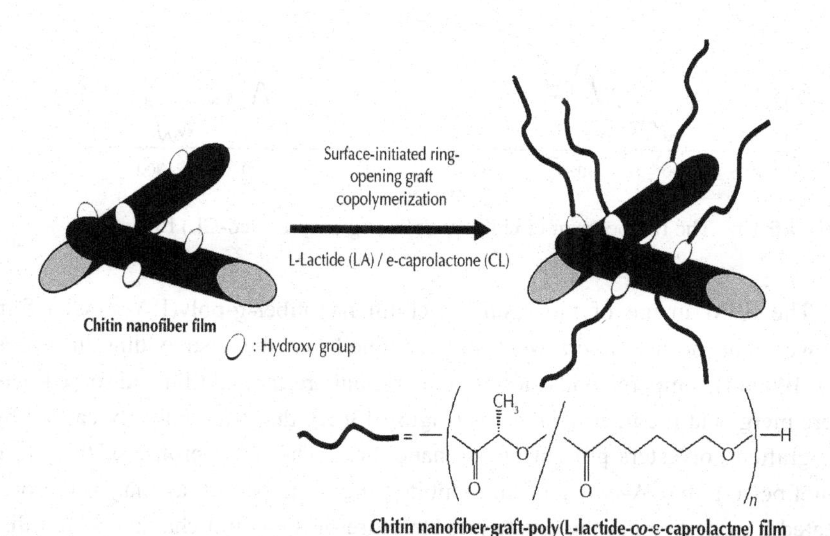

FIGURE 12 Surface-initiated ring-opening copolymerization of LA/CL from chitin nanofiber film.

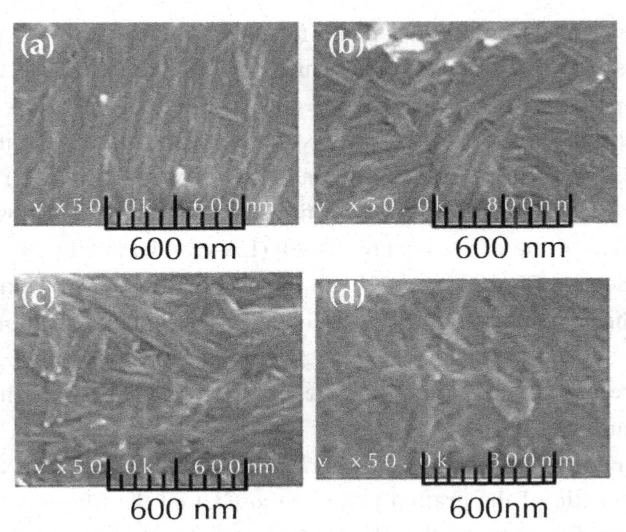

FIGURE 13 The SEM images of (a) chitin nanofiber film, (b) pre-treated chitin nanofiber film, and chitin nanofiber-g-poly(LA-co-CL) films (LA/CL feed molar ratios; (c) 20:80 and (d) 50:50).

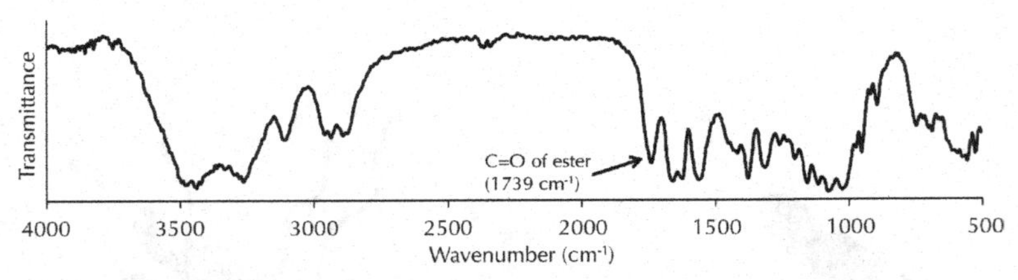

FIGURE 14 The IR spectrum of chitin nanofiber-g-poly(LA-co-CL) film.

The SEM image of the resulting chitin nanofiber-*g*-poly(LA-*co*-CL) film showed that the nanofibers were still remained, but the fiber widths increased (60–100 nm) compared with those of the original pre-treated film and some fibers were merged at the interfacial areas (Figure 13(c)), that was probably caused by the grafted polyesters present on the nanofibers. The XRD profile of the chitin nanofiber-*g*-poly(LA-*co*-CL) film exhibited the same pattern as that of the pre-treated film, indicating the crystalline structure of the chitin chains was not disrupted owing to the occurrence of the graft copolymerization only on surface of the nanofibers. When the surface-initiated graft copolymerization in the various

LA/CL feed molar ratios was performed, the grafting amounts of poly(LA-*co*-CL)s increased with increasing the LA/CL feed molar ratios. Moreover, the LA/CL composition ratios in the polyesters were always higher than the feed ratios. The SEM images of the chitin nanofiber-*g*-poly(LA-*co*-CL) film with the higher grafting amount (20%) obtained by the LA/CL feed molar ratio = 50:50 showed the morphologies where the nanofibers were frequently covered by the grafted polyesters (Figure 13(d)) compared with that of the chitin nanofiber-*g*-poly(LA-*co*-CL) film with the lower grafting amount (12.2%, Figure 13(c)).

The stress-strain curves of the chitin nanofiber-*g*-poly(LA-*co*-CL) films under tensile mode exhibited the larger fracture strain values (4.3–6.2%) than those of the original pre-treated film. Furthermore, the fracture stress values relatively tended to increase with increasing the amounts and the LA/CL composition ratios of the grafted polyesters, whereas the fracture strain values decreased in this order. These data suggested that the mechanical properties of the chitin nanofiber-*g*-poly(LA-*co*-CL) films were strongly affected by the amounts and the LA/CL composition ratios of the grafted polyesters. In comparison with the aforementioned chitin nanofiber PVA blend films (Kadokawa et al., 2011), the present chitin nanofiber-*g*-poly(LA-*co*-CL) films showed much better mechanical properties.

13.5 CONCLUSIONS

This chapter overviewed the preparation of chitin-based nanomaterials such as chitin nanofibers and nanocomposites with synthetic polymers using the ionic liquid. First, dissolution and gelation of chitin with the ionic liquid, AMIMBr, were discussed. Then, the preparation of chitin nanofibers by regeneration from the gel with AMIMBr was described. Immersing the gel in methanol and subsequent sonication gave the chitin dispersion. The SEM image of the dispersion showed the formation of chitin nanofibers. Then, filtration of the dispersion was carried out to give the chitin nanofiber film. The composite film of the chitin nanofibers with PVA was prepared by a similar procedure as that for the chitin nanofiber film. The preparation of chitin nanofiber-*g*-poly(LA-*co*-CL) films by surface-initiated ring-opening graft copolymerization of LA/CL monomers from the chitin nanofiber film was also described. The IR, XRD, and SEM measurements of the obtained film indicated that the graft copolymerization of LA/CL was initiated from hydroxy groups on surface of the chitin nanofibers. The mechanical properties of the composite films were evaluated by tensile testing. The present methods provide new chitin-based nanofibrous and composite materials, leading to the efficient use of chitin as the material source and will have potential to apply in practical material research filed in the future.

KEYWORDS

- **1-Allyl-3-methylimidazolium bromide (AMIMBr)**
- **Chitin**
- **Ionic liquid**
- **Nanofibers**
- **Polysaccharides**

REFERENCES

1. Abdul Khalil, H. P. S.; Bhat, A. H.; Ireana Yusra, A. F. Carbohydr. Polym. 2012, 87, 963–979.
2. Abe, K.; Iwamoto, S.; Yano, H. Biomacromolecules 2007, 8, 3276-2378.
3. Fan, Y.; Saito, T.; Isogai, A. Biomacromolecules 2008, 9, 192-198.
4. Fan, Y.; Saito, T.; Isogai A. Carbohydr. Polym. 2009, 77, 832-838.
5. Feng, L.; Chen, Z. I. J. Mol. Liq. 2008, 142, 1-3.
6. Fujioka, M.; Nagashima, A.; Kenjo, H.; Sakurai, K.; Nishiyama, S.; Noguchi, H.; Ishii, S.; Yoshida, Y. Sen'i Gakkaishi 2005, 61, 282-285.
7. Fujioka, M.; Okada, H.; Kusaka, Y.; Nishiyama, S.; Noguchi, H.; Ishii, S.; Yoshida, Y. Macromol. Rapid Commun. 2004, 25, 1776-1780.
8. Gardner, K. H.; Blackwell, J. Biopolymers 1975, 14, 1581–1595.
9. Goodrich, J. D.; Winter, W. T. Biomacromolecules 2007, 8, 252-257.
10. Grijpma, D. W.; Pennings, A. J. Polym. Bull. 1991, 25, 335-341.
11. Ifuku, S.; Nogi, M.; Abe, K.; Yoshioka, M.; Morimoto, M.; Saimoto, H.; Yano, H. Biomacromolecules 2009, 10, 1584-1588.
12. Ifuku, S.; Nogi, M.; Yoshioka, M.; Morimoto, M.; Yano, H.; Saimoto H. Carbohydr. Polym. 2010, 81, 134-139.
13. Isogai, A.; Saito, T.; Fukuzumi, H. Nanoscale 2011, 3, 71–85.
14. Jayakumar, R.; Prabaharan, M.; Nair, S. V.; Tamura, H. Biotech. Adv. 2010, 28, 142-150.
15. Jayakumar, R,; Tamura, H. Int. J. Biol. Macromol. 2008, 43, 32-36.
16. Jérôme, C.; Lecomte, P. Adv. Drug. Deliv. Rev. 2008, 60, 1056-1076.
17. Kadokawa, J. Preparation of polysaccharide-based materials compatibilized with ionic liquids. In Ionic liquids, application and perspectives; Kokorin, A., Ed.; InTech: Rijeka, 2011, pp 95-114.
18. Kadokawa, J. Preparation of functional ion gels of polysaccharides with ionic liquids. In Handbook of ionic liquids: properties, applications and hazards; Mun, J.; Sim, H., Eds.; Nova Science Publishers: Hauppauge, 2012, pp 455-466.
19. Kadokawa, J.; Takegawa, A.; Mine, S.; Prasad, K. Carbohydr. Polym. 2011, 84, 1408-1412.

20. Kaifu, K.; Nishi, N.; Komai, T. J. Polym. Sci.: Polym. Chem. Ed. 1981, 19, 2361–2363.
21. Kim, J. Y.; Ha, C. S.; Jo, N. J. Polym. Int. 2002, 51, 1123-1128.
22. Kim, H. S.; Kim, J. T.; Jung, Y. J.; Hwang, D. Y.; Son, H. J.; Lee, J. B.; Ryu, S. C.; Shin, S. H. Macromol. Res. 2009, 17, 682-687.
23. Klemm, D.; Heublein, B.; Fink, H.-P.; Bohn, A. Angew. Chem. Int. Ed. 2005, 44, 3358–3393.
24. Kurita, K. Mar Biotechnol. 2006, 8, 203–226.
25. Li, X.; Feng, Q. Polym. Bull. 2005, 54, 47-55.
26. Li, X.; Liu, X.; Dong, W.; Feng, Q.; Cui, F.; Uo, M.; Akasaka, T.; Watari, F.; J. Biomed. Mater. Res. B 2009, 90B, 502-509.
27. Li, J.; Revol, J. F.; Marchessault, R. H. J. Appl. Polym. Sci. 1997, 65, 373-380.
28. Li, J.; Revol, J. F.; Naranjo, E.; Marchessault, R. H. Int. J. Biol. Macromol. 1996, 18, 177-187.
29. Liebert, T.; Heinze, T. BioResources 2008, 3, 576-601.
30. Lou, X.; Detrembleur, C.; Jérôme, R. Macromol. Rapid Commun. 2003, 24, 161-172.
31. Min, B. M.; You, Y.; Kim, J. M.; Lee, S. J.; Park, W. H. Carbohydr. Polym. 2004, 57, 285-292.
32. Minke, R.; Blackwell, J. J. Mol. Biol. 1978, 120, 167–181.
33. Muzzarelli, R. A. A. Chitin nanostructures in living organisms. In Chitin formation and diagenesis; Gupta, S. N., Eds.; Springer: New York, 2011; Chapter 1.
34. Pillai, C. K. S.; Paul, W.; Sharma, C. P. Prog. Polym. Sci. 2009, 34, 641–678.
35. Pinkert, A.; Marsh, K. N.; Pang, S.; Staiger, M. P. Chem. Rev. 2009, 109, 6712-6828.
36. Prasad, K.; Murakami, M.; Kaneko, Y.; Takada, A.; Nakamura, Y.; Kadokawa, J. Int. J. Biol. Macromol. 2009, 45, 221-225.
37. Qin, Y.; Lu, X.; Sun, N.; Rogers, R. D. Green. Chem. 2010, 12, 968-971.
38. Raabe, D.; Romano. P.; Sachs, C.; Fabritius, H.; Al-Sawalmih, A.; Yi, S. B.; Servos, G.; Hartwig, H. G. Mater. Sci. Eng. A-Struct. Mater. Prop. Microstruct. Process. 2006, 421, 43-153.
39. Revol, J. F.; Marchessault, R. H. Int. J. Biol. Macromol. 1993, 15, 329-335.
40. Rinaudo, M. Prog. Polym. Sci. 2006, 31, 603–632.
41. Saito, T.; Kimura, S.; Nishiyama, Y.; Isogai, A. Biomacromolecules 2007, 8, 2485–2491.
42. Saito, T.; Nishiyama, Y.; Putaux, J. L.; Vignon, M.; Isogai, A. Biomacromolecules 2006, 7, 1687–1691.
43. Schiffman, J. D.; Stulga, L. A.; Schauer, C. L. Polym. Eng. Sci. 2009, 49, 1918-1928.
44. Schuerch, C. Polysaccharides. In Encyclopedia of polymer science and engineering, 2nd edition; Mark, H. F.; Bilkales, N.; Overberger, C. G., Eds.; John Wiley & Sons: New York, 1986; Vol. 13, pp 87-162.
45. Seoud, O. A. E.; Koschella, A.; Fidale, L. C.; Dorn S.; Heinze, T. Biomacromolecules 2007, 8, 2629-2647.
46. Setoguchi, T.; Yamamoto, K.; Kadokawa, J. Polymer 2012, 53, 4977-4982.
47. Swatloski, R. P.; Spear, S. K.; Holbrey, J. D.; Rogers, R. D. J. Am. Chem. Soc. 2002, 124, 4974-4975.
48. Vygodskii, Y.S.; Lozinskaya, E.L.; Shaplov, A.S. Macromol. Rapid Commun. 2002, 23, 676–680.

49. Wang, W. T.; Zhu, J.; Wang, X. L.; Huang, Y.; Wang, Y. Z. J. Macromol. Sci., Part B Phys. 2010, 49, 528–541.
50. Welton, T. Chem. Rev., 1999, 99, 2071-2083.
51. Wu, Y.; Sasaki, T.; Irie, S.; Sakurai, K. Polymer 2008, 49, 2321-2327.
52. Zakrzewska, M. E.; Lukasik, E. B.; Lukasik, R. B. Energy Fuels 2010, 24, 737-745.
53. Zhong, C.; Cooper, A.; Kapetanovic, A.; Fang, Z.; Zhang, M.; Rolandi, M. Soft Matter 2010, 6, 5298-5301.

CHAPTER 14

FIRE-RESIST BIO-BASED POLYURETHANE FOR STRUCTURAL FOAM APPLICATION

KHAIRIAH HAJI BADRI and
AMAMER MUSBAH OMRAN REDWAN

CONTENTS

14.1 FIRE-RESIST BIO-BASED POLYURETHANE FOR STRUCTURAL FOAM APPLICATION

14.1.1 POLYURETHANE (PU) AND ITS PROPERTIES

PU is a versatile thermosetting polymer. Its properties are modified either by varying the microstructure or by dispersing either inorganic or organic fillers within the PU continuous matrix (Oertel 1993). PU is a high molecular weight polymer based on the polyaddition of polyfunctional hydroxyl group and isocyanate. There are two types of polyhydroxyl compounds commonly used namely polyester and polyether polyols. Polyester is a high molecular weight substance which contain ester group as the repeating unit in the chain. The mechanical properties and morphological structure of PU depend mainly on the polyol structure, molar mass and its functionality and to a lesser extent, on the nature of the polyisocyanate (Badri et al. 2000). PU foam is formed by gas evolution (either carbon dioxide or chlorofluorocarbon) trapped during the polymerization process between the hydroxyl-containing group compound with polyisocyanate to form urethane chain. The resulted foams would exhibit the relationship between physical and mechanical properties with their chemical composition and density of materials (Oertel 1993).

At present, most polyols used in PU industry are petrochemical-based where crude petroleum oil is used as a starting raw material. However, this material is becoming expensive, rate of depletion is high and they require high processing technology. This necessitates looking at utilizing plants that can serve as an alternative feed stock for monomers in polymer industry. Palm oil (Badri et al. 2000, Badri et al. 2001, Chian & Gan 1998), soybean oil (Guner et al. 2006), and vernonia oil (Kolot & Grinberg 2004) have been converted to polyol to replace petrochemical-based polyol.

The most important commercial PU products are foams that are commonly classified as either flexible or rigid depending on their mechanical performance and cross-link densities. Rigid PU foams are widely used in building insulation and domestic appliances, due to their superior mechanical properties and low density. However, the PU industry is facing environmental challenges due to type of auxiliary blowing agents used during the polymerization process (Zhong et al. 2002). Due to that, n-pentane and hydrochlorofluorocarbon HCFC-141b are of better choices compared to chlorofluorocarbon (CFC). However, uncertainty about the cost and availability of chlorofluorocarbons has led the PU foam industry to focus on pentane as the primary blowing agent, especially in construction industry. Unfortunately, the inherent high flammability of pentane has resulted in PU foams that fail to meet the required regulatory fire tests. Therefore, PU industry has to respond to these challenges by designing additive packages (catalysts, surfactants, cross-linkers, and flame retardants) that can overcome these flammability shortcomings, while providing physical and mechanical properties comparable to those of rigid PU foams blown with HFC or HCFC (Tang et al. 2002).

Polyurethanes, like all organic materials, burn in the presence of oxygen (O_2) under the action of heat. This cause the generation of toxic smoke and reduces O_2 concentration in the surroundings. The application of PU, which is combustible, leads to an increase in fire risk. Flame retardants have been widely used for plastics, rubbers, coatings, and fibers to reduce the risks for applications in areas where safety is essential, such as aircraft, building/construction, public transport, and housings for electrical equipment. In these cases they have to be made incombustible, or at least difficult to ignite and burn. The focus of the present study is to enhance the mechanical and thermal properties and importantly the fire resistivity of PU composites by using phosphite as an additive.

14.1.2 FLAME RETARDANT

Flame retardant is a class of materials that when compounded into plastics, provides a specific reaction during combustion. These reactions cause the initially flammable substances to ignite with more difficulty and will inhibit the propagation compared to the original substrate under laboratory test conditions (Frank 2000). Flame retardant is a substance that can be chemically inserted into polymer molecules or physically blended in polymer after polymerization to suppress, reduce or delay the propagation of a flame through plastic materials. Combustion is supported by three critical elements—energy, O_2, and fuel. For flame retardants to be effective, they must somehow interfere with one or more of those three elements.

Several mechanisms can account for flame retardant functions. Heat absorption can occur through the release of water during the burning process. Char forming can occur that insulates the polymer substrate and also retard the fire by reducing the fuel needed to continue the burning process. Chemical reaction must occur in the flame front for the fire to continue burning. These reaction can be halted or slowed through interference by species (free radicals) generated from flame retardant chemicals during the burning process. Flame retardants are added to polymeric materials, both natural and synthetic, to enhance the flame retardancy properties of the polymers (Innes & Innes 2001). There are two main categories of flame-retardant chemicals, additive flame retardant and reactive flame retardant.

Additive flame retardants are the great majority of the flame retardant. They include halogenated flame retardants (Lyons 1970), organophosphorus and nitrogen-based organic flame retardants (Lewin 2001). All of them are incorporated into the plastic either prior to, during, or more frequently after polymerization reaction. They are used especially in thermoplastics such as acrylonitrile butadiene styrene (ABS), polystyrene (PS), polycarbonate (PC), and thermoplastic elastomers. Sometimes, an inorganic material is added to the compounds to show synergistic effects. The most important one is antimony trioxide (Sb_2O_3) that has no perceptible flame retardant effect on its own. Together with halogen-containing compounds, however, it produces a marked synergistic effect. One of the most important disadvantages of these types of flame retardants (additives) is their volatility and thus, their flame retardancy effect may be gradually lost (Pitts 1972).

A reactive flame retardant is just a small part relative to an additive flame retardant. They are bonded chemically to the polymer chain together with other starting components. This prevents them from bleeding out of the polymer and vaporized, and thus their flame retardancy is retained. They have no plasticizing effect and should not affect on thermal stability of the basic polymer. They are used mainly in thermosets, especially polyesters, epoxy resins, and polyurethanes in which they can be easily incorporated (Rumack et al. 1987). Their high price is the limiting parameter for their wide application. Halogenated phenol and tetrachlorophthalic anhydride are example for reactive flame retardant.

Phosphites are the type of decomposed hydroperoxide secondary antioxidants. The organic phosphite is used widely as affirmer to any polymer type and to keep its physical polymer nature such as its color and molecular weight during processing. Activity such as hydroperoxide decomposition decreased due to its increasing ability in receiving electron and its group size attach to phosphite and blocked aryl phosphite (Schwetlick 1990). Figure 1 shows a type of phosphite antioxidant that is used as a stabilizer for polymer and is used in this study. Irgafos 168 is a phosphite type of antioxidant with low volatility and extreme resistant to hydrolysis. It provides excellent antioxidizing stability to organic polymers, resistance to discoloration and adds long-term protection against thermo-oxidative degradation after processing.

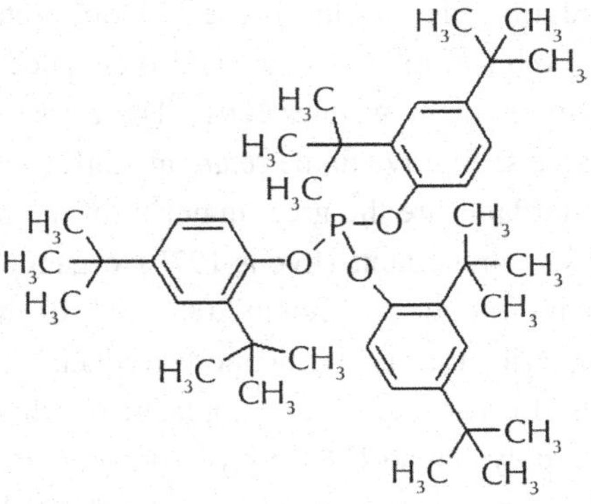

FIGURE 1 The molecular structure of Irgafos 168 [tris (2, 4-di-tert-butylphenyl) phosphite]. (Schwetlick 1990)

This type of phosphite can act as primary antioxidizing agent in the chain scissoring and chain transfer agent in its reaction with alkyl peroxide and then forms hindered aryloxyl radical that can prevent chain auto-oxidation reaction. Nevertheless, hindered aryl phosphate's activity as antioxidant in chain scissoring is lower compared to hindered phenol (Schwetlick 1990). Trivalent phosphorus compounds such as commercialized brand name Irgafos are excellent hydroperoxide decomposers. It is used as flame retardant and works once fire is set. Phosphite flame retardants dissociates into radical species that compete with chain propagating and branching steps in the combustion process.

14.1.3 FIRE RETARDED PU COMPOSITES

The flame retardants are chemical substances used in various products such as plastics, textiles, and furnishing foam to reduce their fire hazards by interfering with the composition of the polymeric materials. The use of polymeric materials as adhesives or as organic composites, for naval, aeronautic or electronic applications, becomes more and more important. However, one of the most important disadvantages of these materials concerns is their thermal and fire resistance behaviors (Aseeva & Zaikov 1985).

To limit the flammability of these materials, it is necessary to incorporate fire retardant (FR) in the polymer. The molecules containing phosphorous groups belong to these fire retarding agents allowing improved material thermal and fire properties. Phosphorus, silicon, and nitrogen compounds are the agents mostly used for replacing halogenated compounds in flame retarding resins, especially epoxies. The halogen free products are attractive and welcome worldwide. Organophosphorus polymers have already been the subject of several works because of the large variety of applications of these products. For example, they can improve adhesive properties and are thus used as adhesion promoters, paints, lacquers, and as adhesives. They are also used for fibers and films with a high mechanical resistance, as an ion exchanger and as lubricants (Spirckel et al. 2002).

Halogen-free flame retardants (HFFR) have also received much attention because of the absence of toxic gases and smokes during combustion compared to halogen-type flame retardants. Intumescent flame retardants, containing phosphorus and nitrogen have become more and more attractive in various industrial applications in recent years. During combustion, the HFFR forms an expanding charred crust to protect the underlying polymeric material from further attack from flame or heating (Zhu & Shi 2001). The action of flame retardants can occur across both or either of the vapor phase and the condensed phase. Halogen-type flame retardants usually follow the vapor phase mechanism while intumescent systems containing phosphorus and nitrogen generally follow the condensed phase mechanism. However, combustion is a complex process, there may be different mechanisms with different flame retardants.

For many years, halogenated hydrocarbons, such as CF_3Br, were used as fire suppressants. However, due to their high ozone depletion potential, they are no longer being manufactured in industrialized countries, as stipulated in the 1990 Montreal Protocol. The search for effective replacements has led to a family of organophosphorus compounds (OPCs) that have shown considerable promise as flame inhibitors. The early work of Twarowski (1993) demonstrated that phosphine (PH_3) accelerated radical recombination in hydrogen oxidation, and subsequent work by Korobeinichev et al. (2004) began to explain how OPCs inhibited hydrogen flames and hydrocarbon flames. Chemically active flame inhibitors alter flame chemistry by catalytic recombination of key flame radicals, especially H and O atoms, and OH radicals. The H atoms are particularly important in flame propagation, since the principal chain branching reaction in hydrogen and hydrocarbon flames is $H + O_2 \rightarrow OH^{\cdot} + O^{\cdot}$. Fast elementary reactions interconnect these small radical species, and removal of any of them through recombination reduces concentrations of all of them correspondingly. Therefore, radical recombination leads to fewer H atoms in the reaction zone, which leads to reduced chain branching and a lower burning velocity in a premixed flame (Korobeinichev et al. 1999).

The need for fire retarding building materials, such as rigid PU foam insulation, has led to the development of many new phosphorus-containing polyether and polyesters. Piechota (1965) has shown the value of phosphorus compounds in such fire retarding systems, and that phosphorus-containing rigid PU foams develop an optimum flame protection at phosphorus content of 1.5%. In general, chemically bond phosphorus compounds are preferred over noncreative additive types. The rigid PU foams containing chemically bond phosphorus flame retarding agents retain essentially all of their phosphorus and most of the initial flame resistance after exposure to various combinations of heat and moisture (Gjurova et al. 1986). The test formulations employing noncreative flame retardants became non-self-extinguishing because of loss of phosphorus during one or more of the aging tests employed. Hydrolytic stability was the main determining factor for retention of phosphorus in the reactive additives. The presence of phosphorus-oxygen-carbon bonds, which are susceptible to hydrolysis in many compounds, has made compounds containing them unsuitable for use under conditions of humidity and heat. This has led to the interest in preparing phosphorus compounds containing stable phosphorus-carbon bonds, especially compounds in which the phosphorus moiety is linked to a carbon-carbon backbone. These compounds, in turn, must also contain suitable groups for further reaction to form polyethers or polyesters useful in preparing polyurethanes (Quinn 1970).

14.1.4 PALM-BASED PU

Palm oil is one of the most widely used plant oils in the world. It is produced from oil palm tree which is grown in mass plantation in tropical countries such as Malaysia. In Malaysia, a ton of its fresh fruit bunches (FFB) yields 200 kg crude palm oil and 40 kg palm kernels which, in turn, yield about 50 wt%, or 20 kg of palm kernel oil (PKO). A hectare of estate can yield 20–24 tons of FFB per year, which in turn yield 4 to 5 tons of palm oil and 400–500 kg of PKO (Tuan Noor Maznee et al. 2001).

There is several research studies carried out that made use of palm oil for PU preparation (Chian & Gan 1998; Tuan Noor Maznee et al. 2001). Palm oil consists of mainly triglycerides (Tri-Gs), thus, it is necessary to introduce hydroxyl groups to a molecule of Tri-G for urethane reaction (Badri 2012). PU which sometimes is also known as urethane or isocyanate polymer has once been of the most important segments in the polymer industry for decades (Healy 1963). The onset of urethane bond dissociation is somewhere between 150°C and 220°C, depending on the type of substituent; the isocyanate and polyol. Saturated hydrocarbons are known to have relatively good thermal and thermo-oxidative resistance compared to polyether and polyester derived from petrochemicals.

The thermal stability of polyurethanes is also very much dependent on the chemical structure of the polyols and diisocyanates used in their formation. Polyols containing aromatic groups show higher thermal stability than that of aliphatic polyols (Sarkar & Adhikari 2001). Javni and coworkers (2000) reported a study on thermal stability of polyurethanes based on vegetable oils such as corn, safflower, sunflower, peanut, olive, soybean, canola, and castor oil in air and also in nitrogen by thermogravimetric analysis (TGA) and Fourier transform infrared (FTIR). In their study, natural oil-based PU had better initial stability in air than the polypropylene oxide-based polyurethane (PPO-PU), while the latter was more stable in nitrogen at initial stage of degradation, when the weight loss at higher conversion is taken as the criterion of stability then oil based PU better thermal stability both in air and nitrogen. The effect of soft segment content and its molecular weight has been studied by Reimann et al. (2003) and Saraf et al. (1985). The results showed that the degree of crosslinking and tensile properties depend mainly on the ratio of soft/hard segments, and are unaffected by variations in the sequence length of the soft segment at a given soft segment content. Polyurethane-nitrolignin (PUNL), a new network polymer, was synthesized from a castor oil based-PU prepolymer and nitrolignin (NL) (Zhang & Huang 2001). The results indicated that PUNL film with 2.8% NL content, was the most miscible, and its tensile strength and elongation at break were 2 times higher than that of PU film. The crosslink densities of PUNL films increased with the increase of NL content until about 3%, similar to the variety of the mechanical properties. Thermogravimetric analysis revealed that the thermal stability of PUNL films was slightly higher than that of PU. The covalent bonds occurred between PU prepolymer and the NL in the PUNL films, forming crosslink networks, which resulted in the enhancement of mechanical properties and thermal stability. NL has a far higher reactivity with PU than nitrocellulose.

14.2 PREPARATION OF PALM-BASED PU FILLED WITH TRIS (2, 4- DI TERT BUTYL PHENYL PHOSPHITE)

14.2.1 2, 4-DITERT-BUTYLPHENYL PHOSPHITE AS THE FLAME RETARDANT

2, 4-ditert-butylphenyl phosphate (commercially known as Irgafos 168) is a hydrolytically stable phosphite processing stabilizer. It is a hydrophobic, high molecular weight (MW = 646 g/mol) compound. It has low volatility and extremely high resistant to hydrolysis (Fischer et al. 1999). It provides excellent antioxidant process stability to organic polymers, resistance to discoloration and adds long-term protection against thermo-oxidative degradation after processing (Ciba Specialty Chemicals 2003). In this study, it undergone sieving procedure and was used without further treatment. An Endocott sieve was used to screen the FR into a size of 45 μm in order to obtain uniformity of particle size and to avoid agglomeration during preparation of the composites. The FR was analyzed using Scanning Electron Microscope X-100 model Philips reveals an average size of 18 μm as shown in Figure 2.

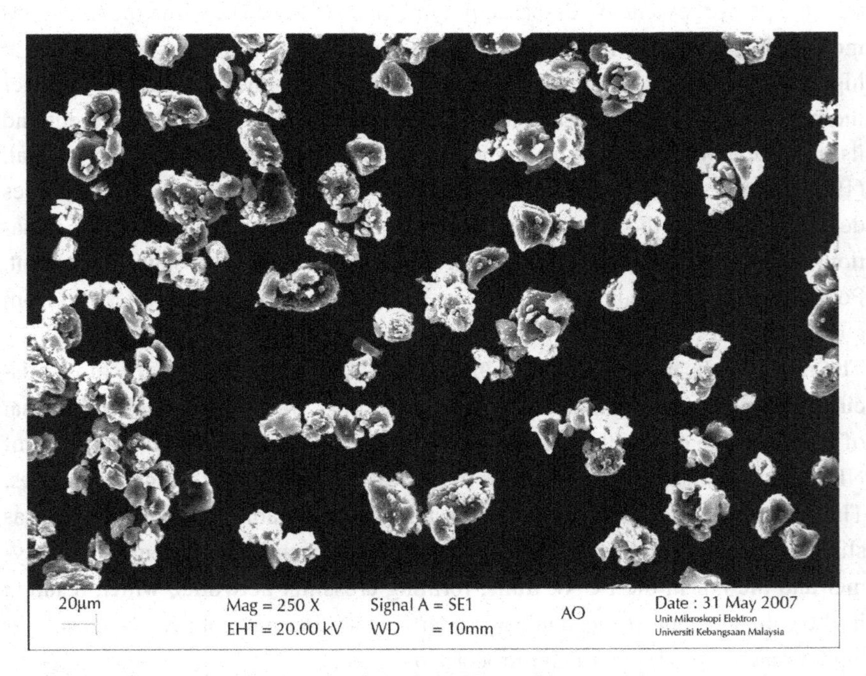

FIGURE 2 The SEM micrograph of Irgafos 168 as a flame retardant.

The melting point was determined using digital melting apparatus model Barnstead, UK. The moisture content was determined by drying of FR in the conventional oven at 105°C for 20 h. It is insoluble in water. The moisture content was 0.33% with melting point of 182°C. The physical properties of Irgafos 168 are tabulated in Table 1.

TABLE 1 Physical properties of Irgafos 168 as a flame retardant

Parameters	Properties
Color	White
State at room temperature	Solid (Powder)
Average particle size, µm	18
Moisture content, %	0.33
Solubility in water	Insoluble
Melting point, °C	182

FTIR Spectroscopy Analysis of FR

The FTIR spectroscopy analysis of the FR indicated the presence of low intensity and broad peak at 3429 cm^{-1} refer to OH (moisture) and sharp peak at 1491 cm^{-1} and 1212 cm^{-1} belong to C=C (aromatic). The P–O–C group respectively as summarized in Table 2. The absorption peaks at 2961 cm^{-1} and 2868 cm^{-1} indicate the CH_3 and CH stretching respectively (Pavia et al. 2001).

TABLE 2 The FTIR spectroscopy analysis of Irgafos 168 as a flame retardant

Functional Group	Wavenumber (cm^{-1})
C=C (aromatic)	1602, 1491
CH_3 (stretching)	2961
CH (stretching)	2868
P–O–C (stretching)	1212, 853
CH and CH_3 deformation	1398

14.2.2 (PKO)-BASED MONOESTER-OH

The patented method described by Badri (2012) is used in preparing the PKO-based monoester-OH (PE) through polycondensation and esterification processes. The synthesis is carried out at temperature range of 185–195°C for 30 min with nitrogen gas blanket. The production is managed by UKM Technology Sdn Bhd through its pilot plant in MPOB-UKM Station, Selangor, Malaysia. The PE is stored in a screw-tight container and used without further purification. The FTIR spectroscopy analysis is carried out using Perkin Elmer spectrum V-2000 spectrometer for the liquid phase PE. Sodium chloride (NaCl) window was used. The PE is scanned between wavenumbers of 4000 to 700 cm^{-1}. The moisture content of PE is determined using Karl Fischer Titrator model Metrohm KFT 701 series following ASTM D4672-91 standard. The PE is a golden yellowish liquid. It has low moisture content (0.09%) and low viscosity (374 cps). These are advantageous in formulating PU system especially when processing of end product is concerned. The physical properties are summarized in Table 3.

TABLE 3 Physical properties of the PKO-based PE

Parameter	Result
Color	Golden yellow
Odor	Odorless
State at room temperature	Liquid
Density at 25°C, g/cm^3	0.992
Viscosity at 25°C, cps	374
Moisture content at 25°C, %	0.29

The FTIR spectrum displays several absorption peaks as tabulated in Table 4. The peak at 3472 cm^{-1} refers to the hydroxyl group O–H from free water molecules overlapping with OH of the carboxylic acid. However, the broad and strong peak at 3390 cm^{-1} refers to the hydroxyl group OH exists in ester chain. Peaks at 2924 and 2854 cm^{-1} indicates the CH_2 and CH_3 scissoring respectively. Peaks at 1739 and 1622 cm^{-1} are carbonyl ester group (C=O) and carbamate (C–N) respectively (Kenkel 1994). The ester synthesized is a substance with hydroxyl functional group terminal. The presence of carbamate group indicates the formation of bond between C of the fatty acid and N of the alcohol amine. Shifting of peaks from 1746 cm^{-1} (C=O for carboxylic acid) to 1739 cm^{-1} indicates the formation of ester linkage.

TABLE 4 The FTIR Analysis of the RBD PKO and PKO-based PE

Functional Groups	Wavenumber	
	RBD PKO	PE
OH (ester)	3472	3390
CH_3 and CH_2	2924, 2854	2924, 2854
C=O	1746	1739
C–N	-	1622

14.2.3 PKO-BASED PU AND ITS FIRE RETARDING COMPOSITES

The PE-based PU was prepared by reacting 90 g crude MDI with 90 g PE-based resin (prepared based on ingredients in Table 5). The mixture was agitated vigorously using a standard propeller with a speed of 2000 rpm until well-mixed (~60 seconds) at room temperature. The reaction time; cream time (CT), gel time (FT), tack-free-time (TFT), and rise time (RT) were noted. These reaction times were important in the preparation of PU composites in identifying the initial reaction time and demolding time.

TABLE 5 Ingredients of the PE-based resin

Ingredients	Part by weight, pbw			
PE	90	90	90	90
Niax L5404	2	2	2	2
Glycerin	10	10	10	10
PMDETA	0.4	0.4	0.4	0.4
TMHDA	0.3	0.3	0.3	0.3
FR (wt% based on resin)	0	2	4	6
Total pbw	102.7	103.9	105.1	106.3

The free-rise density was calculated by dividing the mass of the foam in the cup, with the capacity of the cup as shown in Equation (1):

$$\text{Free-Rise Density (FRD), kg/m}^3 = \frac{\text{(mass of the foam in the cup-mass of the cup), kg}}{\text{capacity of the cup, m}^3} \quad (1)$$

The FRD was set at the range of 37–39 kg/m³ to ensure the resulted molded foam is in the range of 42–45 kg/m³. The above procedures were repeated and the mixture was poured into a pre-waxed mold, covered and screwed tight. The PU was demolded after 30 min. The molded density was calculated using Equation (2).

$$\text{Molded density (MD), kg / m}^3 = \frac{\text{Mass of the molded PU, kg}}{\text{Volume of the mold, m}^3} \quad (2)$$

The molded PU was conditioned for 16 h at 23 ± 2°C before further characterization following standard method BS 4370: Part 1: 1988. The FR-filled PU composites were prepared by using the same procedure, but with the addition of FR (by weight percent of resin) as shown in Table 5.

14.3 PROPERTIES OF PALM-BASED POLYURETHANE AND POLYURETHANE COMPOSITES

14.3.1 FTIR SPECTROSCOPY ANALYSIS

The FTIR Spectroscopy analysis was carried out on Perkin Elmer spectrum V-2000 spectrometer by potassium bromide (KBr) method for the solid sample of PU and its composites. The samples were scanned at wavenumbers ranging from 4000 to 700 cm^{-1}. The differences in peaks of the control PU (0% FR), and the FR-filled composites (2–6%) were observed to identify any chemical changes in the urethane linkage.

Polyurethanes are linear polymers that have a molecular backbone containing carbamate groups. This group, better known as urethane, is produced through a chemical reaction between diisocyanate and polyamide. The FTIR spectra of control PU is shown in Figure 3. The FTIR spectrum of FR-filled PU did not show any significant different or any presence of new peaks compared to the spectrum of the control PU except at one peak. This peak is attributed to the presence of P–O at 853 cm^{-1} and 905 cm^{-1} (as summarized in Table 6). It is not clearly shown when 2% of FR is added but with increasing amount of FR it is obviously shown as indicated in Figure 4. The FR only reacts with the radical that was produced during oxidation reaction of PU (Schwetlick 1990) and thus no indication of chemical bonding between PU and FR.

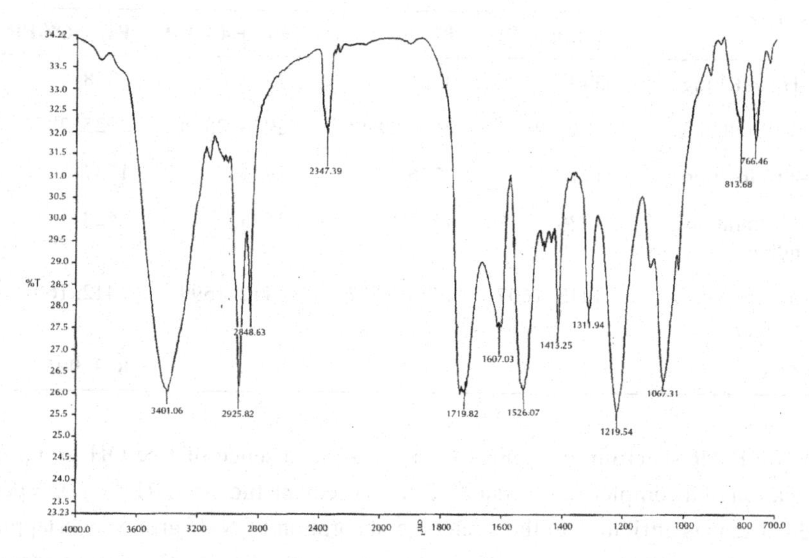

FIGURE 3 The FTIR spectrum of control PU.

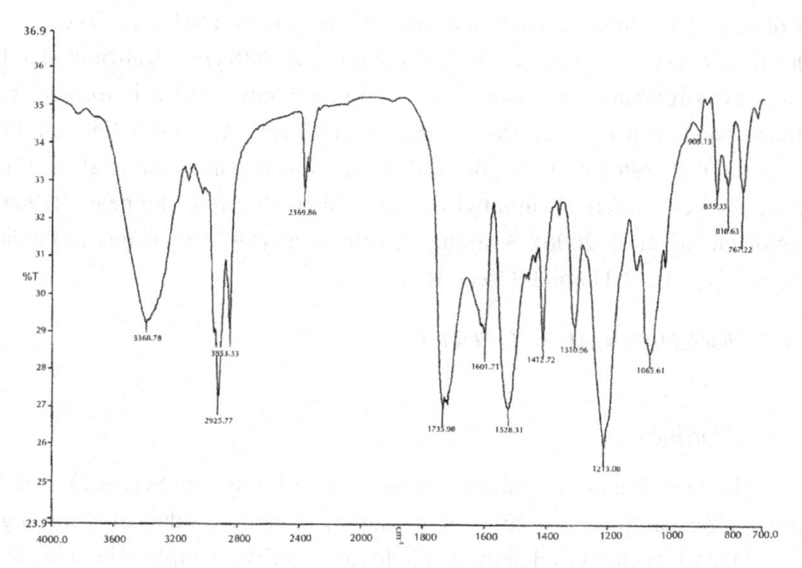

FIGURE 4 The FTIR spectrum for PU composite at 6% FR loading.

TABLE 6 The FTIR Analysis of the control PU and PU composites

Functional Groups	Wavenumber, cm^{-1}			
	Control PU	PU + 2% FR	PU + 4%FR	PU + 6%FR
N-H stretching	3401	3391	3392	3368
C–H Stretching	2925, 2848	2924, 2848	2926, 2848	2925, 2853
C=O Vibration	171	1736	1736	1737
N-H Vibration/ bending	1526	1528	1523	1528
C=C Aromatic stretching	1413, 1607	1413, 1607	1412, 1599	1412, 1601
P–O–C	–	–	853	853, 905

The FTIR spectrum of control PU shows the absence of free OH group; an indication of a complete conversion of PE to urethane moiety (NH–C=O). A peak at 3401 cm^{-1} is attributed to the stretching absorption of N–H group (overlapping with OH peak) (Wang et al. 1999) and the corresponding N–H bending absorption (1526 cm^{-1}) confirmed the presence of the secondary amine that constitutes part of a urethane linkage. Urethane carbonyl absorption (NH–CO–O) occurs at 1700–1760 cm^{-1} (Mortley & Bonin 2007). A band at 2948 cm^{-1} is attributed to the asymmetric stretching vibration of $-CH_3$ and 2848 cm^{-1} which is formed from urethane bond with PE. The sharp bands at 1607 cm^{-1} and 1413 cm^{-1} are both assigned to the stretching vibration of C=C of aromatic ring (Sun et al. 2003). A peak at 1719 cm^{-1} refers to carbonyl vibration and C–N stretching near 1219 cm^{-1} (Chattopadhyay et al. 2006). A strong absorption at 1595 cm^{-1} is due to the aromatic groups of C–NH bond (Czech et al. 2006).

14.3.2 MECHANICAL PROPERTIES

Shore D Hardness

A Portable shore D hardness indentor (Durometer Affri system Series 3300 MRS, Affri Cee-Versea, Italy) is used, with measurement being conducted according to ASTM D2240 standard to determine the hardness of the sample. The boards are cut to samples with dimensions of 130 mm × 130 mm × 3 mm (length × width × thickness). The values of shore D hardness obtained represent mean of five specimen measurements.

A plot of shore D hardness index is shown in Figure 5 indicates the effect of varying amount of FR in PU on shore D hardness index. Hardness is most commonly defined as resistance of polymer material to indentation. Indentation is the pressing motion on a point at the sample with known force (Dowling 1999). Hardness index may give an initial indicator to physical strength of the PU composites. It increased with increasing amount of filler. This result is in close agreement with research carried out by Badri et al. (2006) and Rusu et al. (2001), where composites with filler produced higher hardness index than without filler. The hardness index increased by 8%, 27%, and 38% with 2%, 4%, and 6 % (w/w) of FR loading respectively.

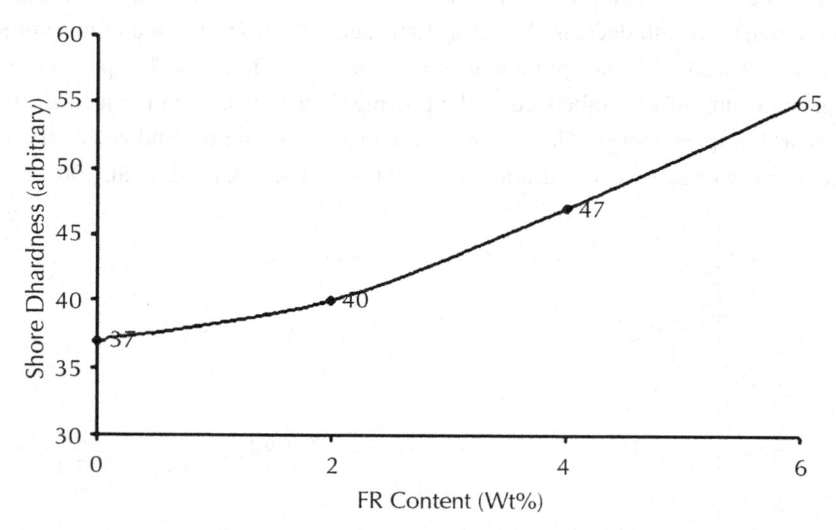

FIGURE 5 Shore D hardness index of the control PU and FR-filled PU composites.

Impact Strength

Impact testing is conducted according to ASTM D256-88 standard. The Izod method is employed, using unnotched samples with dimensions of 63 mm × 13 mm × 3 mm (length × width × thickness) on a Zwick impact tester (Model 5101, Zwick Roell Group, Atlanta, Georgia, USA), with a pendulum energy of 2 J. The impact strengths are calculated by dividing the energy (J) recorded on the tester by cross-sectional area (mm^2) of the specimens. The values of the obtained impact strengths represented the mean of five specimen measurements.

Figure 6 shows the impact strength of the control PU and FR-filled PU composites. The impact strength of materials reflected its ability to resist fracture. Impact strength of the FR-filled PU composites decreased with increasing FR content. This may be attributed to weaken interfacial bonding between FR and PU matrix with higher amount of FR. The highest impact strength is observed for control PU (6.9 k/m^2). The PU composite with 6% FR has the lowest impact strength compared to others, in agreement with research carried out by Liu and Wang (2006) where red phosphorus (RP) and melamine cyanurate (MC)-encapsulated RP (MERP) were used as FR. They observed that impact strength decreased with increasing FR loading, but the situation is much better for MERP flame retardant system due to better compatibility between MERP and the polymer matrix. The impact strength decreased by 9%, 13%, and 17% from 2 to 4 and to 6% FR loading respectively. The optimum loading is observed at 2% FR. The presence of higher loading of FR embedded in the polymer matrix results in reduced ability to absorb impact energy. The FR created matrix discontinuity, and each particle acted as site of stress concentration, and led to micro-crack (Dvir et al. 2003).

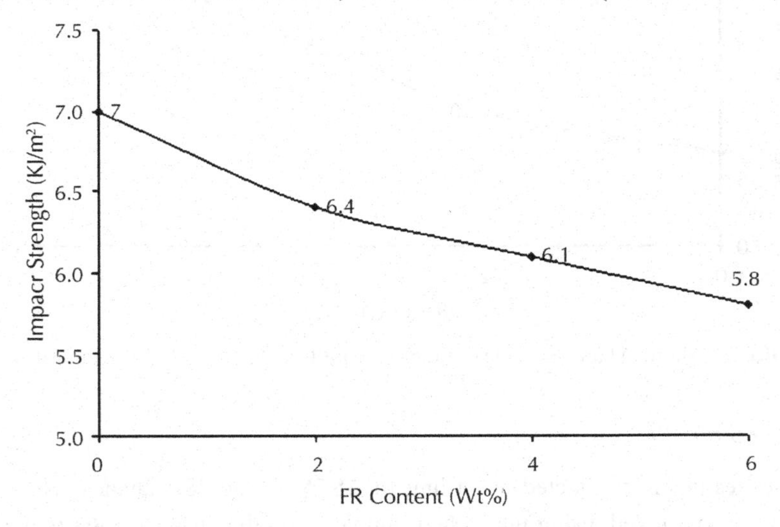

FIGURE 6 Impact strength of the control PU and FR-filled PU composites.

The investigation of the impact strength of the composites revealed that the impact strength of the PU control is higher than with FR -PU composites. This again showed that composites with FR are not able to absorb more energy from the matrix and distributed it efficiently in the composites. The impact strength decreased as the percentage of FR loading increased.

Flexural Strength and Modulus

The flexural strength by means of three points bending testing is conducted according to ASTM D790-86 standard. Samples are cut to dimensions of 130 mm × 13 mm × 3 mm (length × width × thickness). The flexural test is carried out by using a Universal Test Machine (Model 5525, Instron Corporation, Norwood, Massachusetts, USA) at a cross-head speed of 3.1 mm/min. The values obtained represented the mean of five specimen measurements.

Figure 7 and Figure 8 show the flexural strength and modulus of the control PU and the FR-filled PU composites respectively. As the content of FR increased, the flexural strength decreased. The reduction in strength is ascribed to the poor adhesion between FR and PU matrix due to agglomeration of FR particles as exhibited in the SEM micrographs.

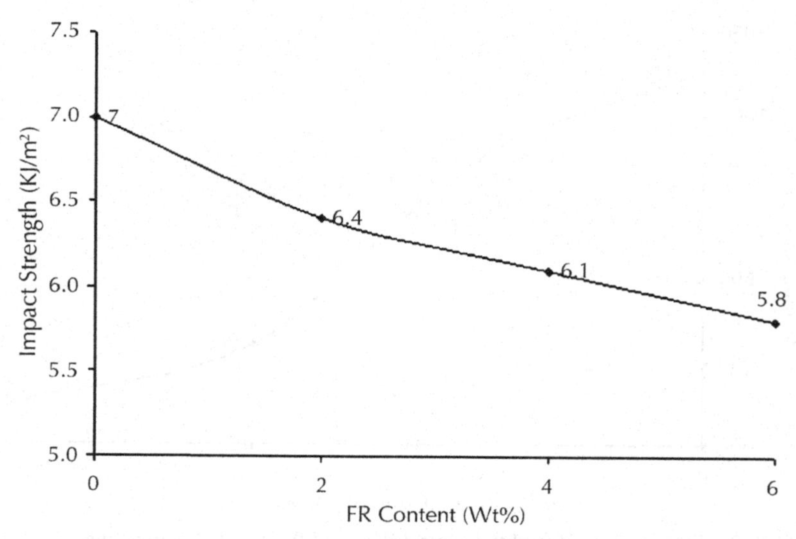

FIGURE 7 Flexural strength of control PU and FR-filled PU composites.

The agglomeration becomes more dominant at higher content of FR and the flexural strength reduced due to uneven dispersion of FR throughout the PU matrix. This result is in agreement with the study reported by Jang et al. (1998) where flexural strength of polymer composite decreased when amount of phosphate as FR was increased.

Maximum flexural strength of polymer composite is observed on control PU and reduced by 6% with addition of 2% FR. Further reduction is observed with higher amount of FR (4% gave a reduction of 25% from the control PU). The highest amount of FR gives the lowest flexural strength as shown in Figure 7 (reduction of 33% from the control PU for 6% FR).

Figure 8 describes flexural modulus of control PU and its composites. Flexural modulus reflects the rigidity of the material. The modulus decreased by 5%, 16%, and 19% respectively when 2%, 4%, and 6% of FR are added to the PU matrix. The lowest modulus is observed at 6% FR with modulus of 827 MPa. Higher loading of FR decreased the wetting properties of the matrix and reduced degree of encapsulation of matrix around FR. As such, the FR is exposed to direct stress and low stress transfer. Parallel to impact strength scenario, this may also lead to weak interfacial bonding between FR and PU matrix (Jang et al. 1998; Rusu et al. 2001).

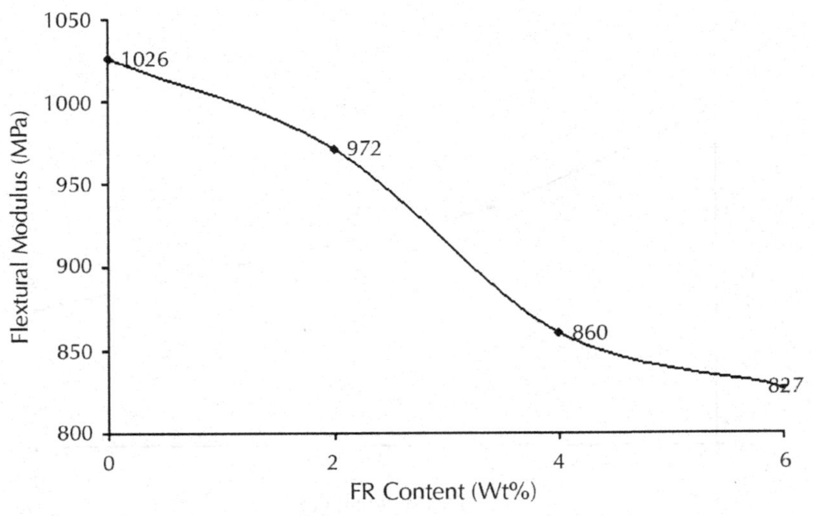

FIGURE 8 Flexural modulus of the control PU and FR-filled PU composites.

14.3.3 THERMAL PROPERTIES

Thermogravimetric Analysis (TGA)

The TGA analysis serves as a diagnostic technique used to determine thermal stability of the material and its fraction of volatile components by monitoring the change in weight that occurs as it is being heated. Thermal stability and weight loss of the PU and its composites was measured using a thermogravimetric analyzer model Shimadzu TGA-50 with temperature range from 30–600°C at a heating rate of 10°C/min under nitrogen gas blanket. Samples are placed in an alumina pan holder in a mass range of 5–15 mg.

The thermogram of FR is shown in Figure 9. It is thermally stable up to 236°C and this is higher than control PU (at 172°C). It undergoes single stage decomposition at 230–295°C with total weight loss of 97%. Thermograms of control PU and FR-filled PU composites with varying FR contents are as tabulated in Table 7. The TGA thermogram of control PU shows two main weight loss stages indicating presence of two thermal degradation temperatures. It is stable at 172°C, a common stability temperature for PU (Hepburn 1991). Initial weight loss of about 14% commences at 172°C to 250°C. Control PU started to degrade at lower temperature (175°C) compared to FR-filled PU composites loaded with FR. There is an insignificant difference in the degradation temperature when FR is added. However, total weight loss of FR-filled PU composites reduced down to 66.0% with addition of 6% FR. The FR plays an important role in boosting the thermal oxidative stability (Shao et al. 1999).

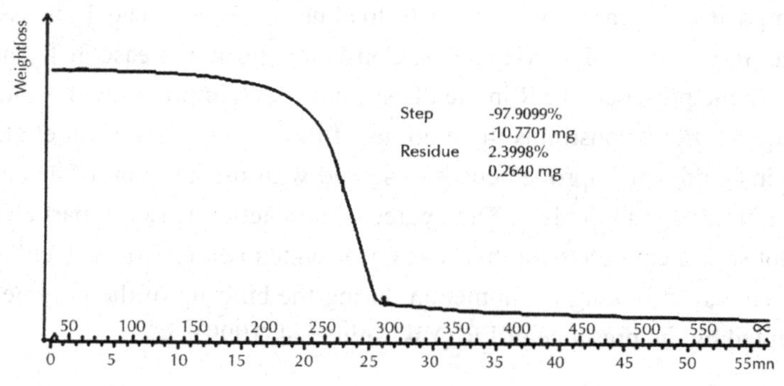

FIGURE 9 The TGA thermogram of the FR (Irgafos 168).

TABLE 7 The TGA results for the control PU and FR-filled PU composites with 2 wt%, 4 wt%, and 6 wt% of FR

Content of FR, wt%	T_{stable}, °C	T_1,°C	Total weight loss, %
0	172	172–495	95.2
2	175	175–496	91.6
4	175	175–497	86.3
6	173	173–500	66.0

Differential Scanning Calorimetry (DSC) Analysis

The thermal behavior of control PU and its composites were determined using a Perkin Elmer Model DSC-7 differential scanning calorimeter interfaced to Model 1020 controller. The samples were analyzed from room temperature to 250°C (as PU decomposed at a temperature more than 250°C) at a heating rate of 10°C/min. the standard aluminum pan is used to analyze about 10 mg samples under nitrogen gas atmosphere.

There is no evidence of the presence of melting and crystallization temperatures peak in the DSC thermogram of the control PU and FR-filled PU composites. The PU is an amorphous polymer (Fedderly et al. 1998). The PU is conveniently described in terms of it characteristic glass transition temperature (T_g) which identify the point where the material changes from rigid to a rubbery behavior (Bernhard 2005). The T_g of control PU is observed at 60°C. The T_g of FR-filled PU composites increased by 3.3% with loading of 2% FR. The T_g is associated with mobility of the polymeric chains. Consequently, an increase in T_g can be attributed to the presence of FR in the PU system which implies that the addition of the flame retardant constrains the motions of the chains (Gregoriou et al. 2005). But an insignificant improvement is observed with the addition of 4% and 6 % FR as summarized in Table 7. The degree of interaction between particles of the FR is not strong enough to modify the activation temperature of molecular movements. The cross-linking phenomenon during the built-up of the polymerization is disrupted by the presence of FR (Yakushin et al. 1999).

The same trend is observed for the specific heat capacity. The intensity of the change in specific heat capacity (ΔC_p) is associated with T_g. Specific heat capacity of control PU is 1392 J/kg·K. Specific heat capacity of the PU composite increased by 5.8 % when 2% FR is added. However, specific heat capacity of FR-filled PU composites decreased by 3.5% and 14% when 4% and 6% FR respectively are added. A significant drop in specific heat capacity (ΔC_p) is observed when more flame retardant is introduced to PU composites, therefore gradual and linear reduction occurred according to addition of FR (Sarier & Onder 2007).

TABLE 8 The T_g and specific heat capacity of control PU and FR-filled PU composites

Samples	Glass transition temperatures (T_g), °C	Specific heat capacity (C_p) J/kg·K
Control PU	60	1392
PU + 2% FR	62	1472
PU + 4% FR	61	1440
PU + 6% FR	57	1190

Bomb Calorimetric Analysis

The effect of FR on network structure of control PU and FR-filled PU composites can be determined by measuring the enthalpy. The enthalpy from combustion was tested with a bomb calorimeter. Bomb calorimeter model IKA C 4000 was used to determine the initial enthalpy. One gram of PU composites is put in the sample cup. The bomb was prepared by putting a wire (5 cm length) between two electrodes. A thread was also tied on the divider of the two wires touching the sample. The bomb was placed inside the chamber and was closed tightly with purging O_2 at a pressure of 30 bar O_2 into the chamber (Oelke & Zuehlke 1969; Arthur 1986). The chamber was transferred to a container containing 1.8 L distilled water. The monitoring is carried out using calorimeter IKA system.

Figure 10 presents the enthalpy of control PU and the FR-filled PU composites. It decreased by increasing the amount of FR. This trend is in good agreement with results reported elsewhere (Toldy et al. 2006) where, they use OPCs and combine them with montmorillonite nanoparticles, and use it to retard flame and incorporate them in epoxy resin.

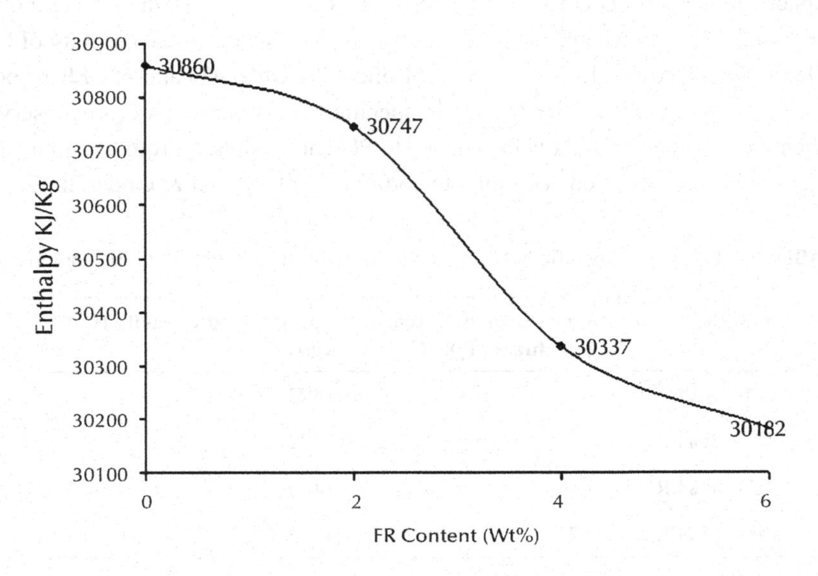

FIGURE 10 Enthalpy of the control PU and FR-filled PU composites.

Enthalpy of control PU is 30860 kJ/kg. When 2% FR is added to the PU matrix, the enthalpy is 30740 kJ/kg. A PU composite with 4% FR showed enthalpy of 30337 kJ/kg. The lowest enthalpy is observed for PU composite with 6% FR (30180 kJ/kg). This is attributed to the fact that less energy is required to break the bond due to the presence of FR.

Fire Resistivity Test

This test is carried out to determine the relative burning characteristics and flame resistance properties. It measured and described the properties of materials, products or assemblies in reactions to heat and flame under controlled laboratory conditions. The test results represented flaming plus glowing time in seconds, for a material under the conditions of the test. ASTM D 5048-90 (procedure B-test of plaque specimens) is conducted as shown in Figure 11.

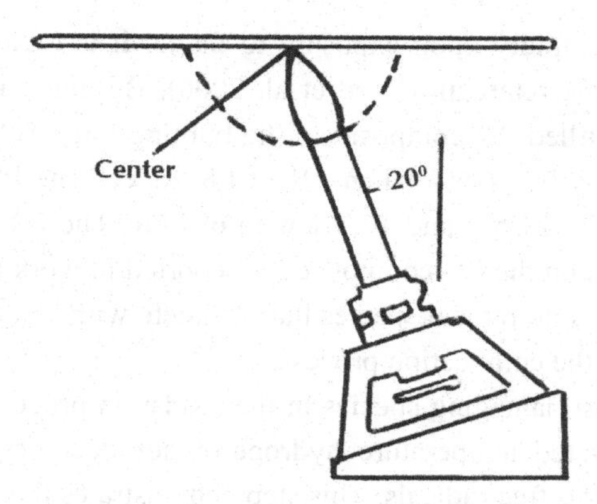

FIGURE 11 Schematic diagram of fire test from ASTM D 5048-90: Procedure B.

Flammability tests are classified based on various characteristics of fire response. In various flammability tests, the fire test is carried out to determine the relative burning characteristic and flame resistance properties. Phosphite used in this study has good flame retardancy and good thermal properties because the hydroperoxide decomposers prevent the split of hydroperoxides into highly reactive alkoxy and hydroxyl radicals (Schkwetlic 1990). Figure 12 shows results of fire test for control PU and its composites with varying amount of FR.

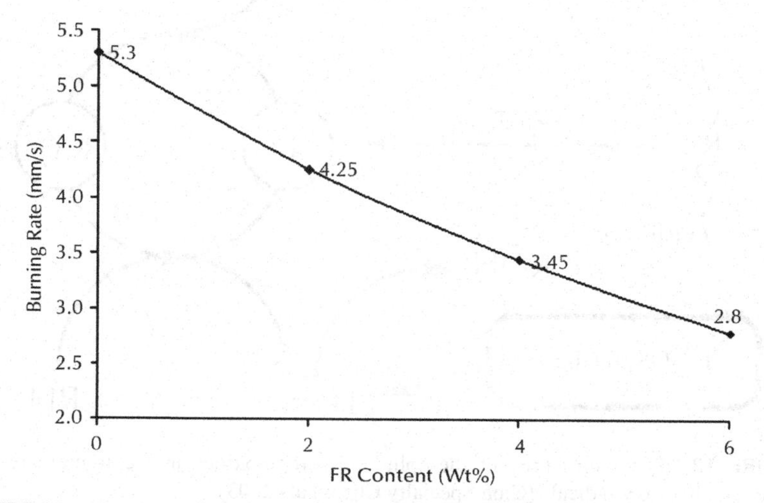

FIGURE 12 Burning rate of control PU and FR-filled PU composites.

The PU burns rapidly upon exposure to flame. It is highly combustible in the absence of flame retardant (Reed et al. 2000). Burning rate for control PU is higher than FR-filled PU composites. The burning rate of PU composites decreased significantly by 26.6% when 2% of FR added. The PU composite with 4% FR decreased by 21.5% and 17.7% with 6% FR. The FR shows excellence fire retarding effect on the PU composites as reported in work (Jang et al. 1998). The FR dissociates into radical species that compete with chain propagating and branching steps in the combustion process.

One of the most damaging species in the oxidation process is the hydroperoxide. Under elevated temperature hydroperoxides decompose via hemolytic cleavage to yield two free radicals. This step demonstrates the catalytic nature of autoxidation. The destruction of hydroperoxides, which continually build up in the polymer, is essential in protecting the polymer (Salamone 1996). Phosphites prevent further formation of free radicals by decomposing unstable hydroperoxides prior to their homolytic cleavage. Instead, the unstable hydroperoxide forms a stable product (Zhu & Shi 2001) as illustrated in Figure 13.

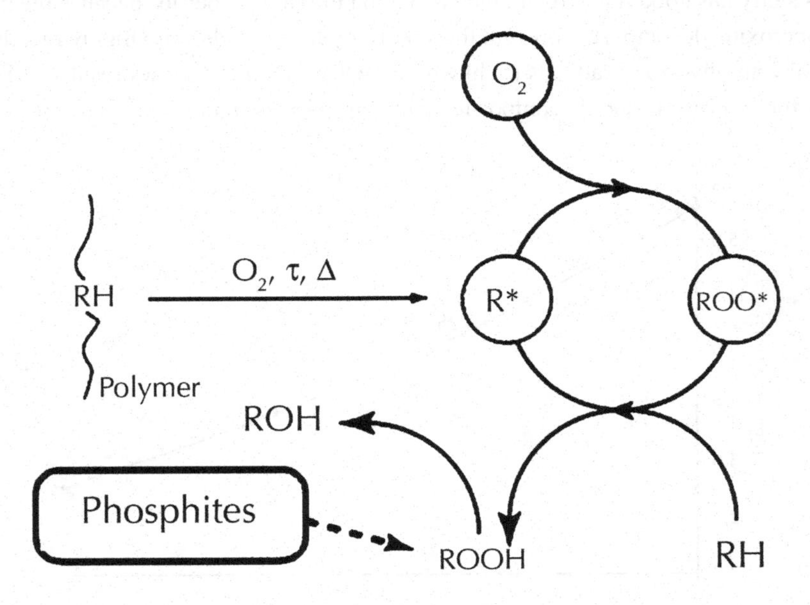

FIGURE 13 Phosphites prevent the split of hydroperoxides into extremely reactive alkoxy and hydroxy radicals (Ciba Specialty Chemicals 2003).

The mechanism is shown below:

Mechanism 1: Phosphite reacts with peroxide

$$
\begin{aligned}
&\dot{R} + O_2 &&\rightarrow R\dot{O}_2 \\
&R\dot{O}_2 + P(OR)_3 &&\rightarrow ROO\dot{P}\,(OR)_3 \\
&ROO\dot{P}\,(OR)_3 &&\rightarrow R\dot{O} + O = P(OR)_3 \\
&R\dot{O} + P(OR)_3 &&\rightarrow RO\,\dot{P}(OR)_3 \\
&RO\,\dot{P}(OR)_3 &&\rightarrow \dot{R} + O = P(OR)_3 \\
&2R\dot{O}_2 &&\rightarrow ROOR + O_2
\end{aligned}
$$

Mechanism 2: Phosphite reacts with hydroperoxide

$$
\begin{aligned}
&R^1OOH + P(OR)_3 &&\rightarrow [R^1O\dot{P}(OR)_3\,H\dot{O}] \\
&[R^1O\dot{P}(OR)_3\,H\dot{O}] &&\rightarrow R^1OH + O = \dot{P}(OR)_3 \\
&R^1OOH + P(OR)_3 &&\rightarrow [HO\dot{P}(OR)_3\,R^1\,\dot{O}] \\
&[HO\dot{P}(OR)_3\,R^1\,\dot{O}] &&\rightarrow R^1OH + O = P(OR)_3
\end{aligned}
$$

Dynamic Mechanical Analysis (DMA)

The DMA is conducted on a dynamic mechanical analyzer model TA Instrumentation (DMA 2980). It is performed using a single cantilever at a frequency of 1 Hz. The sample was cut into a 25 mm × 12.5 mm × 3 mm specimens. The testing temperature ranged from 25°C to 150°C with a heating rate of 5°C /min.

The DMA measures the deformation of a material in response to vibration forces or sinusoidal wave. The storage modulus E′ refers to stiffness of material and tan δ gives the amount of energy dissipated as heat during deformation (Nielsen & Landel 1974). The investigation of dynamic storage modulus and internal friction over a wide range of temperature and frequencies has proven to be very useful in studying the structure of polymers and the vibrational properties in relation to end-use performance (Costa et al. 2005). It is conducted due to (a) T_g determined from tan δ curves is more accurate than by other thermal methods and (b) any decrease in elastic moduli accompanying T_g can be monitored by the storage modulus (Lu et al. 2003).

The storage modulus of the FR-filled PU composites with 0%, 2%, 4%, and 6% FR is shown in Figure 14 (a). The DMA curve indicates a significant decrease in storage modulus E′ around the onset T_g of FR-filled PU composites. Storage modulus of FR-filled PU composites with 2% and 4% FR decreases by 15% and 20% respectively, attributing to poor interaction between FR and PU matrix as supported by SEM micrographs in the following section. Addition of FR decreased storage modulus of the PU composites to a minimum value with loading of 6% FR (975 MPa). The excess amount of FR in PU matrix gives out poor mechanical properties resulting in reduction in flexural and impact strengths. This finding was also obtained by Mat Amin and Badri (2007) where, an increase in the amount of kaolinite in the polymer matrix reduced the mechanical strength of the composite.

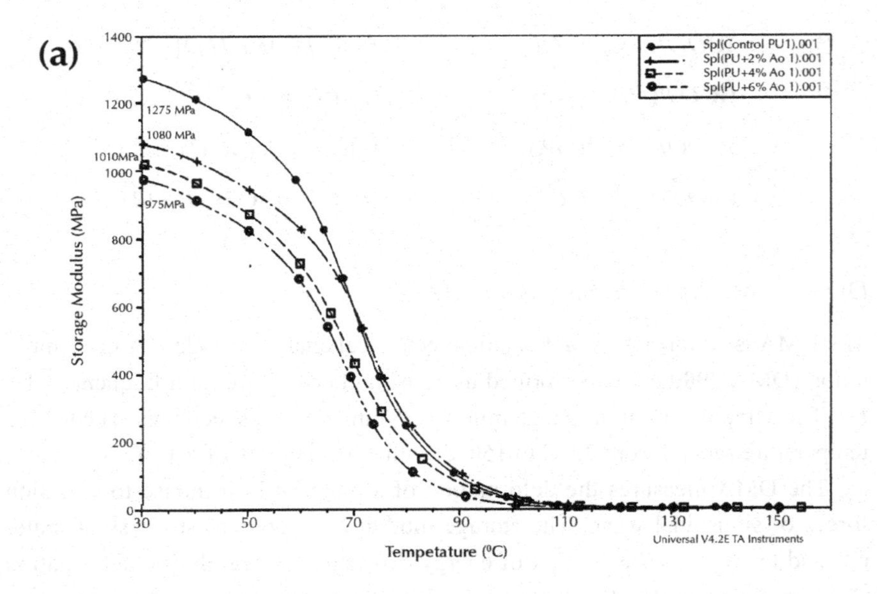

FIGURE 14 *(Continued)*

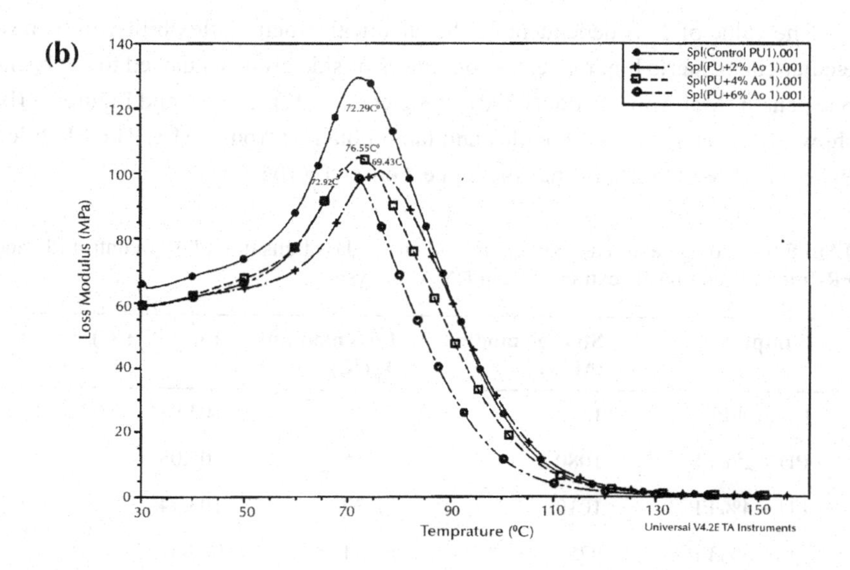

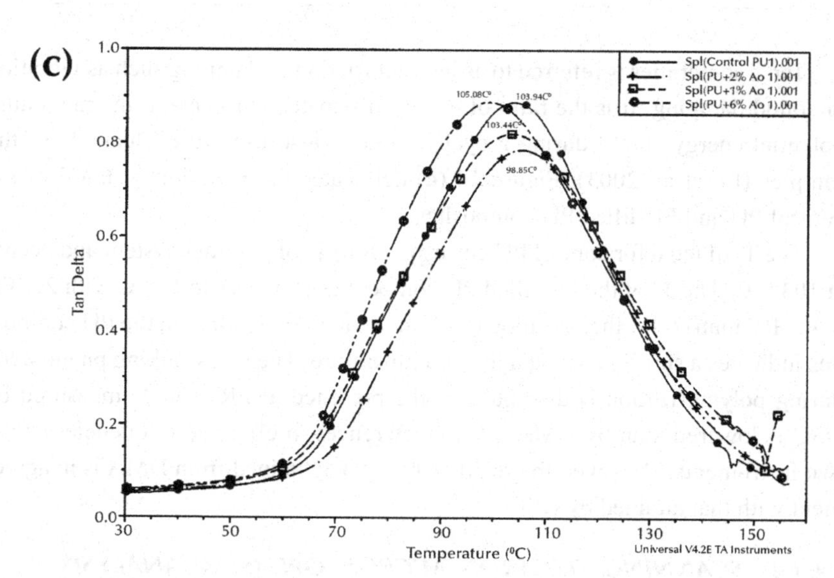

FIGURE 14 The DMA thermograms of control PU and FR-filled PU composites exhibiting (a) storage modulus (b) loss modulus, and (c) tan δ.

The value of T_g is dependent on the chemical structure, flexibility of the molecular chain, steric hindrance, and bulkiness of side groups attached to the backbone chain (Mat Amin & Badri 2007, Song et al. 2005). Table 9 and Figure 14 (b) show the onset T_g of loss modulus and tan δ obtained from DMA. The FR-filled PU composite with 2% FR has the highest onset T_g (105°C).

TABLE 9 Storage and loss moduli and the onset glass transition (T_g) of control PU and FR-filled PU composite extracted from DMA Analysis

Samples	Storage modulus (MPa)	Loss modulus T_g (°C)	Tan δ T_g (°C)
Control PU	1275	72.29	103.94
PU + 2% FR	1080	76.55	105.08
PU + 4% FR	1010	72.29	103.44
PU + 6% FR	975	69.34	98.85

Tan δ sometimes is referred to as internal friction or damping such as vibration or sound damping. It is the ratio of energy dissipated per cycle to the maximum potential energy stored during a cycle. Tan δ is also used to evaluate T_g of the samples (Lu et al. 2003). Figure 14 (c) delineates the vibration of tan δ of the control PU and FR-filled PU composites.

The T_g of tan δ for control PU corresponds to T_g of polymer system and occurs at 103.9°C. The T_g of the FR-filled PU composites increases to 105°C with 2% FR in the PU matrix. Further addition of FR at 4% and 6% loading in the PU composites indicates a shift in tan δ at a lower temperature. The cross-linking phenomena during polymerization is disrupted by the presence of FR. The T_g measured by DSC is lowered than by DMA due to differences in effective frequencies of the two instruments. However, the relationship of T_g obtained from DMA is in agreement with that attained by DSC.

14.3.4 SCANNING ELECTRON MICROSCOPE (SEM) ANALYSIS

A Leo VP SEM-1450 SEM is used to study the interaction and distribution of FR in the PU composites. The observation is made onto impact fractured samples. The samples were coated with gold of thickness 20A with Bio-rad microscience division-SC 500 to prevent electrical charging during examination. The observation is made at 250 times magnification on control PU (0% FR) and its composites with 2%, 4%, and 6% FR.

The SEM micrographs support the findings in the analysis of mechanical properties. Figure 15 (a) shows SEM micrographs for fractured surface of control PU. The control PU has uniform structure with homogeneity in the matrix. SEM micrograph (Figure 15 (b)) of the PU composite with 2% FR indicates uneven distribution of FR particles in the PU matrix. This non-uniformity between FR particles and PU matrix is due to weak interaction between both of them, as indicated by poor mechanical strength of the PU composites. Increase of FR content in the PU matrix by 4% causes initial agglomeration of particles, affecting adversely on dispersion and distribution of FR particles in the PU matrix. This led to further reduction in the mechanical properties of the PU composites; due to poor adhesion between FR and PU matrix as shown in Figure 15 (c). Agglomeration is clearly observed in PU composite with 6% FR as shown in Figure 15 (d). The agglomeration becomes dominant at higher FR content creating a non-uniform distribution of PU composites thus directly affecting the mechanical properties of PU composites (Sumaila et al. 2006). The occurrence of particles agglomeration is responsible for the FR–FR particle contacts instead of PU-FR particle contacts. This shows that there is no physical interaction between FR particles and PU matrix.

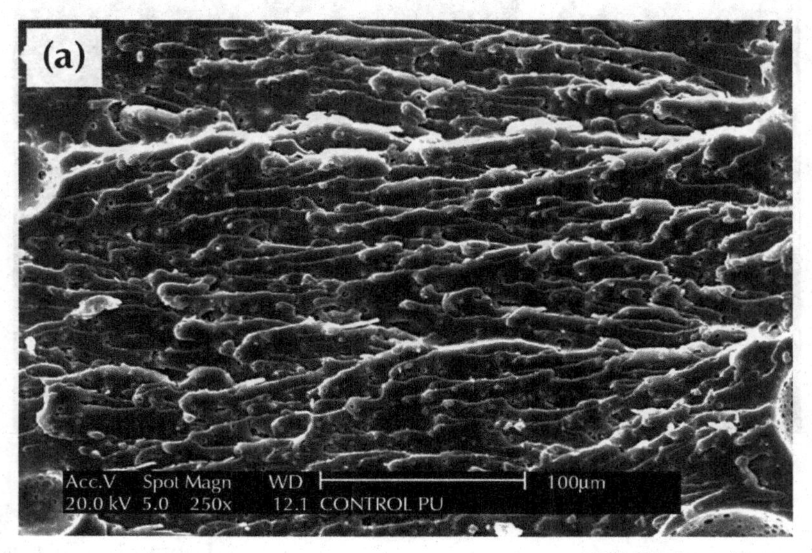

FIGURE 15 *(Continued)*

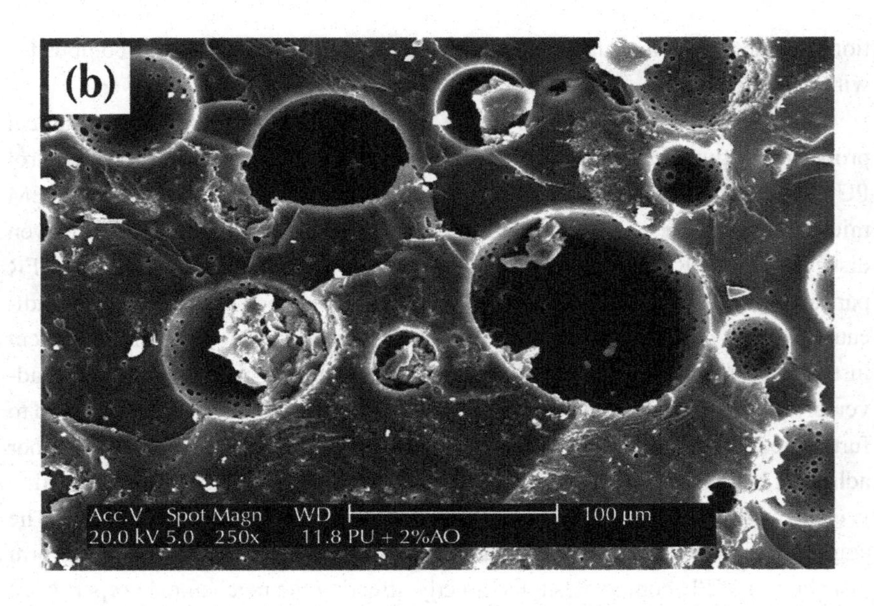

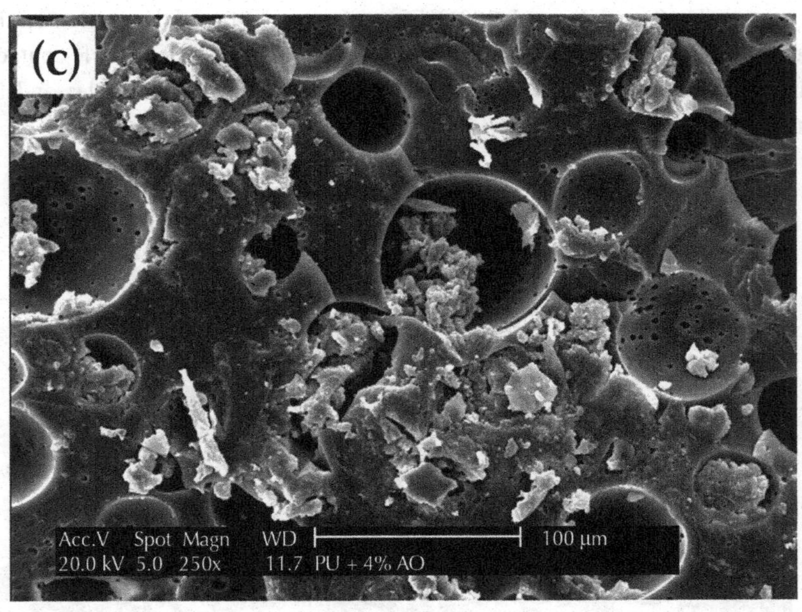

Figure 15 *(Continued)*

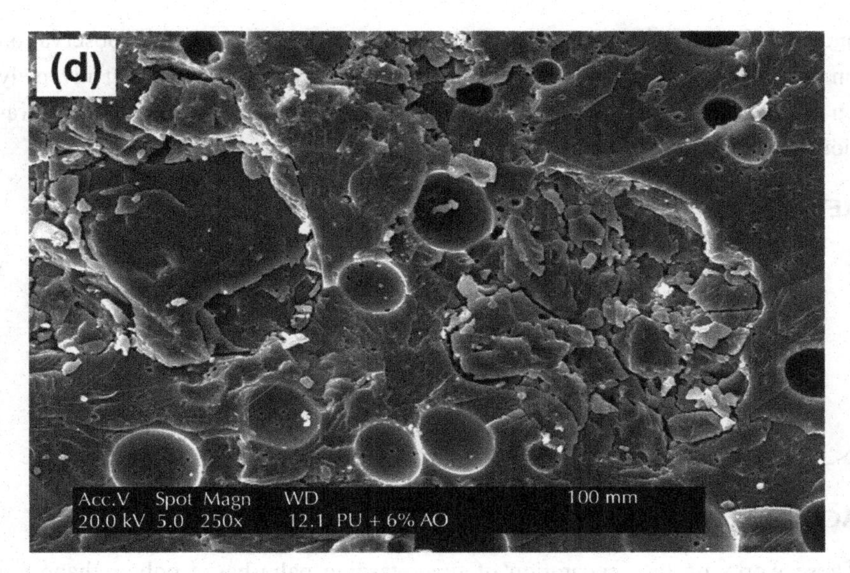

FIGURE 15 SEM micrographs of (a) PU control, (b) PU with 2% FR, (c) PU with 4% FR, and (d) PU with 6% FR.

14.4 CONCLUSION

This study showed the effect of 2, 4-ditert-butylphenyl phosphite (FR) as fire retardant on the mechanical and thermal properties (thermal stability and fire test) of PU. Hardness index obviously increased with increasing amount of FR. However, impact strength of FR-filled PU composites decreased with increasing FR content with the highest observed for control PU and the lowest for PU composite with 6% FR. The optimum loading was observed at 2% FR. The maximum flexural strength was observed on control PU but a reduction with loading of higher amount of FR. The modulus decreased by 5%, 16%, and 19% when 2%, 4%, and 6% of FR were added to the PU matrix respectively. The lowest modulus was observed at 6% FR. Thermal stability of PU composites was improved by the addition of FR, which may be attributed to excellent thermal stability of phosphite as FR agent. The DMA curves indicated that storage modulus E′ decreased significantly around the onset T_g of the PU composites. The fire test indicated lower burning rate (from 5.30 mm/s to 2.80 mm/s) as the percentage loading of FR increased. The enthalpy determined by bomb calorimeter supported the burning test results. Enthalpy of PU composites decreased by 2%, 3%, and 5% when 2%, 4%

and 6% of FR were incorporated to the PU matrix, respectively. The observation on the SEM micrographs indicated agglomeration of particles, affecting adversely on dispersion and distribution of FR particles in PU matrix. This led to deterioration of the mechanical properties of PU composites.

KEYWORDS

- **2, 4-Ditert-butylphenyl phosphite**
- **Burning properties**
- **Dynamic mechanical analysis (DMA)**
- **Flame retardant**
- **Palm-Based Polyurethane**

ACKNOWLEDGMENT

These works on the preparation of fire retarding palm-based polyurethane are based on ranges of polyurethane polyols produced at a larger scale at a pilot plant in Universiti Kebangsaan Malaysia. It was commissioned in November 2011. This is being brought into realization with the support of Universiti Kebangsaan Malaysia under its entities UKM Technology Sdn Bhd and Centre for Collaborative Innovation through a grant No. UKM-IF -1-2010-013. An appreciation is also dedicated to School of Chemical Sciences and Food Technology, Polymer Research Center and Faculty of Science and Technology for research facilities provided. Thanks to Ministry of Higher Education, Ministry of Science, Technology and Innovation for other financial supports in research grants form. Major contributions definitely came from graduates and colleagues of Universiti Kebangsaan Malaysia. For special individuals who initiated this project, Zulkefly Othman and in memory of Haji Badri Haji Zakaria and Hajah Fatimah Haji Ismail, my greatest thanks to you.

REFERENCES

1. Arthur, W. A. *A textbook of physical chemistry,* 3rd ed. Academic Press Inc: Orlando, pp. 155–159 (1986).
2. Aseeva, R. M. and Zaikov, G. E. Flammability of polymeric materials. *Advance Polymer Science*, **70**, 172–229 (1985).
3. Badri, K. H. Process for the production of vegetable oil-based polyurethane polyols. Malaysia Patent MY 145094-A (2012).

4. Badri, K. H. *Biobased polyurethane from palm kernel oil-based polyol*. In Polyurethanes, F. Zafar and E. Sharmin (Eds.), InTech Publication: New York, pp. 447–470 (2012).
5. Badri, K. H., Ahmad, S. H., and Zakaria, S. Development of zero ODP rigid polyurethane foam from RBD palm kernel oil. *Journal of Material Science Letters*, **19**, 1355–1356 (2000).
6. Badri, K. H., Ahmad, S. H., and Zakaria, S. Production of a high-functionality RBD palm kernel oil-based polyester polyol. *Journal of Applied Polymer Science*, **81**(2), 384–389 (2001).
7. Badri, K. H., Bin Othman, Z., and Razali, I. M. Mechanical properties of polyurethane composites from oil palm resources. *Iranian Polymer Journal*, 14, 5 441–448 (2005).
8. Badri, K. Mat Amin, K., Othman, Z., Abdul Manaf, H., and Khalid, K. Effect of filler-to-matrix blending ratio on the mechanical strength of palm-based biocomposite boards. *Polymer International*, **55**, 190–195 (2006).
9. Bernhard, W. *Thermal analysis of polymeric materials*. 3rd ed. Springer: Berlin, (2005).
10. Chattopadhyay, D. K., Aswini, K., Mishra, B., Sreedhar, K., and Raju, K. V. Thermal and Viscoelastic Properties of Polyurethane-Imide/Clay Hybrid Coatings. *Polymer Degradation and Stability*, **91**, 1837–1849 (2006).
11. Chian, K. S. and Gan, L. H. Development of rigid polyurethane foam from palm oil. *Journal of Applied Polymer Science*, **65**, 509–515 (1998).
12. Ciba Specialty Chemicals. Antioxidant free newsletter. http:/www.specialchem4polymers.com/tc. Antioxidant/index [accessed 06/19/07].
13. Costa, M. L.,Almida, S. F., and Rezende, M. C. Hygrothermal effects on dynamic mechanical analysis and fracture behavior of polymeric composites. *Materials Research*, **8**(3), 335–340 (2005).
14. Czech, P., Okrasa, L., Chin, F., Boiteux, G., and Ulanski, J. Investigation of the polyurethane chain length influence on the molecular dynamics in networks crosslinked by hyperbranched polyester. *Polymer*, **47**, 7207–7215 (2006).
15. Dvir, H., Gottlieb, M., Daren, S., and Tartakovsky, S. Optimization of a flame-retarded polypropylene composite. *Composites Science and Technology*, **63**, 1865–1875 (2003).
16. Dowling, N. E. *Mechanical behavior of materials*. 2nd ed. Prentice Hall: Upper Saddle River, New Jersey (1999).
17. Fedderly, J., Compton, E., and Hartmann, B. Additive group contributions to density and glass transition temperature in polyurethanes. *Polymer Engineering and Science*, **38**(12), 2072–2076 (1998).
18. Fischer, K., Norman, S., and Freitag, D. Studies of the behavior of the polymer-additives octadecyl-3-5-di tert butyl 1-4 hydroxyphenyl) proprionate and tri-(2, 4-di-t-butylphenyl) phosphite in the environment. *Chemosphere*, **39**(4), 611–625 (1999).
19. Frank, Y. W. Polymer Additive Analysis by Pyrolysis-Gas Chromatography II. Flame Retardant. *Journal of Chromatography*, **8**, 225–235 (2000).
20. Gjurova, K., Becher, C., Troev, K., and Borisov, G. Thermal behavior of rigid polyurethane foams. *Journal of Thermal Analysis and Calorimetry*, **31**(4), 835–864 (1986).

21. Gregoriou, V. G.,Kandilioti, G., and Bollas, S. T. Chain conformational transformations in syndiotactic polypropylene/layered silicate nanocomposites during mechanical elongation and thermal treatment. *Polymer*, **46**, 11340–11350 (2005).
22. Guner, F. S., Yagc, Y., and Erciyes, A. T. Polymers from triglyceride oils. *Progress in Polymer Science*, **31**, 633–670 (2006).
23. Innes, J. and Innes, A. Flame retardants: Current trends in North America. *Plastic Additives and Compounding*, **3**, 22–26 (2001).
24. Jang, J., Hyuksung, C., Myonghwan, K., and Hyunje, S. The effect of flame retardant on the flammability and mechanical properties of paper-sludge/phenolic composite. *Polymer Testing*, **19**, 269–279 (1998).
25. Javni, I., Petrovic, Z. S., Guo, A., and Fuller, R. Thermal stability of polyurethanes based on vegetable oils. *Journal of Applied Polymer Science*, **77**(8), 1723–1734 (2000).
26. Kolot, V. and Grinberg, S. Vernonia oil-based acrylate and methacrylate polymers and interpenetrating polymer networks with epoxy resins. *Journal of Applied Polymer Science*, **91**(6), 3835–3843 (2004).
27. Korobeinichev, O. P., Sergey, I., Shvartsberg, V. M., and Anatoly, C. The destruction chemistry of organophosphorus compounds in flames—I: quantitative determination of final phosphorus-containing species in hydrogen-oxygen flames. *Combustion and Flame*, **118**, 718–726 (1999).
28. Korobeinichev, O. P., Shvartsberg, V. M., Shmakov, A. G., Bolshova, T. A., Jayaweera, T. M., Melius, C. F., Pitz, W. J., and Westbrook, C. K. Flame inhibition by phosphorus-containing compounds in lean and rich propane flames. *Proceedings of the Combustion Institute*, **30**(2), 2350–2357 (2004).
29. Lewin, M. Synergism and catalysis in flame retardancy of polymers. *Polymer Advanced Technology*, **12**, 215–222 (2001).
30. Liu, Y. and Wang, Q. Melamine cyanurate-microencapsulated red phosphorus flame retardant unreinforced and glass fiber reinforced polyamide 66. *Polymer Degradation and Stability*, **91**, 3103–3109 (2006).
31. Lu, H., Obengb, Y., and Richardson, A. Applicability of dynamic mechanical analysis for CMP polyurethane pad studies. *Materials Characterization*, **49**, 177–186 (2003).
32. Lyons, J. W. *The chemistry and uses of fire retardants*. Wiley-Interscience: New York (1970).
33. Mat Amin, K. A. and Badri, K.H. Palm-based bio-composites hybridized with kaolinite. *Journal of Applied Polymer*, **105**, 2488–2496 (2007).
34. Mortley, H.W. and Bonin, V.T. Synthesis and properties of radiation modified thermally cured castor oil based polyurethanes. *Nuclear Instrument and Method in Physics*, pp. 2841–2842 (2007).
35. Nielsen, L. E. and Landel, R. F. *Mechanical properties of polymers and composites*. 3rd ed. Marcel Dekker, Inc.: New York (1974).
36. Oelke, C. W. and Zuehlke, R. W. *Laboratory physical chemistry*. Van Nostrand Reinhold Company: New York, pp. 204–213 (1969).
37. Oertel, G. *Polyurethane hand book: chemistry-raw material-processing-application-properties*. Hanser Grander publications, Inc: Cincinnati (1993).

38. Pavia, D. L., Lampman, G. M., and Kriz, G. S. *Introduction to spectroscopy*. 3rd ed. Brooks/Cole: Thomson learning Inc (2001).
39. Piechota, H. Some correlations between raw materials, formulation, and flame retardant properties of rigid urethane foams. *Journal of Cellular Plastics,* 1(1), 186–199 (1965).
40. Pitts, J. *Fire flammable*. 2nd ed. New York: John Wiley & Sons (1972).
41. Quinn, E. J. Properties and Stability of Fire-Retardant Rigid Polyurethane Foams From Phoshonopropionate Polyols. *Indian Engineering Chemistry*, **9**, 1 (1970).
42. Reed, C. S., Jonathan, P., and Kiven, T. Polyurethane/ poly[bis (carboxilatophenoxy) phosphazene] blends and their potential as flame retardant materials. *Polymer Engineering Science*, **40**, 2 (2000).
43. Reimann, R.,Mörck, R. Yoshida, H. Hatakeyama, H., and Kringstad, K. P. Kraft lignin in polyurethanes. III. Effects of the molecular weight of PEG on the properties of polyurethanes from a kraft lignin-PEG-MDI system. *Journal of Applied Polymer Science*, **41**(1–2), 3–50 (2003).
44. Rusu, R., Sofian, N., and Rusu, D. Mechanical and thermal properties of zinc powder filled high density polyethylene composites. *Polymer Testing*, **20**, 409–417 (2000).
45. Rumack, M., Marzuk, M., and Georlette, P. *Rubber & plastics news* Publisher: Oklahoma City (1987).
46. Salamon, J. C. *Polymeric material encyclopedia*, flame retardants (overview). CRC Press: Boca Raton, London (1996).
47. Saraf, V., Glasser, W. G., Wilkes, G. L., and McGrath, G. E. Engineering plastics from lignin: IV. Structure property relationships of PEG-containing polyurethane networks, *Journal Applied of Polymer Science*, **30**, 2207 (1985).
48. Sarkar, S. and Adhikari, B. Thermal stability of lignin-hydroxy-terminated polybutadiene–co-polyurethanes. *Polymer Degradation and Stability*, **73**, 169–175 (2001).
49. Sarier, N. and Onder, E. Thermal characteristics of polyurethane foams incorporated with phase change materials. *Thermochimica Acta*, 454, 90–98 (2007).
50. Schwartz, M. M. *Composite Materials Handbook*. McGraw–Hill Book: New York (1984).
51. Schwetlick, K. *Mechanical of antioxidant action of phosphate and phosphate esters*. Science Publisher Ltd: New York (1990).
52. Shao, C., Huang, J., Chen, G., Yeh, J., and Chen, J. Thermal and combustion behaviors of aqueous-based polyurethane system with phosphorus and nitrogen containing curing agent. *Polymer Degradation and Stability*, 65, 359–371 (1999).
53. Song, L. Yuan, H. Y. T., Rui, Z., and Zuyao, W. F. Study on the properties of flame retardant polyurethane/organoclay nanocomposite. *Polymer Degradation and Stability*, **87**, 111–116 (2005).
54. Spirckel, M., Rengier, N., Mortaigne, B., Youssef, B., and Bunel, C. Thermal degradation and fire performance of new phosphonate polyurethanes. *Polymer Degradation and Stability*, 78, 211–218 (2002).
55. Sumaila, M., Ugheoke, B. I., Timon, L., and Oloyede, T. A preliminary mechanical characterization of polyurethane lignocelluloses material. *Leonardo Journal of Sciences* **5**(9), 159–166 (2006).

56. Sun, L., Zhang, Z., and Dang, H. A novel method for preparation of silver nanoparticles. *Materials Letters*, **57**(24–25), 3874–3879 (2003).
57. Tang, Z., Maroto-Valer, M., Andresena, J. M., Miller, J. M., Listemann, M. L., McDaniel, P. L., Morita, D. K., and Furlan, W. R. Thermal degradation behavior of rigid polyurethane foams prepared with different fire retardant concentrations and blowing agents. *Polymer*, **43**, 6471–6479 (2002).
58. Toldy, A., Toth, N., Anna, P., and Marosi, G. Synthesis of phosphorus-based flame retardant systems and their use in an epoxy resin. *Polymer Degradation and Stability*, **91**, 585–592 (2006).
59. Tuan Noor Maznee, T. I., Norin, Z. S., Ooi, T. L., Salmiah, A., and Gan, L., H. Effect of additives on palm-based polyurethane foams. *Journal Oil Palm Resource*, **13**, 7–15 (2001).
60. Twarowski, A. J. The influence of phosphorus oxides and acids on the rate of H + OH recombination *Combustion and Flame*, **94**, 91–107 (1993).
61. Twarowski, A. J. Reduction of a phosphorus oxide and acid reaction set. *Combustion and Flame*, **102**(1–2), 41–54 (1995).
62. Wang, P. S., Chiu, W., Chen, L., Denq, B., Don, T., and Chiu, Y. Thermal degradation behavior and flammability ofpolyurethanes blended with poly (bispropoxyphosphazene). *Polymer Degradation and Stability*, **66**, 307–315 (1999).
63. Yakushin, V. A., Stirna, U. K., Zhmud, N. P. Effect of flame retardants on the properties of monolithic and foamed polyurethanes at low temperatures. *Mechanics of Composite Materials*, **35**, 447–450 (1999).
64. Zhang, L. and Huang, J. Effects of nitrolignin on mechanical properties of polyurethane-nitrolignin films. *Journal of Applied Polymer Science*, **80**(8), 1213–1219 (2001).
65. Zhong, M., Maroto-Valera, M., John, M., Miller, J., Listemann, L. M., McDaniel, P. L., Morita, D. K., and Furlan, W. R. Thermal degradation behavior of rigid polyurethane foams prepared with different fire retardant concentrations and blowing agents. *Polymer*, **43**, 6471–6479 (2002).
66. Zhu, S. and Shi, W. Flame retardant mechanism of hyperbranched polyurethane acrylates used for UV curable flame retardant coatings. *Polymer Degradation and Stability*, **75**, 543–547 (2001).

CHAPTER 15

GRAFT COPOLYMERS OF GUAR GUM VERSUS ALGINATE: DRUG DELIVERY APPLICATIONS AND IMPLICATIONS

ANIMESH GHOSH and TIN WUI WONG

CONTENTS

15.1 GRAFT COPOLYMERS OF GUAR GUM VERSUS ALGINATE

The application of natural polymers in drug delivery system design is a field of ever interest. Broadly, the natural gums are preferred over synthetic polymers due to their non-toxicity, low cost, ready availability, and high regulatory acceptance attributes.

The natural gums are largely safe for oral consumption as they are biodegradable and easily metabolized by the intestinal microflora into individual monomers. Natural gums are widely used as binder and disintegrant in oral solid dosage forms. In oral liquid and topical products, they are used as suspending, thickening and/or stabilizing agents. The use of natural gums is faced with several limitations. These include uncontrolled rates of swelling, excessive thickening, viscosity reduction on storage, and possibility of having microbial contamination. The chemical modification of gums is envisaged to minimize these drawbacks to enable their use in design of specific drug delivery systems (Rana et al., 2011).

The natural gums can be modified *via* carboxymethylation, side chain grafting, cyanoethylation, chemical crosslinking and other processes. Among all gums, guar gum is extensively studied and generally considered as a potential candidate for drug delivery application due to its ability to retard drug release and susceptibility to microbial degradation in large intestine for the purpose of colon-specific delivery. Guar gum is a water-soluble polysaccharide derived from the seeds of *Cyamopsis tetragonolobus*, Leguminosae family. It consists of linear chains of (1→4)-β-D-mannopyranosyl units with α-D-galactopyranosyl units attached by (1→6) linkages (Goldstein et al., 1973) (Figure 1). Guar gum is characterized by a mannose-to-galactose ratio close to 2:1. Its composition constitutes of 80% galactomannan, 12% water, 5% protein, 2% acidic insoluble ash, 0.7% ash, and 0.7% fat. Guar gum is soluble in cold water. It hydrates quickly to produce viscous pseudoplastic solution which exhibits a greater low-shear viscosity than other hydrocolloids (Cheetham and Mashimba, 1990).

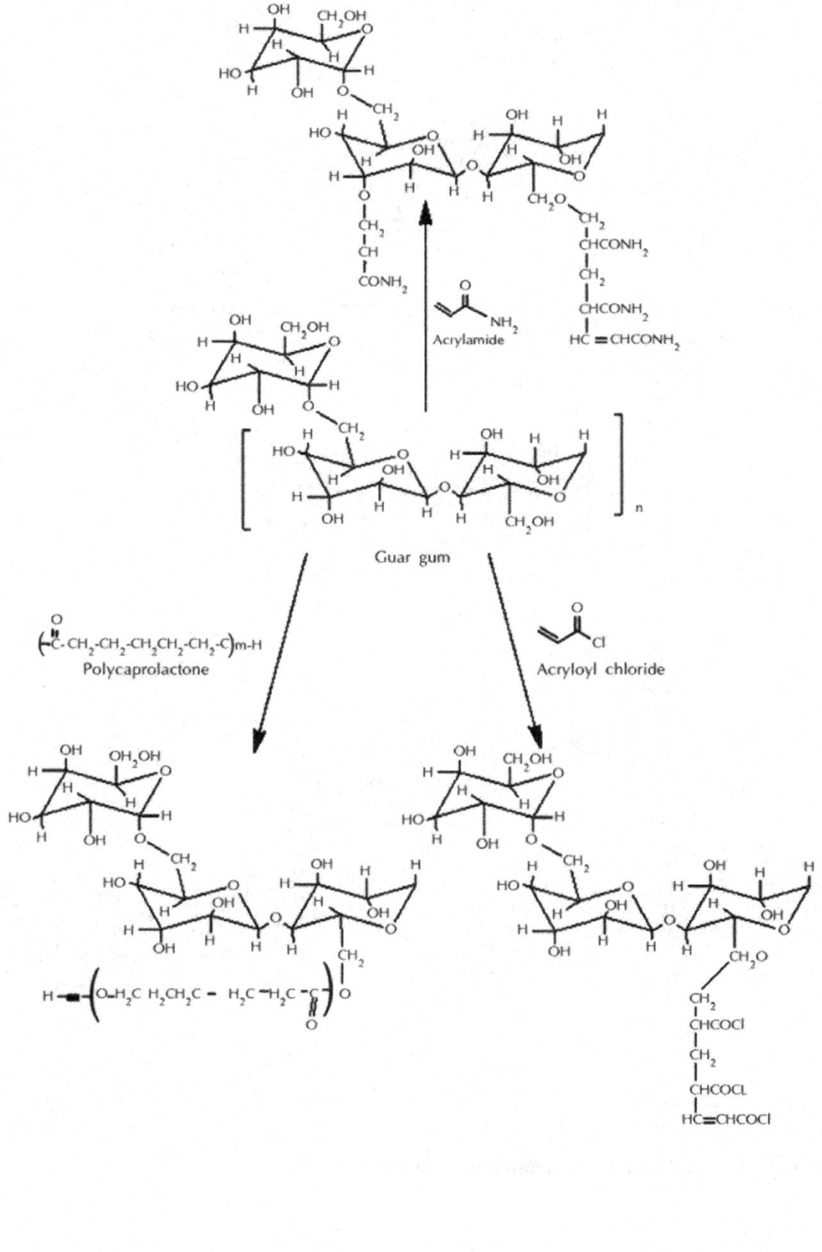

Acrylamide

Guar gum

Polycaprolactone

Acryloyl chloride

FIGURE 1 *(Continued)*

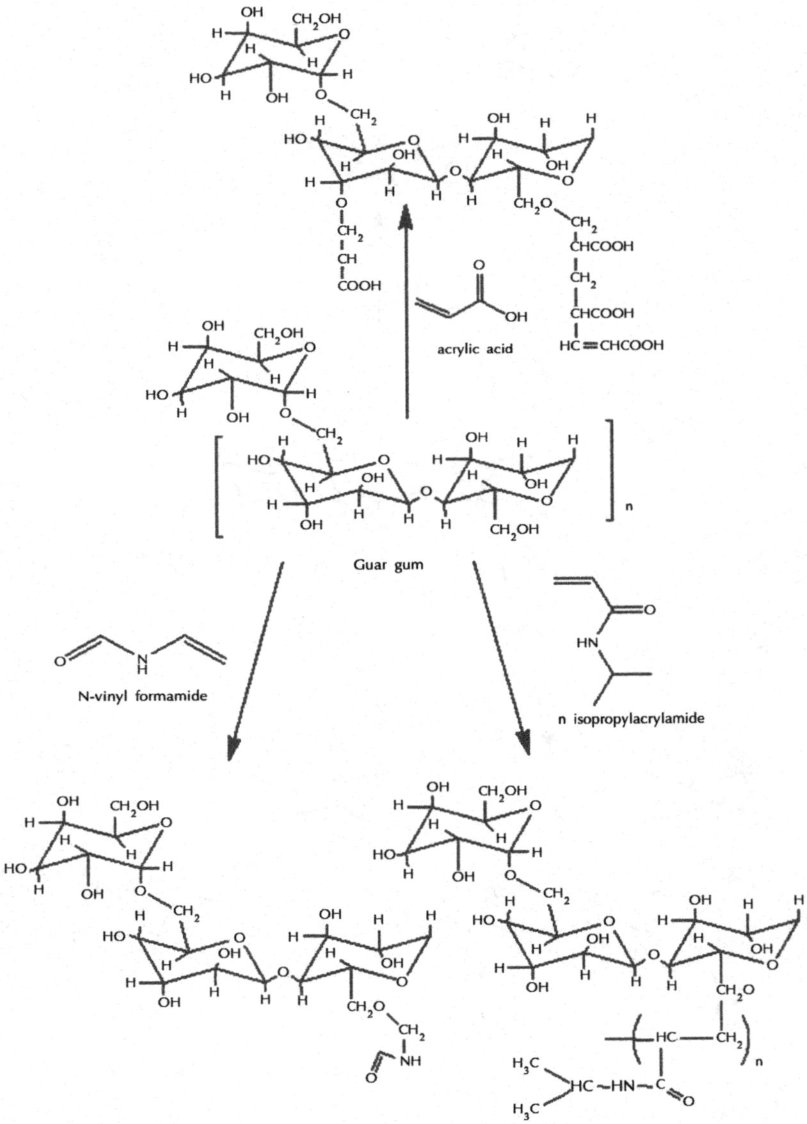

FIGURE 1 Graft copolymerization of guar gum.

Guar gum inherently possesses pharmacological activity. Mosa-Al-Reza et al. (2012) reported that it demonstrates a promising antihyperlipidemic property with a high antiatherogenic potential in both hyperlipidemic and normal rats. Wang et al. (2012) show that sulfated guar gum has antioxidant property. The ingestion of guar gum may modify the propensity of glucose absorption (Blackburn et al, 1984) to give a prolonged feeling of satiety (Leeds, 1987). Guar gum is found to be able to effectively increase the fullness and satiety of oneself in the course of reducing energy intake (Evans and Miller, 1975; Pasman et al., 1997), through decreasing the rate of gastric emptying and small intestinal transit (Harju, 1985; Wilmshurst and Crawley, 1980). However, contradictory outcome has also been reported where guar gum supplementation provides no changes to gastric emptying and/or intestinal transit rate (Meier et al, 1993). Guar gum can be used as a laxative. It is also used to treat diarrhea, irritable bowel syndrome, obesity, and diabetes, as well as, for the prevention of atherosclerosis.

Pharmaceutically, guar gum is widely used as a colon-specific drug carrier in the form of matrix and compression-coated tablets as well as microspheres. Table 1 summarizes type of dosage form, drug, chemical composition, and drug release profiles of some delivery systems made of guar gum.

TABLE 1 Drug delivery systems made of guar gum as the primary drug release modifier

No.	Guar gum quantity in dosage form	Drug	Dosage form	Remarks	Reference
1	97.3%	Dexamethasone	Tablet	72 to 82% of dexamethasone is delivered to colon.	Kenyon et al. (1997)
2	77.2%	Indomethacin	Tablet	*In vitro* drug release without rat caecal medium is 29.2% and this increases to 49.7% and 59.6% with 2% and 4% addition of rat caecal medium into dissolution fluid respectively.	Prasad et al. (1998)

TABLE 1 *(Continued)*

3	55.6% (100 mg guar gum coating over 80 mg core tablet)	Indomethacin	Tablet	*In vitro* cumulative percent drug release with rat caecal medium (2%, 21 hr study) is ≤20 %. *In vivo* scintigraphy study shows intact tablet in small intestine (2 hr), with commencement of disintegration of coat (4 hr), distribution of broken pieces of tablet in ascending colon, hepatic flexure, transverse colon, and splenic flexure (8 hr).	Krishnaiah et al. (1998)
4	80%	5-fluorouracil	Tablet	*In vitro* drug release without rat caecal medium is 50.9%. *In vivo* studies in healthy human volunteers shows maximum plasma drug concentration of 216 ng/ml at 7.6 hr.	Krishnaiah et al. (2003a)
5	75%	Diltiazem	Double coated tablet with inulin and shellac	*In vitro* drug release from uncoated tablet in rat cecal content-free simulated colonic fluid is 60% at 11 hr of dissolution. This increases up to 80% with the use of medium with rat cecal content. Using 2% inulin coating, 28% drug are released after 11 hr of dissolution.	Ravi et al. (2008)

TABLE 1 *(Continued)*

6	65%	Ornidazole	Tablet	*In vitro* drug release without rat caecal medium is 73.4% which increases to 96.8% in the presence of 4 %rat caecal medium.	Krishnaiah et al. (2003b)
7	60%	Rofecoxib	Tablet	*In vitro* drug release without rat caecal medium is 47.3% which increases to 99.5% in the presence of 4% rat caecal medium.	Al-Saidan et al. (2005)
8	50%	Mesalazine	Tablet	*In vitro* drug release without rat caecal medium is 43%which increases to 96% in the presence of 4% rat caecal medium.	Demiröz et al. (2004)
9	44%	Indomethacin	Eudragit FS 30D coated pellets	*In vitro* drug release from double coated pellets is 66.6% at 7 hr of dissolution which increases to 100.2% when drug release is studied in the presence of enzymes. *In vivo* study in Beagle dogs gives maximum plasma drug concentration of 1291.51 ng/ml at 5.41 hr with double coated pellets, and 3296.87 ng/ml at 2.46 hr with single coated pellets.	Ji et al. (2007)

TABLE 1 *(Continued)*

10	20%	Albendazole	Tablet	*In vitro* drug release without rat caecal medium is 20.9% which increases to 43.9% in the presence of 4% rat caecal medium.	Krishnaiah et al. (2001)
11	1%, 1.5%, and 2%	Aceclofenac	Micro-spheres	Microspheres demonstrate similar drug release rates in phosphate buffer solution (pH 7.0) and simulated gastric fluid. A significant increase in drug release (83.2%) is observed in the medium containing 4% rat caecal content which corresponds to *in vivo* analysis.	Ravi et al. (2010)
12	1%, 2%, 3%, and 4%	Methotrexate	Micro-spheres	Microspheres demonstrate similar drug release rates in phosphate buffer solution (pH 7.0) and simulated gastric fluid. A significant increase in drug release (91.0%) is observed in the medium containing 4% rat caecal content.	Chaurasia et al. (2006)

In spite of the guar gum is widely used in native form for drug delivery purpose, its application is complicated by physicochemical and microbial stability concerns. Guar gum can abruptly swell, undergo fast biodegradation in aqueous medium and give premature drug release (Nayak and Singh, 2001; Jumel et al., 1996). To overcome enormous swelling intensity of guar gum, crosslinking of guar gum polymer chains with borax, glutaraldehyde, and trisodium trimetaphosphate shows promising results. The cross-linked guar gum is useful for controlling the release of drugs without its enzymatic digestibility being negated. To achieve site-specific controlled-release attribute *via* modes such as pH or temperature responsiveness, grafting of guar gum has been investigated. The biodegradation propensity of guar gum can also be reduced through grafting with synthetic polymer (Whistler, 1973).

A graft copolymer is a macromolecular chain with one or more species of block connected to the main chain as side chains. Polymer grafting is a convenient method to add new properties to a polymer with minimum loss of the original properties of the substrate. Copolymers are commonly prepared through grafting vinyl or acryl monomers onto the polymer backbone. The chemistry of grafting vinyl/acryl monomers is different from that of non-vinyl/acryl monomers grafting. The latter is executed *via* polycondensation where high temperature and harsh chemical condition are applied. This is unfavorable for preparing graft copolymers of polysaccharides (Maiti et al., 2010). Both hydrophilic and hydrophobic monomers can be grafted onto guar gum (Deshmukh and Singh, 1987; Singh et al., 2004). Grafting is generally considered to result from propagation of radical sites generated on polymeric substrates and the formation opportunity of homopolymers is thus reduced (Nayak and Singh, 2001).

15.2 VINYLIC/ACRYLIC GRAFT COPOLYMERS OF GUAR GUM

Polyvinylic and polyacrylic grafted derivatives of guar gum are mainly achieved by radical polymerization. Graft copolymers of guar gum are prepared by first generating free radicals on the polymer backbone and then allowing these radicals to serve as macroinitiators for the vinyl or acrylic monomer. The chemical and radiation initiator systems are employed to graft copolymerize the monomers onto polysaccharides. The cerium in its tetravalent state is a versatile oxidizing agent. It is frequently used in the graft copolymerization of vinyl monomers onto cellulose and starch. It forms a redox pair with the anhydroglucose units of polysaccharide to yield macroinitiators under slightly acidic conditions (Athawale and Rathi., 1999).

Microwave can be used as an alternative initiator. Sen et al. (2010) report polyacrylamide grafted guar gum synthesized *via* microwave initiation as controlled-release matrix of 5-amino salicylic acid (a drug used for the treatment of ulcerative colitis). The *in vitro* release of drug from the matrix made of polyacrylamide grafted guar gum is dependent on level of grafting which is a function of microwave intensity and duration of treatment. In another study, Toti and

Aminabhavi (2004) develop polyacrylamide grafted guar gum at three different ratios of guar gum to acrylamide (1:2, 1:3.5, and 1:5). The drug release from tablet made of this graft copolymer is diffusion-controlled in case of the copolymer is not hydrolyzed. With copolymer undergoing hydrolyzation, the drug release is swelling-controlled in initial phase of dissolution (simulated gastric medium) followed by diffusion-controlled at prolonged phase of dissolution (simulated intestinal medium). The hydrolyzed polyacrylamide grafted guar gum matrix can have its amide group in graft copolymers converted to carboxylic acid moiety. The matrix is pH sensitive and can be used for intestinal drug delivery.

Soppirnath and Aminabhavi (2002) develop graft copolymer of guar gum with acrylamide and have it cross-linked with glutaraldehyde to form hydrogel microspheres by water-in-oil emulsification method. The microspheres are loaded with two anti-hypertensive drugs, verapamil hydrochloride (water-soluble) and nifedipine (water-insoluble) in investigation of their controlled-release characteristics. The findings indicate that the *in vitro* drug release is dependent on the extent of cross-linking, amount of drug loading, solubility nature of drug molecule, and method of drug loading. The release of drugs is swelling-controlled at the initial stage of dissolution, and is governed by solute diffusion in the later phase. The drug release of such matrix can be faster in simulated intestinal fluid than simulated gastric condition (Soppimath et al., 2001). The rate of drug release may be modulated through changing the matrix swelling extent which is reversible and pulsatile with the changing environmental conditions (pH and ionic strength). The rate of drug release can be further reduced through blending poly (acrylamide) grafted guar gum with chitosan to form interpenetrating polymer network hydrogel microspheres by emulsion cross-linking method using glutaraldehyde as the crosslinking agent (Kajjari et al., 2011). The microspheres are able to encapsulate up to 74% of ciprofloxacin, an antibiotic drug with a plasma half-life of 4 hr, and the release of ciprofloxacin is extended up to 12 hr. Alternatively, guar gum can be grafted with poly (*N*-isopropylacrylamide) and cross-linked with glutaraldehyde into thermosensitive hydrogel (Lang et al., 2008). The drug release is controlled through temperature modulation of the surrounding medium of hydrogel. The drug (sinomenine hydrochloride) is released slowly from the gel at 40°C, and released more rapidly at 35°C.

Tiwari and Prabaharan (2010) develop amphiphilic guar gum grafted with poly (epsilon-caprolactone) using microwave irradiation technique. Under the influence of microwave, guar gum with high poly (epsilon-caprolactone) grafting percentage (>200%) is obtainable within a short reaction time. The drug-release profile of poly (epsilon-caprolactone) grafted guar gum micelles is characterized by initial burst release followed by sustained release phases of hydrophobic model drug, ketoprofen, over a period of 1 to 68 hr. Under physiological conditions, the poly (epsilon-caprolactone) grafted guar gum is hydrolytically degraded into lower-molecular weight fragments within a 7 week period. The micelles derived from this graft copolymer are suitable for use as a controlled-release drug delivery system.

Thakur and co-workers (2009) develop acryloyl guar gum, and it is used as hydrogel carrier and slow-release device for two pro-drugs, L-tyrosine and 3,4-dihydroxy phenylalanine (L-DOPA). The swelling extent of hydrogel is affected by both pH and ionic strength of dissolution medium. It is reversible in response to changing environmental conditions. The highest percentage cumulative release of L-tyrosine is attainable from the use of poly (methacrylic acid) grafted guar gum, while the maximum release of L-DOPA is achievable using poly (acrylic acid) grafted guar gum. Guar gums grafted with poly (2-hydroxyethyl methacrylate) and poly (2-hydroxypropyl methacrylate) however exhibit a higher level of drug release retardation where, 42.3% and 49.1% of drug are retained in matrix even after 12 hr of dissolution.

Huang and co-workers (2007) report polyelectrolyte formed from cationic guar gum and acrylic acid monomer through photo initiated free radical polymerization. The cationic guar gum-poly (acrylic acid) polyelectrolyte was formulated in the form of hydrogel and tablet with ketoprofen embedded in these matrices. The release of drug is favored by intestinal pH with polymer relaxation and erosion serve as the primary mode of drug release. The cationic guar gum-poly (acrylic acid) polyelectrolyte matrix has a great potential for use as colon-specific drug carrier. Other poly (acrylic acid) grafted guar gum has also been prepared by Giri et al. (2012). The poly (acrylic acid) grafted guar gum is formulated with nano-silica to provide slow release characteristics to water-soluble drug encapsulated in membrane dosage form. It has also been blended with poly (vinyl alcohol) into microspheres with an average size of 10 μm by water-in-oil emulsification technique to sustain release of isoniazid, an anti-tuberculosis drug (Sullad et al., 2010). The duration of drug release can be extended to 8 hr from its nascent plasma half-life of 0.5 to 1.6 hr.

Other copolymers of guar gum are designed using N-vinyl formamide, acrylonitrile, methyl acrylate, methyl methacrylate, and ethyl methacrylate as grafted species (Behari et al., 2005; Trivedi et al., 2005). The guar gum employed in grafting may be native or partially carboxymethylated to increase its water solubility for ease of graft copolymerization, or for modulation of drug dissolution in its end application as matrix material. The schematic diagrams of selected graft copolymerization of guar gum are summarized in Figure 1.

15.3 GUAR GUM VERSUS ALGINATE GRAFT COPOLYMERS

A review on physicochemical, biological and drug delivery attributes of alginate graft copolymers has been reported by Wong (2011). Alginate is a water-soluble polysaccharide made of homopolymeric regions of β-D-mannuronic acid (M) blocks and α-L-guluronic acid (G) blocks, interdispersed with regions of alternating structure of G and M acid blocks (Figure 2). It is commonly isolated from brown algae such as *Laminaria hyperborea*, *Ascophyllum nodosum*, and *Macrocystis pyrifera*. This alginate has a M/G ratio of 0.45 to 3.33. The M blocks take the form of extended ribbons, while the G blocks are buckled in shape. The alignment of two G blocks side by side gives rise to diamond configuration cavities which have an ideal dimension for co-operative binding of cations.

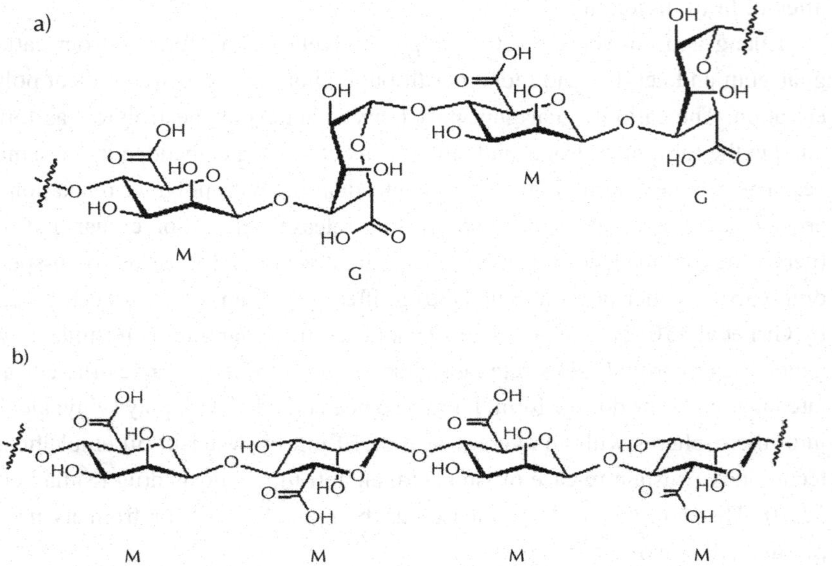

a)

b)

FIGURE 2 *(Continued)*

c)

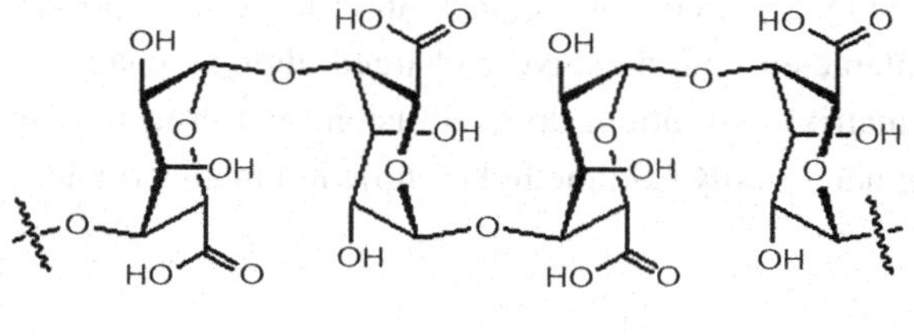

G G G G

FIGURE 2 Chemical structures of alginate with blocks of a) MG, b) MM, and c) GG sequences.

Alginate has pKa values between 3.4 and 4.4 (Wong, 2011). It is practically soluble in water. Alginate precipitates as alginic acid in low pH or gastric medium. In a higher pH medium or intestinal fluid, the alginic acid is converted to soluble viscous layer. Nonetheless, both precipitate and soluble states of alginate could render fast drug release. Similar to guar gum, alginate is non-toxic, biodegradable, and biocompatible. It inherently possesses biological effects, such as anti-cholesterolaemic, anti-hypertensive, anti-diabetic, anti-obesity, anti-microbial, anti-cancer, anti-hepatotoxicity, wound healing, anti-coagulation, and coagulation activities.

Alginate has been used as emulsion thickener, stabilizer, emulsifier, carrier polymer for antigen, enzyme, microbe, animal cell and recombinant gene product, bone and cartilage tissue engineering scaffold, peripheral nerve regeneration implant, wound dressing, dental impression material, anti-heartburn and gastric reflux raft-forming formulation, radioactive and heavy metal absorption agent, and plasma expander (Wong, 2011). Pharmaceutically, alginate has been used as drug carrier in the form of microspheres, microcapsules, gel beads, hydrogel, film, nanoparticles, and tablets. As in the case of guar gum, the alginate has been grafted with poly (*N*-isopropylacrylamide) to strengthen the formed matrix and introduce thermal sensitivity for drug release regulation. The alginate has also been grafted with poly (butyl methacrylate), poly (2-dimethylamino) ethyl methacrylate, poly (methacrylamide), poly (sodium acrylate), and poly (sodium acrylate-co-acrylamide) (Figure 3). Graft copolymerization is largely aimed to retard drug release, with pH and ionic strength sensitivity introduced to the matrix to promote or reduce swelling of dosage form reversibly. This in turn leads to release or retention of drug embedded in the matrix. In addition, cationic character may be incorporated into alginate in order to have positively charged grafted chains interacted with negatively charged alginate chains or drug molecules to prevent matrix dissolution, drug diffusion, and drug release, for example, oxidized alginate-*g*-poly (2-dimethylamino)ethyl methacrylate.

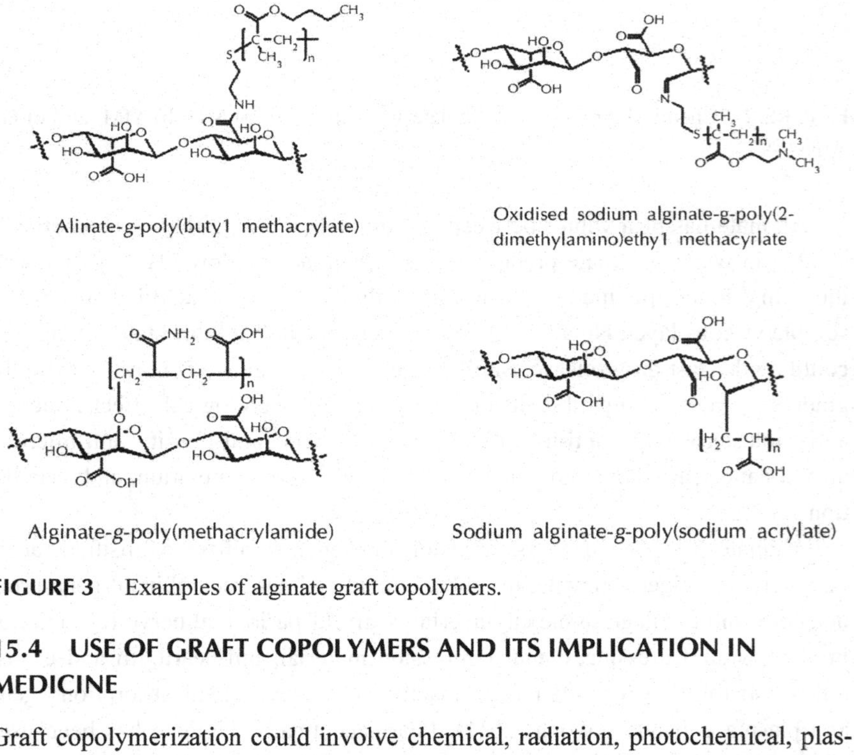

Alinate-*g*-poly(buty1 methacrylate)

Oxidised sodium alginate-g-poly(2-dimethylamino)ethy1 methacyrlate

Alginate-*g*-poly(methacrylamide)

Sodium alginate-*g*-poly(sodium acrylate)

FIGURE 3 Examples of alginate graft copolymers.

15.4 USE OF GRAFT COPOLYMERS AND ITS IMPLICATION IN MEDICINE

Graft copolymerization could involve chemical, radiation, photochemical, plasma-induced, and enzymatic processing. Generally, grafting process is initiated by chemical means. The energy required for grafting may be provided from different sources namely conventional heating or microwave irradiation. Grafting may involve complicated multistep process and hazardous reagents, undesirable chemical modification of the parent polymer, formation of unstable and toxic products, and huge operating costs (Wong, 2011). This hinders commercial use of potential graft copolymers in drug delivery system design. Microwave irradiation demonstrates the concept of green chemistry. The microwave-assisted graft copolymerization requires a very short reaction time and can proceed even in the absence of any redox initiator (Vijan et al., 2012). The interaction between microwave and reagents in a synthesis process is unimpeded by steric hindrance due to large wavelengths of radiation. Unfortunately, the use of microwave in graft copolymerization is much less widespread than redox approach.

15.5 CONCLUSION

The graft copolymerization of guan gum and alginate translates to the formation of "new" excipients. It implies that the synthetic excipients are required to be examined and reviewed from the regulatory standpoints. The impurity-induced toxicity has been one main obstacle in the approval process of excipients for use in pharmaceutical applications. Future direction on guar gum and alginate graft copolymerization research lies in the ability to bulk synthesize functional graft copolymers using the lowest cost, shortest time and meeting the safety requirements of regulatory sectors.

KEYWORDS

- **Alginate**
- **Drug release**
- **Graft copolymerization**
- **Guar gum**

REFERENCES

1. Al-Saidan, S. M., Krishnaiah, Y. S. R., Satyanarayana, V., and Rao, G. S. In vitro and in vivo evaluation of guar gum-based matrix tablets of rofecoxib for colonic drug delivery. *Curr. Drug Deliv.*, **2**(2), 155–163 (2005).
2. Athawale, V. D. and Rathi, S. C. Graft polymerization: starch as a model substrate. *J. Macromol. Sci-Rev. Macromol. Chem. Phys.*, **C39**, 445–480 (1999).
3. Behari, K., Banerjee, J., Srivastav, A., and Mishra, D. K. Studies of graft copolymerization of *N*-vinyl formamide onto guar gum initiated by bromate/ascorbic acid redox pair. *Ind. J. Chem. Tech.*, **12**, 664–670 (2005).
4. Blackburn, N. A., Redfern, J. S., Jarjis, H., Holgate, A. M., Hanning, I., Scarpello, J. H. B., Johnson, I. T., and Read, N. W. The mechanism of action of guar gum in improving glucose tolerance in man. *Clin. Sci.*, **66**, 329–336 (1984).
5. Chaurasia, M., Chourasi, M. K., Jain, N. K., Jain, A., Soni, V., Gupta, Y., and Jain, S. K. Cross-linked guar gum microspheres: A viable approach for improved delivery of anticancer drugs for the treatment of colorectal cancer. *AAPS PharmSciTech*, **7**(3), Article 74 (2006).

6. Cheetham, N. W. H. and Mashimba, E. N. M. Comformational aspects of xanthan-galactomannan gelation-Further evidence from optical-rotation studies. *Carbohydr. Polym.,* **14**, 17–21 (1990).
7. Demiröz, F. T., Acartürk, F., Sevgi, F. T., and Oznur, K. B. In-vitro and in-vivo evaluation of mesalazine-guar gum matrix tablets for colonic drug delivery. *J. Drug Targeting*, **12**(2), 105–112 (2004).
8. Desmukh, S. R. and Singh, R. P. Drag reduction effectiveness, shear stability and biodegradation resistance of guar gum-based graft copolymers. *J. Appl. Polym. Sci.*, **33**, 1963–1975 (1987).
9. Evans, E. and Miller, D. S. Bulking agents in the treatment of obesity. *Nutr. Metab.*, **18**, 199–203 (1975).
10. Giri, A., Bhunia, T., Mishra, S., Goswami, L., Panda, A. B., Pal, S., and Bandyopadhyay, A. Acrylic acid grafted guar gum-nanosilica membranes for transdermal diclofenac delivery. *Carbohydr. Polym.,* http://dx.doi.org/10.1016/j.carbpol.2012.08.035 (2012)
11. Goldstein, A. M., Alter, E. N., and Seaman, J. K., Guar gum, in: R. L. Whistler (Ed.), Industrial Gums, Polysaccharides and their derivatives, Academic Press, New York, pp. 303–321 (1973).
12. Harju, E. Increases in meal viscosity caused by addition of guar gum decrease postprandial activity and rate of emptying of gastric contents in healthy subjects. *Panminerva Med.,* **27**, 125–128 (1985).
13. Huang, Y., Yu, H., and Xiao, C. pH-sensitive cationic guar gum/poly (acrylic acid) polyelectrolyte hydrogels: Swelling and in vitro drug release. *Carbohydr. Polym.*, **69**, 774–783 (2007).
14. Ji, C., Xu, H., and Wu, W. In vitro evaluation and pharmacokinetics in dogs of guar gum and Eudragit FS30D-coated colon-targeted pellets of indomethacin. *J. Drug Targeting*, **15**(2), 123–131 (2007).
15. Jumel, K., Harding, S. E., and Mitchell, J. R. Effect of gamma irradiation on the macromolecular integrity of guar gum. *Carbohydr. Res.*, **282**, 223–236 (1996).
16. Kajjari, P. B., Manjeshwar L. S., and Aminabhavi, T. M. Novel interpenetrating polymer network hydrogel microspheres of chitosan and poly(acrylamide)-grafted-guar gum for controlled release of ciprofloxacin. *Ind. Eng. Chem. Res.,* **50**, 13280–13287 (2011).
17. Kenyon, C. J., Nardi, R. V., Wong, D., Hooper, G., Wilding, I. R., and Friend, D. R. Colonic delivery of dexamethasone: a pharmacoscintigraphic evaluation. *Alimentary Pharmacol. Ther.,* **11**(1), 205–213 (1997).
18. Krishnaiah, Y. S. R., Muzib, Y., Indira Rao, G., Srinivasa Bhaskar, P., and Satyanarayana, V. Studies on the development of colon targeted oral drug delivery systems for ornidazole in the treatment of amoebiasis. *Drug Dev. Ind. Pharm.*, **10**(2), 111–117 (2003).
19. Krishnaiah, Y. S. R., Nageswara Rao, L., Bhaskar Reddy, P. R., Karthikeyan, R. S., and Satyanarayana, V. Guar gum as a carrier for colon specific delivery: influence of metronidazole and tinidazole on in vitro release of albendazole from guar gum matrix tablets. *J. Pharm. Pharm. Sci.*, **4**(3), 235–243 (2001).

20. Krishnaiah, Y. S. R., Satyanarayana, S., Rama Prasad, Y., and Narasimha Rao, S. Evaluation of guar gum as a compression coat for drug targeting to colon. *Int. J. Pharm.*, **171**(2), 137–146 (1998).
21. Krishnaiah, Y. S. R., Satyanarayana, V., Dinesh Kumar, B., Karthikeyan, R. S., and Bhaskar, P. In vivo pharmacokinetics in human volunteers: oral administered guar gum-based colon-targeted 5-fluorouracil tablets. *Eur. J. Pharm. Sci.*, **19**(5), 355–362 (2003).
22. Lang, Y., Jiang, T., Li, S., and Zheng, L. Study on physicochemical properties of thermosensitive hydrogels constructed using graft-copolymers of poly(*N*-isopropylacrylamide) and guar gum. *J. Appl. Polym. Sci.*, **108**, 3473–3479 (2008).
23. Leeds, A. R. Dietary fiber: mechanism of action. *Int. J. Obes. Relat. Metab. Disord.*, **11**(1) 3–7 (1987).
24. Maiti, S., Ranjit, S., and Sa, B. Polysaccharide-based graft copolymers in controlled drug delivery. *Int. J. Pharm. Tech. Res.*, **2**, 1350–1358 (2010).
25. Meier, R., Beglinger, C., Schneider, H., Rowedder, A., and Gyr, K. Effect of a liquid diet with and without soluble fiber supplementation on intestinal transit and cholecystokinin release in volunteers. *J. Parent. Enteral Nutr.*, **17**, 231–235 (1993).
26. Mosa-Al-Reza, H., Saeed, S., Sadat, D. A., and Marziyeh, A. Comparison of the beneficial effects of guar gum on lipid profile in hyperlipidemic and normal rats. *J. Med. Plants Res.*, **6**, 1567–1575 (2012).
27. Nayak, B. R. and Singh, R. P. Synthesis and characterization of grafted hydroxypropyl guar gum by ceric ion induced initiation. *Eur. Polym. J.*, **37**, 1655–1666 (2001).
28. Pasman, W. J., Westerterp-Plantenga, M. S., Muls, E., Vansant, G., van Ree, J., and Saris, W. H. M. Effect of one week of fibre supplementation on hunger and satiety or energy intake. *Appetite,* **29**, 77–87 (1997).
29. Prasad, Y. V., Krishnaia, Y. S. R., and Satyanarayana, S. In vitro evaluation of guar gum as a carrier for colon-specific drug delivery. *J. Controlled Release*, **51**, 281–287 (1998).
30. Rana, V., Rai, P., Tiwary, A. K., Singh, R. S., Kennedy, J. F., and Knill, C. J. Modified gums: Approaches and applications in drug delivery. *Carbohydr. Polym.*, **83**, 1031–1047 (2011).
31. Ravi, P., Rao Kusumanchi, R. M., Mallikarjun, V., Babu Rao, B., and Raja Narender, B. Formulation and evaluation of guar gum microspheres of aceclofenac for colon targeted drug Delivery. *J. Pharm. Res.*, **3**, 1510–1512 (2010).
32. Ravi, V., Mishra, S. T., and Kumar, P. Influence of natural polymer coating on novel colon targeting drug delivery system. *J. Mater. Sci.: Mater. Med.*, **19**(5), 2131–2136 (2008).
33. Sen, G., Mishra, S., Jha, U. and Pal, S. Microwave initiated synthesis of polyacrylamide grafted guar gum (GG-*g*-PAM)-Characterizations and application as matrix for controlled release of 5-amino salicylic acid. *Int. J. Biol. Macromol.*, **47**, 164–70 (2010).
34. Singh, V., Tiwari, A., Tripathi, D. N., and Sanghi, R. Microwave assisted synthesis of guar-*g*-polyacrylamide. *Carbohydr. Polym.*, **58**, 1–6 (2004).
35. Soppimath, K. S., Kulkarni, A. R., and Aminabhavi, T. M.. Chemically modified polyacrylamide-*g*-guar gum-based crosslinked anionic microgels as pH-sensitive

drug delivery systems: preparation and characterization. *J. Controlled Release*, **75**, 331–45 (2001).
36. Soppirnath, K. S. and Aminabhavi, T M.. Water transport and drug release study from cross-linked polyacrylamide grafted guar gum hydrogel microspheres for the controlled release application. *Eur. J. Pharm. Biopharm.*, **53**, 87–98 (2002).
37. Sullad, A. G., Manjeshwar, L. S., and Aminabhavi, T. M. Novel pH-sensitive hydrogels prepared from the blends of poly(vinyl alcohol) with acrylic acid-graft-guar gum matrixes for isoniazid delivery. *Ind. Eng. Chem. Res.*, **49**, 7323–7329 (2010).
38. Thakur, S., Chauhan, G. S., and Ahn, J. Synthesis of acryloyl guar gum and its hydrogel materials for use in the slow release of l-DOPA and l-tyrosine. *Carbohydr. Polym.*, **76**, 513–520 (2009).
39. Tiwari, A and Prabaharan, M. An amphiphilic nanocarrier based on guar gum-graft-poly(epsilon-caprolactone) for potential drug-delivery applications. *J. Biomater. Sci. Polym. Ed.*, **21**(6), 937–949 (2010).
40. Toti, U. S. and Aminabhavi, T. M. Modified guar gum matrix tablet for controlled release of diltiazem hydrochloride. *J. Controlled Release*, **95**, 567–77 (2004).
41. Trivedi, J. H., Patel, N. K., and Trivedi H. C. Grafting of vinyl monomers onto sodium salt of partially carboxymethylated guar gum: Comparison of their reactivity. *Polym. Plast. Tech. Eng.*, **44**, 407–425 (2005).
42. Vijan, V., Kaity, S., Biswas, S., Isaac, J., and Ghosh, A. Microwave assisted synthesis and characterization of acrylamide grafted gellan: application in drug delivery. *Carbohydr. Polym.*, **90**, 496–506 (2012).
43. Wang, J., Zhao, B., Wang, X., Yao, J., and Zhang, J. Structure and antioxidant activities of sulfated guar gum: homogeneous reaction using DMAP/DCC catalyst. *Int. J. Biol. Macromol.*, **50**, 1201–1206 (2012).
44. Whistler, R. L. Industrial gum. New York: Academic press, p. 6 (1973).
45. Wilmshurst, P. and Crawley, J. C. W. The measurement of gastric transit time in obese subjects using ^{24}Na and the effects of energy content and guar gum on gastric emptying and satiety.*Br. J. Nutr.*,**44**, 1–6 (1980).
46. Wong, T. W. Alginate graft copolymers and alginate-co-excipient physical mixture in oral drug delivery. *J. Pharm. Pharmacol.*, **63**(12), 1497–1512 (2011).

CHAPTER 16

THERMAL PROPERTIES OF POLYHYDROXYALKANOATES

YOGA SUGAMA SALIM, CHIN HAN CHAN, KUMAR SUDESH, and SENG NEON GAN

CONTENTS

16.1 INTRODUCTION TO POLYHYDROXYALKANOATES

Polyhydroxyalkanoates (PHAs), optically active polyesters with (*R*) absolute configuration and chiral centers, can be synthesized by variety of Gram-positive and Gram-negative microorganisms. They serve as energy reserve in the cytoplasmic fluid of bacteria, or as a sink for excess reducing equivalents for microorganisms (Doi, 1990). Poly(3-hydroxybutyrate) (PHB) is an example of homopolymer PHAs, which was first isolated by a French microbiologist in 1925 from *Bacillus megaterium* using chloroform extraction (Lemoigne, 1925; 1926). Following the discovery of PHB, copolymer containing 3-hydroxyvalerate (PHBV) (Doi et al., 1988), 4-hydroxybutyrate (PHBB) (Doi et al., 1988) and 3-hydroxyhexanoate (PHBHHx) (Doi et al., 1995) have been subsequently discovered. The synthesis of PHAs, in general, starts from production of hydroxyacyl-CoAs from carbon substrate and precursors, followed by the polymerization of hydroxyacyl-CoAs by PHAs syntheses into PHAs. To date, there are at least 140 types of PHAs monomers successfully identified (Steinbüchel and Valendin, 1995). Depending on the carbon substrates used and biochemical pathways of the microorganisms, the molecular structure and copolymer composition of PHAs can be judiciously altered to create polymeric materials with specific physical properties ranging from semicrystalline plastic to elastic materials, which resemble rubber. The PHAs are known to exhibit biodegradability (Jendrossek and Handrick, 2002; Mukai and Doi, 1995; Numata et al., 2009) in soil (Mergaert et al., 1993; Sang et al., 2002), sludge (Briese et al., 1994), seawater (Gonda et al., 2000) and tropical mangrove ecosystem (Sridewi et al., 2006). It has been long known that PHAs degrading microorganisms tend to attack the amorphous region of PHAs leading to a reduction of molecular weight (Doi et al., 1989), and a recent study of PHAs soil degradation by Boyandin and co-workers also shows the same tendency (2012).

The PHAs also exhibit biocompatibility to osteoblast (Zhao et al., 2007), fibroblast (Sun et al., 2007), stem cells (You et al., 2011; Xu et al., 2010), and can be used for tarsal repair (Zhou et al., 2010), thus, they serve as alternative and environmentally friendly polymeric materials for many applications including environmental and medical applications. Current studies are devoted to the development of better genetically engineered bacteria (Nomura et al., 2004; Hiroe et al., 2012; Yee et al., 2012; Bhubalan et al., 2011) producing higher amount of PHAs (da Cruz Pradella et al., 2012; Berezina, 2012) from renewable cheaper carbon sources such as fatty acids, sugars or industrial wastes (Sudesh, 2013; Chaudhry et al., 2011; Tripathi et al., 2012). In fact, the choice of chemicals plays a major role to determine the efficiency of recovery process as well as the production cost in industrial scale. Kunasundari and Sudesh summarized various recovery processes from literatures and discussed the strategies to reduce production cost of PHAs in large scale production (2011). The challenge of PHAs recovery lies within the choice of solvents and chemicals with a certain level of purity, which depends on the types of final application, without compromising the molecular weight of extracted PHAs.

During storage over a certain period of time, example 1–2 years, PHAs undergo secondary crystallization and the molecular weight of the PHAs may gradually decrease. From the standpoint of product development, the knowledge of crystallization kinetics is of critical need in regulating the size, orientation/allignment and structure types (banded, fibrillar, shish kebab, and so on) of the spherulites of PHA crystals, which in turn, affects the mechanical properties of the materials. The requirement for crystallization of a semicrystalline polymer is restricted to a temperature range between its glass transition temperature (T_g) and equilibrium melting temperature (T_m^0).

16.2 THEORETICAL BACKGROUND OF ISOTHERMAL CRYSTALLIZATION

The crystallization of polymer in bulk as well as in solution is initiated by nucleation followed by growing of spherulites (Mandelkern, 2002). A common fundamental approach to study isothermal crystallization kinetics is the heuristic Avrami phase transition theory (Avrami, 1939; 1940; 1941).

16.2.1 CRYSTALLIZATION KINETICS UNDER ISOTHERMAL CONDITION FOR SEMICRYSTALLINE POLYMER

The Avrami model was originally derived for the study of kinetics of crystallization and growth of a simple metal system, and further extended to the crystallization of polymer. Avrami assumes the nuclei develop upon cooling of polymer and the number of spherical crystals increases linearly with time at a constant growth rate in free volume. The Avrami equation is given as follow:

$$X(t) = 1 - \exp\left[-K^{1/n}(t - t_0)\right]^n \quad (1)$$

where, $X(t)$ is relative crystallinity at time t, t_0 is an induction time, $K^{1/n}$ is an overall rate constant of isothermal crystallization (min^{-1}); and n is Avrami exponent. Induction time is defined as the initial time in which the polymer crystals start to grow. Selection of induction time for Avrami fitting is important to achieve reasonable value of $K^{1/n}$ and n (Lorenzo et al., 2007). The Avrami exponent depends not only on the structure of nuclei/crystal but also on the nature of the nucleation (Avrami, 1940).

Rearrangement of Equation (1) arrives at Equation (2), and the calculation of estimated half-time crystallization from Avrami plot is shown in Equation (3).

$$\lg\left[-\ln\left(1-X\right)\right]=\lg K^{1/n}+n\lg\left(t-t_0\right) \tag{2}$$

$$t^*_{0.5}=\frac{(\ln 2)^{1/n}}{K^{1/n}} \tag{3}$$

$K^{1/n}$ and n can be extracted from the intercept and the slope of Avrami plot, lg[–ln(1–X)] versus lg(t–t_0), respectively. The prime requirement of Avrami model is the ability of spherulites of a polymer to grow in a free space. Besides, Avrami equation is usually only valid at low degree of conversion, where impingement of polymer spherulites is yet to take place. The rate of crystallization of polymer can also be characterized by reciprocal half-time $(t^*_{0.5})^{-1}$. The use of Avrami model permits the understanding on the kinetics of isothermal crystallization as well as non-isothermal crystallization. However, in this chapter the discussion of the kinetics of crystallization is limited to isothermal conditions.

Basic principle underlying the isothermal crystallization of PHAs will be emphasized in this chapter. Crystallization of polymer melts is often accompanied by a heat release in the system. This can be measured using differential scanning calorimeter (DSC). Following the isothermal crystallization of PHAs, the influence of thermal treatment of a semicrystalline polymer on the mechanical properties is also given.

16.2.2 SPHERULITE GROWTH RATES FOR SEMICRYSTALLINE POLYMER

Radial spherulite growth rates (G) in semicrystalline polymer can be analyzed using Lauritzen–Hoffman approach, which was developed in 1973 and 1976 (Lauritzen and Hoffman, 1973; Hoffman et al., 1976). G is dependent on crystallization temperature, expressed as:

$$G = G_0 \exp\left(-\frac{\Delta F^*}{RT_c}\right)\exp\left(-\frac{\Delta\Phi}{k_B T_c}\right) \tag{4}$$

G_0 is a pre-exponential constant assumed to be constant or proportional to T_c, ΔF^* is the activation energy responsible for the transport of molecular chains across liquid-solid interface. $\Delta\Phi$ is the free energy of formation of a surface nucleus with critical size. R and k_B is gas constant and Boltzman constant, respectively. The quantity G is influenced by the type, molecular weight, composition, and so on of the semicrystalline polymer.

16.3 CRYSTALLINITY AND MELTING BEHAVIOR

16.3.1 CRYSTALLINITY AND MELTING BEHAVIOR OF PHAS BASED ON DSC

The DSC is a powerful tool to characterize important thermal features of PHAs materials such as apparent melting temperature (T_m), enthalpy of melting (ΔH_m), crystallization temperature (T_c) and T_g. In recent years, there have been numerous reports on the determination of crystallinity of PHAs using DSC, X-ray diffraction, vibrational spectroscopy particularly Fourier transform infrared and so on, however in this chapter we focus on the determination of crystallinity using DSC. The crystallinity of PHAs or any other polymer can be regulated through heating-cooling-heating processes, and these processes affect the final mechanical properties. The first heating is usually done to eliminate thermal history to which the polymer is stored. Upon cooling, polymer crystals are formed at preselected crystallization temperatures. Once the crystallization process completes, the polymer is re-heated in order to determine the T_m and ΔH_m. For semicrystalline polymer processing, the polymer is usually heated and followed by cooling, for examples: injection molding, hot-pressing, extrusion and so on. To determine the crystallinity of PHAs, melting enthalpy of a 100% crystalline PHAs is often used as a reference value. The ratio of melting enthalpy corresponds to the crystalline lamellae, which is obtained from first or second heating process, and melting enthalpy of 100% crystalline material yields crystallinity (X^*).

The PHB has the highest melting point (170–180°C) in the class of PHAs due to its high degree of crystallinity (X^{*} = 55–65%) (Barham et al., 1984), depends on the molecular weight, biosynthesis pathway, and thermal treatment. Its copolymers such as PHBV, PHBB, and PHBHHx have lower crystallinity, lower melting points and wider processing window for processing. Figure 1 summarizes the crystallinity of (a) PHBV, (b) PHBB, and (c) PHBHHx in identical thermal conditions unless otherwise stated. Owing to the complexity and lack of studies on the 100% crystallinity materials for PHA copolymers such as PHBB and PHBHHx, all the crystallinity calculations used are based on the 100% crystallinity of PHB materials (146.6 J g^{-1}), as determined by Barham et al. (1984). The fundamental issue here is to understand the crystallinity with increasing comonomer compositions. The PHBV can crystallize in either PHB or poly(hydroxyvalerate) (PHV) lattice depending on the compositional distribution. The transformation of PHB lattice to PHV lattice is between 40–45 mol% of 3HV as indicated by the dash curve in Figure 1(a). Exceeding this composition, the 3HV crystals dominantly prevail due to isodimorphism (Bluhm et al., 1986; Bloembergen et al., 1986). The open marker in Figure 1 indicates first heating cycle starting from room temperature or lower to above the melting temperature of PHAs while the solid marker indicates the second heating after isothermal crystallization at preselected crystallization temperature. There have not been many reports on the crystallinity of PHAs with high composition of second monomer determined after isothermal crystallization by using DSC. The crystallinity of PHBB is illustrated in Figure 1(b). Here, it also behaves in similar manner to that of PHBV. The crystallinity reduces drastically to approximately 15% when 40 mol% of 4HB present. The crystallinity of PHBB increases gradually from 65 to 100 mol% of 4HB. The corresponding polymer crystal in this region is expected to come from the fractions of 4HB. Studies of the crystallinity of PHBHHx show a slightly different trend at high composition of 3HHx. The crystallinity does not increase with increasing monomer up to 100 mol% of 3HHx, as illustrated in Figure 1(c) (Doi et al., 1995; Zini et al., 2007; Tsuge et al., 2004; Asrar et al., 2002; Qu et al., 2006; Shimamura et al., 1994; Cheng et al., 2008; Feng et al., 2002; Cai and Qiu, 2009; Yu et al., 2010). This indicates that dominant amorphous longer side chains of 3HHx disrupt the crystalline region of 3HB in the copolymer. The side chain of the random comonomer units in PHAs does affect the crystallinity of the PHAs at the same comonomer composition (mol%) for the isodimorphised comonomer crystal. For instance, at 80 mol% of second comonomers in PHBV, PHBB, and PHBHHx, a reduction of crystallinity is observed at 30%, 20%, and 10%, respectively.

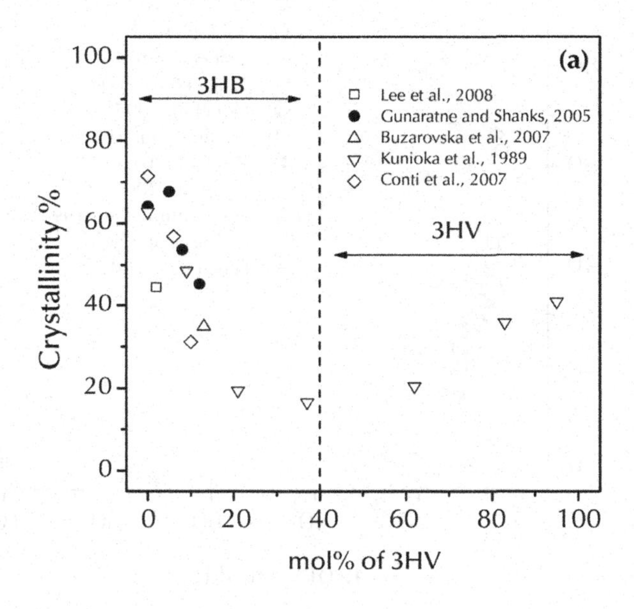

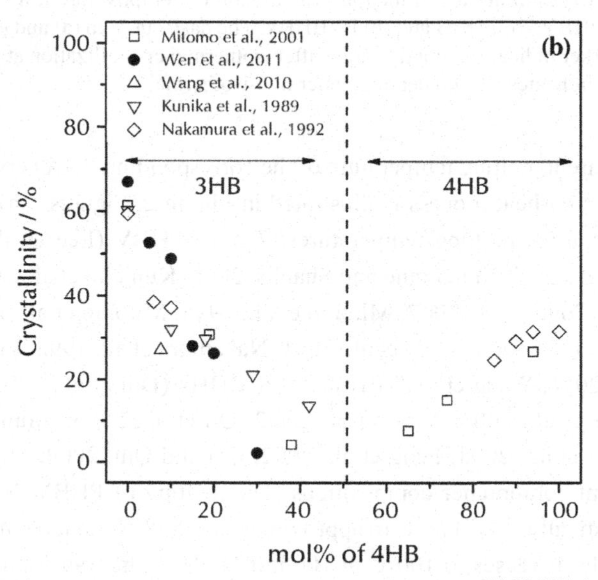

FIGURE 1 *(Continued)*

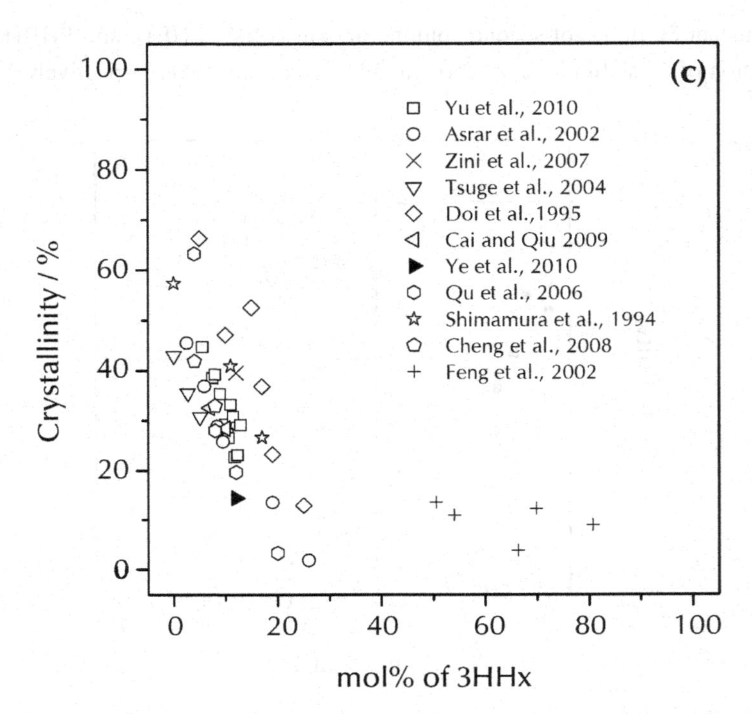

FIGURE 1 Crystallinity as a function of comonomer composition in a series of PHAs samples: (a) PHBV, (b) PHBB and (c) PHBHHx. The dash curve in (a) and (b) is for visual aid. Solid marker indicates second heating after isothermal crystallization at preselected T_c. Open marker indicates second heating under non-isothermal.

The apparent melting temperature of the corresponding PHAs crystals formed during heat-cool-heat process is illustrated in Figure 2. Figures 2(a), (b), and (c) show the apparent melting temperatures, T_ms, of PHBV (Lee et al., 2008; Buzarovska et al., 2007; Gunaratne and Shanks, 2005; Kunioka et al., 1989; Yamada et al., 2001; Conti et al., 2007; Mitomo et al., 1995; Mitomo et al., 1999), PHBB (Kunioka et al., 1989; Mitomo et al., 2001; Nakamura et al., 1992; Lu et al., 2011; Wen et al., 2011; Wang et al., 2010) and PHBHHx (Doi et al., 1995; Zini et al., 2007; Tsuge et al., 2004; Asrar et al., 2002; Qu et al., 2006; Shimamura et al., 1994; Cheng et al., 2008; Feng et al., 2002; Cai and Qiu, 2009; Yu et al., 2010) with different comonomer compositions. The melting of PHBV, in Figure 2(a), reduces drastically from 175°C to approximately 70–90°C up to 40 mol% of 3HV and gradually increases to 100°C at 100 mol% 3HV. The transition of dominant crystal lattice from 3HB to 3HV indicates that the 3HV units co-crystallized in the sequence of 3HB units and accommodates the repeating monomer unit of 3HB as part of its crystal structure. This phenomenon is an evident of isodimorphism in PHBV. In the case of PHBB, the T_ms of PHBB reduce from 175°C, for PHB, to approximately 50°C at and above 40 mol% of 4HB. The trend of T_ms of PHBHHx also greatly reduces with the increase in 3HHx content. Both PHBB and PHBHHx have multiple melting peaks. It is evident that there are multiple traces of melting peaks when second comonomers present, as shown in the example in Figure 2(b) with the vertical dash curve for PHBB with 38 mol% of 4HB at 157°C, 118°C, and 54°C. Multiple melting peaks observed in PHAs may result from the presence of crystals with different stability and lamellar thickness (isomorphism and polymorphism), melting-crystallization-remelting mechanism during heating-cooling process, molecular weight differences and/or physical aging (Watanabe et al., 2001).

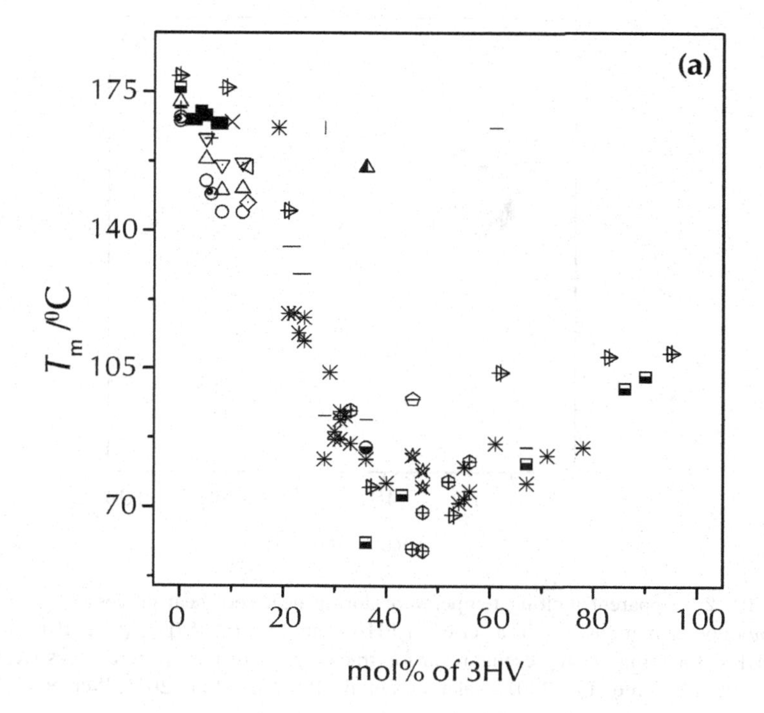

FIGURE 2 *(Continued)*

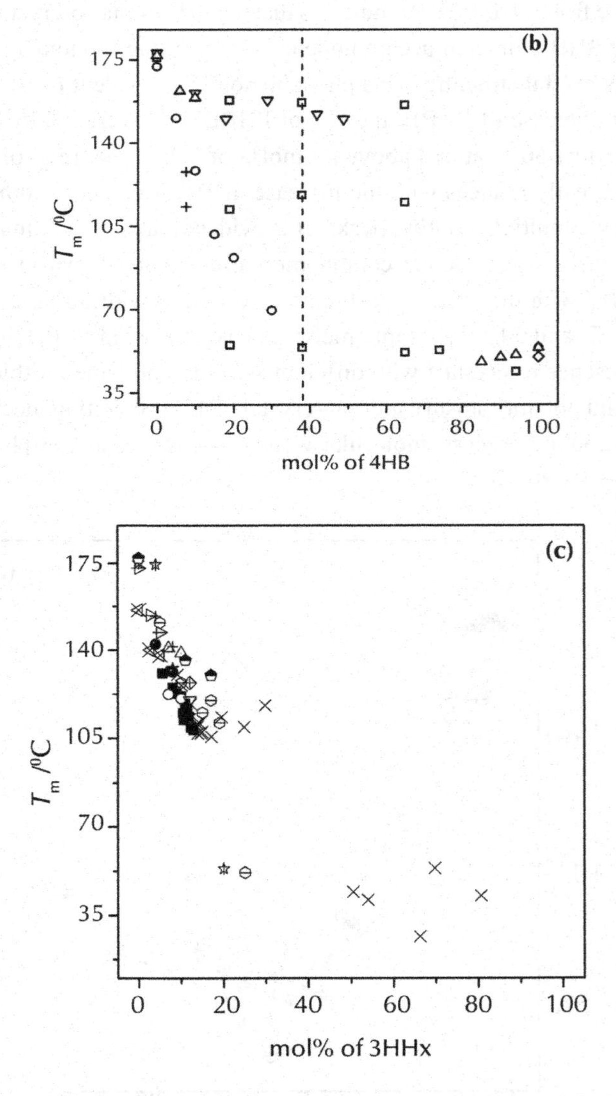

FIGURE 2 Apparent melting temperature during heat-cool-heat process as a function of comonomer composition in a series of PHAs samples: (a) PHBV, (b) PHBB, and (c) PHBHHx. Each marker represents results extracted from respective references (Kunioka et al., 1989; Mitomo et a., 2001; Nakamura et al., 1992; Lu et al., 2011; Wen et al., 2011; Wang et al., 2010). The dash curve in (b) is meant for visual aid.

Although multiple melting points of PHAs in most literatures are reported after non-isothermal crystallization (as shown in Figure 2), similar observation of multiple melting points of PHAs also can be observed after isothermal crystallization. An example of multiple melting peaks (Peaks I and II) is observed in PHB8HHx (number '8' denotes mol% of 3HHx) crystallized isothermally at 112°C (Figure 3). The dominant peak is peak I at 135°C, which is postulated to be the melting of PHB8HHx crystals formed during isothermal crystallization at 112°C. Confirmation of melting peak corresponds to the primary crystal can be done by heating the polymer sample at various heating rates after isothermal crystallization. Besides, the intensity of lower T_m changes and it shifts with increasing T_c. Meanwhile the intensity of higher T_m gradually disappears at higher isothermal T_c. This shows that a more perfect crystal is allowed to form at higher T_c, therefore, first melting peak corresponds to the melting temperature of primary crystallites formed during isothermal crystallization after cooling from the melt. Chen et al. (2005), in the study of isothermal crystallization of PHB15HHx, also observed multiple melting peaks.

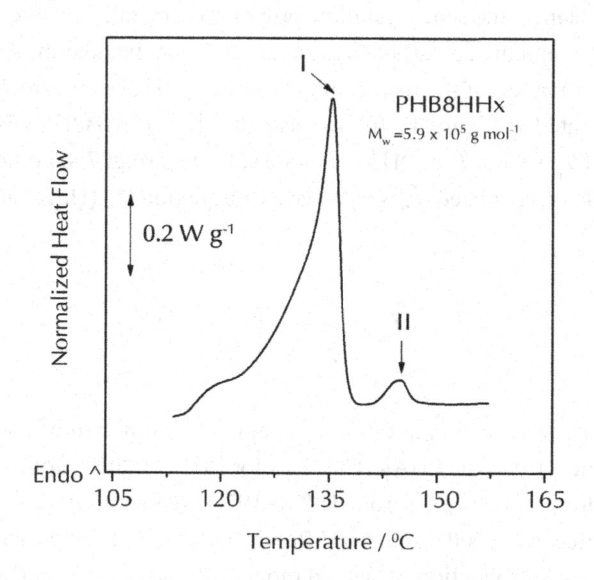

FIGURE 3 Multiple melting peaks exhibited by PHB8HHx, which was isothermally crystallized at 112°C.

The first melting peak is attributed to the melting of primary lamella formed during crystallization while the second peak is due to the recrystallization process. In addition, there is a minor and broad endothermic peak located at 60–85°C in PHB15HHx after isothermally crystallized at 48–60°C for a certain interval of time. They suggest that this peak is due to the pre-melting behavior of intermediate crystals containing crystalline and amorphous state. Conversely, Ding et al. (2011) and Hu et al. (2007) suggest that this peak is due to the melting of lamella formed during secondary crystallization.

16.3.2 EQUILIBRIUM MELTING TEMPERATURES (T_m^0) OF PHAS

The T_m^0 of a semicrystalline polymer is defined as the melting temperature of most stable and perfect crystal. It can be determined by a step-wise annealing procedure as proposed by Hoffman and Weeks (1962). When a semicrystalline polymer with long chain molecules such as PHAs is crystallized near to its T_m^0, the crystallization rate is so slow that it is not practical to study the kinetics of isothermal crystallization and the subsequent corresponding apparent melting temperature. Hence, the semicrystalline polymer is crystallized at temperature far away from T_m^0 and the corresponding apparent T_m can be determined. Extrapolation of linear function of the results of T_m versus T_c (dash curve) to $T_m = T_c$ yields T_m^0 (as illustrated in Figure 4). In this case, the T_m^0 of PHB8HHx is 160°C. The data points of T_m versus T_c of PHAs shall be sufficient (e.g. 7–8 points) in order to obtain reliable extrapolated T_m^0s by means of Equation (5) (Hoffman and Weeks, 1962).

$$T_m = \frac{1}{\gamma} T_c + \left(1 - \frac{1}{\gamma}\right) T_m^0 \tag{5}$$

Figure 5 shows a compilation of T_m^0 values as a function of mol% of comonomer content for various PHAs. The T_m^0s for PHB slightly vary from one to the other literatures, which range from 187 to 197°C (Chan et al., 2004; Don et al., 2006; El-Shafee et al., 2001), but the difference is within the experimental error of reported values. Introduction of second monomer, such as 5.7 mol% and 7 mol% of 3HV, slightly reduces the T_m^0s to 184°C (Wang et al., 2010) and 182°C (Chun and Kim, 2000), respectively (Inspection of the results can be guided by the dotted linear curve in Figure 5).

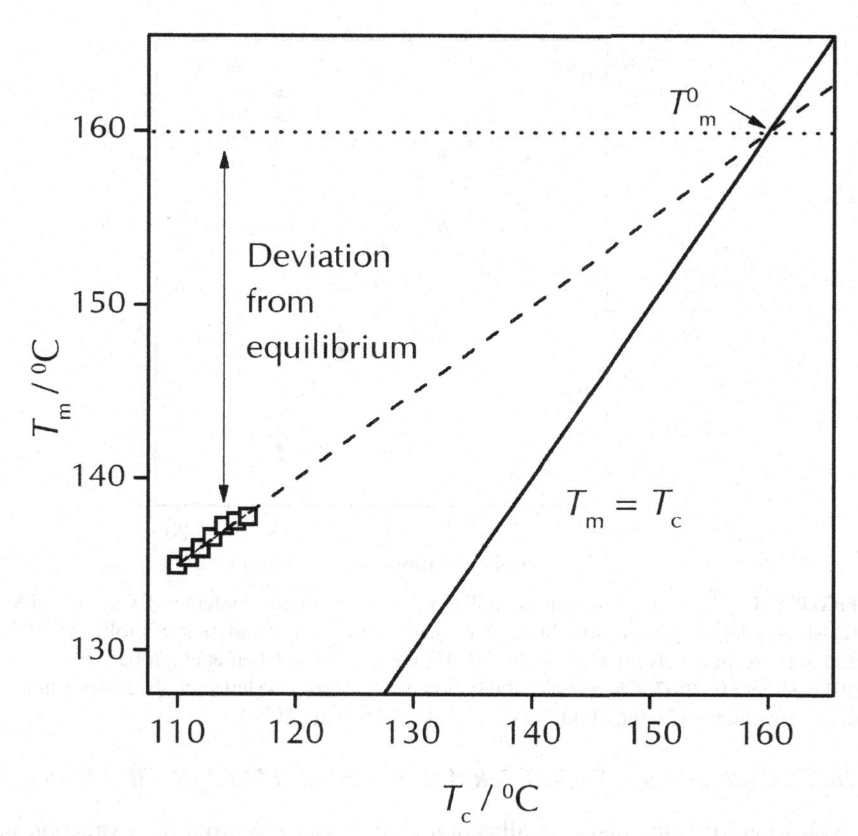

FIGURE 4 Extrapolation of T^0_m from the apparent melting temperature as a function of isothermal crystallization of PHB8HHx.

Similar trend of depression of T^0_m of PHBHHx is also found along the dash curve. The dash curve shows a linear decrease in T^0_ms with increasing comonomer content of 3HHx (Cai and Qiu, 2009; Chen et al., 2005; Ding et al., 2011; Hu et al., 2007; Ye et al., 2010; Lim et al., 2006). The copolymers of PHBB also show linear depression of T^0_ms as a function of mol% of 4HB content (Lu et al., 2011). It is apparent that existence of second monomers such as 3HV, 3HHx, and 4HB disturb the conformational crystalline chain of PHB, allowing the chains to be more flexible with higher free volume, higher mobility and a reduced T_g (Ishida et al., 2005).

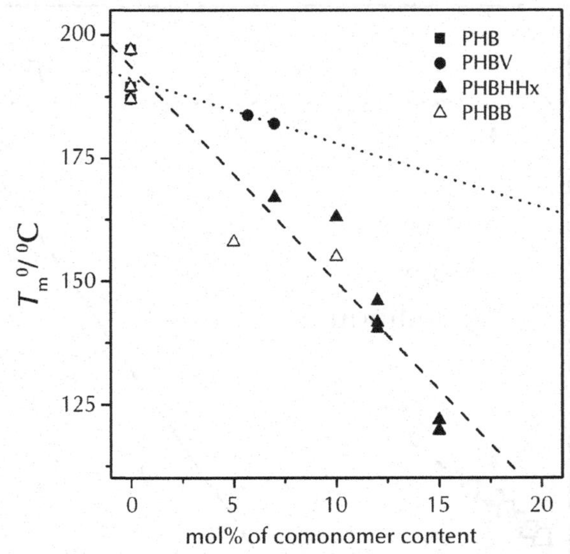

FIGURE 5 T_m^0 s as a function of mol% of the comonomer contents of various PHAs. (Dash and dotted curves show linear depression of T_m^0 s as a function of mol% of 3HHx and 3HV, respectively) (Cai and Qiu, 2009; Lu et al., 2011; Chen et al., 2005; Ding et al., 2011; Hu et al., 2007; Chan et al., 2004; Don et al., 2006; El-Shafee et al., 2001; Wang et al., 2010; Chun and Kim, 2000; Ye et al., 2010; Lim et al., 2006).

16.3.3 KINETICS OF ISOTHERMAL CRYSTALLIZATION OF PHAS

Evaluation of isothermal crystallization of PHAs begins from the extraction of area integral of exothermic crystallization peaks as a function of time, as illustrated in Figure 6(a). The integration of 50% area of crystallization peak gives half-time of crystallization, $t_{0.5}$. One of the important parameters used to describe the crystallization kinetics is $t_{0.5}$, which is defined as the time for 50% of the crystallinity to develop. This figure shows an example of PHB12HHx, which was isothermally crystallized at 110°C, and the half-time of crystallization is 7.0 min. Figure 6(b) illustrates the plots of relative crystallinity [$X(t)$] versus crystallization time (t-t_0), for PHB8HHx isothermally crystallized at 112°C. In general, all the PHAs show similar pattern of crystallization transformation curves. They exhibit similar 'sigmoidal-shape' of isotherms as in Figure 6(b). There is a close relationship between molecular weight of a polymer and its crystallization rate. Not unexpectedly, the crystallization transformation/rate also changes with respect to the comonomer contents.

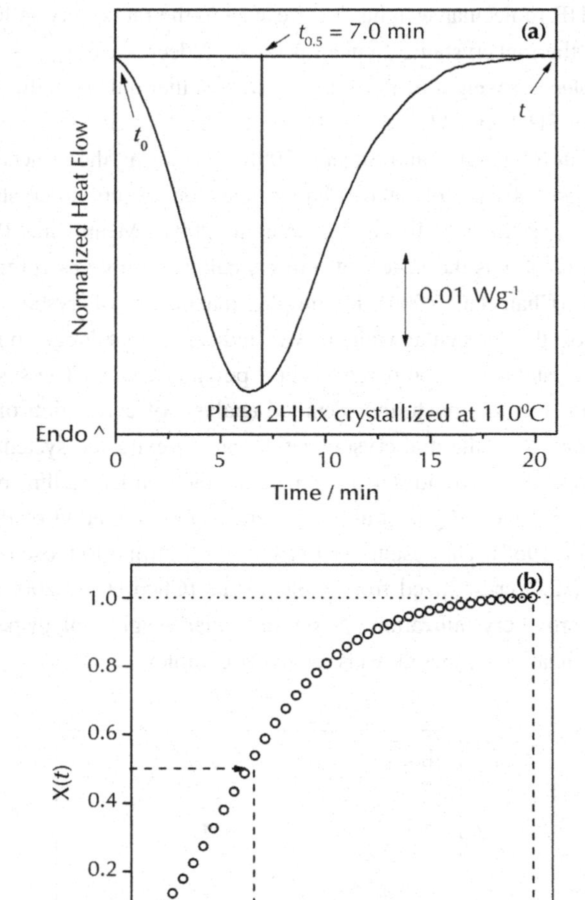

FIGURE 6 (a) Half-time of crystallization from the crystallization peak for PHB12HHx crystallized isothermally at 110°C. (b) Plots of relative crystallinity versus crystallization time during isothermal crystallization of PHB8HHx crystallized at 112°C.

The overall crystallization rate can be expressed in terms of reciprocal half-time of crystallization, $t_{0.5}^{-1}$. Quantity $t_{0.5}^{-1}$ is plotted against the crystallization temperature of PHB with different M_ws, as illustrated in Figure 7. This figure shows an example of different PHB molecular weights with respect to the rate of crystallization at corresponding isothermal crystallization temperature. A decrease of $t_{0.5}^{-1}$ is observed with increasing molecular weight of PHB at the same isothermal crystallization temperature. The $t_{0.5}$ for PHB with $M_w = 230{,}000$ g mol^{-1} and $M_w = 350{,}000$ g mol^{-1} at 110°C is about 2.5 min (Gunaratne and Shanks, 2006). Increasing the molecular weight of PHB to $M_w = 540{,}000$ g mol^{-1}, the polymer takes longer time to crystallize, and $t_{0.5}$ increases almost six folds to 14 min (Chan et al., 2004). Meanwhile, PHB with low $M_w = 3{,}900$ g mol^{-1} has the fastest overall crystallization rate as compared to PHB with higher M_w (Chan et al., 2004). This implies that the rate of crystallization of PHB is dependent on the molecular weight, with indication that longer polymer chains are harder to crystallize as compared to short polymer chains. The suggestion from Figure 7 is rather crude, no further detailed analysis of correlation of M_w of other PHAs to the rate of isothermal crystallization can be extracted systematically from literatures. However, this trend is in agreement to other semicrystalline polymers such as linear poly(ethylene) (Ergoz et al., 1972) and poly(tetramethyl-*p*-silphenylene siloxane) (Magill, 1968). The results for PHB are also similar to those of syndiotactic poly(styrene) (*s*PS) crystallized from molten state (Chen et al., 2009). The dependency of isothermal crystallization rate on molecular weight, polydispersity and isothermal crystallization temperatures is relatively complex.

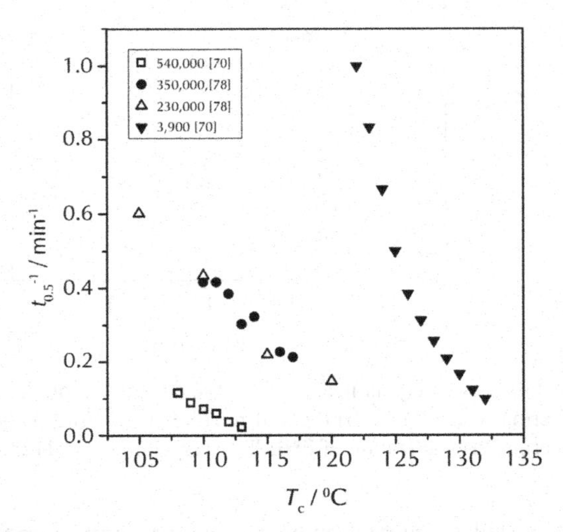

FIGURE 7 Dependence of overall isothermal crystallization rate on average weight of molecular weight (g mol^{-1}) of PHB at corresponding crystallization temperature (Chan et al., 2004; Gunaratne and Shanks, 2006).

It is interesting at this point to examine the influence of different types or content of the comonomer on the rate constant ($K^{1/n}$) of isothermal crystallization in PHAs samples. A summary of $K^{1/n}$ values for PHB, PHBV, PHBB, and PHBHHx is given in Figure 8. Generally, it is shown that the rate constant of isothermal crystallization reduces exponentially for PHB (Lu et al., 2011; An et al., 1997), PHBV (Peng et al., 2003), PHBHHx (Chen et al., 2005), and PHBB (Lu et al., 2011) with increasing crystallization temperatures (c.f. solid curve). Rate constants of isothermal crystallization reduce with the increment of comonomer content in PHAs, illustrated by the vertical dash line. It is also evident that, for instance, at high undercooling ($T_m^0 - T_c$) for PHB (indicated by arrow), the rate constants reduce. When PHAs are crystallized at high undercooling, which is closer to its T_g, the mobility of PHAs chains to the growth front of PHAs crystal is restricted thus the rate constants reduce. This effect is also seen in the radial growth rate of PHAs described later. In most semicrystalline polymers, the rate constant of isothermal crystallization decreases exponentially with increasing crystallization temperatures at low undercooling temperature.

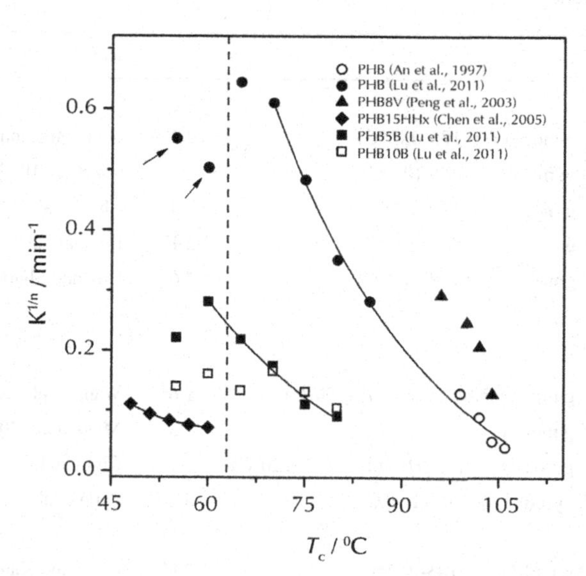

FIGURE 8 Rate constants of PHAs isothermal crystallized at various crystallization temperatures. Solid curves show exponential decrease of the rate constant of isothermal crystallization of PHB [(•) Lu et al., 2011; (○) An et al., 1997], (■) PHB5B (Lu et al., 2011), and (♦) PHB15HHx (Chen et al., 2005). Arrow indicates reduction of the rate of isothermally crystallization at high undercooling (compare to text).

From the Avrami model [Equation (1)], the exponent n can be determined. Table 1 shows n values for different PHAs from different investigators. According to Mandelkern, n value plays a key role in the analysis of overall crystallization kinetics. Although in a few cases (2004), the n value is affected by the crystallization temperature, the revelation of the crystallization mechanism based on only DSC result remains a challenge. All the n values reported in PHAs have non-integral values, indicating nature of instantaneous or non-instantaneous nucleation. The PHB with $M_w = 2.3 \times 10^5$ g mol^{-1} has average n values of 2.2 (Gunaratne and Shanks, 2006). At higher molecular weight of PHB, $M_w = 2.9 \times 10^5$ and 3.5×10^5 g mol^{-1}, n values appear to be lower at 1.7 and 1.4 respectively. With further increase of M_w to 4.0×10^5 g mol^{-1} and 5.4×10^5 g mol^{-1}, n values increase. This increment is expected to influence the growth and nucleation mechanism of PHB. Here, it is very challenging to judge the influence of molecular weight on the n values due to the differences in the experimental conditions (e.g. different undercoolings) by different research groups.

TABLE 1 Influences of PHAs molecular weight and comonomer composition on the Avrami exponent, n

PHAs	n	Ref
PHB		
$M_w = 2.3 \times 10^5$ g mol^{-1} ($T_c = 105$–$120°C$)	2.2*	Gunaratne and Shanks, 2006
[a]$M_w = 2.9 \times 10^5$ g mol^{-1} ($T_c = 99$–$106°C$)	1.7*	An et al., 1997
$M_w = 3.5 \times 10^5$ g mol^{-1} ($T_c = 120$–$130°C$)	1.4*	Chan et al., 2004
$M_w = 4.0 \times 10^5$ g mol^{-1} ($T_c = 55$–$85°C$)	2.1*	Lu et al., 2011
$M_w = 5.4 \times 10^5$ g mol^{-1} ($T_c = 108$–$113°C$)	2.6*	Hay and Sharma, 2000
PHBV		
8mol% 3HV		
$M_w = 2.5 \times 10^5$ g mol^{-1} (*6mol% 3HV*) ($T_c = 90°C$)	1.6	Wang et al., 2010
$M_w = 4.2 \times 10^5$ g mol^{-1} ($T_c = 95°C$)	2.3	Miao et al., 2008
$M_w = 4.7 \times 10^5$ g mol^{-1} (*7mol% 3HV*) ($T_c = 105$–$120°C$)	2.9*	Chun and Kim, 2000
[b]$M_w = 5.8 \times 10^5$ g mol^{-1} ($T_c = 96$–$104°C$)	3.1*	Peng et al., 2003
12mol% 3HV		
$M_w = 2.4 \times 10^5$ g mol^{-1} ($T_c = 105$–$112°C$)	2.8*	Chan and Kammer, 2009
$M_w = 4.5 \times 10^5$ g mol^{-1} (*14mol% 3HV*) ($T_c = 70°C$)	2.6	Qiu et al., 2005

TABLE 1 *(Continued)*

PHBB		
5mol% 4HB (M_w = 3.9 × 10^5 g mol^{-1})		
T_c = 55°C	2.1	Lu et al., 2011
T_c = 60°C	1.8	
T_c = 65°C	2.2	
T_c = 70°C	2.2	
T_c = 75°C	1.8	
T_c = 80°C	2.0	
10mol% 4HB (M_w = 4.0 × 10^5 g mol^{-1})		
T_c = 55°C	1.9	Lu et al., 2011
T_c = 60°C	1.9	
T_c = 65°C	1.9	
T_c = 70°C	2.0	
T_c = 75°C	2.0	
T_c = 80°C	2.0	
PHBHHx		
7mol% 3HHx (M_w = 2.4 × 10^5 g mol^{-1})		
T_c = 95°C	2.6	Cai and Qiu, 2009
T_c = 100°C	2.7	
T_c = 105°C	2.6	
T_c = 110°C	2.9	
10mol% 3HHx (M_w = 2.4 × 10^5 g mol^{-1})		
T_c = 100°C	2.4	Cai and Qiu, 2009
T_c = 105°C	3.1	
T_c = 110°C	3.2	
T_c = 115°C	3.1	
15mol% 3HHx (M_w = 7.5 × 10^5 g mol^{-1})		
T_c = 48°C	2.8	Chen et al., 2005
T_c = 51°C	2.8	
T_c = 54°C	2.7	
T_c = 57°C	2.6	
T_c = 60°C	2.6	
18mol% 3HHx (M_w = 2.4 × 10^5 g mol^{-1})		
T_c = 85°C	2.6	Cai and Qiu, 2009
T_c = 90°C	2.6	
T_c = 95°C	2.5	
T_c = 100°C	2.7	

[a] obtained by measuring the viscosity in ethylene dichloride at 30°C
[b] obtained by measuring the viscosity in chloroform at 30°C
* average result

If we compare other copolymer systems in PHBB and PHBHHx, it is clear that crystallization temperatures influence the growth and nucleation processes. In general, the higher the isothermal crystallization temperatures, the higher the n values. Interestingly, within the same molecular weight in PHBB sample, the average n values decrease with increasing mol% of 4HB.

The activation energy of PHAs is calculated using Hoffman–Arrhenius' approach (Hoffman, 1982). The Hoffman's Arrhenius-like model is described as:

$$\frac{1}{t_{0.5}} = \frac{1}{\tau}\exp\left[-\frac{\Delta g^*}{k_B T_c}\right] \tag{6}$$

and

$$\Delta g^* = L\alpha\beta\frac{\sigma^2 T_m^0}{\Delta H_m \rho \Delta T_c} \tag{7}$$

where, Δg^* is the free energy barrier that must be overcome before a new stable phase can develop. ΔT_c is the undercooling ($T_m^0 - T_c$). All the abbreviation of temperature independent quantities is given by B, and τ is a constant that does not depend on temperature. It is thus convenient to define the reciprocal half-time of isothermal crystallization ($t_{0.5}^{-1}$), as:

$$\frac{1}{t_{0.5}} = \frac{1}{\tau}\exp\left[-\frac{B}{\Delta T_c}\right] \tag{8}$$

When $\frac{\Delta T_c}{T_m^0}$ is constant or approximately constant, the activation energy, B^*, can be rewritten as:

$$B^* \equiv \frac{B}{\left(1-\frac{\Delta T_c}{T_m^0}\right)} \tag{9}$$

By multiplying the B^* energy of activation with gas constant, R, one can obtain the activation energy in the unit of J mol^{-1} or kJ mol^{-1}.

The activation energy of isothermal crystallization in PHB8HHx is illustrated in Figure 9. There is a linear relationship from the plot of reciprocal undercooling, ΔT_c^{-1}, versus natural logarithm of reciprocal half-time of crystallization, $\ln(t_{0.5}^{-1})$, as represented by Equation (8). The activation energy of PHB8HHx calculated from the slope of Equation (8) using Equation (9) yields 1.7 kJ mol^{-1}. By applying the same approach, Figure 10 summarizes the activation energy of PHBHHx from various investigators with different comonomer compositions. The activation energy of PHBHHx is essentially dependent on the comonomer contents in the range of temperature studied. As a consequence, a linear decrease in activation energy of isothermal crystallization is observed with increasing 3HHx content, as illustrated in Figure 10. Longer side chain of 3HHx contributes to larger free volume, slower rate of isothermal crystallization, thus having lower energy densities to crystallize.

This is in agreement to the results of rate constant obtained from Avrami isothermal crystallization model. Meanwhile, the analysis of activation energy of isothermal crystallization of PHAs has been investigated from different approach. It has been proposed that the plot of ln V_c versus $(RT)^{-1}$, according to $V_c = Ae^{-\Delta E/RT}$ (Vilanova et al., 1985), yields Arrhenius' activation energy. R is a gas constant (8.314 J mol^{-1} K^{-1}). For a given polymer such as PHB, the activation energy is 83 kJ mol^{-1} (An et al., 1997). By adding one methyl and ethyl side chain to the backbone of homopolymer, it was found that the activation increases to 110 kJ mol^{-1} for PHBV (Peng et al., 2003), and 127 kJ mol^{-1} for PHBHHx (Chen et al., 2005). This increment is related to the obscurity of polymer crystals to crystallize due to the bigger size of side groups.

The limitation to this approach is that it does not consider the effect of undercooling on the rate of isothermal crystallization. Nevertheless, the Hoffman-Arrhenius' model using Equation (6) to Equation (9) describes the activation energy based on both nucleation and growth as a function of the undercooling ($T_m^0 - T_c$). Generally, at low undercooling, the growth rate of PHB increases with lower T_c until it reaches maxima. Conversely, at high undercooling, the movement of polymer to the growth front of the crystals is restricted when the T_c is closer to T_g (compare to Figure 11), in this case, the growth rate of PHB decreases with lower T_c (Mandelkern, 2004).

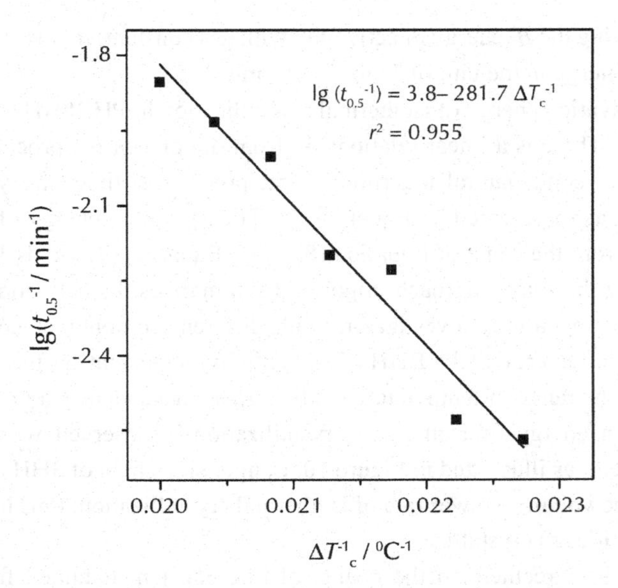

FIGURE 9 Hoffman–Arrhenius plot of PHB8HHx crystallized at various undercooling. Solid curve represents regression function after Equation (8).

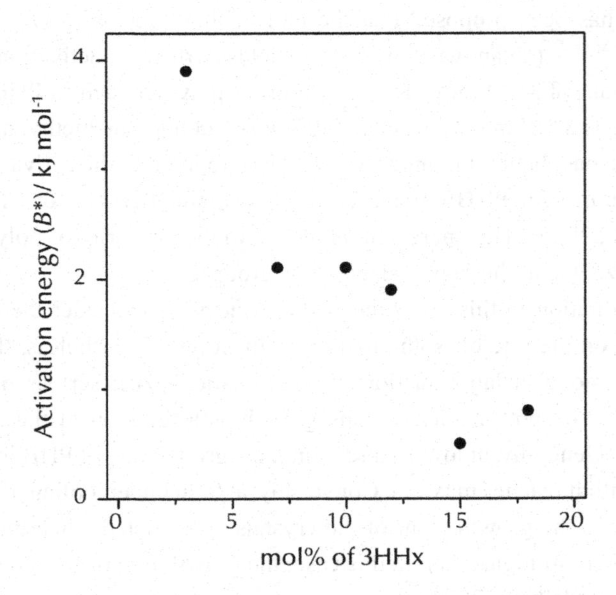

FIGURE 10 Activation energy of isothermal crystallization [calculated after Equation (8)] of PHBHHx as a function of comonomer composition.

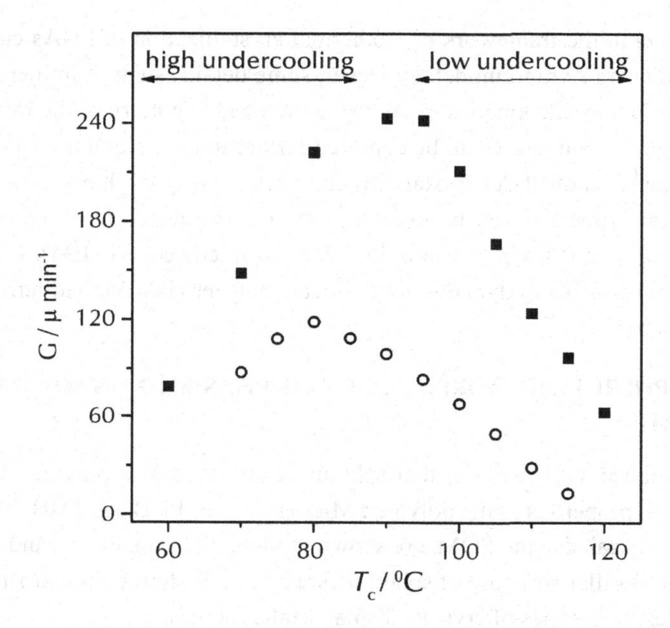

FIGURE 11 Radial growth rate of spherulite of (■) PHB (Cimmino et al., 1998) and (○) PHB8V (Peng et al., 2003) isothermally crystallized at different crystallization temperatures.

The radial growth rate of these crystallites can be calculated from Equation (4). Figure 11 shows the radial growth rates (G) for two PHAs, that is PHB and PHBV, estimated using POM. There is an increase in the radial growth rate of PHAs crystals, both PHB and PHB8V, at high undercooling until it reaches a maximum rate, and progressively decreases at low undercooling. It is clear that quantity G of PHB reaches its peak crystals growth at approximately 90°C (Cimmino et al., 1998), while PHB8V reaches its peak crystals growth at approximately 80°C (Peng et al., 2003). Chan and co-workers showed that PHBV undergoes self-seeding nucleation and it requires higher undercoolings for a substantial rate of crystallization (2011), unlike poly(ethylene oxide) and poly(ethylene terephthalate) (Chan et al., 2011). When PHAs are crystallized near to T_g (at high ΔT_c), the growth of PHAs crystals are slower and nucleation are favorable, thus small crystal seeds forms. The radial growth rate of the spherulites versus T_c of a semicrystalline polymer is often in agreement to the reciprocal half-time of crystallization versus T_c using DSC.

In general, the framework of isothermal crystallization of PHAs can be well described using Avrami model. However, some details, such as influence of molecular weight on the kinetics of crystallization and geometry of the rate-controlling nucleus, retain interest to be explored further in the field of microbial PHAs. Further analyses of PHAs crystals structure related to growth rate of crystal and mechanical properties can be explored using microscopes [e.g. scanning electron microscope (SEM), transmission electron microscope (TEM), etc.], X-ray diffractometer (XRD), dynamic mechanical analyzer (DMA), and tensile meter, respectively.

16.4 SPHERULITIC MORPHOLOGY OF PHAS AND ITS MECHANICAL PROPERTIES

It was noted in early session that spherulitic structure of a polymer affects the mechanical properties of the polymer. Micrographs of PHB and PHB12V crystals structure, revealed using POM, are shown in Figure 12. The PHBV and PHBHHx show fine fibrillar structure of spherulites while PHB shows circular ring banded spherulites, both types of crystals display Maltese cross.

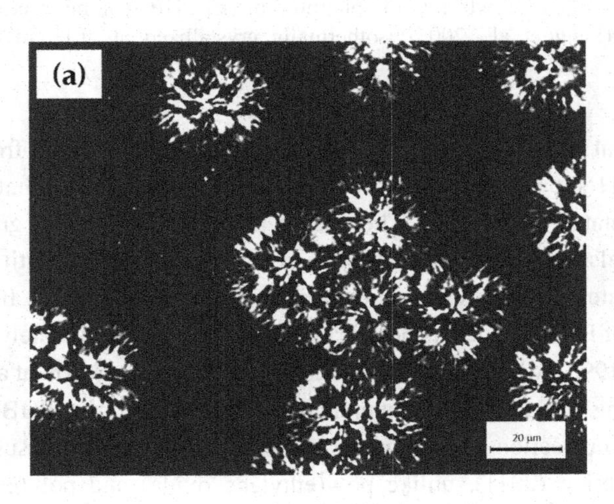

FIGURE 12 *(Continued)*

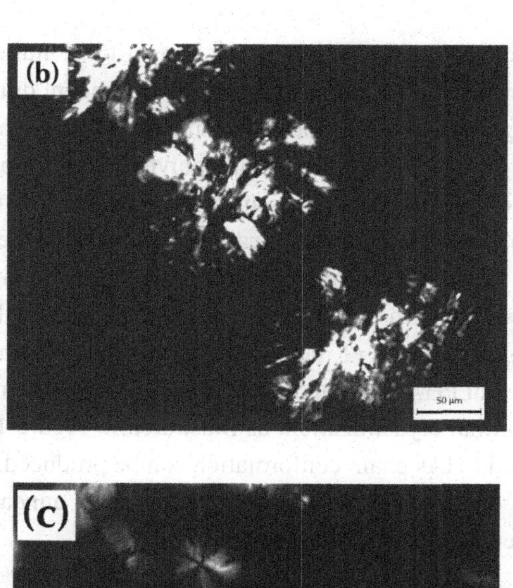

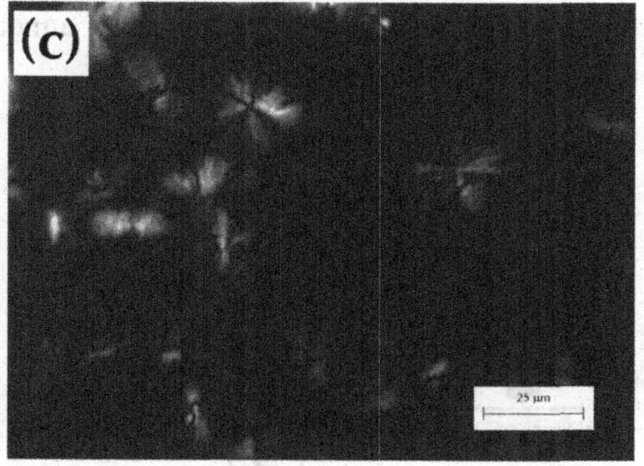

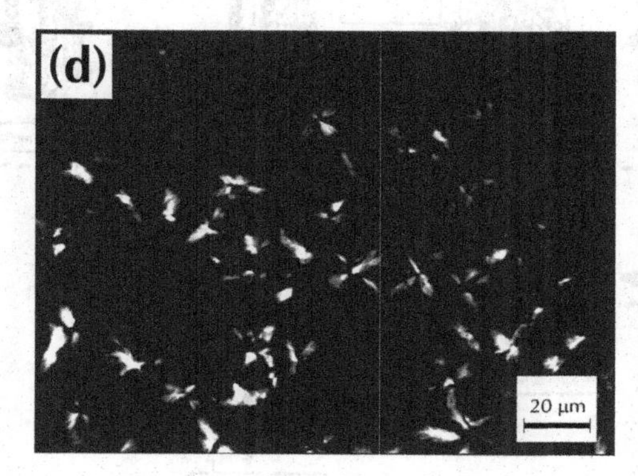

FIGURE 12 Examples of growing PHA spherulitic morphology at predetermined crystallization temperature. (a) PHB ($M_w = 3.5 \times 10^5$ g mol^{-1}) at $T_c = 129$ °C, 50× (b) PHB ($M_w = 3{,}900$ g mol^{-1}) at $T_c = 130$ °C, 50× (c) PHB12V ($M_w = 2.4 \times 10^5$ g mol^{-1}) at $T_c = 120$ °C with 40×, and (d) PHB8HHx ($M_w = 5.9 \times 10^5$ g mol^{-1}) at $T_c = 110$ °C with 20×.

The mechanical properties of PHAs are known to be influenced by type of crystals formation. The first study of temperature dependent drawing method that produces ultra-high modulus polymer was studied for poly(ethylene) (Capaccio et al., 1976). The drawing of semicrystalline polymer leads to physically orientation of random coiled amorphous region of the polymer into relatively aligned or stretched form (β-form), and thus great improvement of mechanical strength can be observed. The first experimental observation of planar zigzag or β-form of PHB was made by Yokouchi et al. (1973), followed by Orts et al. (1990). Another form of PHAs crystal, α-form (2_1 helix with shish kebab structure), is one of the common structures of PHAs crystals and it can be produced from one-step annealing without isothermal crystallization, as illustrated in Figure 13(a). The β-form with fully extended PHAs chain conformation can be produced using a combination technique of one- or two-step drawing followed by annealing (Iwata et al., 2005; Furuhashi et al., 2004), as summarized in Table 2.

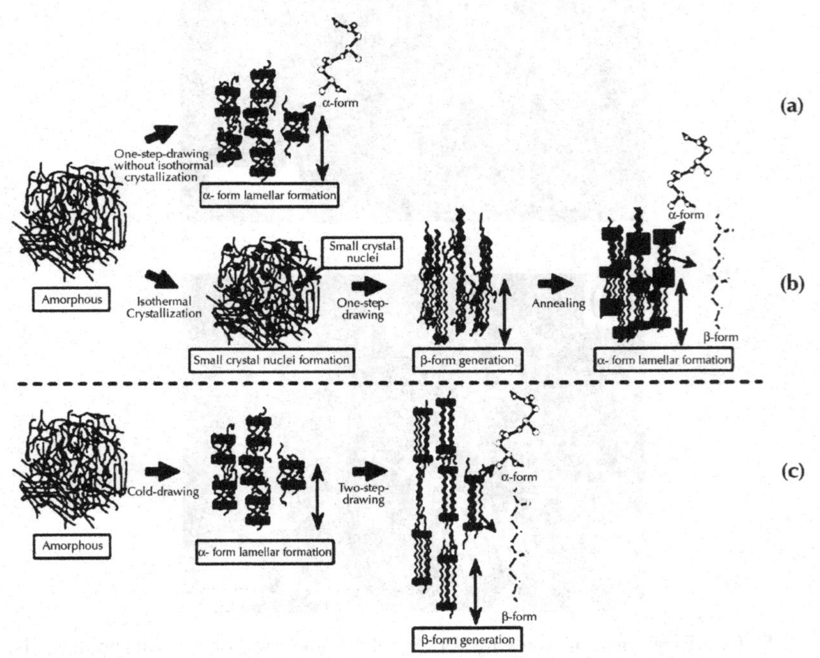

FIGURE 13 Mechanism to produce planar zigzag conformation (β-form) in high-strength PHB8V fibers by different drawing methods: (a) one-step-drawing without isothermal crystallization, (b) one-step-drawing after isothermal crystallization, and (c) cold-drawing and two-step-drawing. The vertical arrows indicate the drawing direction.

Source: Reprinted (adapted) with permission from Tanaka et al. (2006). Copyright (2006) American Chemical Society.

The β-form contributes to higher tensile strength because the amorphous region between the lamella is fully extended, as illustrated in Figure 13(b) and 13(c). There are several factors influencing the tensile strength of semicrystalline polymer in drawing method, such as the speed of melt-spinning, drawing ratio, applied strain, molecular weight of polymer, crystallization temperature, and annealing condition. In cold one-step drawing, the molten polymer is quenched into ice water and annealed at temperature near to its T_g while in the cold two-step drawing method, the polymer is quenched, annealed, kept at room temperature and further drawing is applied before annealing.

TABLE 2 Tensile strength of PHB prepared by different drawing methods

M_w (g mol^{-1}) of PHB	Tensile strength (MPa)	Conditions	Type of Crystal(s)	Ref
-	18	none	α-form	Li et al., 2011
-	330	high-speed melt spinning and drawing followed by annealing (drawing ratio ~7)	α-form β-form	Schmack et al., 2000
-	310	two-step annealing process of melt-spun fibers	α-form β-form	Yamane et al., 2001
4.2×10^5	416	combination of cold-drawing and two-step drawing methods followed by annealing under tension	α-form β-form	Furuhashi et al., 2004
7.0×10^5	740	a four-times drawn fiber crystallized for 72 h	α-form β-form	Iwata, 2005
11.0×10^6	1320	combination of cold-drawing and two-step drawing methods at room temperature (after 60× drawing)	α-form β-form	Iwata et al., 2004

Kusaka et al. shows that the β-form PHB obtained using combination of drawing method has tensile strength at 1.3 GPa (1999), while Tanaka et al. reported that PHB8V has a tensile strength of 1 GPa (2006). There is a remarkable increment of tensile strength when ultra high molecular weight of PHB is used in the drawing as compared to lower molecular weight. The property of aligned β-form can be retained for several months and no cracks were observed in PHAs, suggesting a suppression of secondary crystallization (Fischer et al., 2004). In addition to annealing thermal treatment, Martinez–Salazar et al. in their study show that there is a thermal expansion and cracking of PHB and PHBV after isothermal crystallization (1989). The Bragg *d*-spacings of PHBV having different molar compositions, which are obtained from X-ray diffractometer, show a gradual and linear increment with increasing crystallization temperatures. This implies that when PHB is cooled slowly, larger size of PHB spherulites are formed thus it weakens the mechanical properties of PHB.

16.5 SUMMARY

The PHAs are semi-crystalline naturally occurring biodegradable aliphatic polyesters. The spherulitic morphology formation in PHAs is strongly coined by thermal behavior/treatment of/on PHAs. Properties are influenced by the crystallite structures of PHAs, which can be reflected in the thermal properties of the blends, for example the crystallization and melting behavior. Avrami model is used to describe the isothermal crystallization kinetics of PHAs in this chapter. An understanding on the thermal properties of the semi-crystalline PHAs is deemed important for fine adjustment of structure-property relationships. Thermal properties influence development of morphologies that in turn strongly affect the mechanical behavior of PHAs, which is highlighted in the last part of the chapter. Extensive study needs to be carried out in order to give satisfactory structure-property relationships to PHAs with different comonomer, comonomer content, different molecular weight, and so on.

KEYWORDS

- **Arrhenius' activation energy**
- **Differential scanning calorimeter (DSC)**
- **Hoffman–Arrhenius' model**
- **Radial growth rates**
- **Semicrystalline polymers**

ACKNOWLEDGMENT

The authors are indebted to Hans–Werner Kammer (Martin-Luther Universität Halle, Germany) for his contributions to this chapter. They also gratefully acknowledge financial support of Research Excellent Fund from Universiti Teknologi MARA [600-RMI/DANA 5/3/RIF (636/2012)], Long-Term Research Grant from Ministry of Higher Education, Malaysia (Kenaf: Sustainable Materials in Automotive Industry) and Internal Grant of Postgraduate Research Fund (UM.TNC2/IPPP/UPGP/638/PPP).

REFERENCES

1. An, Y., Dong, L., Xing, P., Zhuang, Y., Mo, Z., and Feng, Z. *Eur. Polym. J.*, **33**, 1449–1452 (1997).
2. Asrar, J., Valentin, H. E., Berger, P. A., Tran, M., Padgette, S. R., and Garbow, J. R. *Biomacromol.*, **3**, 1006–1012 (2002).
3. Avrami, M. *J. Chem. Phys.*, **7**, 1103–1112 (1939).
4. Avrami, M. *J. Chem. Phys.*, **8**, 212–224 (1940).
5. Avrami, M. *J. Chem. Phys.*, **9**, 177–184 (1941).
6. Barham, P. J., Keller, A., Otun, E. L., and Holmes, P. A. *J. Mater. Sci.*, **19**, 2781–2794 (1984).
7. Berezina, N. *Biotechnol. J.*, **7**, 304–309 (2012).
8. Bhubalan, K., Chuah, J. A., Shozui, F., Brigham, C. J., Taguchi, S., Sinskey, A. J., Rha, C. K., and Sudesh, K. *Appl. Environ. Microbiol.*, **77**, 2926–2933 (2011).
9. Bloembergen, S., Holden, D. A., Hamer, G. K., Bluhm, T. L., and Marchessault, R. H. *Macromol.*, **19**, 2865–2871 (1986).
10. Bluhm, T. L., Hamer, G. K., Marchessault, R. H., Fyfe, C. A., and Veregin, R. P. *Macromol.*, **19**, 2871–2876 (1986).
11. Boyandin, A. N., Prudnikova, S. V., Filipenko, M. L., Khrapov, E. A., Vassil'ev, A. D., and Volova, T. G. *Appl. Biochem. Microbiol.*, **48**, 28–36 (2012).
12. Briese, B. H., Jendrossek, D., and Schlegel, H. G. *FEMS Microbiol. Lett.*, **117**, 107–111 (1994).
13. Buzarovska, A., Bogoeva-Gaceva, G., Grozdanov, A., Avella, M., Gentile, G., and Errico, M. *J. Mater. Sci.*, **42**, 6501–6509 (2007).
14. Cai, H. and Qiu, Z. *Phys. Chem. Chem. Phys.*, **11**, 9569–9577 (2009).
15. Capaccio, G., Crompton, T. A., and Ward, I. M. *Polym.*, **17**, 644–645 (1976).
16. Chan, C. H. and Kammer, H. W. *Polym. Bull.*, **63**, 673–686 (2009).
17. Chan, C. H., Kammer, H. W., Sim, L. H., and Winie, T. *Int. J. Pharm. Pharm. Sci.*, **3**, 1–6 (2011).
18. Chan, C. H.; Kummerlöwe, C.; Kammer, H. W. Macromol. Chem. Phys. 2004, 205, 664-675.
19. Chaudhry, W. N., Jamil, N., Ali, I., Ayaz, M. H., and Hasnain, S. *Anal. Microbiol.*, **61**, 623–629 (2011).
20. Chen, C. M., Hsieh, T. E., and Ju, M. Y. *J. Alloys Compd.*, **480**, 658–661 (2009).

21. Chen, C., Cheung, M. K., and Yu, H. F. *Polym. Int.*, **54**, 1055–1064 (2005).
22. Cheng, M. L., Lin, C. C., Su, H. L., Chen, P. Y., and Sun, Y. M. *Polym.*, **49**, 546–553 (2008).
23. Chun ,Y. S. and Kim, W. N. *Polym.*, **41**, 2305–2308 (2000).
24. Cimmino, S., Iodice, P., Martuscelli, E., and Silvestre, C. *Thermochim. Acta*, **321**, 89–98 (1998).
25. Conti, D. S., Yoshida, M. I., Pezzin, S. H., and Coelho, L. A. F. *Fluid Phase Equilib.*, **261**, 79–84 (2007).
26. da Cruz Pradella, J. G., Lenczak, J. L., Delgado, C. R., and Taciro, M. K. *Biotechnol. Lett.*, **34**, 1003–1007 (2012).
27. Ding, C., Cheng, B., and Wu, Q. *J. Therm. Anal. Calorim.*, **103**, 1001–1006 (2011).
28. Doi, Y. *Microbial Polyesters*, VCH Publisher, New York, (1990).
29. Doi, Y., Kanesawa, Y., Kawaguchi, Y., and Kunioka, M. *Makromol. Chem. Rapid Commun.*, **10**, 227–230 (1989).
30. Doi, Y., Kitamura, S., and Abe, H. *Macromol.*, **28**, 4822–4828 (1995).
31. Doi, Y., Kunioka, M., Nakamura, Y., and Soga, K. *Macromol.*, **21**, 2722–2727 (1988).
32. Doi, Y., Tamaki, A., Kunioka, M., and Soga, K. *Appl. Microbiol. Biotechnol.*, **28**, 330–334 (1988).
33. Don, T. M., Chen, P. C., Shang, W. W., and Chiu, H. J. *Tamkang J. Sci. Eng.*, **9**, 279–290 (2006).
34. El-Shafee, E., Saad, G. R., and Fahmy, S. M. *Eur. Polym. J.*, **37**, 2091–2104 (2001).
35. Ergoz, E., Fatou, J. G., and Mandelkern, L. *Macromol.*, **5**, 147–157 (1972).
36. Feng, L., Watanabe, T., Wang, Y., Kichise, T., Fukuchi, T., Chen, G. Q., Doi, Y., and Inoue, Y. *Biomacromol.*, **3**, 1071–1077 (2002).
37. Fischer, J. J., Aoyagi, Y., Enoki, M., Doi, Y., and Iwata, T. *Polym. Degrad. Stab.*, **83**, 453–460 (2004).
38. Furuhashi, Y., Imamura, Y., Jikihara, Y., and Yamane, H. *Polym.*, **45**, 5703–5712 (2004).
39. Gonda, K. E., Jendrossek, D., and Molitoris, H. P. *Hydrobiologia*, **426**, 173–183 (2000).
40. Gunaratne, L. M. W. K. and Shanks, R. A. *Eur. Polym. J.*, **41**, 2980–2988 (2005).
41. Gunaratne, L. M. W. K. and Shanks, R. A. *J. Therm. Anal. Calorim.*, **83**, 313–319 (2006).
42. Hay, J. N. and Sharma, L. *Polym.*, **41**, 5749–5757 (2000).
43. Hiroe, A., Tsuge, K., Nomura, C. T., Itaya, M., and Tsuge T. *Appl. Environ. Microbiol.*, **78**, 3177–3184 (2012).
44. Hoffman, J. D. *Polym.*, **23**, 656–670 (1982).
45. Hoffman, J. D., Davis, G. T., and Lauritzen, J. I. The rate of crystallization of linear polymers with chain folding. *In Treatise on Solid State Chemistry*; N. B. Hannay, (Ed.), Plenum Press, New York, (1976).
46. Hoffman, J. D. and Weeks, J. J. *J. Res. Natl. Bur. Stand. Sec. A. Phys Chem.*, **66**, 13–28 (1962).
47. Hu, Y., Zhang, J., Sato, H., Noda, I., and Ozaki, Y. *Polym.*, **48**, 4777–4785 (2007).
48. Ishida, K., Asakawa, N., and Inoue, Y. *Macromol. Symp.*, **224**, 47–58 (2005).
49. Iwata, T. *Macromol. Biosci.*, **5**, 689–701 (2005).

50. Iwata, T., Aoyagi, Y., Fujita. M., Yamane, H., Doi, Y., and Suzuki, Y. *Macromol. Rapid Commun.*, **25**, 1100–1104 (2004).
51. Iwata, T., Fujita, M., Aoyagi, Y., Doi, Y., and Fujisawa, T. *Biomacromol.*, **6**, 1803–1809 (2005).
52. Jendrossek, D. and Handrick, R. *Annu. Rev. Microbiol.*, **56**, 403–432 (2002).
53. Kunasundari, B. and Sudesh, K. *Express Polym. Lett.*, **5**, 620–634 (2011).
54. Kunioka, M., Tamaki, A., and Doi, Y. *Macromol.*, **22**, 694–697 (1989).
55. Kusaka, S., Iwata, T., and Doi, Y. *Int. J. Biol. Macromol.*, **25**, 87–94 (1999).
56. Lauritzen, J. I. and Hoffman, J. D. *J. Appl. Phys.*, **44**, 4340–4352 (1973).
57. Lee, W. H., Loo, C. Y., Nomura, C. T., and Sudesh, K. *Bioresour. Technol.*, **99**, 6844–6851 (2008).
58. Lemoigne, M. *Ann. Inst. Pasteur.*, **39**, 144–173 (1925).
59. Lemoigne, M. *Bull. Soc. Chim. Biol.*, **8**, 770–782 (1926).
60. Li, S. Y., Dong, C. L., Wang, S. Y., Ye, H. M., and Chen, G. Q. *Appl. Microbiol. Biotechnol.*, **90**, 659–669 (2011).
61. Lim, J. S., Noda, I., and Im, S. S. *J. Polym. Sci. Part B Polym. Phys.*, **44**, 2852–2863 (2006).
62. Lorenzo, A. T., Arnal, M. L., Albuerne, J., and Muller, A. J. *Polym Test.*, **26**, 222–231 (2007).
63. Lu, X., Wen, X., and Yang, D. J. *Mater. Sci.*, **46**, 1281–1288 (2011).
64. Magill, J. H. *J. Polym. Sci. Part B Polym. Lett.*, **6**, 853–857 (1968).
65. Mandelkern, L. *Crystallization of Polymers*, Second Edition–Vol. 2. Kinetics and mechanisms; Cambridge University Press, Cambridge, (2004).
66. Mandelkern, L. *Crystallization of Polymers*, Second Edition–Vol. 1. Equilibrium Concepts; Cambridge University Press, Cambridge, (2002).
67. Martinez-Salazar, J., Sanchez-Cuesta, M., Barham, P. J., and Keller, A. *J. Mat. Sci. Let.*, **8**, 490–492 (1989).
68. Mergaert, J., Webb, A., Anderson, C., Wouters, A., and Swings, J. *Appl. Environ. Microbiol.*, **59**, 32333238 (1993).
69. Miao, L., Qiu, Z., Yang, W., and Ikehara, T. *React. Funct. Polym.*, **68**, 446–457 (2008).
70. Mitomo, H., Morishita, N., and Doi, Y. *Polym.*, **36**, 2573–2578 (1995).
71. Mitomo, H., Takahashi, T., Ito, H., and Saito, T. *Int. J. Biol. Macromol.*, **24**, 311–318 (1999).
72. Mitomo, M., Hsieh, W. C., Nishiwaki, K., Kasuya, K., and Doi, Y. *Polym.*, **42**, 3455–3461 (2001).
73. Mukai, K. and Doi, Y. *Prog. Indus. Microbiol.*, **32**, 189–204 (1995).
74. Nakamura, S., Doi, Y., and Scandola, M. *Macromol.*, **25**, 4237–4241 (1992).
75. Nomura, C. T., Tanaka, T., Gan, Z., Kuwabara, K., Abe, H., Takase, K., Taguchi, K., and Doi, Y. *Biomacromol.*, **5**, 1457–1464 (2004).
76. Numata, K., Abe, H., and Iwata, T. *Materials*, **2**, 1104–1126 (2009).
77. Orts, W. J., Marchessault, R. H., Bluhm, T. L., and Hamer, G. K. *Macromol.*, **23**, 5368–5370 (1990).
78. Peng, S., An, Y., Chen, C., Fei, B., Zhuang, Y., and Dong, L. *Eur. Polym. J.*, **39**, 1475–1480 (2003).

79. Qiu, Z., Yang, W., Ikehara, T., and Nishi, T. *Polym.*, **46**, 11814–11819 (2005).
80. Qu, X. H., Wu, Q., Liang, J., Zou, B., and Chen, G. Q. *Biomater.*, **27**, 2944–2950 (2006).
81. Sang, B. I., Hori, K., Tanji, Y., and Unno, H. *Appl. Microbiol. Biotechnol.*, **58**, 241247 (2002).
82. Schmack, G., Jehnichen, D., Vogel, R., and Tändler, B. *J. Polym. Sci. Part B Polym. Phys.*, **38**, 2841–2850 (2000).
83. Shimamura, E., Kasuya, K., Kobayashi, G., Shiotani, T., Shima, Y., and Doi, Y. *Macromol.*, **27**, 878–880 (1994).
84. Sridewi, N., Bhubalan, K., and Sudesh, K. *Polym. Degrad. Stab.*, **91**, 2931–2940 (2006).
85. Steinbüchel, A. and Valentin, H. E. *FEMS Microbiol. Lett.*, **128**, 219–228 (1995).
86. Sudesh, K. Polyhydroxyalkanoates from Palm Oil: Biodegradable Plastics, *Springer*, Heidelberg, New York, Dordrecht, and London (2013).
87. Sun, J., Dai, Z., Zhao, Y., and Chen, G. Q. *Biomater.*, **28**, 3896–3903 (2007).
88. Tanaka, T., Fujita, M., Takeuchi, A., Suzuki, Y., Uesugi, K., Ito, K., Fujisawa, T., Doi, Y., and Iwata, T. *Macromol.*, **39**, 2940–2946 (2006).
89. Tripathi, A. D., Yadav, A., Jha, A., Srivastava, S. K. *J. Polym. Environ.*, **20**, 446–453 (2012).
90. Tsuge, T., Saito, Y., Kikkawa, Y., Hiraishi, T., and Doi, Y. *Macromol. Biosci.*, **4**, 238–242 (2004).
91. Vilanova, P. C., Ribas, S. M., and Guzman, G. M. *Polym.*, **26**, 423–428 (1985).
92. Wang, L., Wang, X., Zhu, W., Chen, Z., Pan, J., and Xu, K. *J. Appl. Polym. Sci.*, **116**, 1116–1123 (2010).
93. Wang, X., Chen, Z., Chen, X., Pan, J., and Xu, K. *J. Appl. Polym. Sci.*, **117**, 838–848 (2010).
94. Watanabe, T., He, Y., Fukuchi, T., and Inoue, Y. *Macromol. Biosci.*, **1**, 75–83 (2001).
95. Wen, X., Lu, X., Peng, Q., Zhu, F., and Zheng, N. *J. Therm. Anal. Calorim.*, **109**, 959–966 (2011).
96. Xu, X. Y., Li, X. T., Peng, S. W., Xiao, J. F., Liu, C., Fang, G., Chen, K. C., and Chen, G. Q. *Biomater.*, **31**, 3967–3975 (2010).
97. Yamada, S., Wang, Y., Asakawa, N., Yoshie, N., and Inoue, Y. *Macromol.*, **34**, 4659–4661 (2001).
98. Yamane, H., Terao, K., Hiki, S., and Kimura, Y. *Polym.*, **42**, 3241–3248 (2001).
99. Ye, H. M., Wang, Z., Wang, H. H., Chen, G. Q., and Xu, J. *Polym.*, **51**, 6037–6046 (2010).
100. Yee, L. N., Chuah, J. A., Chong, M. L., Phang, L. Y., Raha, A. R., Sudesh, K., and Hassan, M. A. *Microbiol. Res.*, **167**, 550–557 (2012).
101. Yokouchi, M., Chatani, Y., Tadokoro, H., Teranish, K., and Tani, H. *Polym.*, **14**, 267–272 (1973).
102. You, M., Peng, G., Li, J., Ma, P., Wang, Z., Shu, W., Peng, S., and Chen, G. Q. *Biomater.*, **32**, 2305–2313 (2011).
103. Yu, F., Nakamura, S., and Inoue, Y. *Polym.*, **51**, 4408–4418 (2010).
104. Zhao, Y., Zou, B., Shi, Z., Wu, Q., and Chen, G. Q. *Biomater.*, **28**, 3063–3073 (2007).

105. Zhou, J., Peng, S. W., Wang, Y. Y., Zheng, S. B., Wang, Y., and Chen, G. Q. *Biomater.*, **31**, 7512–7518 (2010).
106. Zini, E., Focarete, M. L., Noda, I., and Scandola, M. *Compos. Sci. Technol.*, **67**, 2085–2094 (2007).

CHAPTER 17

REPLACING PETROLEUM-BASED TACKIFIER IN TIRE COMPOUNDS WITH ENVIRONMENTAL FRIENDLY PALM OIL-BASED RESINS

SIANG YIN LEE and SENG NEON GAN

CONTENTS

17.1 REPLACING PETROLEUM-BASED TACKIFIER IN TIRE COMPOUNDS WITH ENVIRONMENTAL FRIENDLY PALM OIL-BASED RESINS

In the tire industry, a tackifier is used in many places. During the forming process, some parts such as the tread and side wall of a tire, are attached together by the tackifier. In the construction of high performance tire, for example the racing car tire, the incorporation of a suitable tackifier could improve the road gripping properties.

Conventionally, common tackifiers used in the tire industry are phenolic and petroleum resins derived from the non-sustainable petroleum. For example, Japanese Patent Laying-Open No. 2005-248056 proposes a rubber composition including 0.1–50 parts by weight of a rosin ester resin as a tackifier for a tire tread. This rosin ester resin has a softening point of 125°C or higher and acid value of 20 or less with respect to 100 parts by weight of a rubber component. It was found that the grip performance and abrasion resistance have been improved without deteriorating abrasion appearance of the tire. Although rosin and terpene resins have also been used as tackifiers, the tack strength of these materials is not satisfactory, and very often some enhancements are made by copolymerizing with petroleum-based monomers. However, this approach is not so preferable in terms of environmental load.

In order to reduce environmental load, the use of biomass material such as plant oil has also been considered. However, improved effects of sufficient tackiness are hardly obtained especially for practical use as a tackiness-imparting additive in the tire application. Hence, there are some rooms for improvement in terms of the tackiness of these materials. Therefore, an alkyd resin which is a kind of biomass material, having both tackiness-imparting and reversion resistance improving effects, has been considered to be added to rubber composition. It was expected thatthe addition of alkyd resin can impart good properties for a rubber composition for use in tires. However, if a large amount of alkyd resin is added to achieve good reversion resistance, the abrasion resistance may sometimes be deteriorated, and in addition, the costs may be increased. Moreover, a hardness change during thermal aging may sometimes be increased. Therefore, it is desired to provide a rubber composition with low environmental load and with high tackiness and reversion resistance, but also relatively cheap.

Recently, a patent application was filed on a method of chemical modifications of natural rubber and epoxidized natural rubber by alkyds made from palm oil and its fatty acids(Gan, 2006). Through variation of chemical structure, composition, and proportion of alkyd relative to the rubber, the physical properties of the rubber and epoxidized rubber could be enhanced for specific applications. For example, the palm oil-based alkyds had been applied in the modification of the shear and peel strengths of natural rubber based pressure sensitive adhesives(Lee et al., 2013). Furthermore, the researchers from University of Malaya of Malaysia and Sumitomo Rubber Industries Ltd. (SRI R&D) of Japan were the first to investigate the use of alkyd as tackifier in tire compound during the period of 2005–2011(Hattori et al., 2005; Hattori et al., 2007; Hattori et al., 2011). The objectivewas to reduce the petroleum-based constituents by increasing the content of biomass material in alkydand at the same time improve the tire performance without sacrificing the rubber tackiness and physical properties.Thus, the use of palm oil-based resins as tackifiers in dry rubber compounds for tire applications is discussedin this chapter.

17.2 ALKYD RESINS

The alkyd resins are polyfunctional oil-modified polyesters, synthesized by reacting polybasic acids with polyhydric alcohols together with a vegetable oil, animal oil or its derivatives, *via* a stepwise polymerization process(Patton, 1962). Ever since the alkyds were first introduced 80 years ago by Kienle in 1927(Kienle, 1930), they have enjoyed a consistent annual growth (Patton, 1962). Alkyd was the first synthetic polymer being applied in surface coatings industry(Lambourne, 1987)and till now it is one of the widely used synthetic resins in the paint and surface coating industry.Alkyd-based coating is well known for its fast dryness, good corrosion protection, high gloss, and ease of application even over poorly treated surface. Besides, alkyds tend to be lower in cost yet give coatings that exhibit fewer film defects during application(Weiss, 1997). Since the alkyd reaction is one of the most versatile resin-forming reactions, the properties of alkyds can be modified through physical or chemical blending with other polymers (Patton, 1962). Examples of polymers that commonly used to do physical blending with alkyds are nitrocellulose, urea-formaldehyde, melamine-formaldehyde, chlorinated rubber, and chlorinated paraffin,whereas, the polymers commonly used to modify alkyds chemically are styrene, phenolics, silicones, epoxies, isocyanates, and formaldehyde. Whether the modification is physical or chemical, the properties and performance of the blended system will be enhanced than that of any individual resin, especially in flexibility, toughness, adhesion, and durability of the coatings.

A polyhydric alcohol is an alcohol with more than one hydroxyl group (–OH). Table 1 lists some examples of common polyhydric alcohols used for preparation of alkyd. These polyhydric alcohols may be used alone or in combination of at least two kinds.Among these, ethylene glycols, glycerol, and pentaerythritol are the most common polyhydric alcohols used in alkyd formulation. As ethylene glycols are only bifunctional, they are mainly used to formulate unmodified resins which are thermoplastics that are linear and cannot be cross-linked except in the case of unsaturated polyesters. Longer chain glycols, such as propylene glycol and diethylene glycol will only provide the finished resins with even greater flexibility. Therefore, they are usually used together with glycerol in an alkyd formulation. If ethylene glycols are used with a drying oil *via*alcoholysis process, then the glycerol present in the oil becomes the determining factor in the formation of resin.

In order to obtain an alkyd resin with suitable molecular weight, a trivalent alcohol, especially glycerol is preferable in common alkyd formulation due to its trifunctionality. With this trifunctionality, glycerol is able to add more branching in the alkyd macromolecule, providing a three-dimensional polymeric network and thus results in a high molecular weight resin. However, there is a difference in reactivity of the hydroxyl groups in glycerol, where the terminal hydroxyl groups ($–CH_2OH$) called primary or α-hydroxyl groups are alike the central group (–CHOH) contains a secondary or β-hydroxyl group. All the three hydroxyl groups react at about the same rate with fatty acids at 180–280°C, howeverthe α-hydroxyl groups react more rapidly than the β-hydroxyl groups with dibasic acids, such as phthalic (Taylor et al., 1972).

TABLE 1 Some of the common polyhydric alcohols used for preparation of alkyd

Bivalent alcohol	Formula
Ethylene glycol	$HO(CH_2)_2OH$
Propylene glycol	$CH_3CH(OH)CH_2OH$
1,3-butylene glycol	$CH_3CH(OH)CH_2CH_2OH$
Pentanediol	$HO(CH_2)_5OH$

TABLE 1 *(Continued)*

Neopentyl glycol	$HOCH_2C(CH_3)_2CH_2OH$
Diethylene glycol	$HO(CH_2)_2O(CH_2)_2OH$
Triethylene glycol	$HO(CH_2)_2O(CH_2)_2O(CH_2)_2OH$
Dipropylene glycol	$CH_3CH(OH)CH_2OCH_2CH(OH)CH_3$
Trivalent alcohol	**Formula**
Glycerol	$HOCH_2CH(OH)CH_2OH$
Trimethylolethane	$CH_3C(CH_2OH)_3$
Trimethylolpropane	$CH_3CH_2C(CH_2OH)_3$
Polyvalent alcohol	**Formula**
Pentaerythritol	$C(CH_2OH)_4$
Dipentaerythritol	$(CH_2OH)_3CCH_2)_2O$
Sorbitol	$HOCH_2(CHOH)_4CH_2OH$

Pentaerythritol, $C(CH_2OH)_4$, is made by condensing formaldehyde and acetaldehyde in the presence of alkali and water. It is a solid material of high melting point (253°C), which has four primary reactive hydroxyl groups. It then forms a much more complex resin with phthalic anhydride than with glycerol. Hence, only medium-long and extra long oil length alkyds, with the content of fatty acids about 65–70%, are commercially available in order to avoid rapid gelation during manufacturing (Taylor and Marks, 1972). However, its reactivity can be reduced, either by partial replacement with glycol (functionality 4 plus functionality 2 is equivalent to average functionality 3), or more easily by the use of larger proportions of modifying fatty acids. Because the four hydroxyl groups are equivalent in reactivity, pentaerythritol behaves more uniformly than glycerol when esterifies, and the course of the esterification can be more readily predicted. Besides, the product molecules have a more symmetrical structure. However, commercial pentaerythritols are not pure, containing up to 15–20% of dipentaerythritols which may affect the properties of the final alkyds as predicted. Owing to the excessive reactivity of pentaerythritol compared to glycerol, the manufacturing of short and medium drying oil alkyds is almost impossible.

Overall, an amount ratio of polyhydric alcohol in the alkyd resin is an important factor to obtain the polymer with desired properties when condensation reaction is carried out. For example, if the amount is less than 5% by weight, the molecular weight of the finished alkyd is difficult to be improved. However, if the amount exceeds 40% by weight, there will be a lot of the unreacted polyhydric alcohols remained in the system and thus causing an excessive increase in hygroscopic property.

On the other hand, a polybasic acid is an acid having a basicity of at least two. Examples of polybasic acids are saturated dibasic acids such as succinic acid, adipic acid, azelaic acid, and sebacic acid, unsaturated dibasic acids are such as terephthalic acid, maleic acid, fumaric acid, and isothalic acid, whereas, anhydrides of saturated dibasic acids or unsaturated dibasic acids are phthalic anhydride, maleic anhydride, glutaric anhydride, and succinic anhydride. The saturated dibasic acids are usually for the manufacture of plasticizing alkyds. For example, the alkyds synthesized using adipicorsebacic acids arepredominantly linear and less rigid (Figure 1).

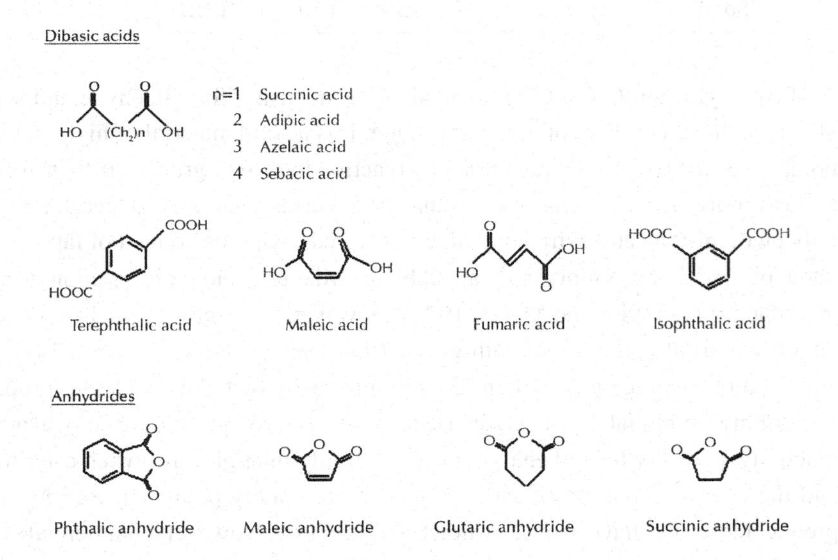

FIGURE 1 Adipic or sebacic acids used for synthesizing alkyds.

Among the polybasic acids, unsaturated polybasic acids or their anhydrides are preferable, from the reason that they are obtainable at a low cost and could form hydrogen bond among molecules and thus providing higher tackiness to the final alkyd. By far, the most widely used unsaturated polybasic acid is phthalic anhydride. It has the advantage that the first esterification reaction proceeds rapidly by opening the anhydride ring and thus reduces the reaction time. Phthalic anhydride also has a relatively low melting point, where the crystals can melt, dissolve and react readily in the reaction mixture at 131°C. The shorter time at lower reaction temperature can thus reduce the extent of side reactions of the polyol components.

The maleic acid, or more commonly maleic anhydride, is usually used together with phthalic anhydride in the formulation of drying oil alkyds. Besides its difunctionality, maleic anhydride also provides additional functionality arising from its double bond, which enables it to form the polyunsaturated alkyds. In normal alkyd manufacture, the use of maleic anhydride is usually restricted to 1–10% of the phthalicanhydridecontent (Taylor and Marks, 1972). Such replacement is sufficient to produce improvements in color, processing time and water resistance when compared with maleic-free alkyds. However, the replacement of phthalic anhydride by maleic anhydride in excessive quantities may cause gelation during synthesis.

The fumaricacid is treated similarly to maleic anhydride and derived from vegetable.It is less toxic than maleic anhydride and thus more preferable than maleic anhydride in the alkyd formulation in consideration of reducing environmental load.Since a carboxyl group is in a *trans* position, fumaric acid forms hydrogen bond among molecules and thereby becomes possible to impart higher tackiness than maleic anhydride. Besides, fumaric acidexhibits less loss by sublimation and is claimed to give a faster reaction leading to the formation of faster drying alkyd of better color.However, compare to maleic anhydride, fumaric acid requires a higher processing temperature for solubility and liberates an additional mode of water during processing.

Theterephthalic acid is made by oxidizing *p*-xylene or di-isopropyl benzene. It is used almost entirely in glycol-terephthalic ester resins for the production of synthetic fibers and films, with great strength, elasticity and relatively good chemical resistance.

Isophthalic acid has a much higher melting point (354°C) than phthalic anhydride and thus is more difficult to use since it does not easily melt into the reaction mixture. High temperatures are required for longer times than with phthalic anhydride and hence more dimerization of fatty acids occurs with isophthalic acid, resulting in the final resin of higher viscosity. The longer reaction time at higher temperature also leads to greater extent of side reactions of the polyol components. However, polyesters of isophthalic acid are more resistant to hydrolysis than those of phthalic anhydride in the pH range of 4 to 8, the most important range for exterior durability of coating.

In practical, the favorable amount for polybasic acids in the alkyd resin formulation is 10–50% by weight, where the condensation is difficult to precede if the amount of ratio is less than 10% by weight; while the tackiness of finished alkyd tends to be low if the amount ratio exceeds 50% by weight. Furthermore, the excess amount of acid group may increase the extent of side reaction during synthesis which can cause gelation if the reaction proceeds at high temperature for longer time.

On the other hand, vegetable or animal oils are othermodifiers in an alkyd formulation. Examples of vegetable oils are palm oil, palm kernel oil, soybean oil, olive oil, colza oil, sesame oil, wood oil, castor oil, and linseed oil, whereas, animal oils are such as fish oil, whale oil, beef oil, mutton oil, and hoof oil. Natural resins such as rosin, amber, and shellac, and synthetic resins such as ester gum, phenol resin, carbon resin, and melamine resin, also have been reported as a modifier in the alkyd formulation.

From the recent trend of emphasizing environmental friendly, oils derived from resources other than petroleum, either vegetable or animal, should be utilized in alkyd formulation. These oils are generally available at low cost and that tackiness can be imparted without excessively increasing hardness. An amount of oil in the alkyd resins is preferably 10–85% by weight. If an amount ratio is less than 10% by weight, it tends that hardness is excessively increased when an alkyd resin is compounded into a rubber. On the other hand, an alkyd resin having a suitable molecular weight is hardly obtained if an amount ratio exceeds 85% by weight.

There are many ways of classifying alkyds. Generally, alkyd resin is classified in term of oil length and oil type. The oil length refers to the percentage by weight of the oil portion in the final alkyd, where long oil length alkyds contain more than 55% oil, medium oil length alkyds have between 45–55% oil, and short oil alkyds content less than 45% oil portion (Earhart, 1949; Moore, 1950). Short oil length alkyds show resin-like properties which have higher viscosity, tackiness and impart the hardness of a coating; whereas, the long oil length alkyds show oil-like properties, where they are softer, having higher flowability and form coating with less tackiness. The medium oil length alkyd corresponds to an even weight mixture of resin-like and oil-like properties. The final structures and properties of an alkyd would depend on the overall formulation as well as the type of vegetable oil being used.

Another classification is to classify alkyds into oxidizing and non-oxidizing types (Wicks et al., 1999). An oxidizing alkyd contains drying (unsaturated) or semi-drying oils or fatty acids and often have an oil length in excess of 45%. Some authors classify oils as drying, semidrying and non-drying based on their iodine value (Rheineck and Austin, 1968). Although iodine values can serve as satisfactory quality control specifications, they are sometime not useful and can be misleading in defining a drying oil or for predicting reactivity. A useful empirical relationship is that non-conjugated oils that have a drying index greater than 70 are considered as drying oils (Greaves, 1948).

In fact, the naturally occurring oils consist of triglycerides, tri-esters of glycerol, and fatty acids. Some triglycerides are drying oils, but many are not. Alkyds formulated from drying oils are able to air-dry at ambient temperature, when the oils react with oxygen, cross-link and form solid films. The reactivity of drying oils with oxygen results from the presence of diallylic groups, (that is, two double bonds are separated by some methylene groups, $–CH=CHCH_2CH=CH–$) or conjugated double bonds.

A non-oxidizing alkyd contains non-drying (saturated) oils or fatty acids and therefore the alkyd is not capable of forming coherent film by air oxidation. Hence, the non-oxidizing alkyd is used as polymeric plasticizer or as hydroxyl-functional resin, which is cross-linked by melamine-formaldehyde or urea-formaldehyde resins or by isocyanatecross-linkers. The oil length for non-oxidizing alkyds is usually formulated below than 45% (Wicks et al., 1999).

Another classification is by distinguishing the unmodified with modified alkyds. Modified alkyds contain other monomers in addition to polyhydric alcohols, polybasic acids and fatty acids. Examples are styrenated alkyds (Sheehter and Wynstra, 1602; Bhow and Payne, 1950; Cowan, 1954) and silicone alkyds (Von-Fischer and Bobalek, 1953; Hileset al., 1955; Karimet al., 1960).

In term of alkyd synthesis, alkyds that are formulated from an identical chemical composition will exhibit different properties and performance depending on their preparation processes. There are various procedures in preparing alkyd resins (Anonymous, 1962; Patton, 1962; Kaska and Lesek, 1991). In general, alkyds can be prepared from two processes, that is fatty acid and alcoholysis (or monoglyceride).

In the fatty acid process, alkyds are prepared from fatty acids in three processes. The first process can be performed in a single step where all the raw materials, polyhydric alcohol, fatty acid, and polybasic acid, are added together at the start of the reaction and the esterification is carried out simultaneously in range of 220–250°C (Wicks et al., 1999). Thus, there is a free competition among the –COOH groups with the –OH groups. An alkyd with more branching is synthesized.The second process can be conducted in two steps. The polyhydric alcohol (glycerol) and polybasic acid are first reacted together at 180°C until a soft clear viscous resin is obtained. Then, warm fatty acids are added and the reaction temperature is raised to 180–220°C (Wicks et al., 1999). In this case, the –COOH groups of polybasic acid are deliberately reacted with the primary –OH groups of glycerol to form glyceryl phthalate before fatty acid is added. Consequently, the –COOH groups of fatty acid are forced to esterify with the leftover –OH groups of glyceryl phthalate. Hence, the alkyd's backbone is linear. While, the third process was carried out under a step-by-step esterification of the fatty acids, which involves a stepwise addition of fatty acids, from 40–90%, at 180–250°C (Kraft, 1957). By withholding a part of this chain-terminating ingredient, the polymer is predominantly linear and of high molecular weight.

On the other hand, in the alcoholysis (monoglyceride) process, alkyds are prepared from glyceride oil in a two-stage procedure. The first stage in the process is to convert the polyhydric alcohol and glyceride oil into a single homogeneous monoglyceride phase at 230–250°C in the presence of a catalyst. At the second stage, this monoglyceride in turn provides a solvent for the polybasic acid added next to react under esterification in order to complete the alkyd reaction at 180–250°C. This process is usually applied when phthalic anhydride is used as the polybasic acid since phthalate anhydride is not soluble in the oil but is soluble in the glycerol. Hence, transesterification of oil with glycerol must be carried out as a separate step before phthalate anhydride is added or otherwise glyceryl phthalate gel particles would form early in the first stage of the process.

Goldsmith has proposed a plausible explanation for the difference of the alkyds' characteristics prepared by different processes (Goldsmith, 1948). He indicated that the differences were due to the differential rates of reaction between –OH and –COOH groups of raw materials, depending on their specific location on the parent molecules. Comparisons of the proposed chemical structures can be made by considering these alkyds are synthesized from the same raw materials and formulations of fatty acid, PA and glycerol.

In the fatty acid preparation process, there is a free competition between the –COOH groups of fatty acids and phthalate anhydride, with the –OH groups of glycerol, assumed that all added at the beginning. Compare to phthalate anhydride, the long hydrocarbon chain in fatty acid creates steric hindrance and acts as a structural block in order to esterify with the nearby hydroxyl groups. Therefore, the –COOH groups of fatty acid that lag behind in esterification with the primary –OH groups in glycerol have to settle for connections with secondary –OH groups. These can be seen from the alkyd structures as shown:

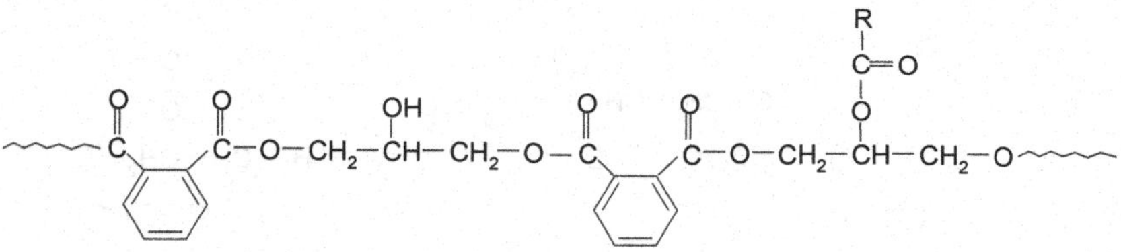

where, short oil length alkyds that are formed by fatty acid procedure, utilizing 1 mole of fatty acid, 2 moles of phthalate anhydride and 2 moles of glycerol.

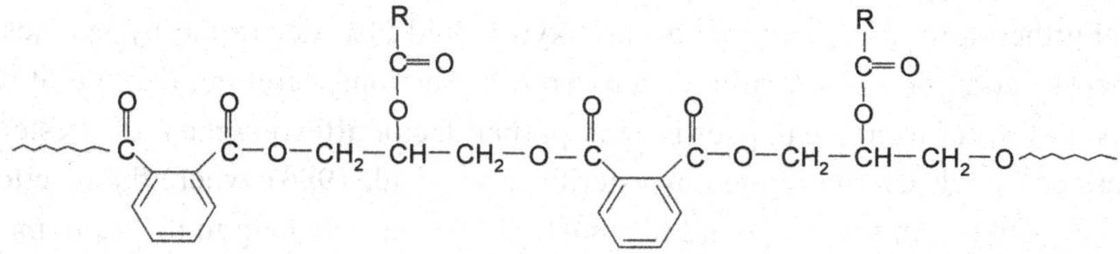

where, long oil length alkyds that are formed by fatty acid procedure, utilizing 1 mole of fatty acid, 2 moles of phthalate anhydrideand 2 moles of glycerol.

On the other hand, in the alcoholysis (monoglyceride) process, the competition is more rigged, where the –COOH groups of fatty acid are deliberately reacted with the primary groups of glycerol before any PA is added. Thus, the –COOH groups of the diacid PA are placed at a competitive disadvantage and is forced to react with the leftover –OH groups. An example of structures of a short oil length and long oil length alkyds synthesized from alcoholysis(monoglyceride) procedure are shown as follows:

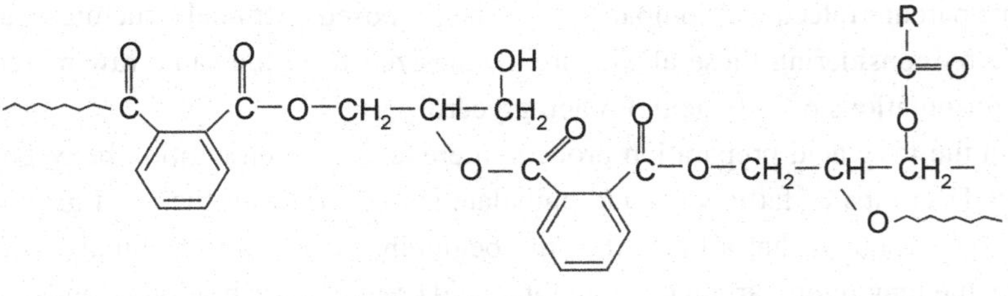

where, short oil length alkyds that are formed by alcoholysis(monoglyceride) procedure, utilizing 1 mole of fatty acid, 2 moles of phthalate anhydrideand 2 moles of glycerol.

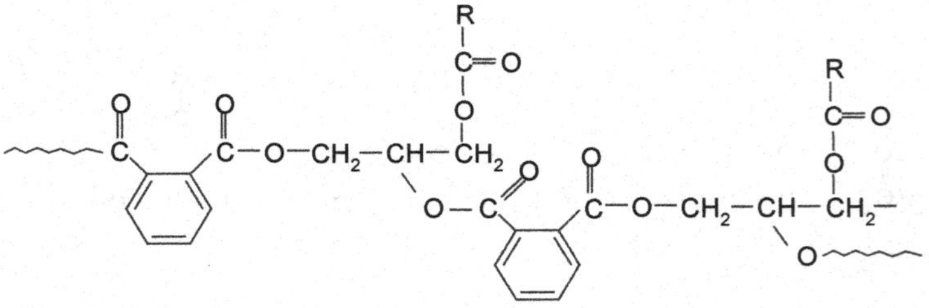

where, long oil length alkyds that are formed by alcoholysis(monoglyceride) procedure, utilizing 1 mole of fatty acid, 2 moles of phthalate anhydride and 2 moles of glycerol.

Furthermore, the structures of an alkyd would also determine by synthesis process variation. Esterification is a reversible reaction; therefore, the rate of removal of water from the reactor is an important factor affecting the rate of esterification. Alkyds can be formed by esterification (Paul, 1986), where the reaction occurs between the –COOH and the –OH groups either belong to the same molecule or in two different compounds according to the reaction as follows:

$$R\text{-}COOH + R'\text{-}OH \leftrightarrow RCOOR' + H_2O$$

For the alkyds synthesized from anhydrides, the reaction involves two distinct steps. In the first step, esterification proceeds rapidly by opening the anhydride ring, which happens at a lower temperature and caused the formation of half-esters. There is no water of reaction evolved. On further heating, the second step continues and long chain molecules are produced which contain free hydroxyl groups in excess. This step happens at a slower rate and water of reaction is releasedFigure 2 (Paul, 1986).

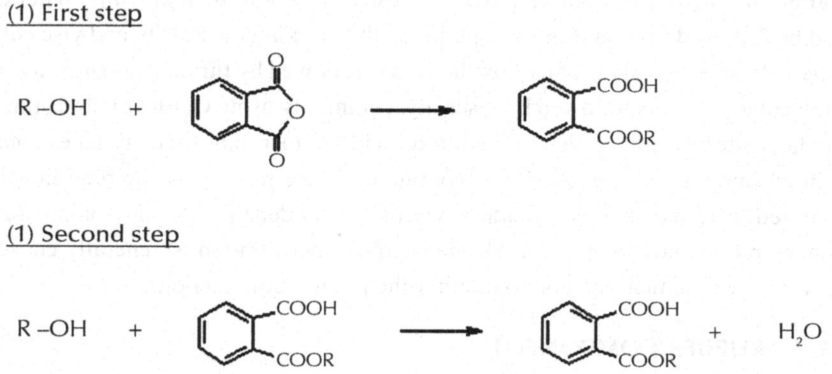

(1) First step

R–OH + [phthalic anhydride] → [benzene ring with COOH and COOR]

(1) Second step

R–OH + [benzene ring with COOH and COOR] → [benzene ring with COOH and COOR] + H_2O

FIGURE 2 Chemical reaction during the synthesis of alkyds using anhydrides.

Esterification can be conducted either in the absence of solvent (fusion cook) or with the presence of solvent (solvent cook).In a fusion cook, a flow of inert gas is usually maintained through the reactor, to prevent ingress of air and to remove water of reaction. There is a considerable loss of volatile reactants and phthalate anhydride sublimation (in the reflux condenser) during a fusion cook, particularly at higher temperature. Therefore, fusion cook is generally applied in the formulation of long oil length alkyds in order to reduce the effect of unanticipated losses on the predicted functionality. On the other hand, the presence of solvent in a solvent cook facilitates the removal of the water of reaction, allowing a shorter processing time being achieved without heating to excessively high temperature, where a typical temperature ranges for solvent process is around 200–240°C. The solvent vapor blanket serves as an inert atmosphere, preventing ingress of air, thus enabling light color products to be polymerized with a minimum of inert gas usage. Furthermore, the accumulation of sublimed solid monomers, mainly phthalate anhydride, in the reflux condenser, can be returned into the reactor together with solvent reflux.

In fact, the acidity and viscosity of the final alkyds can be affected by the progress of reactionduring esterification, for example, the reaction temperature and reaction time.Hence, the choice of a reaction temperature is important in an alkyd synthesis, where the temperature must allow the reaction to be carried out within a reasonable time period to reduce the operation cost, yet not so elevated to cause destructive decomposition, discoloration and an excessive loss of volatile material during the reaction. In fact, the progress of reaction in an alkyd preparation can be followed and controlled by periodically checking the acidity and viscosity, where the disappearance of carboxylic acid is followed by titration and increase in molecular weight is followed by viscosity, commonly using Gardner bubble tubes or rheometer. However, determination of acid number and viscosity takes some time in sampling as the reaction is continuing in the reactor. While time limit is required for viscosity determination, viscosity also depends strongly on solution concentration and temperature. Hence, acid number titration is generally chosen as a more convenient method to monitor the progress of reaction.

17.3 RUBBER COMPONENT

Natural rubber (NR) is produced from rubber tree (*Heveabrasiliensis)*, which is originally grown in South America. NR is a polyolefin with a precise *cis*-1,4-isoprene molecule and the weight-average molecular weight ($\overline{M}_w$) (Subramaniam, 1972; Subramaniam, 1975)is generally between the limits of 1.6 and 2.3 × 10^6. However, NR in its initial form is not a useful material. It does not have good mechanical properties, and is easily oxidized by oxygen and ozone.In a process known as rubber compounding, the weak rubber can be chemically transformed into useful and versatile materials for diverse applications ranging from adhesives, latex products, foams, and tires to bridge bearings. The main ingredients in compounding are chemicals which can introduce certain amounts of cross-linking reactions between the rubber molecules. Filler and antioxidant are also important ingredients for enhancing the hardness and extending the service life span.Other minor additives include processing aids, plasticizers, blowing agents and pigments. A simple compounding formulation may contain as many as six or more ingredients.

Vulcanization is the most common method employed in rubber compounding. It converts rubber from a linear polymer to a three-dimensional network by cross-linking the polymer chains. The first commercial method for vulcanization has been attributed to Charles Goodyear. His process (heating natural rubber with sulfur) was first used in Springfield, Massachusetts in 1841 (Coran, 2005). Ever since, vulcanization has remained the cornerstone of compounding, and sulfur is still the principle vulcanizing ingredient for linking the C=C double bonds, because of its low cost, fast reactions, and minimal interference with other compounding ingredients. For more than 100 years, there have not been many significant changes in rubber compounding, in terms of new variations in the formulations.

Vulcanization, thus, is a process of chemically producing junctures by the insertion of cross-links between polymer chains. A cross-link may be a group of sulfur atoms in a short chain, a carbon to carbon bond, a polyvalent organic radical, an ionic cluster, or a polyvalent metal ion. The process is usually carried out by heating the rubber, mixed with vulcanizing agents in a mold under pressure. In fact, vulcanization increases elasticity while decreases plasticity by increasing the retractile force and reducing the amount of permanent deformation remaining after removal of the deformation force(Coran, 2005). According to the theory of rubber elasticity, the retractile force to resist a deformation is proportional to the number of network-supporting polymer chains per unit volume of elastomer (Flory, 1953). A supporting polymer chain is a linear polymer segment between network junctures. An increase in the number of junctures or cross-links gives an increase in the number of supporting chains. In an unvulcanized linear high polymer (above its melting point), only molecular chain entanglements constitute junctures.

Sulfur vulcanization always carried out in the presence of sulfur, together with activators and organic accelerators. The activators, usually zinc oxide and stearic acid, increase the number of cross-links and stiffness, but amount used in sulfur vulcanization are not critical provided that certain minimum levels are exceeded. The accelerator increases the rate of cure and the efficiency of sulfur cross-linking.

In a normal sulfur vulcanization system, the temperatures required are in the range of 120–160°C when the sulfur-sulfur bonds in the elemental S_8sulfur rings start to break (Munk, 1989). In any case, the primary sites of attack are hydrogen on the α carbon on the double bond, producing the structure as shown. As a result, the physical properties of the cross-link networks, such as resilience, stiffness and resistance to creep, are generally increased.

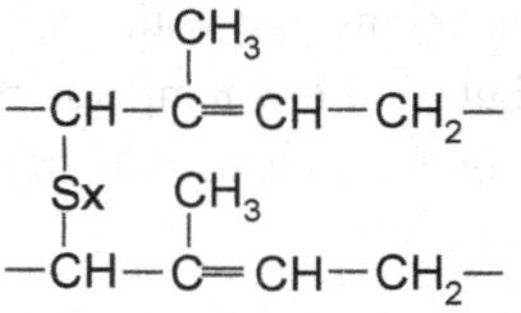

$$
\begin{array}{l}
\quad\quad\quad\;\; CH_3 \\
\quad\quad\quad\;\;\; | \\
-CH-C{=}CH-CH_2- \\
\quad\;\; | \\
\quad\; Sx \quad CH_3 \\
\quad\;\; | \quad\quad\;\; | \\
-CH-C{=}CH-CH_2-
\end{array}
$$

where, x = 2, 4 ,6, 8.

In fact, the physical properties obtained depend on the types of cross-link formed and the extent of main-chain modification by side reactions. This is usually being largely determined by the vulcanization system, although cure time and temperature also have an important effect. Generally, there are three types of typical sulfur vulcanization systems, namely, conventional vulcanization, efficient vulcanization and semi-efficient vulcanization.

A conventional vulcanization system contains high sulfur level, 2-3.5 phr and low level of accelerator, typically 0.3-1 phr. At optimum cure, the vulcanizates contain mostly polysulfidiccross-links with a relatively high level of chain modification. These polysulfide networks generally have good tensile, tear, and fatigue properties and excellent resistance to low-temperature crystallization. However, they are susceptible to reversion, that is loss of properties on overcure and to oxidative ageing. Other drawbacks include high compression set and high rates of secondary creep and stress relaxation at elevated temperatures, due to exchange reactions between polysulfidiccross-links.

Low sulfur levels, typically 0.25-0.7 phr, with high accelerator levels, typically 2.5-5 phr will give mainly monosulfidiccross-links with lower level of chain modification. They are therefore known as efficient vulcanization systems. These systems give vulcanizates that are more resistant to reversion, oxidative ageing and high temperature compression set due to the higher thermal stability of monosulfidiccross-links. However, compared to conventional vulcanizationvulcanizates, efficient vulcanizationvulcanizates have a lesser extent of resistance to tearing, fatigue and wear, and lower tensile strength. Besides, their resistance to low temperature crystallization is poor and their rubber-to-metal bonding may be difficult. Therefore, efficient vulcanization systems are used for the vulcanization of thick articles and those used at elevated temperatures.

For semi-efficient vulcanization systems, intermediate sulfur level of 1-2 phr and 2.5-1 phr of accelerator are often used. The vulcanizates have physical properties intermediate between those of conventional vulcanization and efficient vulcanizationvulcanizates. In fact, they give some improvements in reversion, ageing resistance and compression set compared with conventional vulcanizationvulcanizates, but resistance to fatigue and low temperature crystallization is impaired. However, they have higher scorch safety, particularly when sulfenamide accelerators are used in the system.

Examples of the rubber component that may be used in the tire compounds are natural rubber (NR), modified natural rubber, styrene-butadiene rubber (SBR), butadiene rubber (BR), isoprene rubber (IR), butyl rubber (IIR), ethylene propylene dienemonomer (EPDM) rubber, acrylonitrile butadiene rubber (NBR), chloroprene rubber (CR), and epoxidized natural rubber (ENR). These rubbers may be used alone or in combination. Among these, for a tread and a sidewall, SBR and BR are preferable because of excellent abrasion resistance, fatigue resistance and flex crack growth resistance. As for SBR, an amount of combined styrene in SBR is preferably at least 10% and at most 60%. It tends that either sufficient grip performance is hardly obtained or abrasion resistance is deteriorated when the amount of combined styrene is either less than 10% or more than 60% when used for a tread. As for BR, it is preferable that to contain at least 90% of cis-1,4 bond. By compounding such BR, flex crack growth resistance and aging resistance performance can be improved when utilized in a tread of tires particularly for trucks and buses and a sidewall of tires including for general passenger automobiles. In the case where the rubber composition is used for sidewalls of pneumatic tires, SBR is not usually used.

Among modified rubbers, ENR is preferable in the tire compounds from reasons that necessary grip performance can be obtained for use as a tread, a sea-island structure having a suitable size is formed with NR, flex crack growth resistance can be improved, and ENR can be comparatively inexpensively available compared with other modified natural rubbers. Although ENR has a disadvantage of causing reversion more easily compared with NR, ENR is preferable since larger modifying effects of physical properties can be obtained especially in the case of high-temperature vulcanization by compounding an alkyd resin into ENR.

Commercially available ENR may be used or NR may be epoxidized as ENR. The method of epoxidizing NR may include chlorohydrin method, direct oxidation method, hydrogen peroxide method, alkyl hydroperoxide method, and peroxide method. As the peroxide method, an example is a process of reacting organic peracid such as peracetic acid or performic acid with NR. An epoxidization ratio in ENR is preferably to be in the range of 3% to 80% by mole. When an epoxidization ratio is less than 3% by mole, an effect of modification tends to be small but when the ratio exceeds 80% by mole, polymers tends to cause gelation. The partial conversion of double bonds to epoxide groups has led to improvements of NR properties such as oil resistance, low gas permeability, good wet grip, and high damping characteristics (Gelling, 1991). ENR contains both epoxide and unsaturated sites, which could be utilized for chemical modifications. While the double bonds can be vulcanized by sulfur and peroxide, the epoxide groups provide sites for interaction with compounds having other functional groups (Loo, 1985; Baker et al., 1986; Gan, 1989; Hashim and Kohjiya, 1994).

In the case of compounding non-modified NR, an amount of such NR is preferably at most 85% by weight to avoid deterioration of flex crack growth resistance and ozone resistance. In addition, an amount of such NR is preferably at most 80% by weight when used in a rubber composition for a tread, and also an amount of such NR is preferably 15–85% by weight when used in a rubber composition for a sidewall. However, it is more preferably that an amount ofNR and/or modified NR is at most 50% by weight in the rubber component in order to reduce the petroleum resources ratio and further reducing environmental load, and in addition, the tackiness-imparting effect can be achieved well, because of the use of tackifiers, for example, alkyd resins that show good compatibility with NR and modified NR.

When a modified natural rubber is compounded, an amount of a modified natural rubber is preferably more than 30% by weight in order to obtain significant effect of improvement in grip performance. On the other hand, an amount of a modified natural rubber is preferably 15–85% by weight when used for a sidewall in order to allow the formation of an appropriate sea-island structure with other rubber as NR, and thereby improve crack resistance property of the tire compound.

17.4 PROPERTIES OF PALM OIL-BASED ALKYD MODIFIED TIRE COMPOUNDS

Several applications of joint patents between University of Malaya (UM) of Malaysia and Sumitomo Rubber Industries Ltd. (SRI R&D) of Japan on the application of palm oil-based alkyd in the rubber composition for tire and pneumatic tire had been filed(Hattori et al., 2005; Hattori et al., 2007; Hattori et al., 2011). As described in these patents, alkyd was compounded as a tackifier in rubber compositionand has shown an improvementin tackiness,hardness as well as lowered the rolling resistance of thetire, compared with rubber composition compounding with a petroleum resin or a terpeneresin. Besides, alkyd has also shown good controls on reversion which is frequently observed when a rubber is vulcanized with sulfur for a short time at a high temperature.

Alkyds synthesized from palm kernel oil involves two stages, where the first stage was the alcoholysis of palm kernel oil in the presence of an alkali catalyst at 180–200°C, follows by second stage the condensation reaction of monoglycerideswith a dicarboxylic acid (phthalic anhydride and fumaric acid) in a medium of xylene at 210–220°C (Figure3).The formulation of palm oil-based alkyds is shown in Table 2.

These new environmental friendly tackifiers with varying percentage of fumaric acid in the total of polybasic acid content were synthesized. The incorporation of fumaric acid provides additional double bondin the polyunsaturated alkyds which is then serves as sites for cross-linking reactions. In addition, these fumaric acid modified alkyds are expected to form more hydrogen bonds among molecules due to its *trans*-oriented carboxyl groups lies in fumaric acid molecules and thus resulting higher tackiness than the unmodified alkyds (Figure 4).

TABLE 2 Alkyd resins synthesized from palm kernel oil (PKO)

Alkyd code*	**Composition (% by weight)**				**Ratio of resources other than petroleum (% by weight)**
	PKO	**Glycerol**	**PA**	**FA**	
AlkFA0	58.1	16.4	25.6	0	74.5
AlkFA15	58.6	16.5	21.9	3.0	78.1
AlkFA25	58.9	16.6	19.4	5.1	80.6
AlkFA35	59.3	16.8	16.9	7.0	83.1

where, PKO= palm kernel oil, PA= phthalic anhydride, and FA= fumaric acid.

* Alkyd code was based on the percentage of FA in the total of polybasic acid content.

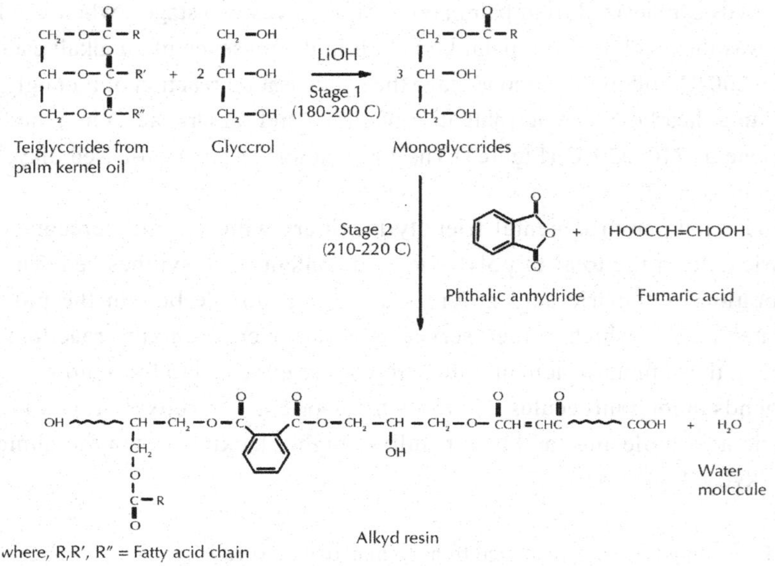

Teiglyccrides from palm kernel oil + 2 Glyccrol → LiOH, Stage 1 (180-200 C) → 3 Monoglyccrides

Stage 2 (210-220 C): + Phthalic anhydride + Fumaric acid (HOOCCH=CHOOH)

→ Alkyd resin + H_2O (Water molccule)

where, R,R′, R″ = Fatty acid chain

FIGURE 3 A plausible reaction mechanism in the preparation of palm oil-based alkyd.

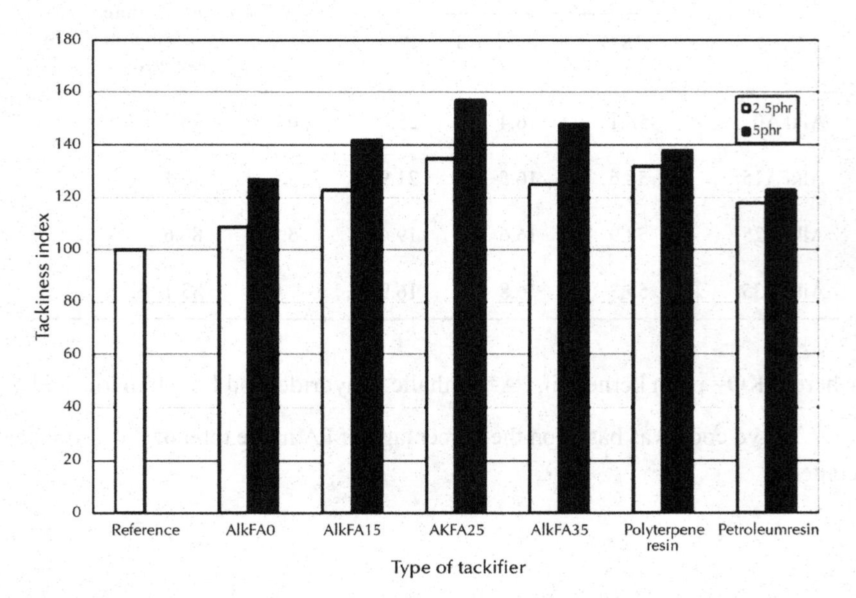

FIGURE 4 Tack strengths imparted to the rubber compound by different tackifiers.

Various properties of the obtained alkyds are displayed in Table 3. Basically, the finished alkyds have three fundamental functional groups, that is hydroxyl, carboxylic and olefinic groups. It was postulated that only the hydroxyl and carboxylic groups in alkyds might react with the epoxy groups of ENR at ambient temperature. Hence, it was important to determine the free hydroxyl and carboxylic groups remained in those alkyds prior to the manufacture of unvulcanized rubber composition. Both the free hydroxyl and carboxylic groups could be determined from the hydroxyl number and acid number titrations, respectively.

TABLE 3 Properties of palm kernel oil alkyd resins

Alkyd code	Acid value/ mg KOH g^{-1}	Hydroxyl value/ mg KOH g^{-1}	Softening point (°C)	Molecular weight	
				M_n	M_w
AlkFA0	20	84	4	1092	1641
AlkFA15	19	84	4	1118	1797
AlkFA25	17	82	5	1264	2359
AlkFA35	15	82	5	1247	2239

An acid value is referred to an amount (mg) of potassium hydroxide added to neutralize 1 g of a resin when the resin is dissolved in an organic solvent such as toluene. From Table 2, four alkyds were formulated with an acid value between 15-20 mg KOH g^{-1} where these values falls within the preferable range (5-60 mg KOH g^{-1}) of acid value of an alkyd. It is important to obtain an acid value of alkyd resin at least 5 in order to synthesize alkyds at a low cost with lower acidity. However, an acid value of alkyd is preferably at most 60 to enable reasonable high cross-linking density in order to prevent the deterioration of an abrasion appearance or lowering of hardness during running.

The hydroxyl value is referred to an amount (mg) of potassium hydroxide added in order to neutralize acetic acid bonded to 1 g of acetylated resin. A hydroxyl group value of the alkyd resin is preferably at least 50 and at most 100. Sufficient tackiness is hardly obtained if the hydroxyl group value is less than 50 while rolling resistance is deteriorated when the value exceeds 100. All the four palm oil-based alkyds were synthesized with a hydroxyl value between 82–84 mg KOH g^{-1}, containing 1.5×10^{-3} mol g^{-1} of –OH groups. Compared to the –COOH concentrationranges from 2.7×10^{-4} mol g^{-1} to 3.6×10^{-4} mol g^{-1} of –COOH groups, these alkyds were invariably formulated with excess hydroxyl groups. In fact, these alkyds contain both hydroxyl and carboxylic groups. Reactions occur mainly *via* the carboxylic groups. At high temperatures of about 170°C, similar to the conditions of vulcanization, the hydroxyl groups can undergo dehydrogenation in order to generate –C=C– groups which could then involve with cross-linking reactions. These reactions appear to account for the improvements of tire compound properties, such as better wet grip and lower rolling resistance.

The softening point is the temperature at which a material softens or becomes less viscous beyond some arbitrary softness or viscous.A softening point of the alkyd resin is preferably at least –40°C or at most 30°C in order to obtain sufficient tackiness and to avoid excessive increase in hardness and rolling resistance which then will further lower its wet grip performance.

In Table 3, the molecular weight indicates a molecular weight calculated in terms of polystyrene measured by the gel permeation chromatography (GPC). Preferably, the number average molecular weight (M_n) of the alkyd resin is within 250–5000, while the weight average molecular weight (M_w) of the alkyd resin is within 450–10000. If $M_n$orM$_w$is less than the lower range of the preferred limit, it tends to cause bleeding in alkyd resin or rubber foamsdue to the volatilization of alkyd resin during vulcanization. However, hardness is excessively increased, grip performance, particularly wet grip performance is lowered, and fuel efficiency is deteriorated when M_n or M_w exceeds its higher limit.

In the rubber composition for a tire, the content of alkyd resin is preferably within a range of 0.5–10 parts by weight with respect to 100 parts by weight of the rubber component. If the alkyd resin is more than 0.5 parts by weight, both the suppressing reversion and tackiness imparting effects are particularly good. Conversely, reduced abrasion resistance and cost increase can be prevented well if the content of alkyd resin less than 10 parts by weight. For a rubber composition for a tread, the amount of the alkyd resin is preferably 0.5–10 parts by weight and the amount is preferably 1 to 10 parts by weight when used for a composition for a sidewall.

Figure 4 matches the tack strength performance of palm kernel oil alkyd resins as a tackifier in rubber compound againsta reference (the same ENR compound without alkyd), polyterpene (a β-pinene resin with softening point of around 30°C) and petroleum resin (C5 type aliphatic hydrocarbon resin withsoftening point around 100°C), two of the very common tackifiers in the tire industry. These tackifiers were mixed into the ENR50 compound at 2.5 phr and 5 phr, in combination with the other standard vulcanizing chemicals in a Banbury Mixer (BR type) at 77 rpm maintained at 20°C using chiller water. The tackiness was evaluated in the unvulcanized state by PICMA tack tester from Toyo Seiki Seisaku-Sho Ltd., at a separating speed of 30 mm min^{-1} and a measurement time of 2.5s. It was shown that the performance of AlkFA15 could already match the polyterpene, petroleum resin and reference and the performances of AlkFA25 and AlkFA35 were superior to the other tackifiers.

With reference to the alkyd compositions in Table 2, phthalic anhydride was the only petroleum based chemical, while the other constituents could be considered as natural materials. As portion of the phthalic anhydride was replaced by fumaric acid, the natural content of the alkyd has increased, and the tack strength has also improved. Figure 5 is an empirical fit of the tack strength with the fumaric acid content in the alkyd. The tack strength achieved a maximum for AlkFA25 which contained around 5% by weight of fumaric acid in the alkyd.

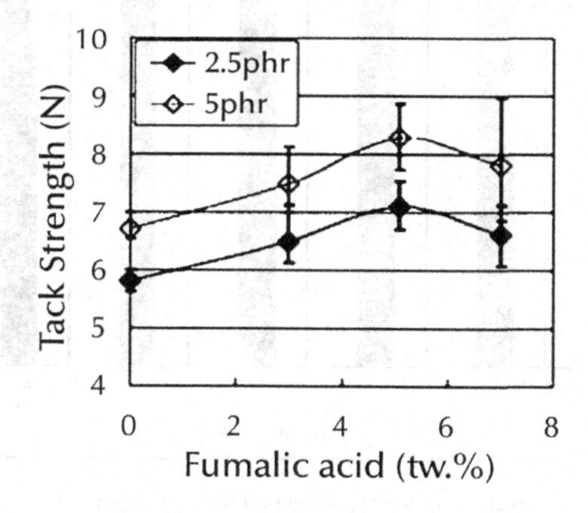

FIGURE 5 Empirical fit of the tack strength with the fumaric acid content.

Figure 6 displays the effect of tackifier on abrasive resistance of rubber compound as the tread rubber of tire. The rubber compound was vulcanized at 170°C for 12 min and then was then evaluated as tread rubber in tire under the conditions of loading of 2.5 kg, a slip ratio of 40%, a temperature of 20°C, and a measuring time for 2 min using a Lambourn abrasion tester manufactured by Iwamoto Seisakusho Co., Ltd. The abrasive wear of the tire tread of passenger cars and trucks travelling on common roads is characterized by its abrasion when there is a direct contact of a vehicle with the road. From Figure 6, the incorporation of polyterpene or petroleum resin had cause the abrasive resistance to be reduced slightly. The reduction of abrasive resistance was much less with the alkyd tackifiers. In fact, AlkFA15 and AlkFA25 showed almost no reduction of abrasive resistance at 2.5 phr. In comparison to the results at 5 phr, higher tackifier generally lowers the abrasive resistance and may cause faster wear out of the rubber. The lowering of abrasive resistance is around 4% for the polyterpene or the petroleum resin, but less than 2% for the alkyds. This shows that higher tackifier generally lowers the abrasive resistance and may cause faster wear out of the rubber.

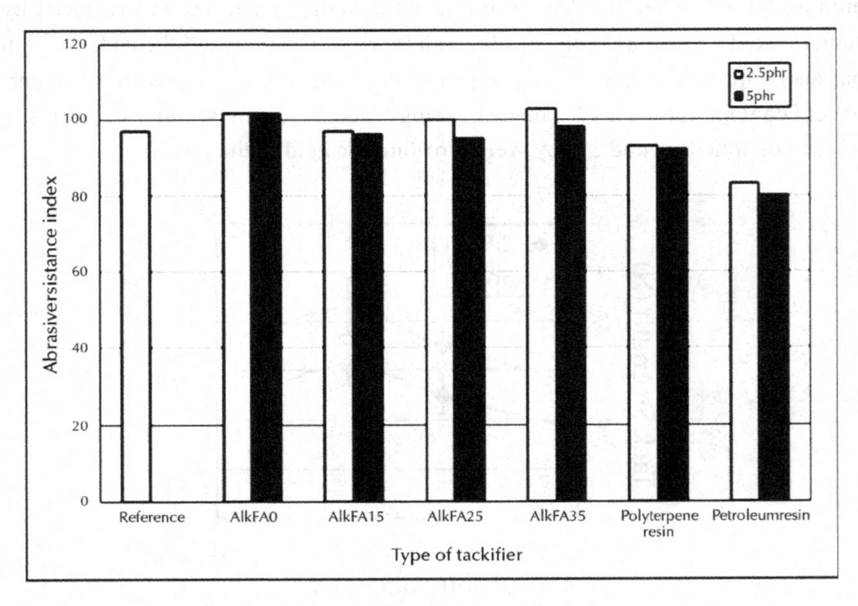

FIGURE 6 Effect of tackifier on abrasive resistance of rubber.

The WET μ and Tan δ were measured in the evaluation of a rubber compound as the tread rubber of tire for the various rubber compounds containing the different tackifiers at the same level of 2.5 phr. The WET μ, evaluated by a flat-belt type friction tester, refers to the coefficient of friction between the wet surfaces, and would reflect the tire performance in road gripping under wet conditions. On the other hand, the Tan δ would be related to the rolling resistance of the tire compound and was measured by viscoelasticity tester.In general, as shown in Table

4, incorporation of a tackifier can improve road gripping of the tread rubber as indicated by a higher value of WET . AlkFA15 and AlkFA25 showed better performance than the polyterpene and petroleum resin. All the alkyd tackifiers were able to reduce the rolling resistance as reflected by the smaller Tanδ. The best road gripping was achieved by AlkFA25, while the best reduction of rolling resistance was achieved by AlkFA35. In fact, the lower the rolling resistance the better the tire performance in fuel saving.

TABLE 4 Effect of incorporation of 2.5phr of tackifier on tread rubber

Tackifier	Reference	Polyterpene	Petroleumresin	AlkFA15	AlkFA25	AlkFA35
Wet μ	1.36	1.40	1.42	1.47	1.50	1.38
Tan δ	0.13	0.13	0.14	0.09	0.09	0.08

17.5 CONCLUSION

The palm oil-based alkyds could be used as tackifiers to replace polyterpene and petroleum resin in ENR rubber compound. They could provide better tack strength, impart better road gripping and reduce rolling resistance in tread rubber. However, they should be applied at low level to avoid reducing the abrasive resistance of the rubber.

KEYWORDS

- **Alkyd-based coating**
- **Esterification**
- **Isophthalic acid**
- **Pentaerythritol**
- **Polymerization process**

REFERENCES

1. Anonymous. *The Chemistry and Processing of Alkyd Resins*, Monsanto Chemical Co (now Solutia, Inc), New York, (1962).
2. Baker, C. S. L., Gelling, I. R., andAzemi, S. J. *J Nat Rubb Res*, **1**(2), 135 (1986).
3. Bhow, N. R. and Payne, H. F. *IndEngChem*, **42**, 700(1950).

4. Coran, A. Y. In *The Science and Technology of Rubber* (Chap. 7).J. E.Mark,B. Erman, and F. R.Eirich(Eds.), Elsevier Academic Press, New York(2005).
5. Cowan, J. C. *J Am Oil Chemists' Soc*, **31**, 529 (1954).
6. Earhart, K. A. *IndEngChem*, **41**, 716 (1949).
7. Flory, P. J. *Principles of Polymer Chemistry* (Chap. 11). Cornell Univ Press, Ithaca, New York(1953).
8. Gan, S. N. andBurfield, D. R. *Polymer*,**60**, 1903 (1989).
9. Gan, S. N. *Modification of Rubber by Alkyds*. Malaysia Patent, PI20060622, UM, 15-Feb-2006.
10. Gelling, I. R. *J Nat Rubb Res*, **6**, 184 (1991).
11. Goldsmith, A. H. *IndEngChem*, **40**, 1205 (1948).
12. Greaves, J. H. *Oil Colour Trades J*, **113**, 949 (1948).
13. Hashim, A. S., andKohjiya, S. *J PolymSci,PolymChem Ed*, **32**, 1149 (1994).
14. Hattori, T.,Terakawa, K., Ichikawa, N.,Sakaki, T.,Gan, S. N., and Lee, S.Y. *Rubber compositionfor tire, andpneumatic tire using the same*. Japan Patent, JP2007119629A, Joint patent of UM & SRI, 28-Oct-2005.
15. Hattori T., Choong D. H., Gan S. N., and Lee S. Y.*The effect on tackiness and cure of rubber compounds by alkyd resins synthesized from vegetable oil*. 14th Polymer Material Forum, the Society of Polymer Science, Japan, 15-Nov-2005, [Preprint of 14th Polymer Material Forum (1PD19)].
16. Hattori, T.,Terakawa, K., Ichikawa, N.,Sakaki, T.,Gan, S. N., and Lee, S.Y. *Rubber composition for tire and pneumatic tire using the same*. Malaysia Patent, PI 20070671, Joint patent of UM & SRI, 30-April-2007.
17. Hattori, T.,Terakawa, K., Ichikawa, N.,Sakaki, T.,Gan, S. N., and Lee, S. Y. *Rubber compositionfor tire andpneumatic tire using the same*.Malaysia Patent, PI 20070672, Joint patent of UM & SRI, 30-April-2007.
18. Hattori, T.,Terakawa, K., Ichikawa, N.,Sakaki, T.,Gan, S. N., and Lee, S.Y. *Rubber composition, rubber compositionfor tire, and pneumatic tire using the same*.US Patent, US 2007/0100061 A1, Joint patent of UM & SRI, 3-May-2007.
19. Hattori, T.,Terakawa, K., Ichikawa, N.,Sakaki, T.,Gan, S. N., and Lee, S.Y. *Rubber compositionfor tire and pneumatic tire using the same*.European Patent, EP 1782966 (A1), Joint patent of UM & SRI, 9-May-2007.
20. Hattori, T.,Terakawa, K., Ichikawa, N.,Sakaki, T.,Gan, S. N., and Lee, S.Y. *Rubber compositionfor tire, and pneumatic tire using the same*.China Patent, CN1958669A, Joint patent of UM & SRI, 9-May-2007.
21. Hattori, T.,Terakawa, K., Ichikawa, N.,Sakaki, T.,Gan, S. N., and Lee, S.Y. *Rubber compositionfor tire, and pneumatic tire using the same*.China Patent, CN1958669B, Joint patent of UM & SRI, 15-June-2011.
22. Hattori, T.,Terakawa, K., Ichikawa, N.,Sakaki, T.,Gan, S. N., and Lee, S.Y. *Rubber compositionfor tire and pneumatic tire using the same*.European Patent, EP 2363304 (A2), Joint patent of UM & SRI, 7-Sept-2011.
23. Hiles, R. C., Golding, B., and Shreve, R. N. *IndEngChem*, **47**, 1418 (1955).
24. Karim, A. F., Golding, B., and Morgan, R. A. *J ChemEng Data*, **5**, 117 (1960).
25. Kaska, J. andLesek, F. *Prog Org Coat*, **19**, 283 (1991).
26. Kienle, R. H. *IndEngChem*, **22**, 590 (1930).
27. Kraft, W. H. *Am Paint J*, **28**, 41 (1957).

28. Lambourne, R. *Paint and Surface Coatings, Theory and Practice*(Chap 1).Horwood Limited, New York(1987).
29. Lee, S. Y.,Gan, S. N., Hassan, A., Tan, I. K. P.,Terakawa, K., and Ichikawa, N.*The Adhesion Properties of Natural Rubber Pressure-Sensitive Adhesives Using Palm Kernel Oil-Based Alkyd Resins as a Tackifier*. Composite Interfaces(2013).
30. Loo, C. T. *Vulcanization of Epoxidised Natural Rubber with Dibasic Acids*.ProcIntRubbConf 1985; Rubber Research Institute of Malaysia, Kuala Lumpur, 368.
31. Moore, C. G. *JAm Oil Chemists's Society*, **27**, 510 (1950).
32. Munk, P. *Introduction to Macromolecular Science* (Chap. 2). John Wiley & Sons, New York, (1989).
33. Paul, S. *Surface Coatings, Science and Technology*, John Wiley & Sons, New York, **69** (1986).
34. Patton, T. C. *Alkyd Resin Technology*; Interscience Publishers, New York, Chap. 1(1962).
35. Patton, T. C. *Alkyd Resin Technology*; Interscience Publishers, New York, Chap. 2(1962).
36. Patton, T. C. *Alkyd Resin Technology*; Interscience Publishers, New York,Chap. 7(1962).
37. Rheineck, A. E. and Austin, R. O.In *Treatise on Coatings*, R. R. Myers andJ. S. Long(Eds.), Marcel Dekker, New York, pp. 181–248 (1968).
38. Sheehter, L. andWynstra, J. *IndEng Chem.*, **47**, 1602 (1955).
39. Subramaniam, A. *RubbChemTechnol*, **45**, 346 (1972).
40. Subramaniam, A. Molecular Weight and Other Properties of Natural Rubber: A Study of Clonal Variations, ProcIntRubbConf 1975; Rubber Research Institute of Malaysia, Kuala Lumpur, 3.
41. C. J. A.Taylor and S.Marks(Eds.)*Convertible Coatings*, Part 3; Chapman & Hall, London, Chap. 5 (1972).
42. VonFischer, W. andBobalek, E. G. *Organic Protective Coatings*, Reinhold: New York, Chap. 16 (1953).
43. Weiss, K. D. *ProgPolymSci*, **22**, 203 (1997).
44. Wicks, Z. W., Jones, F. N., and Pappas, S. P. *Organic Coatings*, John Wiley & Sons: New York, Chap. 15 (1999).

CHAPTER 18

MISCIBILITY, THERMAL PROPERTIES, AND ION CONDUCTIVITY OF POLY(ETHYLENE OXIDE) AND POLYACRYLATE

LAI HAR SIM, SITI ROZANA BT. ABD. KARIM, and
CHIN HAN CHAN

CONTENTS

18.1 MISCIBILITY, THERMAL PROPERTIES, AND ION CONDUCTIVITY OF POLY(ETHYLENE OXIDE) AND POLYACRYLATE

In the 20th century, after World War II, the great demand for materials in the construction, engineering, automobile, as well as domestic sectors have exhausted natural resources for materials. The dire need to replace conventional natural materials by synthetic materials to cater for the fast pace of industrial development and to meet the demand for cheaper yet good quality materials give rise to the era of plastics and polymeric materials. In the 1950s, the production of synthetic polymers flourished rapidly and became the most important industry in the world (Utracki, 2002). In view of its great demand, polymer technology has invariably diversified from single polymer to combinations of polymers, incorporation of additives, and reinforcements. Apart from acquiring synergistic effects, blending of polymers aims at widening, and modifying the range of properties of existing polymers such that new materials with the desired combination of properties for specific usage can be developed. Polymer blends are the physical mixtures of two or more chemically different homopolymers or copolymers. Blending is, in recent years, a well-established, easy processing, and cost-effective way for obtaining tailor-made polymeric systems (Utracki, 1989). In addition, it offers the means for industrial and/or municipal plastic waste recycling. The performance of polymer blends depends very much on the properties of the polymeric components, phase behavior, and the blend morphology. Optimization of the polymer blend performance can be done through the control of blend morphology *via* molecular structures of the components, blend compositions, and processing conditions (Paul and Newman, 1979, Utracki, 1991: Dumolin, 2002).

The study of polymer-polymer miscibility has attracted great interest both commercially and academically because of its effect on the glass transition temperature (T_g), a basic property of polymer, which plays an important role in determining the overall physical properties of polymer blends (Olabisi et al., 1979). A miscible binary blend refers to a molecularly homogeneous system consisting of mixed segments of the two blending constituents at equilibrium whereas distinct phases of two constituents coexist in an immiscible binary system. Polymers, especially those with high molecular weights, are generally immiscible. Therefore, most of the multicomponent polymer systems commercially utilized today are two-phase blends. Poly(vinyl chloride) (PVC)/isotactic poly(methyl methacrylate) (*i*-PMMA) (Forger, 1977) and poly(butadiene-*co*-styrene) (PBS)/PVC (Grundmeier et al., 1977) are two of the numerous important examples of commercial immiscible blends which possess improved heat distortion temperatures, flame-retardant properties, notched impact strength, and excellent mechanical properties compared to either of the blending components. The drawback of two-phase blends is the interfacial adhesion between the respective phases governing the ultimate mechanical properties attainable with the blends. However, compatibility between two immiscible polymers can be improved through grafting and the use of block copolymers (Longworth, 1967: Robeson et al., 1973).

Poly(ethylene glycol) (PEG) and poly(ethylene oxide) (PEO) are polymers of ethylene oxide of different chain lengths and molecular weights (M_w). Historically, PEG refers to polyether with M_w < 100,000 g mol^{-1} while PEO is the long chain PEG with M_w >100,000 g mol^{-1} (Bailey and Koleske, 1976; Hamon et al., 2001). These polymers of ethylene oxide are thermoplastic polyether synthesized by anionic polymerization of ethylene oxide monomers. While PEG and PEO with different molecular weights find use in different applications and have different physical properties like viscosity due to chain length effects, their chemical properties are nearly identical. PEG is widely used in pharmaceutical applications, cosmetics (tooth paste, lotions, and so on), and medical research due to its non-toxic and biodegradable characteristics whereas, PEO with high stability, viscoelasticity, amphiphilic, and high corrosion resistance, is always chosen as water soluble resin, soil stabilizer, adhesives, and so on. On the whole, PEO is the most extensively investigated polyether both industrially and academically especially as polymer blends and polymer electrolytes because of its ability to form miscible blends with numerous amorphous polymers and to solvate high concentrations of many inorganic salts, respectively (Hamon et al., 2001; Rajendran et al., 2002a). To date, PEO is still being actively studied to develop its potentiality as solid polymer electrolyte (SPE) in lithium ion rechargeable batteries, fuel cells, electrochromic windows, and smart windows (Pitawala et al., 2007). The miscibility, thermal properties, and morphology of the blends of amorphous PMMA with a crystallizable component like PEO have been well documented (Hamon et al., 2001; Li and Hsu, 1984; Martuscelli et al., 1986; Silvestre et al., 1987; Marco et al., 1990; Talibuddin et al., 1996).

Polyacrylate, a thermoplastic polymer, synthesized by polymerization of acrylate monomers, consists of acrylic acid (AA) esters shown in Figure 1 as the general structure. Methacrylate monomer differs from the acrylate monomer in that the H atom at the vinyl oxygen is substituted by a methyl group ($-CH_3$). The wide variety of side chains and the variations in tacticity contribute to polyacrylate, a wide range of M_w with each range having specific applications. In recent years, the consumption of polyacrylates such as poly(methyl acrylate) (PMA), poly(ethyl acrylate) (PEA), poly-2-ethylhexyl acrylate (PEHA), poly(butyl acrylate) (PBA), poly(aryl methacrylate) (PArM) and PMMA in the manufacturing industry as fiber modifier, coating material, and raw material for the formulation of paints and adhesive, is rising tremendously (Gaborieau et al., 2007; Huang et al., 2002). Among the broad variety of available acrylic polymers, PMMA is the most extensively investigated in the last decade. Owing to its crystal clear transparency, excellent weatherability, moderate stiffness, and toughness (Ulrich, 1982), PMMA is widely used in consumer goods, in medicine as dental cements, and bone reconstruction (Hanake, 2004), in aircraft industry as glazing material (Kennedy, 1993). Besides, its affinity for cations like lithium ions has in the past decade attracted intensive research on its potential applications as polymer electrolytes (Rajendran et al., 2002b).

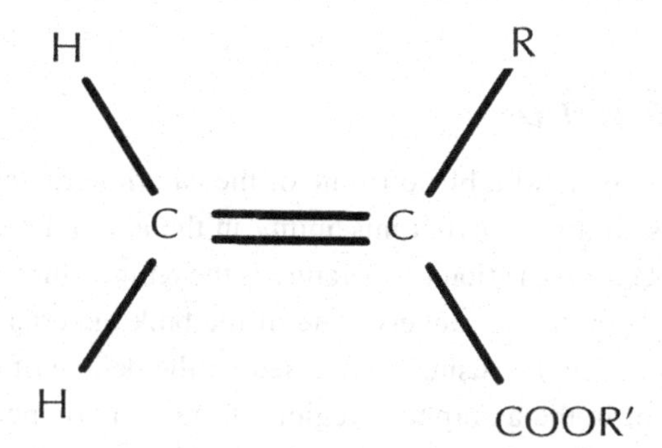

FIGURE 1 General formula for acrylate and methacrylate monomer with R'=H and CH_3, respectively.

18.2 MISCIBILITY

Miscibility refers to the molecular mixing of the components down to the level adequate to yield macroscopic properties expected of a single phase system (Olabisi et al., 1979). Mutual miscibility or immiscibility of the two components in the melt and in the amorphous state of a crystallizable binary blend plays a vital role in the blend's microstructure, crystallization behavior, and degree of dispersion of one component in the matrix of another as well as inter-phase adhesion (Silvestre et al., 1996).

The most commonly used and widely accepted experimental methods are the calorimetric determination of single composition dependant (T_g) and equilibrium melting temperature (T_m^o). Microscopy methods such as polarized optical microscopy (POM) (Kryszewski et al., 1973), scanning electron microscopy (SEM) (Walters and Keyte, 1965), and transmission electron microscopy (TEM) (Smith and Andries, 1974) have been frequently used as a preliminary indication of the extent of miscibility in polymer blends. Scattering methods like small angle neutron scattering (SANS) (Kruse et al., 1976) and small angle X-ray scattering (SAXS) (Fischer et al., 1976) enable the elucidation of the chain conformation of the blends. Spectroscopic techniques such as nuclear magnetic resonance (NMR) and Fourier transform infrared spectroscopy (FTIR) are useful tools in demonstrating miscibility by providing a valuable insight into the nature of the specific interactions between the components in polymer blends. Each of these techniques used for detecting miscibility has its own strength and limitation with resolutions ranging from segmental level to micrometer domains (Utracki, 2002; Gedde, 1995).

18.2.1 FOX EQUATION

The determination of T_g of a blend is one of the calorimetric techniques used to elucidate the miscibility or partial miscibility in the amorphous phase of binary polymer blends. Glass transition temperature is the temperature at which the transition from the glassy to the rubbery state of the bulk material takes place. The establishment of miscibility using T_g is based on the degree of dispersion of the second component in the amorphous region of the first component and that the size of the disperse phase domain is ≤ 15 nm (Silvestre et al., 1996; Shultz and Young, 1980). It is noteworthy that blends which exhibit a T_g are miscible whereas phase-separated binary blends demonstrate two T_gs corresponding to those of the neat components throughout the whole composition range. Fox equation (Fox, 1956) as given in Equation (1) is widely used to predict the monotone dependence of T_g on composition of binary blends. According to this equation, the T_g of a miscible binary blend can be determined, to a good approximation, by the following relationship:

$$\frac{1}{T_g} = \frac{W_1}{T_{g1}} + \frac{W_2}{T_{g2}} \tag{1}$$

where T_g, T_{g1}, and T_{g2} in Kelvin are the glass transition temperatures of the blend, pure component 1 and pure component 2, while W_1 and W_2 are weight fractions of component 1 and component 2, respectively. However, this method is reliable in ascertaining polymer-polymer miscibility if only the content of the second component is greater than 10%, and the T_gs of the two neat constituents must differ by at least 20 °C (Utracki, 1989; Olabisi et al., 1979).

18.2.2 FLORY-HUGGINS MODEL

From the thermodynamic viewpoint, miscibility is determined by the molar Gibb's free energy of mixing (ΔG_{mix}) which in turn is governed by the combinatorial molar entropy of mixing (ΔS_{mix}) and molar enthalpy of mixing (ΔH_{mix}) as shown in the thermodynamic relationship below where T is the absolute temperature:

$$\Delta G_{mix} = \Delta H_{mix} - T\Delta S_{mix} \tag{2}$$

For binary polymer mixtures in which the blending components possess polar functional groups and specific intermolecular interactions such as dipole-induced dipole, dipole-dipole, hydrogen bonding, ion-dipole, and so on which are exothermic processes operate between the components, thus achieving a negative ΔH_{mix}. On the other hand, ΔS_{mix} is obviously positive and unfavorable as the molecular chains transform from a random state to an ordered arrangement. Since ΔS_{mix} has an infinitesimally small contribution to favorable mixing of polymers especially those of high molecular weight. Therefore, intermolecular interactions between components of polymer mixtures, in some cases, play a deciding role in thermodynamic miscibility. As a consequence, polymer-polymer miscibility is temperature dependent. The thermodynamic miscibility can be summarized as:

$$\Delta G_{mix} \approx \Delta H_{mix} < 0 \text{ since } S_{mix} \approx 0$$

The Flory-Huggins relation as given in Equation (3) is used to describe thermodynamically polymer-polymer miscibility (Huggins, 1942; Flory, 1942; Utracki, 1962; Koningsveld, 1967; McMaster, 1973):

$$\Delta G_{mix} = RT \ \frac{\Phi_1}{N_1}\ln\Phi_1 + \frac{\Phi_2}{N_2}\ln\Phi_2 + \chi_{12}\Phi_1\Phi_2 \tag{3}$$

where Φ and N_i are the volume fraction and the degree of polymerization of component "i", χ_{ij} is the segmental binary interaction parameter or commonly known as the Flory-Huggins interaction parameter while R is the gas constant. The first two logarithmic terms in Equation (3) give the combinatorial entropy of mixing (ΔS_{mix}) which, as mentioned earlier, is insignificantly small especially for high molecular weight polymers (that is, N_i is relatively large). Therefore, the miscibility or immiscibility of the blend mainly depends on the last term of the equation—the Van Laar-type expression, which represents the exothermic ΔH_{mix}. On the whole, most binary polymer blends, according to the Flory-Huggins equation, achieve thermodynamic miscibility through specific interaction ($\Delta H_{mix} < 0$) characterized by a large negative value of χ_{ij} that increases with temperature.

18.2.3 HOFFMAN-WEEKS EQUATION

Another calorimetric technique in determining the miscibility of semicrystalline-amorphous and semicrystalline-semicrystalline blends is the progressive suppression of T_m^o of one component by ascending content of the second component in the blend. T_m^o of a polymer is defined as the melting point of an assembly of large crystals, with negligible surface effects, and in equilibrium with the polymer liquid. The crystallization of a polymer can only proceed in a temperature below T_m^o on the ground that crystal nucleation is greatly inhibited at the proximity of T_m^o. The T_m^o of a polymer can be determined experimentally by step-wise annealing procedure after Hoffman-Weeks (Hoffmann and Weeks, 1962; Groeninckx et al., 2002). According to the step-wise procedure, the sample is isothermally crystallized in a range of crystallization temperatures (T_c). Subsequently, the sample is allowed to crystallize at the same range of T_cs for a reasonable period of time, for example, five half-times ($5t_{0.5}$), and the corresponding melting temperatures (T_ms) are obtained from the peak of the endotherms. Half time of crystallization ($t_{0.5}$) is the time taken for 50% of the crystallinity of the crystallizable component to develop. The use of $5t_{0.5}$ serves to impose equivalent thermal history to the sample at all T_cs (Chan et al., 2004; Tan et al., 2006). Figure 2 shows a linear plot of the apparent or experimental T_m versus T_c, normally known as the Hoffman-Weeks plot. Extrapolation of the Hoffman-Weeks plot to intersect with the theoretical linear curve of $T_m = T_c$ yields the value of T_m^o.

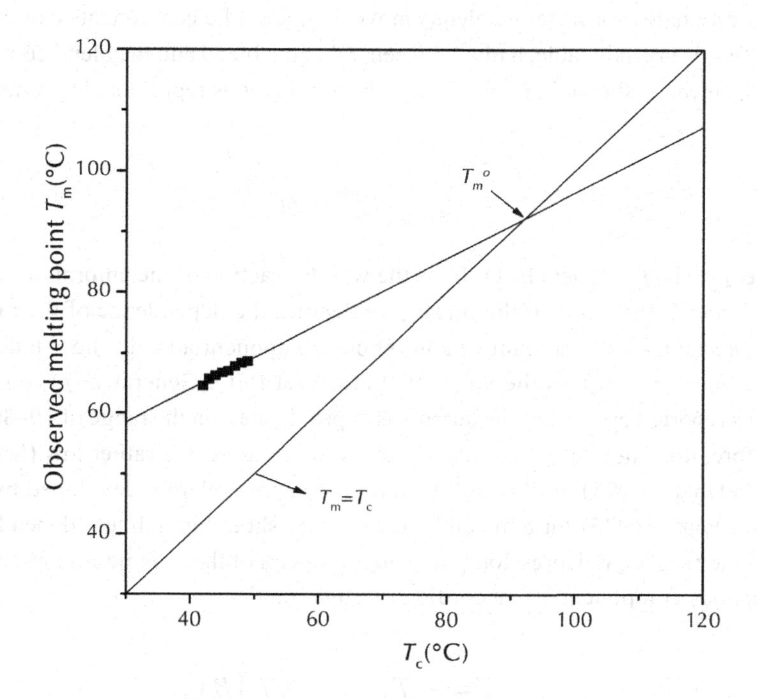

FIGURE 2 Hoffmann-Weeks plot. The intercept point is the T_m^o.

For perfect crystals of finite sizes and no occurrence of recrystallization, the linear relationship between T_m and T_c can be described by Equation (4) (Hoffmann and Weeks, 1962):

$$T_m = \langle 1 - \frac{1}{\gamma} \rangle T_m^0 + \frac{1}{\gamma} T_c \quad (4)$$

The stability parameter, $\frac{1}{\gamma}$ which is a morphological factor, assumes values between 0 and 1. $\frac{1}{\gamma} = 0$ implies that $T_m = T_m^o$ for all T_c which means the crystals are most stable. On the other hand, the crystals are inherently unstable when $\frac{1}{\gamma} = 1$ which results in $T_m = T_c$.

Additionally, for miscible blends in which one of the components (for example, PEO) is crystallizable, a plot between T_m^o of the blend and the blend compositions is linear as shown in Figure 3. The linear relation is represented by Equation (5).

$$T_m^o = T_m^{PEO} - \beta W_2 \tag{5}$$

where T_m^{PEO} is T_m^o of neat PEO, W_2 is the weight fraction of the amorphous component and β, the slope of the linear plot denotes the dependence of the rate of T_m^o depression on the amount of amorphous component added. The *y*-intercept of the linear plots gives the value of T_m^o for neat PEO. Generally, T_m^o for neat PEO as reported by most of the authors is approximately in the range of 70–80 °C, therefore, the value of T_m^o at ≈ 64 °C observed in Figure 3 is rather low (Katime and Cadenato, 1995). Following the thermodynamic relation developed by Nishi and Wang (1975) for a miscible blend, there should be a linear dependence of the melting point depression (ΔT_m) on the square of the volume fraction of the amorphous component v_2^2 according to Equation (6):

$$\Delta T_m = -\left(T^{PEO} V_1 / \Delta H_1\right) B\, v_2^2 \tag{6}$$

where V_1 and ΔH_1 are the molar volume and ΔH_m for the crystallizable component, respectively. The parameter B is related to the binary Flory-Huggins interaction parameter χ_{12} as shown in Equation (7):

$$B = RT\chi_{12} / V_2 \tag{7}$$

where V_2 and χ_{12} denote the molar volume of the amorphous component and the Flory-Huggins interaction parameter, respectively.

Based on Equation (6), parameter B can be evaluated as $\frac{\Delta H_1}{V_1}$ gives the latent heat of fusion for 100% crystalline component per unit volume. The value of χ_{12} can then be calculated by substituting the value B into Equation (7).

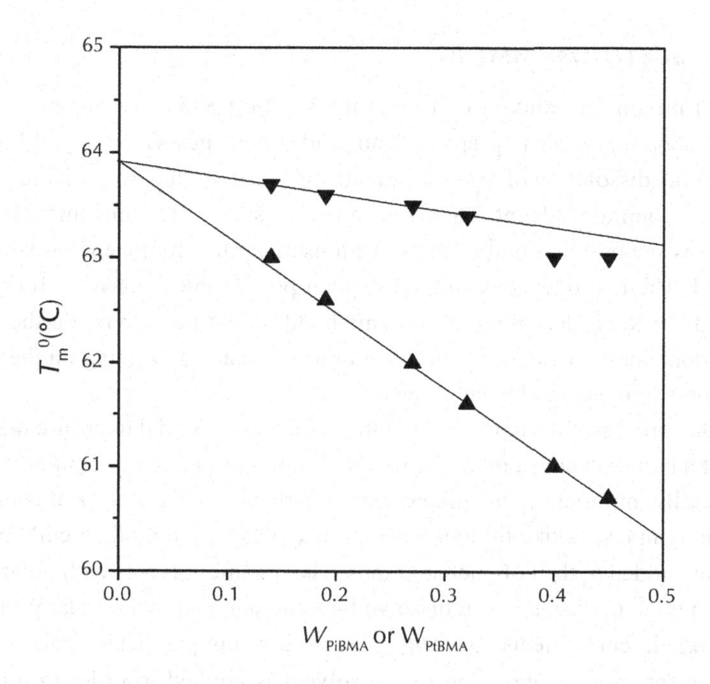

FIGURE 3 T_m^o of PEO/PiBMA and PEO/PtBMA as a function of weight fractions of (▲) PiBMA and (▼) PtBMA, respectively. The *y*-intercept point for both the linear plots denotes T_m^o of neat PEO. Modified from Katime and Cadenato, (1995).

18.3 PREPARATION OF BLENDS AND COMPOSITES

Blends and composites can be processed by solvent-free techniques (Dey et al., 2009; Wang et al., 2010a; Wang et al., 2010b; Smith and Hashemi, 2011) like hot press, extrusion, injection molding, and so on, the solution cast method (Kumar et al., 2011; Zhou, et al., 2011; Takahashi and Tadokoro, 1973; Zhu et al., 2000; Mohan et al., 2007; Chaurasia et al., 2011; Kim et al., 2012) depending on the type of fillers used and the form of polymer matrix. While organically polymer hybrids favor the melt-processed techniques, most of the composite polymer electrolytes (CPE) are prepared by the solution cast technique in which the ionic salt and the polymer are added to sonically dispersed ceramic fillers in a common organic solvent. The common blend and composite preparation methods are described in detail.

18.3.1 SOLUTION MIXING

Solution mixing or solution casting is the simplest and most common method used in laboratory scale to prepare blends and composites. Generally, this process involves the dissolution of a certain quantity (in ratio) of the two polymer components in a common solvent. The mixture is then stirred at a convenient temperature below the boiling point of the solvent used, until a homogeneous solution is obtained. The homogeneous solution is then poured into a glass or Teflon Petri dish and left to air dry either in the fume hood or in a glove box. Further drying can be done in an oven, a vacuum oven or a desiccator depending on the characterization parameters to be determined.

In this process, the choice of a common solvent is crucial to ensure the attainment of a homogeneous polymer solution. A good inspection of common solvent is especially important in the preparation of polymer electrolytes and composites when inorganic salts like lithium perchlorate ($LiClO_4$) and inorganic fillers, silica (SiO_2) are added to the polymer solutions. The organic solvents with polar groups will be a good choice as it can dissolve both the polymers with polar groups and the inorganic components. In solution casting technique, besides choosing the right solvent, slow evaporation of the solvent is applied in order to produce a smooth and dry thin film without phase boundary or bubbles forming on the film. Since this method requires a large amount of organic and usually toxic solvents to be removed by evaporation, the usage for this method is usually limited to laboratory scale work. However, this technique is applied to prepare thin membranes, surface layers, and paints on the industrial scale (Wiley, 2011).

18.3.2 MELT MIXING

Melt mixing is the most widely used method to prepare polymer blends and composites both on the laboratory or industrial scale where large polymer sample size is available (Wiley, 2011; Robenson, 2007). Generally, this process involves the mixing of the blend components in the molten state using an extruder or a batch mixer. Most of the commercial plastics in the market are produced by melt mixing thermoplastics. There are many different types of extruders available in the market but one type of extruder can be used to process different types of polymer blends. The advantage of this method is the capability of processing a large sample size without the use of organic solvents. Meanwhile, if the blends are prepared in solvents, then the solvent has to be volatilized before the melt mixing process (Robenson, 2007). Nevertheless, the drawback of this process is the high energy consumption leading to the high cost of production. Furthermore, mixing the blend components at high temperature may cause chemical changes in the molecular level of the polymer chain resulting in the retardation of some of the useful properties such as crystallization rate and morphology of the polymer blend.

18.3.3 HOT ISOSTATIC PRESS (HIP) METHOD

The HIP or hipping or hot-press method as it is normally known is a technique that involves two important parameters—temperature and pressure. In this technique, the sample is subjected to a very high temperature (up to 600 °C) and pressure so as to produce a thin film with uniform thickness (Wiley, 2011). This technique is a well known industrial manufacturing process to produce materials with minimum porosity and defects, thus improving the strength of a material. Although this process is normally used to press hard materials such as steel, alloy, and ceramics, it is also frequently employed in the preparation of polymer blends (Wiley, 2011). However, to prepare polymer blends using this technique, the polymer components have to be physically mixed by either solution casting or melt mixing until a homogeneous slurry forms before pressing it under two stainless steel blocks at elevated temperature (Agrawal et al., 2011). The thin films produced by hot-press method do show some improvement in properties especially in ion conductivity and mechanical property over those produced by solution casting method. Besides, hot-press technique promises a rapid, least expensive, and free-solvent preparation which is environmental friendly.

18.4 INSTRUMENTATION

The common instruments used in the characterization of the polymer blends are given as follows:

18.4.1 THERMAL GRAVIMETRIC ANALYSIS (TGA)

The TGA is a branch of thermal analysis which examines the mass loss of a material as a function of temperature, time, and sometimes environment. The main usage of this instrument is to characterize the decomposition temperature (T_d) or to investigate the thermal stability of materials under a variety of conditions. In addition, it is also used to study the kinetics of the physicochemical processes of materials.

The sample size used in TGA is very small (approximately 8-10 mg). Sample is weighed then mounted on the crucible of the microbalance before heated isothermally or in dynamic temperature depends on whether thermal stability is studied as a function of time or temperature, respectively, under nitrogen atmosphere. The amount of material decomposed at the end of a set range of temperature or time interval is recorded in a thermogravimetry curve. This curve is a plot of percent mass loss of a material (Δm) as a function of temperature (T) or time (t) (Hatakeyama and Quinn, 1999). The value of T_d is extracted as the onset temperature or the temperature at the inflection point of the thermogravimetry curve as depicted in Figure 4. The T_d of a polymer blend is an important parameter which provides information not only on the polymer thermal transformation after blending process but also on the processing temperature range to which degradation of the polymer blend can be avoided.

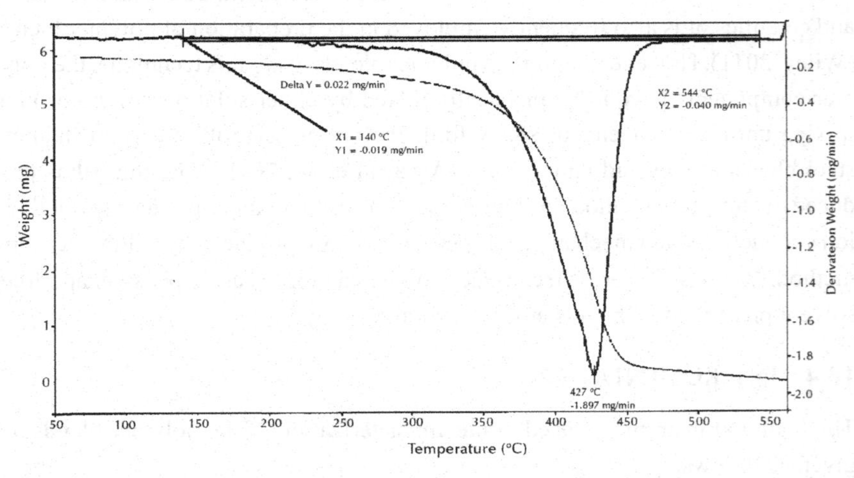

FIGURE 4 The thermogravimetric curve for neat polyacrylate (PAc) after heating to 600 °C. T_d = 427 °C.

18.4.2 DIFFERENTIAL SCANNING CALORIMETRY (DSC)

The DSC is a thermo-analytical technique in which the difference in the amount of heat required to increase the temperature of a sample and reference are measured as a function of temperature (Hatakeyama and Quinn, 1999). This technique is useful to determine the thermal properties such as T_g, melting, and crystallization behavior of polymers. Different cooling systems can be used. It is of good practice to determine the T_d of a sample before subjecting it to thermal analysis using a DSC. In this way the furnace in a DSC can be protected from contamination by any volatile residue of a degradation process in the case where a sample is heated beyond its T_d. The calorimetric processes applied to determine the miscibility of a blend are the T_g and the melting behavior as described earlier in the chapter. The value of T_g can be taken as the inflection point or the midpoint of the difference in heat capacity (∂C_p) at the reheating cycle from the DSC thermogram of the sample. In Figure 5, the two distinct T_gs of an immiscible blend of PEO/PAc 80/20 are estimated as the midpoint of ∂C_p from a DSC trace of the second heating cycle of the sample.

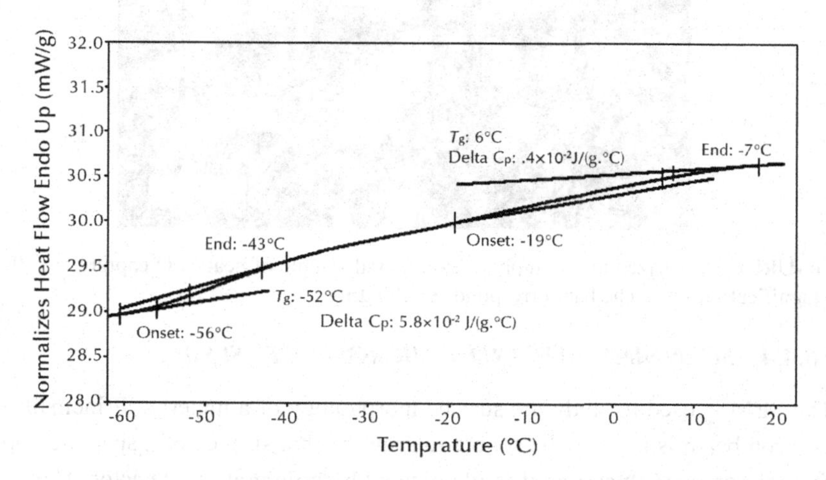

FIGURE 5 DSC trace of reheating cycle for the immiscible PEO/PAc 80/20 blend at annealing time (t_a) = 1 min. Glass transition temperatures for PEO and PAc are $T_g^{PEO} = -52$ °C and $T_g^{PAc} = 6$ °C.

18.4.3 POLARIZED OPTICAL MICROSCOPE (POM)

The POM is a conventional optical microscope (OM) added with a pair of polarizer such that the microstructure of the material or crystallites can be observed as it interacts with the polarized light. It is the most basic and widely used microscopic method to examine the features and internal surface of a material especially polymer and liquid crystals in micrometer level. The POM gives information on the grain size, grain boundary, and also the multiple phases which exist in a material

(Smith and Hashemi, 2011). It can reveal images and fine details of specimens at a range of magnifications from 2x up to 2000x (Sawyer and Grubb, 1996) depending on the type of POM used. The POM micrograph of an as-prepared sample (a completely dried film after solution casting) of neat PEO is shown in Figure 6.

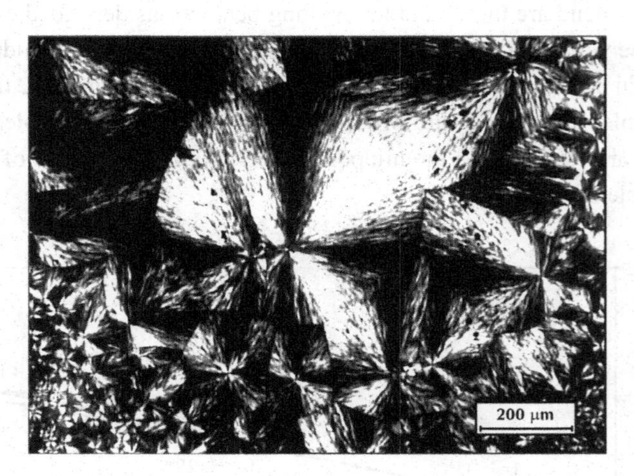

FIGURE 6 Polarized micrograph of as-prepared sample of neat PEO captured at 30 °C. Magnification—5x. The bar corresponds to 200 μm.

18.4.4 SCANNING ELECTRON MICROSCOPE (SEM)

The SEM is used to study the surface morphology of a material. Principally, an electron beam is used as the source to scan the thin surface of a specimen up to few micrometers thick and it is illuminated by a light at the detector. The SEM inspects the topographic of a sample in high range of resolution (from 20x up to 1,00,000x) and a much larger depth of field (nano scale) that leads to the production of 3D images of the samples as compared to POM. For nonconductive materials like polymers, a conductive coating on the surface or the use of low accelerating voltage is required. Besides, SEM can give more information and better understanding on the surface changes such as fracture, defect, and crystal structure of a material (Sawyer and Grubb, 1996) compared to POM.

18.4.5 TRANSMISSION ELECTRON MICROSCOPE (TEM)

The TEM is an electron optical instrument that uses electron beam with accelerating high voltages (100 kV–300 kV). It is required to be operating in vacuum since air scatters electrons. As it uses short wavelength of electrons, high resolution (0.5

nm) of image is possible. The TEM can probe bulk, thin layer of specimen until < 0.2 μm of thickness and it can be magnified up to 2×10^6 in magnitude (Sawyer and Grubb, 1996). Sample used for TEM must be of very thin layer and flat parallel surfaces in order to allow the electrons to pass through the thin specimen and because of this, the sample preparation for TEM is critical. The specimen of a sample must be properly prepared by cutting out a thin section around 3000-500 micrometer from the thick sample using electric discharge machining (EDM) and rotating wire saw before the thin layer specimen is further cut to reduce the thickness up to 50 μm using a microtome (Smith and Hashemi, 2011).

18.4.6 X-RAY DIFFRACTION (XRD) SPECTROSCOPY

The XRD is an example of X-ray wave interference. It is a non-destructive and fast technique used to study the structure of any state of matter with an X-ray beam. When a monochromatic incident beam of X-ray of wavelength λ strikes a set of planes (*hkl*) of a crystalline material, interference of the diffracted beams produces various diffraction patterns with different angles recorded as diffraction peaks. This phenomenon is known as XRD (Smith and Hashemi, 2011). The position, width and intensity of the diffraction peaks together with the wavelength of the X-ray beam applied and the set of planes obtained from the instrument provide important information on the crystal lattice structure of a material. For polymeric materials, SAXS, and wide-angle X-ray diffraction (WAXS) are usually employed to determine the degree of crystallization, crystallite size, type of crystal orientation, phase composition, and so on (Patnaik, 2004). The specimen used for XRD is usually solid samples grind to a fine powder (200–300 mesh) and shaped into thin rod after mixing with colloidal binder or tampered into a uniform capillary. Meanwhile, liquid samples must be converted into crystalline derivatives (Patnaik, 2004).

18.4.7 DYNAMIC MECHANICAL ANALYZER (DMA)

The DMA is a technique to examine the viscoelasticity or rheological properties of a material especially polymers. Rheology is the study of deformation and flow of materials. Principally, DMA can be described as applying an oscillating force to a sample and analyzing the response of the sample to the force (Fang, 2009). By definition, the applied force is the stress (σ) and the response of the materials to the stress is known as strain (γ) (Menard, 2008). The mechanical and rheological properties of polymers normally investigated by DMA or a rheometer are tensile strength, elongation break, tensile modulus, storage modulus (G'), and loss modulus (G''). The storage and loss moduli depend not only on the temperature but also on the frequency applied. Stiffness, elasticity, and modulus are important mechanical properties of polymers owing to its wide application as automotive, engineering, and construction materials. The plots of storage and loss moduli as functions of frequency for epoxidized natural rubber (ENR) with 50 mol% epoxidation (ENR-50) determined using DMA are shown in Figure 7.

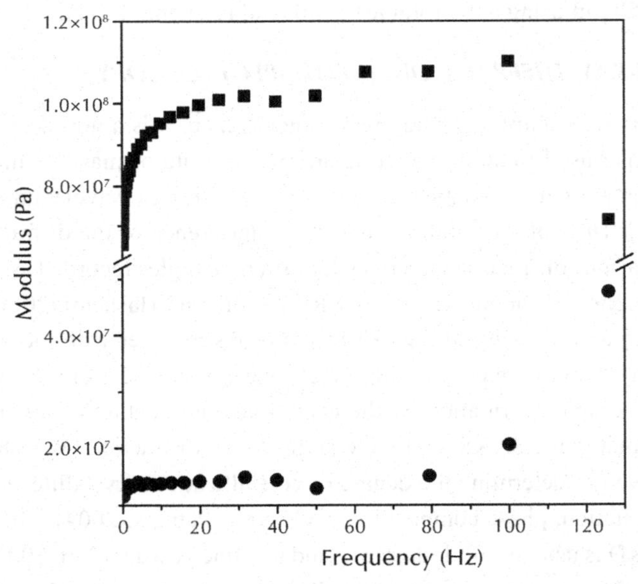

FIGURE 7 The plot of storage and loss modulus of 50 mol% of ENR-50 as a function of frequency determined using DMA. (■) and (●) represent the storage and loss modulus of ENR-50, respectively.

18.5 POLY(ETHYLENE OXIDE) (PEO)

The regularity of the linear macromolecular chains of PEO, as illustrated in its chemical structure shown in Figure 8, enables the polymer to acquire a high degree of crystallinity (X_c) of ~70–80%. The relative degree of crystallinity (X_c) for PEO applied in this book chapter is evaluated from the enthalpy of fusion (ΔH_m) of PEO extracted from the area of the melting curve of PEO in the reheating cycle of the DSC trace of the sample after Equation (8):

$$X_c = \frac{\Delta H_m}{W_{PEO} \ \Delta H_{ref}} \times 100\ \% \tag{8}$$

where $\Delta H_{ref} = 188.3\ J\ g^{-1}$, is the enthalpy of fusion of 100% crystalline PEO (Cimmino, 1990) and W_{PEO} denotes the weight fraction of PEO in the blend. For neat PEO, $W_{PEO} = 1$.

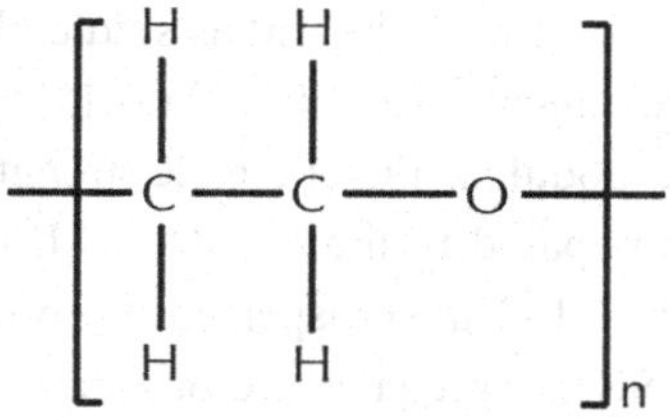

$$\left[-\mathrm{CH_2}-\mathrm{CH_2}-\mathrm{O}- \right]_n$$

FIGURE 8 Chemical structure of PEO.

It is well known that PEO spherulites have a Maltese cross extinction pattern and a very fine spherulite texture as observed under POM (Ketabi and Lian, 2012; Choi and Kim, 2004; Xi et al., 2005; Choi, 2004) (refer to Figure 9). Meanwhile, under SEM, the image of pure PEO shows waving smooth and uniform structure without any phase separation (Johan et al., 2011).

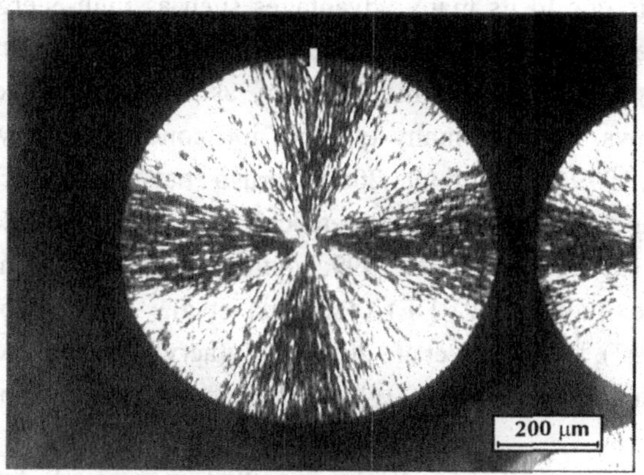

FIGURE 9 Micrograph of neat PEO captured after crystallization at $T_c = 45\ °C$, the arrow showing the Maltese cross birefringence (Sim et al., 2011).Magnification—5x. The bar corresponds to 200 µm.

In the last decade, nanocomposite has been widely investigated in the search for advanced polymeric materials with improved physical, mechanical, and electrical properties. The distribution of the nanofiller and its interaction with the macromolecular chain govern the mechanical and physical properties of the polymer nanocomposites. From the SEM micrograph, one can observe the exact location of the nano-sized fillers either within the PEO spherulite or in the inter-spherulitic regions or the amorphous region of PEO (Johan et al., 2011; Yue et al., 2009; Joykumar and Bhat, 2004). The addition of 5 wt% fumed silica to the PEO/10 wt% sodium montmorillonite (Na^+-MMT) hybrid as studied by Burgaz, (2011) shows significant property improvements. The PEO/Na^+-MMT/SiO_2 nanocomposite has higher storage and elastic modulus, higher tensile strength, good thermal stability, and lower crystallinity compared to the PEO/Na^+-MMT hybrid and neat PEO. The nano-silica aggregates (1–2 μm) dispersed uniformly within the spherulite while both the clay and SiO_2 aggregates are observed in the inter-sphrulitic region. The higher density of SiO_2 aggregates in the amorphous phase coupled with the physical cross-link sites generated by the strong H-bonding interaction between the ether oxygen of PEO and the surface silanols of the hydrophilic SiO_2 contribute mainly to the extra reinforcement and stiffness in the PEO/Na^+-MMT/SiO_2 system. The disruption in the crystallization of PEO caused by the clay particles and the H-bonding interaction between SiO_2 and PEO, lead to the formation of smaller nonisotropic crystallites.

According to Chandra and Chandra, (1994), SPE plays an important role in solid state ionic due to its many advantages such as high-energy density, leak proof, good electrolyte-electrode contact, volumetric stability, solvent-free condition, easy handling, and mouldability, as well as wide electrochemical stability windows. To date, many types of ion conducting polymers *viz.* PEO (Armand et al., 1979; Kim and Oh, 2000; Caruso et al., 2002; Rocco et al., 2002), PMMA (Floriańczyk et al., 1991; Appetecchi, et al., 1995; Bohnke et al., 1993a; Bohnke et al., 1993b; Song et al., 1999; Vondrak et al., 2004; Uma et al., 2005), poly(vinylidene fluoride) (PVDF) (Tsuchida et al., 1983; Choe et al., 1995; Jacob et al., 1997), PVC (Han et al., 2002), poly(acrylonitrile) (PAN) (Song et al., 1999; Abraham and Alamgir. 1990) have been investigated as hosts for polymer electrolytes.

Owing to its relatively simple structure that can provide fast ion transport (Labreche et al., 1996) and the high solvating capability of a wide variety of metallic salts, PEO is extensively studied as a model system in polymer blends and polymer electrolyte (Hamon et al., 2001; Rocco et al., 2002; Rocco et al., 2004; Mahendran et al., 2005). The major drawback of PEO as SPE is its conductivity, which is much too low (<< 10^{-4} S cm^{-1}) for practical application in solid-state batteries and electrochemical devices. Efforts to enhance the ionic conductivity by improving the chain mobility can be realized *via* various approaches such as using plasticizers (Song et al., 1999; Deepa et al., 2002a; Kumar and Sekhon, 2002; Kumar and Sekhon, 2008); blends (Borkowska et al., 1993; Rajendran et al., 2001; Osman et al., 2005; Chan and Kammer, 2008; Kato and Kawaguchi, 2012; Katime and Cadenato, 1995); copolymers (Wieczorek and Stevens, 1997; Ko et al., 2004; Krivorotova et al., 2010); block copolymer (Ma and Dezmazes, 2004); grafts (Li et al., 2006; Li et al., 2009), inter penetrating network (Munich-Elmer and Jannasch, 2006) and in recent years, the addition of organic and inorganic filler materials (Kryszewski et al., 1973; Burgaz, 2011; Pandey et al., 2008; Dey et al., 2008; Stefanescu et al., 2010).

Generally, conductivity of a polymeric electrolyte is determined by the concentration of charge carrier and its mobility. Ion dissociation is markedly influenced by the lattice energy of the salt (Murata et al., 2000; Othman et al., 2007) as well as the solvating ability of the polymer matrix. Ion transport is associated with the amorphous phase of a polymer electrolyte (Berthier et al., 1983). Segmental motion in polymeric chain is another important criteria governing ion conductivity because it affects the mobility of both cations and anions by controlling the local free volume and viscosity of the environment surrounding charge-transporting ions (Croce et al., 1993; Stallworth et al., 1995). When an inorganic salt is added to PEO, the cations from the salt form polymer-salt complexes with the ether oxygen of PEO, hence, change the straight or bent lamellae to randomly oriented lamellae (Gupta and Rhee, 2012). Restricted crystallization of PEO under the influence of $LiClO_4$ was also observed by Sim et al., (2011) as the volume filled morphology of as-prepared sample of salt-free neat PEO (see Figure 6) changed to the morphology of scantily dispersed aggregates of PEO spherulites as shown in Figure 10. Besides, the PEO spherullite became coarse in texture and irregular in shape with dark spots representing the amorphous regions within the spherulite due to the entrapment of salt as observed in Figure 11 (Sim et al., 2011). However, addition of inorganic salt is often associated with an increase in the T_g of the polymer resulting in restricted segmental motion of the chain (Bamford et al., 2003; Marzantowicz et al., 2005; Guilherme et al., 2007; Suthanthiraraj and Sheeba, 2007; Sim et al., 2011). Therefore, low molecular weight organic solvents such as ethylene carbonate (EC), propylene carbonate (PC), and PEG and so on are often added as plasticizers to enhance ion transport through an increase in the amorphous region of the polymer electrolyte (Bandara et al., 1998; Nicotera et al., 2002).

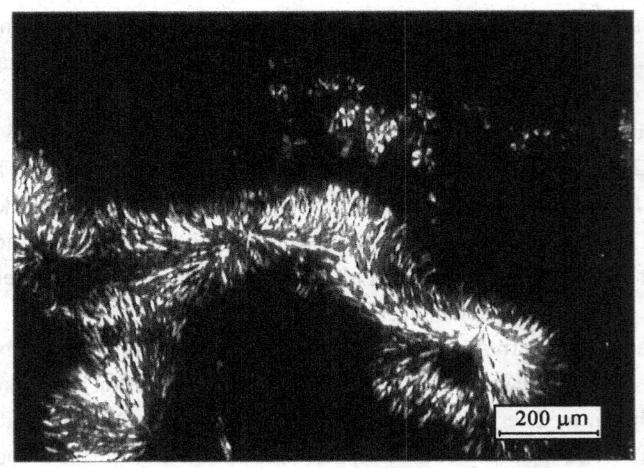

FIGURE 10 Micrograph of as-prepared sample of PEO doped with 15 wt% $LiClO_4$ (Sim, 2009). Magnification: 5x. The bar corresponds to 200 μm.

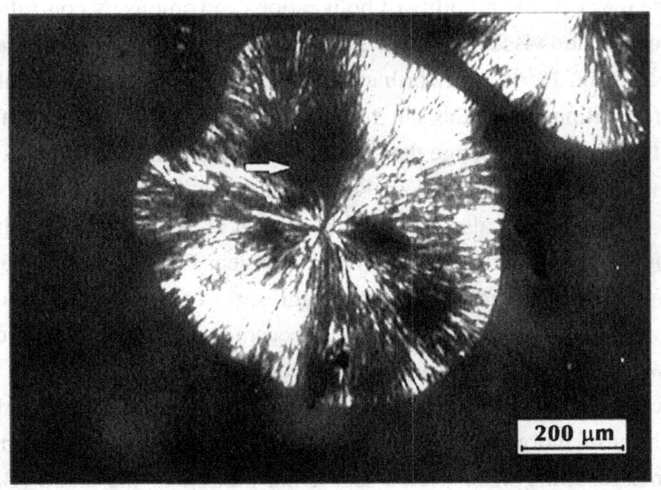

FIGURE 11 Micrograph of PEO doped with 15 wt% $LiClO_4$ captured after crystallization at T_c = 45 °C. The arrows show the entrapment of salt within the spherulite (Sim et al., 2011). Magnification: 5x. The bar corresponds to 200 μm.

Plasticized polymer-salt complexes are classified as gelled polymer electrolytes (GPE), fabricated by incorporating a large quantity of liquid plasticizer such as low molecular weight polyethylene glycols or aprotic organic solvent to the polymer matrix-salt complex that is capable of forming a structurally stable gel like polymer host.

The most widely studied and representative of all gelled electrolytes are the PEO-, PMMA-, PAN- and PVdF-based polymer electrolytes. Ion transport in gelled electrolytes occurs through a continuous conduction path of the plasticizer-rich phase quite similar to that of a liquid electrolyte and that the polymer host merely acts as a matrix of structural stability. This is especially so with polymer matrix having high solvating ability like PEO (Appetecchi et al.,1995; Sim et al., 2011; Suthanthiraraj and Sheeba, 2007). High ambient ion conductivity, no doubt, is a favorable characteristic of gelled electrolytes but the corrosive organic solvent tends to react with the metal electrodes and the volatility of the solvent raises a safety concern. Plasticized polymer electrolytes normally possess poor mechanical properties and interfacial instability (Vondrak et al., 2004; Othman et al., 2007; Croce et al., 1994).

Another approach extensively applied in recent years to improve the ion conductivity (σ), lithium ion transference number (t^+), mechanical properties, and the electrode-electrolyte interfacial stability of a polymer electrolyte is the addition of inorganic or ceramic fillers into the polymer-salt complexes (Capiglia et al., 1999; Kim et al., 2003; Chen-Yang et al., 2008; Croce et al., 2001; Rahman et al., 2009; Shen et al., 2009; Zhang et al., 2011; Munichandratah et al., 1995; Wieczorek, 1992). Micro and nano-sized inorganic filler such as silicone oxide (SiO_2), alumina (Al_2O_3), ceria (CeO_2), and so on are incorporated into PEO-salt complex in an effort to improve the mechanical, thermal stability, and ion conductivity of PEO-based polymer electrolytes. The effect of nano-fillers on the thermal properties of the PEO-based polymer complex varies with the type of nano-particles as well as the polymer-salt complex host matrix.

Croce et al., (1999) dispersed nano-sized ceramic powders, TiO_2 and Al_2O_3, individually into PEO-$LiClO_4$ complex and annealed the solution cast membrane at 97 °C (that is above the melting temperature, 70 °C, of PEO). The ceramic fillers with their large surface area cross-link the PEO segments via Lewis acid-base type reactions, thus, preventing local PEO chain reorganization resulting in freezing the highly disorder configuration of PEO at ambient temperature. The nanocomposite is found to retain this amorphous state for as long as a month. Both the

T_g and crystallinity of the PEO-based nanocomposite are reduced under such high degree of disorder. The network structure accounts for the large enhancement of the Young's modulus and of the yield point stress in the nanocomposite polymer electrolyte.

Besides, the ceramic fillers provide surface conducting pathways as a result of its Lewis acid type interaction with the PEO chains, hence, promoting ion transport which leads to enhanced ion conductivity. The T_g of the polymer host matrix PEO-ammonium perchlorate (NH_4ClO_4) investigated by Dey and Karan, (2008) arise in a narrow range between –41 °C and –46 °C while the T_m of PEO-salt complex is lower than that of pure PEO upon addition of nano-CeO_2 resulting in enhanced ion conductivity of the polymer nanocomposite (see Table 1).

Pitawala et al., (2007) studied the combined effect of both plasticizer and nano-ceramic filler on the thermal behavior and conductivity of $(PEO)_9$-lithium trifluoromethanesulphonate ($LiCF_3SO_3$ or LiTf) composite polymer electrolyte. The formula $(PEO)_9LiCF_3SO_3$ denotes the chemical composition of the polymer-salt complex in which "9" is the molar ratio of (ethylene oxide (EO)/$LiCF_3SO_3$). According to their work, addition of 15 wt% Al_2O_3 lowered the T_m and T_g of $(PEO)_9$LiTf from 58 °C and –44 °C to 51 °C and –50 °C, respectively and the conductivity was raised by two orders of magnitude from 4×10^{-7} to 2×10^{-5} S cm^{-1} as shown in Table 1. On the other hand, the effects on the T_m, T_g, and ion conductivity (σ) of $(PEO)_9$LiTf brought about by the addition of 50 wt% plasticizer alone are smaller as compared to the addition of nano-fillers. Nevertheless, the combined effect of both the EC and Al_2O_3 on the thermal and conductivity properties of the $(PEO)_9$LiTf is commendable as the reported values of T_m, T_g, and σ of $(PEO)_9$ LiTf incorporated with 50 wt% EC and 15 wt% Al_2O_3 are 49 °C, –56 °C and 2×10^{-4} S cm^{-1}. The major contribution for the enhancement of ion conductivity comes from the structural modification of the polymer host by the plasticizer and the nano-fillers.

Besides providing a new conduction path, plasticizer also enhances salt dissociation and lowers the T_g of PEO. It is suggested that the filler disrupts the stacking of the PEO crystalline lamella, thus, lowers the T_g by increasing the volume fraction of the amorphous phase of the polymer host (Pitawala et al., 2007, Johan et al., 2011; Burgaz, 2011).

TABLE 1 Different parameters (T_g, T_m, ΔH_m, χ_c) obtained from DSC studies for PEO-based polymer electrolytes and CPE

PEO-based polymer electrolytes and composites	T_g (°C)	T_m (°C)	ΔH_m (J/g)	χ_c (%)	Ref.
PEO	–54*	71	155	83	
PEO/NH_4ClO_4	–46	63	65	35	Dey et al., 2008
$PEO/NH_4ClO_4/CeO_2$ (5 wt%)	–44	60	69	37	
$PEO/NH_4ClO_4/CeO_2$ (10 wt%)	–47	66	46	24	
$PEO/NH_4ClO_4/CeO_2$ (20 wt%)	–41	61	68	36	
$PEO/LiCF_3SO_3$	–44	58			
$PEO/LiCF_3SO_3$/50 wt% EC	–48	57			Pitawala et al., 2007
$PEO/LiCF_3SO_3$/15 wt% Al_2O_3	–50	51			
$PEO/LiCF_3SO_3$/50 wt% EC/15 wt% Al_2O_3	–56	49			

* From Sim et al., (2009).

In view of some contrasting reports where fillers are found to have insignificant or negative effect on the conductivity of CPE (Panero et al., 1992; Kumar et al., 1994; Chandra et al., 1995), Choi et al., (1997) did a thorough study on the effect of twenty two ceramic fillers with sizes ranging from 0.023 to 13 μm on the thermal and electrical properties of $(PEO)_{16}LiClO_4$. The ceramic fillers used are SiO_2 (0.007 and 0.014 μm), iron oxide (Fe_2O_3) (0.023 and 1 μm), titanium oxide (TiO_2) (0.032 and 5 μm), Al_2O_3 (0.037 and 0.05 μm), boron nitride (BN) (1 μm), silicon carbide (SiC) (1 and 13 μm), silicon nitride (Si_3N_4) (1 μm), tungsten carbide (WC) (1 μm), barium titanate ($BaTiO_3$) (2 μm), molybdenum sulphide (MoS_2) (2 μm), zirconium oxide (ZrO_2) (3 μm), boron carbide (B_4C) (5 μm), lead titanate ($PbTiO_3$) (5 μm), aluminium nitride (AlN) (10 μm), titanium boride (TiB_2) (10 μm), and calcium silicate ($CaSiO_3$) (25 μm). Ramifications from their work showed that the addition of all the twenty two ceramic fillers caused a decrease in the values of T_m and T_g of the polymer complex $(PEO)_{16}LiClO_4$, attributed to the formation of smaller crystallites and higher flexibility of the macromolecular chain, respectively. On the other hand, higher ΔH_m for the CPE as compared to the polymer-salt complex reflects an increase in the volume fraction of the crystalline phase which is contradictory to reports that inert fillers enhance the formation of amorphous regions leading to higher ion conductivity (Munichandratah et al., 1995; Quartarone et al., 1998; Croce et al., 1999; Yuan and Ahao, 2006). Because of the competitive effect of increased segmental motion of the polymer chain and increased crystallinity, the ion conductivity of all the CPE studied in this work is either slightly enhanced or reduced at room temperature (~30 °C) as shown in Figure 12.

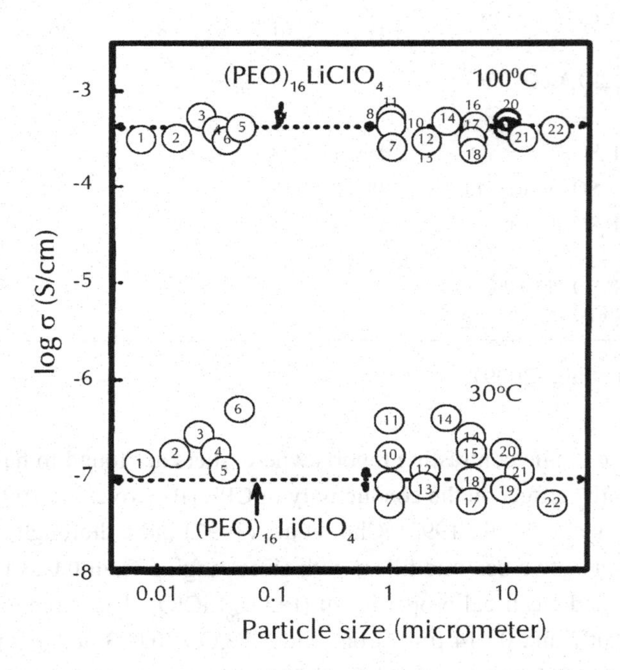

FIGURE 12 Conductivity versus particle sizes of twenty two ceramic fillers added to $(PEO)_{16}LiClO_4$ complex at 30 °C and 100 °C. The numbers in the figure denote the names and the particle sizes (given in the text) of the ceramic powders. For fillers with more than one particle size, the numbers are given in order of increasing particle size. (1, 2) SiO_2,(3,8) Fe_2O_3, (4, 17, 18) TiO_2, (5, 6) Al_2O_3, (7) BN, (9, 21) SiC, (10) Si_3N_4, (11) WC, (12) $BaTiO_3$, (13) MoS_2, (14) ZrO_2, and (15) B_4C (16) $PbTiO_3$ (19) AlN (20) TiB_2 (22) $CaSiO_3$. Reproduced from Choi and co-workers, copyright (1997), by permission of Elsevier Ltd.

The effective enhancement of physical, mechanical, and electrical properties of CPE is mainly governed by the modification of the microstructure of the polymer chain caused by the nature, grain size distribution, particle concentration, and type of the fillers used (Croce et al., 2001; Choi et al., 1997; Quartarone et al., 1998; Wieczorek et al., 1989). The addition of an optimum content of filler has a beneficial effect on the mechanical and electrical properties of a CPE while high concentration could result in the aggregation of the inorganic filler (Joykumar and Bhat, 2004; Jeon et al., 2006; Marcinek et al., 2005) or formation of well-defined crystallite region (Kumar et al., 1994; Rajendran and Uma, 2000) which acts as insulators leading to obstruction in ion conduction.

It has also been suggested that the conductivity enhancement by one or more order(s) in magnitude for most of the nanocomposite polymer electrolytes (Dey et al., 2009; Johan et al., 2011; Pandey et al., 2008; Dissanayake et al., 2003) is attributed to the transient hydrogen bonding of migrating ions with the Lewis-acid-base- type O–OH surface groups on the nano-sized ceramic filler (Johan et al., 2011; Dissanayake et al., 2003) which forms the ceramic-polymer chain network, providing ceramic surface conducting pathway for ion-transport (Croce et al., 1999). Besides, the large surface area of the nano-sized fillers prevents local reorganization of the PEO chains at ambient temperature leading to the formation of the highly disordered amorphous phase and the formation of ceramic filler-PEO chain network which favors ion transport (Johan et al., 2011; Croce et al., 1999; Appetecchi et al., 2000).

On the whole, besides high ion conductivity, wide electrochemical stability, and compatibility with the lithium metal electrode are important parameters to ensure good performance in rechargeable lithium batteries. Therefore, ceramic filler is superior to plasticizer in ensuring the good performance of rechargeable batteries because of the absence of any liquid and the interfacial stabilizing action of the dispersed filler which assures high electrochemical stability.

18.6 POLYACRYLATE

Most polyacrylates, especially the atatic and syndiotactic polyacrylates, are amorphous due to the irregular array of pendant ester groups, which prevent the parallel alignment of the polymer backbones. The T_gs of the different polyacrylates vary from subambient temperature to values above 100 °C. The T_g and the melting temperature for some polyacrylates are listed in Table 2.

TABLE 2 List of T_g and T_m values for some polyacrylates

Types of polyacrylates	T_g (°C)	Ref.	T_m (°C)	Ref.
Poly(arylalkyl methacrylates) [a]	10–56	Krause et al., 1965 Chen et al., 1994	-	
Poly(aryl methcrylates) [a]	75–155	Krause et al., 1965 Porter, 1995 Gargallo and Russo, 1975 Gargallo et al., 1996 Lewis, 1968	-	
Poly(ethyl methacrylates)	63–65 (*atactic*)	Porter, 1995 ICI, 1996	-	
	120 (*syndiotactic*)	ICI, 1996	47	Wiley, 1948
	8 (*isotactic*)	ICI, 1996	-	
Poly(ethyl acrylate)	-24 (*conventional*) -24 (*syndiotactic*) -25 (*isotactic*)	Li and Hsu, 1984	-	
Poly(butyl acrylate)	-43	Brandrup et al., 2009	47	Brandrup et al., 2009
Poly(methyl acrylate)	14	Wu et al., 2005	-	
Poly(methyl methacrylates)	105 (*conventional*) 115/160 (*syndiotactic*) 43–45 (*isotactic*)	Olabisi and Simba, 1975 Shetter, 1962 / Karev and MacKnight, 1968 Shetter, 1962 Karev and MacKnight, 1968 Thompson, 1966	-	

[a] The T_g values are depending on the substitution on the vinyl group of methacrylates.

PMMA is known for its hardness and transparency in nature. The mechanical properties of pure PMMA can be altered dramatically by the addition of foreign materials such as plasticizer, inorganic salt, and nano-particles as well as when subjected to temperature, frequency changes or an applied stress. Studies on the effects of nano-particles on the mechanical properties of PMMA, showed that the addition of nano-materials such as SiO_2, clay and fillers enhance the physical properties of the polymer (Krückel et al., 2012; Münstedt et al., 2010; Fischer et al., 2012). Research done by Kruckel et al., (2012) revealed that carbon black filled PMMA composites show pronounced improvement in viscoelastic properties such as higher storage modulus, lower percolation threshold, and higher conductivity than PMMA filled with carbon fiber. This is attributed to the high specific surface area as well as the electrical conductivity of the carbon black filler. Similar behavior was observed by Münstedt et al., (2010) when they incorporated nano-silica (SiO_2) into PMMA. According to Münstedt and his co-workers, a good understanding on the relationship between the properties of the nano-particles and their influence on the rheology of the nanocomposites renders an insight into the dispersion and interaction between the particles and the molecules of the composite matrix which is the key to resolve issues with respect to the effect of the nano-particles on the material property enhancement.

Creep and creep-recovery experiments were applied (Jacob et al., 1997) to investigate interactions between particles and matrix molecules at low particle concentrations and to get an insight into the long relaxation or retardation times. This method has an advantage over the dynamic-mechanical experiments as molecular processes can be investigated at very low particle concentration and at very long relaxation time so that a new cycle which begins with each frequency change can reach its steady state to deliver time-independent data. The linear steady-state recoverable compliance obtained increases by six orders in magnitude with an addition of maximum particle concentration of 2.1 vol% due to long relaxation time caused by the interaction of nano-particles with the surface molecules of the matrix, resulting in retardation in molecular mobility. Investigation of the viscoelastic properties at long experimental time using the creep and creep-recovery method allows one to get a deeper understanding on the particle-matrix interaction. Additionally, based on the sensitive influence of the surface area on the steady-state elastic compliance, one can predict the quality of particle dispersion in the composite matrix (Münstedt et al., 2010). Other than storage modulus, incorporation of clay into PMMA causes a significant enhancement on the tensile modulus as well as the toughness of the polymer chain, but results in a reduction on the elongation at break (Drzewinski, 1993).

The PMMA is well documented to form miscible blends with a variety of semi-crystalline polymers such as PEO via dipole-dipole interactions with its carboxyl group (–COO^-) (Kern, 1957; Martuscelli et al., 1987; Ito et al., 1987; Drzewinski, 1993; Drzewinski, 1994; Cimmino et al., 1998). Besides, PMMA is also extensively studied as polymer host in polymer electrolytes especially as gel electrolytes. Adding plasticizer like EC and PC improves the ion conductivity at 25 °C (σ_{25}) of PMMA-Li salt GPE up to values as high as 10^{-3} S cm^{-1} as shown in Table 3 (Deepa et al., 2002b; Deepa et al., 2004; Agnihotry et al., 2004; Ali et al., 2007). The enhancement in conductivity is attributed to the physicochemical properties such as the high dielectric constant (ε) of the plasticizer which enhances the dissociation of the lithium salt (Tobishima and Yamaji, 1984). The dielectric constants at 25 °C of EC and PC are 85 and 69, respectively. Besides, plasticizer reduces the viscosity of the polymer. Consequently, the macromolecular chains become more flexible and the free volume of PMMA increases leading to a fast ion transport through the polymer matrix, thus, increasing the ion conductivity of the GPE. However, to our knowledge, PMMA, due to its high T_g has never been applied as SPE as it requires plasticizers to provide the liquid like conducting pathway.

TABLE 3 Variations of ion conductivity at 25 °C (σ_{25}) as a function of concentrations of the plasticizer, PC for PMMA doped with LiTF, LiTFSI, and $LiClO_4$

Plasticizer (PC) concentration (wt%)	σ_{25} (S cm^{-1})		
	*LiTF (Ali et al., 2007) (× 10^{-6})	*LiTFSI (Deepa et al., 2004) (× 10^{-3})	$LiClO_4$ (Deepa et al., 2002) (× 10^{-3})
0.0	1	9 × 10^{-4}	3 × 10^{-5}
0.1	2	1	2
0.2	2	2	3
0.4	3	3	6
0.8	3	4	8

TABLE 3 *(Continued)*

1.0	5	4	8
1.2	6	4	8

* The conductivity is measured in solution form.

18.7 BLENDS OF PEO AND POLYACRYLATE

18.7.1 GENERAL

Polymer blending is the most convenient and cost effective way of designing new polymeric materials with properties not attainable by single polymers, tailored for specific applications (Rajendran et al., 2002c; Tan and Johan, 2011). Polymer blends can be divided into different categories depending on the degree of miscibility of the two blending components. PEO is known to exhibit miscibility with many amorphous polymers via hydrogen bonding between its ether oxygen atom and proton donors such as poly(vinyl alcohol) (PVA) (Pedrosa et al., 1995), oligoesters (OE) and polyester resins (PER) (Zheng et al., 1989), poly(vinyl phenol) (PVPh) (Sotele et al., 1997) and poly(methyl vinyl ether-maleic acid) (PMVE-MAc) (Rocco et al., 2001). However, when blended with non-proton donors, miscibility in the binary PEO-based blend system can also be achieved *via* relatively weak dipole-dipole interactions (Silvestre et al., 1987; Martuscelli et al., 1984; Ramana Rao et al., 1985 Shafee and Ueda, 2002). The weak intermolecular interactions or strong hydrogen bonds in miscible PEO-based blend tend to hinder the crystallization of PEO, resulting in suppressing its crystallinity and lowering its melting temperature.

The thermal behavior and the spherulitic morphology of PEO/PMMA have been widely explored by many authors (Li and Hsu, 1984; Martuscelli et al., 1986; Silvestre et al., 1987; Talibuddin et al., 1996; Martuscelli et al., 1987; Schantz, 1997). Other than the extensively documented PEO/PMMA system, no other acrylate polymers like poly(propyl methacrylate (PPMA), poly(butyl methacrylate) (PBMA), poly(cyclohexyl methacrylate) (PCMA), and so on are recognized for their miscibility with PEO, and until recently, poly(phenyl methacrylate) (PPhMA) (Woo et al., 2000), poly(benzyl methacrylate) (PBzMA) (Mandal et al., 2000), poly(n-butyl methacrylate) (PnBMA) (Shafee and Ueda, 2002), PMA

(Pfefferkorn et al., 2011), poly(iso-butyl methacrylate) (PiBMA) (Katime and Cadenato, 1995), poly(tert-butyl methacrylate) (PtBMA) (Katime and Cadenato, 1995), and poly(4-vinylphenol-*co*-2-hydroxyethyl methacrylate) (PVPh-HEM) (Pereira and Rocco, 2005) are found to be miscible with PEO. Owing to the vast differences both in the surface structure and T_gs of PEO and PMMA (T_gs for PEO and PMMA are –52 and 105 °C, respectively), PEO/PMMA blend forms a complex system (Mu et al., 2012). The two polymers are marginally miscible in the liquid state *via* weak intermolecular interactions but are immiscible in the solid state at room temperature as PEO crystallizes below 65 °C and PMMA is in its glassy state below 105 °C. Table 4 shows a compilation of experimental details for PEO/PMMA blends reported in previous literature.

TABLE 4 Some of the PEO/PMMA blends investigated in the past and present decades

Polymer Blend Components	Other Component	Properties Reported	References
PEO/PMMA	$LiClO_4$-DEP/DMP/DOP	Morphology and conductivity	Rajendran et al., 2002
PEO (M_w = 107,000 g mol^{-1})/PMMA (M_w = 218,000 g mol^{-1})		Rheological properties	Kato and Kawaguchi, 2012
PEO (M_w = 17,000 and 35,000 g mol^{-1}) /PiBMA and PtBMA		Thermal properties	Katime and Cadenato, 1995
PEO (M_w = 600,000 g mol^{-1})/PMMA (M_w = 35,000 g mol^{-1})	$LiClO_4$-MnO_2-EC	Conductivity, thermal properties, and crystal structure	Tan and Johan, 2011
PEO (M_w = 6,000 g mol^{-1})/PMMA (M_w = 4,200 g mol^{-1})	Au films	Morphology, intermolecular interaction, and thermal properties	Wang et al., 2003
PEO (M_w = 100,000 g mol^{-1})/PMMA (M_w = 110,000 g mol^{-1})		Morphology, thermal, and crystallization behavior	Martuscelli et al., 1984

TABLE 4 *(Continued)*

PEO (M_w = 6,000–990,000 g mol^{-1})/PMMA (M_w = 1,000–525,000 g mol^{-1})		Thermal and crystallization behavior	Alfonso and Russel, 1986
PMMA (*i*PMMA, *a*PMMA, *s*PMMA)[a]/PEO (oligomers; M_n = 400–20,000 g mol^{-1})		Thermal properties	Hamon et al., 2001
PEO (M_w = 101,200 g mol^{-1})/PMMA (M_w = 4900–101,000 g mol^{-1})		Morphology	Okerberg and Marand, 2007
PEO (M_w = 200,000 g mol^{-1})/*a*PMMA(M_w = 90,000 g mol^{-1})		Thermal properties and molecular structure	Schantz, 1997
PEO (M_w = 145,000 g mol^{-1})/*d*-PMMA (M_w = 129,000 g mol^{-1}) and *h*-PMMA[b] (M_w = 125,000 g mol^{-1})		Crystallization behavior	Ito and Russel, 1987
PEO (M_w = 144,000 g mol^{-1})-Poly (n-butyl methacrylate) (PnBMA) (M_w = 33,700 g mol^{-1})		Miscibility and thermal properties	Shafee and Ueda, 2002
PEO (M_w = 100,000 g mol^{-1})/PMMA (M_w = 120,000 g mol^{-1})	Ammonium salt-OMMT	Morphology, rheology, and crystal structure	Kim et al., 2005
d-PMMA-*d*-PEO/*h*-PEO [b]		Mechanical and crystallization behavior	Schawhn et al., 2012
PEO(M_w = 300,000 g mol^{-1})/PAc (M_w = 170,000 g mol^{-1})	$LiClO_4$	Thermal properties, conductivity and morphology	Sim et al., 2011

[a] indicate the stereoregularity of PMMA; *i*PMMA (isotactic), *a*PMMA (atactic), *s*PMMA (syndiotactic).

[b] organicity modified PEO and PMMA; d (deuterio) and h (protio).

18.7.2 FACTORS AFFECTING THE MISCIBILITY OF PEO/PMMA BLENDS

Studies by several authors have revealed that miscibility of PEO/PMMA blends is greatly influenced by the tacticity of PMMA, mainly attributed to the effect of polymer chain conformation and flexibility of PMMA stereoisomers on the phase behavior of blends (Martuscelli et al., 1986; Silvestre et al.,1987; Martuscelli et al., 1984; Ramana Rao et al., 1985; Schantz, 1997; Alfonso and Russel, 1986). Blends of PEO and atactic PMMA (*a*PMMA) are among the most studied semicrytalline/amorphous system found to be miscible by SAXS, DSC, POM, and ^{13}C NMR (Hamon et al., 2001; Martuscelli et al., 1986; Martuscelli et al., 1984; Alfonso and Russel, 1986; Hoffman, 1979; Martuscelli et al., 1980; Cimmino et al., 1990; Chen et al., 1998; Martuscelli et al., 1983a). Silvestre et al., (1987) and Cimmino et al., (1988) have concluded in their work that *i*PMMA is more preferred than syndiotactic PMMA (*s*PMMA) to form miscible blends with PEO. This conclusion is in disagreement with the findings reported by Ramana Rao et al., (1985), Marco et al., (1990), and Hamon et al., (2001).

On examining the effect of molecular weight of the amorphous *i*PMMA on the miscibility of PEO/*i*PMMA blends, Marco et al., (1990) found that the PEO (6×10^5 g mol^{-1})/ *i*PMMA (1.4×10^6 g mol^{-1}) blends were incompatible. However, compatibility of the blends occurs when the molecular weight of *i*PMMA is reduced to 7×10^5 g mol^{-1} and 2×10^5 g mol^{-1}. Therefore, it is important to note that blends whose components are incompatible could possess a "window of compatibility" attributed to the molecular weights of the components, in addition to other factors like blend composition, polarity of solvents and thermal procedures (Marco et al., 1990; Martuscelli et al., 1984). Cortazar et al., (1982), in their studies on the thermodynamic mixing of the components in the blends of PEO ($M_w = 4.0 \times 10^6$ g mol^{-1})/PMMA and PEO ($M_w = 3.7 \times 10^5$ g mol^{-1})/PMMA, observed a depression in the melting points of the blends and the Flory-Huggins interaction parameter χ_{12} determined were –0.131 and –0.136, respectively. These findings establish the miscibility of the two components in the molten state, and furthermore reveal that the M_w of PEO has no influence on the binary interaction parameter of the blends.

18.7.3 MISCIBILITY OF PEO/PMMA BLENDS

Glass Transition Temperature

The strongest evidence indicating miscibility of a binary blend is the observation of a T_g. Over the entire composition range, a single T_g is observed for PEO/PMMA blend, suggesting miscibility. However, it is found that T_g of the blends with compositions 50–70 wt% of PMMA content deviates from the theoretical Fox relation and occurs very close to that of the neat PEO. This discrepancy in the quantitative assessment of the T_g is ascribed to the overlapping of the T_g of the blends at ~40–60 °C with the melting point of the PEO (Martuscelli et al., 1986; Alfonso and Russel, 1986).

In our study on the PEO/PMMA system, a single T_g is observed for all blend compositions from PEO/PMMA 10/90 to 90/10 but the variation of T_g as a function of weight fraction of PEO deviates largely from that calculated using the Fox equation [unpublished result]. Nevertheless, the trend of the variation is similar to that exhibited by the PEO/PPhMA and PEO/PBzMA blends (Woo et al., 2000; Mandal et al., 2000). On the other hand, two distinct T_g's corresponding to the two neat constituents are observed throughout the entire composition range for the immiscible PEO/PAc system (Sim et al., 2011).

The polyacrylate used by Sim et al., (2011) is a thermoplastic acrylic copolymer synthesized (Gan, 2005) by free-radical polymerization of six monomer units added in semi batches to the reactor, as described in detail by Zhou et al., (2004). Schematic representation of the chemical structure of PAc showing part of the random distribution of the six monomer units *viz.* styrene, methyl methacrylate (MMA), butyl acrylate (BA), AA, 2-hydroxy ethylacrylate (2HEA), and isobutyl methacrylate (iBMA) in the backbone of the copolymer is shown in Figure 13 (Sim et al., 2009). The alphabets a–f denote mole fraction of 0.16, 0.17, 0.39, 0.19, 0.06, and 0.03, respectively of each monomer unit in the copolymer. Figure 14 displays the difference in the T_g results between a miscible and an immiscible blend systems.

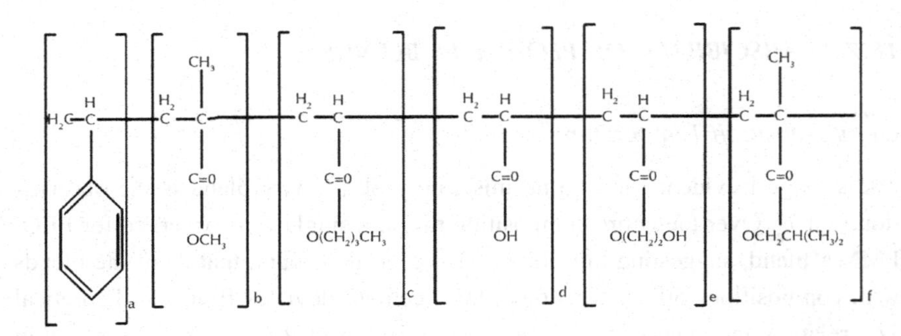

FIGURE 13 Schematic representation of the chemical structure of PAc. Composition of each monomer unit in the copolymer is given in mole fraction—(a) 0.16, (b) 0.17, (c) 0.39, (d) 0.19, (e) 0.06, and (f) 0.03. Modified from Sim et al., (2009).

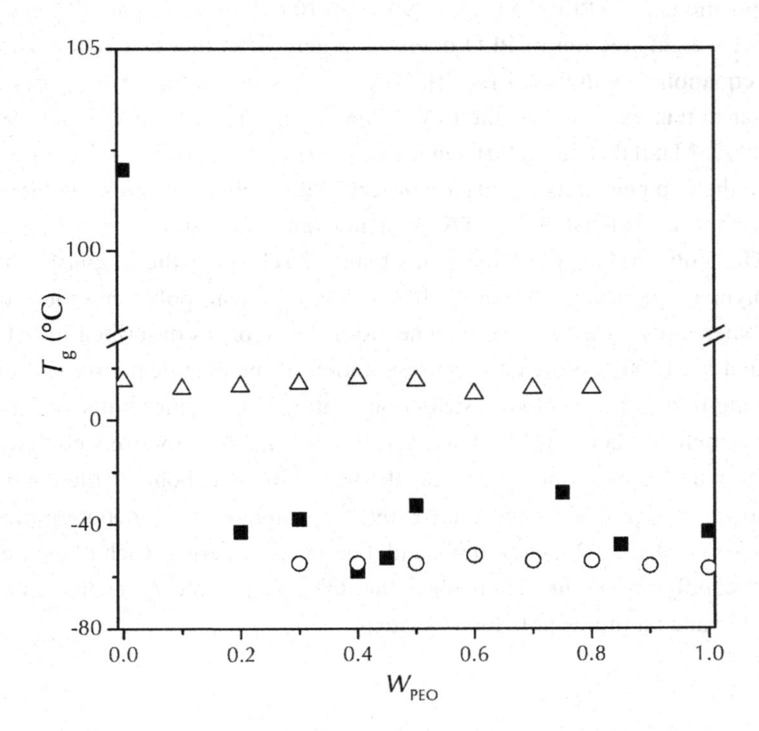

FIGURE 14 Glass transition temperature (T_g) versus weight fractions of PEO (W_{PEO}). (■) T_g of miscible PEO/PMMA blend (Martuscelli et al., 1986), T_g of (○) PEO and (Δ) PAc for the immiscible PEO/PAc blend (Sim et al., 2011).

Melting and Crystallization Behavior

In most studies, miscibility has been determined by depression of the T_m and/or T_m°, existence of a T_g, reduction in crystallinity (X_c), spherulite growth rate (G), and isothermal crystallization rate constant (K_A) as well as changes in morphology (Chan et al., 2004; Rocco et al., 2001; Dreezen et al., 1999; Zhong and Guo, 2000; Guo et al., 2001; Lü et al., 2004). Successive depression of T_m° of PEO by ascending polyacrylate content in blends of PEO and various types of amorphous polyacrylates are depicted in Table 5. Depression of T_m° is ascribed to the decrease in the rate of PEO crystallization as the carboxyl group of polyacrylate interacts with the carbon atom of the C-O-C group of PEO in the miscible PEO/polyacrylate blends.

TABLE 5 T_m° of some PEO/polyacrylate blends

% PEO	**T_m° (°C)**				
	PEO/PMMA (Martuscelli et al., 1984)	**PEO/PiBMA (Katime and Cadenato, 1995)**	**PEO/PtBMA (Katime and Cadenato, 1995)**	**PEO/Pn-BMA (Shafee and Ueda, 2002)**	**PEO/PMA (Pfeffer-korn et al., 2011)**
100	75	64	64	74	62
90	74	64	64	73	61
80	73	63	64	72	60
70	72	63	63	73	59
60	71	62	63	71	59
50	-	60	61	-	59

Martuscelli et al., (1984) studied the crystallization and thermal behavior of PEO (1.0×10^5 g mol^{-1})/PMMA (1.1×10^5 g mol^{-1}), and found that the values of G, K_A as well as T_m° reduced proportionally according to the PMMA content, and hence, when combined with a value of –0.35 for χ_{12}, corroborated the miscibility of the PEO/PMMA blend (Calahorra et al., 1982). However, Colby, (1989) and

Zawada et al., (1992) designate the PEO/PMMA system as marginally miscible because of the weak dipole-dipole interactions between the blending components with value of $\chi_{12} \cong 0$, implying that the blend can be immiscible depending on the temperature and composition range. Meanwhile, the study by Dionisio et al., (2000) on the effect of PEO on the relaxation process of PMMA using the dielectric relaxation spectroscopy (DRS) complemented with calorimetric analysis gives further insight into the mixing mechanism of the PEO/PMMA blend. In addition, the presence of a singular broad transition in the blend with 20 wt% PEO, obtained using DRS and DSC, is associated with the merging of the α process into a complex αβ relaxation process and the presence of concentration fluctuations, depicting that the blend lies in the border of miscible/immiscible region.

The immiscible blends, however, demonstrate inconsistent variations of T_m^o. Figure 15 depicts the variations of T_m^o as a function of W_{PEO} for a miscible blend of PEO/PMMA (Martuscelli et al., 1984) and an immiscible blend of PEO/PAc (Sim et al., 2011).

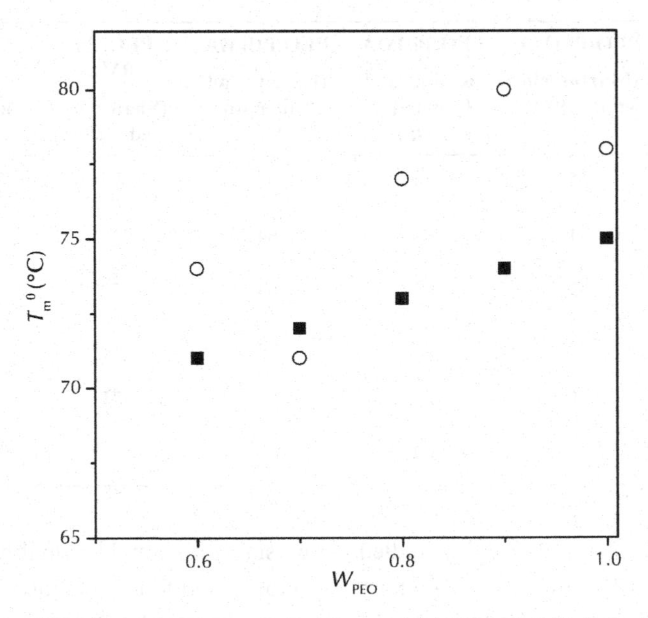

FIGURE 15 T_m^o as functions of W_{PEO} for blends of (■) PEO/PMMA (Martuscelli et al. 1986), (○) PEO/PAC (Sim et al., 2011).

Morphology

Investigation into the morphology of isothermally crystallized PEO/PMMA blend by Martuscelli et al., (1986) shows that the as-prepared sample with PEO content as low as 60 wt% is volume filled with spherulites without observing any separate domains of the amorphous component suggesting that the PMMA diluent is incorporated, during crystallization process, into the inter-lamellar regions of PEO spherulite resulting in the formation of a mixed homogeneous amorphous phase as shown in Figure 16 (a). However, when PEO/PMMA 60/40 blend is subjected to isothermal crystallization at crystallization temperature (T_c) = 46 °C, PEO spherulite adopts a coarse texture and irregular shape with the absence of a Maltese cross birefringence pattern as shown in Figure 16 (b) (Martuscelli et al., 1984). Coarsening of the crystalline lamellae and the disappearance of the birefringence are due to the presence of PMMA in the inter-lamellar regions of PEO spherulite which interrupts the crystallization process. Besides, the morphology of the after melt PEO/PMMA 60/40 demonstrates that the spherulites are scantily dispersed in a wide amorphous region (Martuscelli et al., 1984). Disruption in the crystallization process of PEO in an immiscible blend only takes place at higher content of the v_2^2 added. Figure 16 (c) depicts that the PEO spherulites in the immiscible PEO/PAc 40/60 blend are interspersed with large amorphous domains in the inter-lamellae as well as the inter-spherulitic regions implying incomplete rejection of PAc at the crystallite growth front during the crystallization process (Sim, 2009). As a result, the presence of a small amount of PAc in the inter-lamella domain of PEO/PAc 50/50 blend, after isothermal crystallization at T_c = 45 °C, causes the PEO spherulite as shown in Figure 16 (d) to adopt a slightly irregular and coarse spherulitic morphology but with the Maltese cross clearly visible. No crystal is observed for the PEO/PAc 40/60 blend after isothermal crystallization at 45 °C.

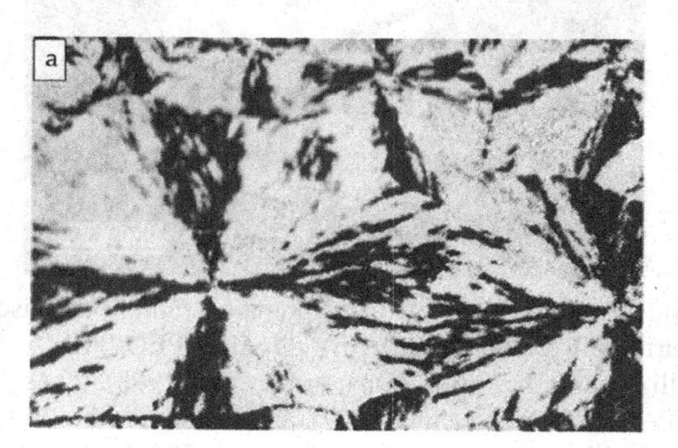

FIGURE 16 *(Continued)*

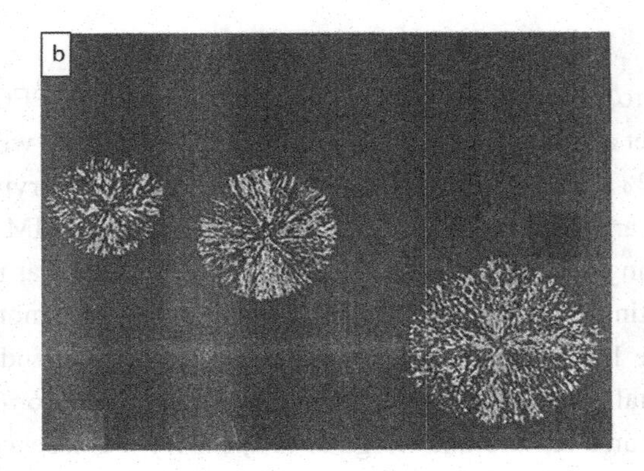

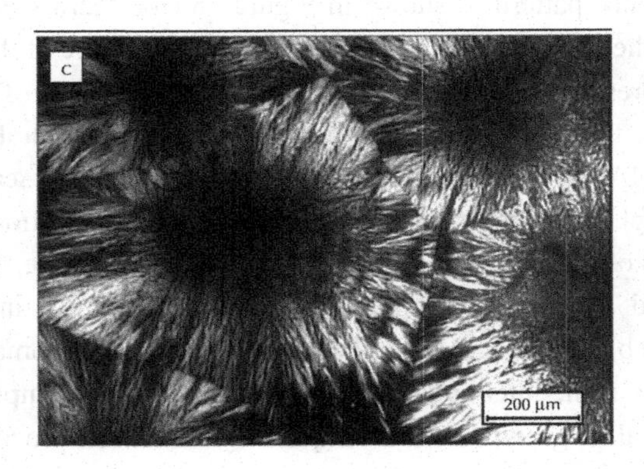

FIGURE 16 Optical micrographs of (a) as-prepared sample of miscible PEO/PMMA 60/40 blend (Martuscelli et al., 1986), (b) miscible PEO/PMMA 60/40 blend after isothermal crystallization at T_c = 46 °C (Quartarone et al., 1998), (c) as-prepared sample of immiscible PEO/PAc 50/50 blend (Sim, 2009), and (d) immiscible PEO/PAc 50/50 blend after isothermal crystallization at T_c = 45 °C (Sim et al., 2011).

Spectroscopic Technique

The spectroscopic techniques are particularly useful in examining the degree of phase separation and microstructure changes induced by blending (Lü et al., 2004; Zhong and Guo, 1998; Hsieh et al., 2001; Baskaran et al., 2006). Li and Hsu, (1984) analyzed the crystallization behavior of PEO/PMMA blends using FTIR and calorimetric techniques. On examining the FTIR spectra, it was reported that the crystalline band of 1360 and 1340 cm^{-1} doublet had been replaced by the amorphous 1349 cm^{-1} band with increasing PMMA content, indicating a significant reduction in crystallization rate leading to attenuation in the crystallinity of PEO (see Figure 17). The microstructure of PEO is changed from the trans-state to the gauche form attributed to the interaction between oxygen atoms and CH_2 groups. Wang et al., (2003) in their studies on crystalline structure of PEO/PMMA blend film display weak interaction between the ether oxygen of PEO and the polar methoxy group of PMMA in the blend.

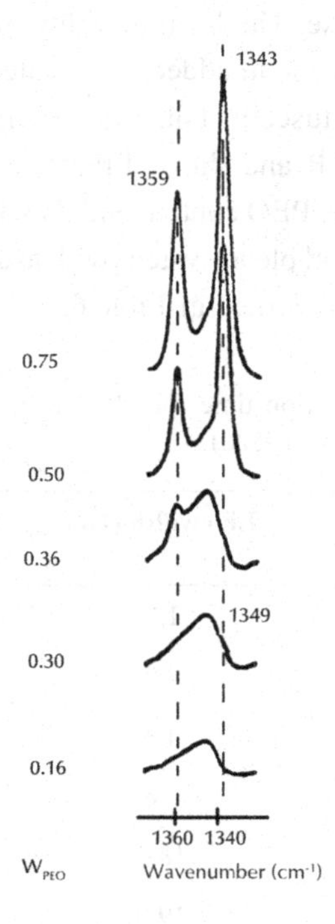

FIGURE 17 Infrared spectra (IR) of the CH_2 wagging mode of PEO for PEO/PMMA blend. Modified from Li and Hsu, (1984).

The existence of weak dipole-dipole interactions between PEO and PMMA is also confirmed by applying vibrational spectroscopy (Ramana Rao, 1985), SAXS, and SANS (Ito et al., 1987). Recently, FTIR spectroscopy has been widely employed to substantiate ramifications on the miscibility of binary blends studied by other techniques (Guo et al., 2001; Baskaran et al., 2006; Ramesh et al., 2007). Most of the authors have concluded that the PEO/PMMA blend is miscible both in the melt (Martuscelli et al., 1983a; Martuscelli et al., 1983b) and in the amorphous phase (Parizel et al., 1997; Straka et al., 1995). Further evidence on the miscibility, molecular structure, and local dynamics of PEO/*a*PMMA blend is demonstrated by ^{13}C and ^{1}H NMR spectroscopy (Marco et al., 1990; Pfefferkorn et al., 2011; Martuscelli et al., 1983a; Parizel et al., 1997; Straka et al., 1995; Lartigue et al., 1997). Schantz, (1997) and Straka et al., (2002) who carried out solid-state ^{13}C NMR measurement on melt-mixed blends of PEO/*a*PMMA concluded that all the blends with PEO weight fraction $\leq 30\%$ were completely homogeneous in a length scale of 20–70 nm with part of the PMMA and PEO chain intimately mixed in the amorphous phase. The local mobility of amorphous PEO decreases in proportion to the PMMA content added as revealed by the relaxation behavior for protons and carbons. Martuscelli et al., (1983a) investigated the miscibility of PEO/*a*PMMA using ^{13}C NMR and showed that the linewidth of PMMA signal was strongly dependent on the PEO content and that the PEO mobility was greatly reduced leading to a more rigid blend system with ascending PMMA content. The results obtained at 90 °C are recorded in Table 6.

TABLE 6 Linewidth and relaxation time for PEO in blends of PEO/*a*PMMA at 90 °C using ^{13}C NMR (Martuscelli et al., 1983a)

PEO/*a*PMMA blend (% PEO)	Linewidth (Hz)	Relaxation time (s × 10^{-3})
100	1.7	840
90	4.0	502
80	4.8	488
70	5.6	461
60	18.0	345
50	19.0	240

18.7.4 MISCIBLE BLENDS OF PEO AND OTHER POLYACRYLATES

Other than PEO/PMMA system, the miscibility of blends of PEO/PiBMA and PEO/PtBMA (Katime and Cadenato, 1995) is verified by the suppression of T_m^o with ascending amorphous component content and negative χ_{12} values of –0.106 and –0.034, respectively. The reduction in T_m^o is due to the formation of imperfect PEO crystallites as a result of disrupted crystallization caused by the presence of the amorphous component. Meanwhile, the isothermal crystallization kinetics of the PEO/PiBMA and PEO/PtBMA blends follow the Avrami equation (Avrami, 1939) implying that the rate of crystallization of PEO in the blends decreases progressively with ascending amorphous PiBMA and PtBMA content. The Avrami equation applied under isothermal crystallization conditions is presented as Equation (9):

$$X_t = 1 - \exp\left[K_A t^n\right] \quad (9)$$

where X_t is the normalized crystallinity given as the ratio of degree of crystallinity at time t and the final degree of crystallinity that is at $t \rightarrow \infty$. K_A is the overall rate constant of crystallization. The Avrami exponent (n) is a numerical value which depends on the type of nucleation and the geometry of crystal growth (Schultz, 1981). The Avrami plot is a plot of log [-ln (1 – X_t)] versus log t after Equation (9). The linear domain of the Avrami plots of PiBMA at T_c = 42 °C and 46 °C are demonstrated in Figure 18.

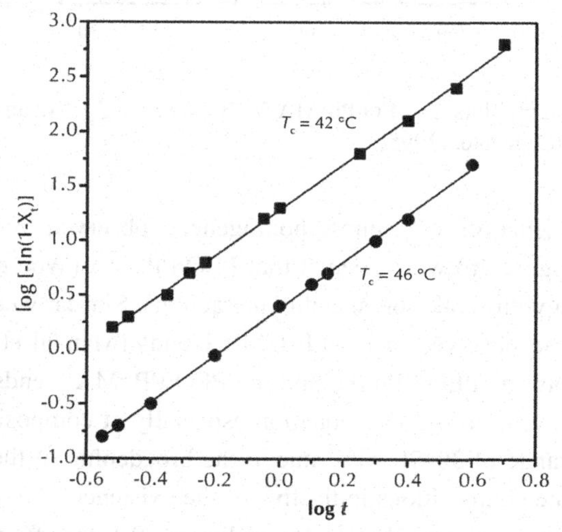

FIGURE 18 Avrami plots for PEO/PiBMA 50/50 blends at T_c = 42 and 46 °C. Modified from Katime and Cadenato, (1995).

The y-intercept obtained from the linear plot between equilibrium melting point depression (ΔT_m^{o}) and the square of the volume fraction of the amorphous component (v_2^2) after Equation (6) (described in detail in Miscibility) developed by Nishi and Wang, (1975) for both the blends is close to zero as shown in Figure 19 suggesting that entropy effect has insignificant influence on the χ_{12}.

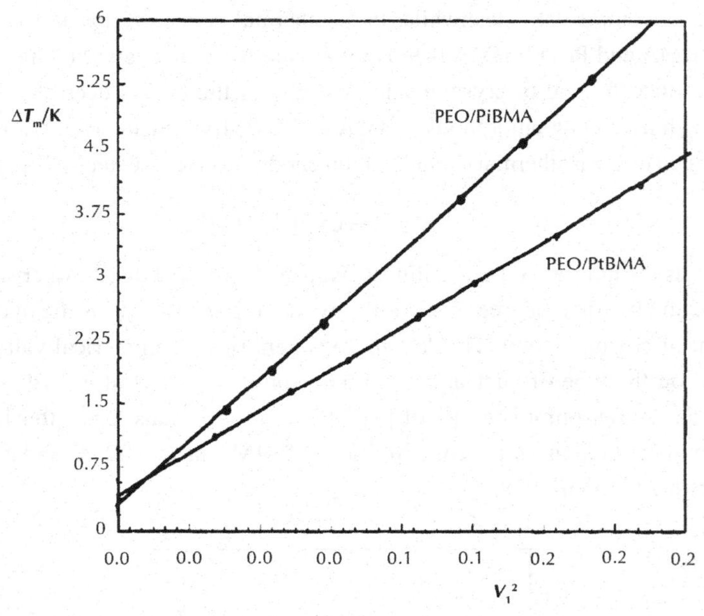

FIGURE 19 Avrami plots for PEO/PiBMA 50/50 blends at T_c = 42 and 46 °C. Modified from Katime and Cadenato, (1995).

Besides evidence of a T_g, phase homogeneity observed in POM and SEM micrographs (Figure 20) which depict that PEO/PPhMA (Woo et al., 2000) is a miscible system with weak non-specific interactions. Similar weak dipole-dipole interaction is also observed in PEO/PBzMA blends (Mandal et al., 2000). The single T_gs of both the PEO/PPhMA and the PEO/PBzMA blends exhibit a large negative deviation from the Fox equation especially at compositions with PEO content in the range of 30–70 wt%, due to the broadening of the blend samples with intermediate compositions indicative of the existence of a certain degree of intermolecular aggregation. The low χ_{12} value of –0.1 at 60°C recorded for the

PEO/PBzMA blend system correlates with the IR results which show that miscibility of the PEO/PPhMA and the PEO/PBzMA blends depends only on the weak physical van der Waals interactions.

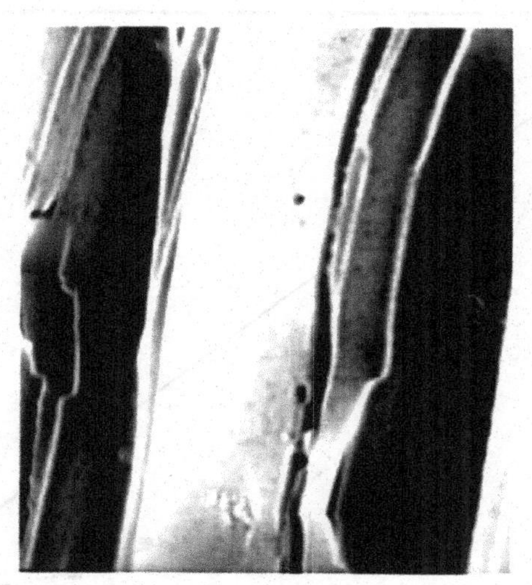

FIGURE 20 SEM microgoraph for PEO/PPhMA 30/70 showing a smooth and homogeneous surface morphology of the miscible blend. Reproduced from Woo et al., copyright (2000), by permission of Elsevier Ltd.

The presence of a single composition dependent T_g and a slight decrease in T_m^o with increasing content of PnBMA affirm the miscibility of the PEO/PnBMA blend (Mandal et al., 2000). Nevertheless, the χ_{12} value ≈ 0 indicates low interaction strength between the blending components. The crystallization and melting behavior of the blends show that the blends exhibit a lower critical solution temperature type (LCST) demixing at temperature > 85 °C. In the case of the PEO/PnBMA blend, the T_g data fits Equation (10), the Gordon-Taylor equation, which is given as:

$$T = \frac{w_1 T_{g1} + K w_2 T_{g2}}{w_1 + K w_2} \tag{10}$$

where w_1 is the weight fraction of polymer component and the parameter K is the ratio of the differences of the coefficient of expansion ($\Delta\alpha$) near the glassy and the rubbery states. The plot of T_g versus W_{PEO} for the PEO/PnBMA blend is shown in Figure 21.

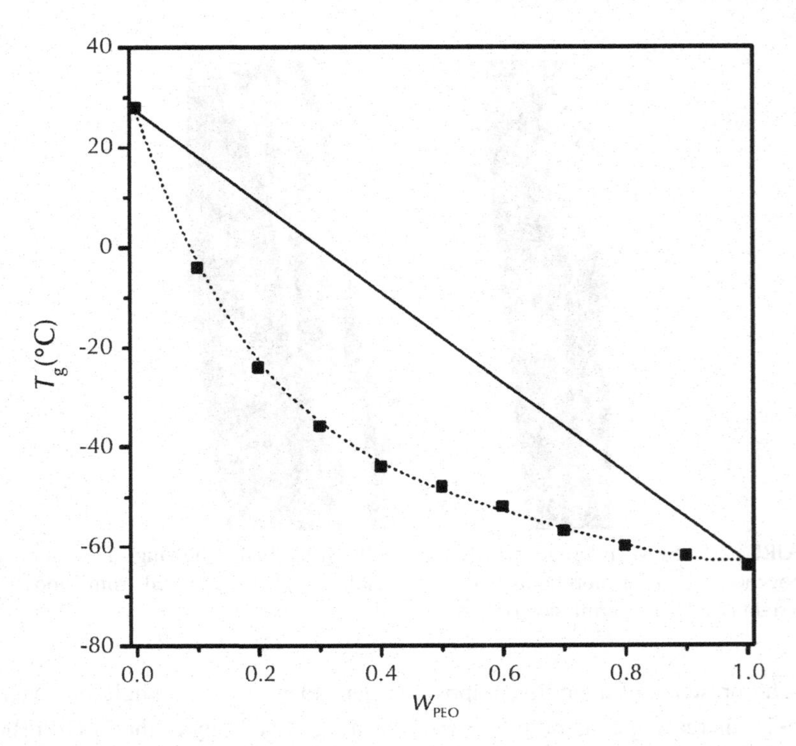

FIGURE 21 Glass transition temperature versus weight fractions of PEO for PEO/PnBMA blends. The solid and dotted line represent theoretical T_g-composition curves according to Fox and Gordon-Taylor equations, respectively. Modified from Mandal et al., (2000).

Pfeffercorn et al., (2011) in their investigation on the crystallization and melting behavior of PEO/PMA blends using DSC and temperature-resolved small-angle X-ray scattering (TR-SAXS) reveal that when the blends are heated after isothermal crystallization, PEO lamellae thickened from once-folded to extended chain crystals prior to melting. The maximum T_m^o depression and χ_{12} values for the blends are found to be approximately 2 K and –0.04, respectively implying miscibility with weak attractive interactions.

On the whole, values of χ_{12} evaluated from T_m^o depression study for all the PEO/polyacrylate systems investigated are rather small and close to 0. In addition, the absence of specific interactions in the PEO/polyacrylate systems including the extensively studied PEO/PMMA system as evident from the IR spectra concludes that the overall phase behavior of PEO and polyacrylate blends is miscibility with weak dipole-dipole intermolecular attractions (Rao et al., 1985). In contrast, χ_{12} values between –0.5 and –2.5, significant suppression of crystallinity is observed between the components of PEO/PVPh-HEM blends (Pereira and Rocco, 2005). The Miscibility of the PEO/PVPh-HEM blends is obviously associated with strong hydrogen bonds due to the presence of phenyl and hydroxyl groups in the acrylate polymer. It is interesting to note that the single T_gs exhibit by all the miscible PEO/polyacrylate blends deviate from the Fox equation suggesting that the intermolecular homogeneity is far from covalent-bonded copolymers.

The nanocomposites consisting of nano-sized organoclay or inorganic fillers dispersed in a polymer blend matrix have recently attracted much attention due to their tremendous improvements in thermal stability, physical, and mechanical properties (Kim et al., 2005; Kim et al., 2000; Lim et al., 2002). The PEO/PMMA/organically modified montmorillonite (OMMT) nanocomposite prepared by solution casting method has been demonstrated, using XRD, to form an intercalated hybrid structure with some exfoliated forms (Kim et al., 2005; Lim et al., 2002). In addition, PMMA is found to have better affinity for OMMT than PEO and that the layered silicate is more homogenously dispersed in the blend system than in PEO polymer system, based on results from XRD, Flory-Huggins interaction parameter as well as images from TEM. The intercalation of OMMT in the PEO/PMMA blend plays an important role in the enhancement of the mechanical and elastic properties of the blends.

18.8 BLENDS OF PEO/POLYACRYLATE WITH INORGANIC SALT

18.8.1 GENERAL

Of all the novel approaches to prepare polymer electrolytes, polymer blending is widely employed and known to be very promising (Rocco et al., 2002; Rocco et al., 2004; Chan and Kammer, 2008; Li and Khan, 1993).The main advantages of blending are simplicity of preparation and easy control of physical properties *via* compositional manipulation. PEO-based electrolyte is the earliest and the most extensively studied system. Because of its inherently high degree of crystallinity

(~70–80%), solvent free PEO-salt complexes suffer a serious drawback in conductivity. Therefore, the main focus in improving the ion conductivity of this electrolyte system is to suppress its crystallinity by blending with a compatible non-crystalline polymer such as PMVE-MAc (Rocco et al., 2002), poly(bisphenol A-co-epichlorohydrin) (PBE) (Rocco et al., 2004), PMMA (Rajendran et al., 2001; Florjanczyk et al., 2004), poly(ethylene glycol)methacrylate (PEGMA) (Marco et al., 1996), and so on. Apart from miscible binary blend electrolytes, immiscible blend electrolytes between PEO and poly(2-or 4-vinylpyridine) (P2VP or P4VP) (Li and Khan, 1993), PPO (Acosta and Morales, 1996), PVA (Mishra and Rao, 1998], ENR (Chan and Kammer, 2008; Glasse et al., 2002), and PAc (Sim et al., 2011) do exhibit comparatively good ion conductivity in the range of 10^{-7}–10^{-5} S cm^{-1}. It is noteworthy to recognize that in most of the PEO-based polymer electrolytes, PEO provides the percolating pathway for the transport of ions in the blends (Rocco et al., 2002; Rocco et al., 2004; Chan and Kammer, 2008; Sim et al., 2011; Mishra and Rao, 1998).

18.8.2 GLASS TRANSITION TEMPERATURE

It is generally observed that incorporation of an inorganic salt to a blend elevates the single T_g of a miscible blend (Rocco et al., 2002; Rocco et al., 2004; Rajendran et al., 2001) as well as the two T_g's of an immiscible binary system (Chan and Kammer, 2008; Sim et al., 2011; Acosta and Morales, 1996). This is attributed to the formation of complexes between the cation of the salt and the partially negatively charged polar group of the polymer component (Rocco et al., 2002; Rocco et al., 2004; Florjanczyk et al., 2004). Ion-dipole interaction between cation and macromolecular ligand in a semicrytalline/amorphous blend can be investigated on the basis of DSC, FTIR, POM, SEM, and X-ray dispersive studies.

For the marginally miscible PEO/PMMA 25/75 and 75/25 blends investigated by authors, a single T_g higher than that of the pristine salt-free blend prevails throughout the range of $LiClO_4$ added (1–12 wt%) to the blends. Variations of T_g values as a function of salt content added for the PEO/PMMA 25/75 and 50/50 blends are shown in Figure 22. Phase separation begins to occur when 10 wt% $LiClO_4$ is added to the 50/50 blend as shown in Figure 22, most probably attributed to the preferential formation of the PEO-Li^+-PEO or PMMA-Li^+-PMMA complexes in which the ion-dipole interactions between Li^+ ions and ether or carbonyl oxygen of PEO and PMMA, respectively are stronger than the weak dipole-dipole interaction between the two blending components.

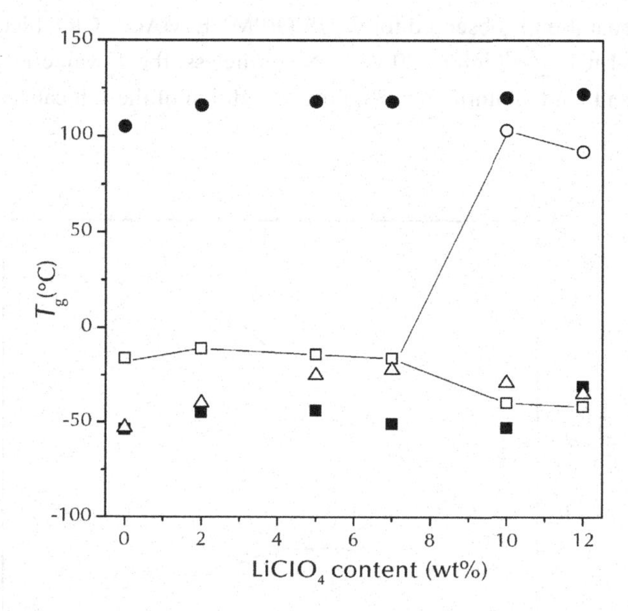

FIGURE 22 Variations of T_g as functions of W_{PEO}. (■, ●) T_g's for the neat PEO and neat PMMA, respectively. T_gs for the (Δ) PEO/PMMA 25/75 blend, (□) 50/50 blend before phase separation. T_gs of (□, ○) PEO and PMMA, respectively, for the PEO/PMMA 50/50 blend after phase separation (unpublished results).

Similar behavior, passing from a miscible system at lower salt content to an immiscible one at higher salt content is also observed in the PEO/PBE/$LiClO_4$ blend (Rocco et al., 2004). Figure 23 depicts the dependence of T_g values of the PEO/PBE/$LiClO_4$ blend on the concentration of salt. Phase separation of the blend with the presence of two distinct T_gs is observed when the salt content ≥ 15 wt%. Results from FTIR indicate the formation of three types of complexes at different salt content. At low salt content, complexes of the type PEO-Li^+-PBE prevail in which compatibilization is facilitated by Li^+ ions forming ion-dipole interactions with both PEO and PBE in place of the original hydrogen bond in the PEO/PBE salt-free blend. In some immiscible blends, the presence of an optimum amount of salt enables the two otherwise immiscible components to become compatible (Rocco et al., 2002; Mani and Stevens, 1992; Molnar and Eisenberg, 1992; Gao et al., 1992; Wang et al., 2005). At high salt content, phase separation occurs when the polymers preferentially forming complexes of the type PEO-Li^+-PEO and PBE-Li^+-PBE (Li and Khan, 1993; Wieczoreck et al., 1998). On the contrary, no phase separation is observed for the PEO/PMVE-MAc/$LiClO_4$ blends with the addition of $LiClO_4$ as high as 20 wt%. Nevertheless, the T_g value increases with ascending salt content until 10 wt%, further addition of the salt causes a decrease in T_g value.

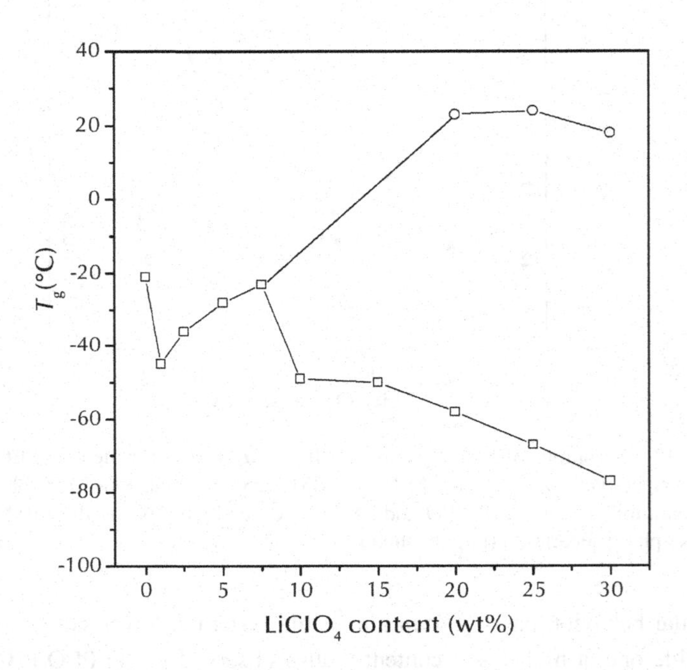

FIGURE 23 Variations of T_g for PEO/PBE blend with the addition of $LiClO_4$. (□) indicates the T_g of the blend before phase separation, (□,○) T_gs for PEO and PBE, respectively for PEO/PBE blend after phase separation. Modified from Rocco et al., (2004).

Chan and Kammer, (2008) and Sim et al., (2011) studied on the immiscible PEO/ENR/$LiClO_4$ and PEO/PAc/$LiClO_4$ systems reported that the two T_gs of the immiscible blends estimated under constant salt content were observed to be higher than their respective values in the pristine salt-free systems. Figure 24 depicts the variations of T_g as a function of weight fraction of PEO (W_{PEO}) for the salt-free and salt-added PEO/ENR and PEO/PAc systems. The salt content in the PEO/ENR/$LiClO_4$ and PEO/PAc/$LiClO_4$ blends are 12 and 15 wt%, respectively. A close examination of the plots in Figure 24 concludes that the difference in the T_g value of PEO between the salt-free and salt-added PEO/ENR blends (ΔT_g^{PEO}) is very much higher than that of ENR ($\Delta T_g^{PEO} >>> \Delta T_g^{ENR}$) but only moderately higher than the difference in the T_g value of PAc between the salt-free and salt-added PEO/PAc blends ($\Delta T_g^{PEO} > \Delta T_g^{PAc}$) implying that the salt is preferably dissolved in PEO as compared to ENR and PAc in the two immiscible blends (Chan and Kammer, 2008; Sim et al., 2011). Higher dissolution of the salt in PAc than in ENR could be due to the presence of more polar carbonyl groups in PAc. Qualitatively, the distribution of the salt and $LiClO_4$ in the two blending components of the phase separated PEO/ENR/$LiClO_4$ and PEO/PAc/$LiClO_4$ blends can be evaluated from Equation (11):

$$K_D = \frac{mass\ of\ Li\ in\ PEO}{mass\ of\ Li\ in\ ENR\ or\ PAc} \approx \frac{\Delta T_g^{PEO}}{\Delta T_g^{ENR}\ \text{or}\ \Delta T_g^{PAc}} \tag{11}$$

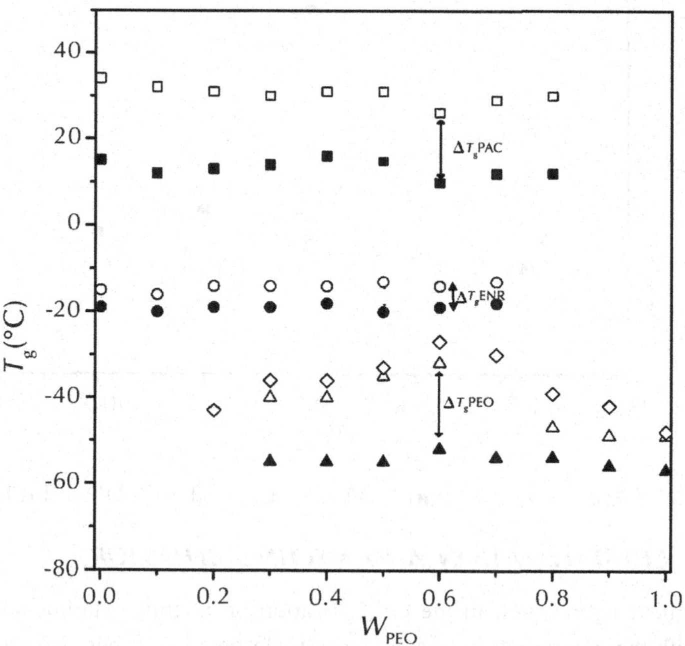

FIGURE 24 Glass transition temperatures as functions of weight fractions of PEO in salt-free and salt-added PEO/ENR and PEO/PAc blends. Solid markers denote the salt-free polymer components—(▲) PEO, (●) ENR, and (■) PAc. The salt-added components are represented by the open markers—(◊, ○) PEO and ENR added with 12 wt% $LiClO_4$, (Δ, □) PEO and PAc added with 15 wt% $LiClO_4$. The doubled headed arrow shows the difference in T_g values between each salt-free and salt-added polymer component for the two blends.

Plots of the distribution coefficient (K_D) of $LiClO_4$ as functions of W_{PEO} for the PEO/ENR/$LiClO_4$ and PEO/PAc/$LiClO_4$ blend systems are depicted in Figure 25. The difference in the ΔT_g^{PEO} and K_D values of the two amorphous polymers, ENR and PAc, are reflected in the conductivity results of the two immiscible blend systems to be discussed later.

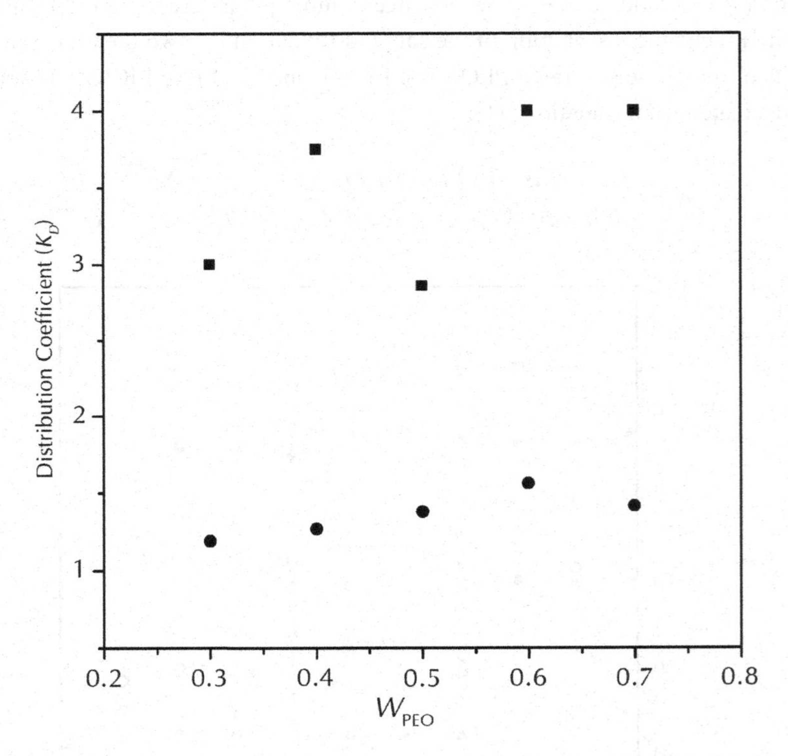

FIGURE 25 K_D of $LiClO_4$ for the (■) PEO/ENR/$LiClO_4$ and (●) PEO/PAc/ $LiClO_4$ blends.

18.8.3 CRYSTALLIZATION AND MELTING BEHAVIOR

The frequent interruption in the crystallization of a semicrystalline/amorphous blend with the incorporation of salt not only elevates its T_g, but also reduces the crystallization kinetics and G of the crystallizable component in both miscible and immiscible blends (Rocco et al., 2002; Rocco et al., 2004; Chan and Kammer, 2008; Florjanczyk et al., 2004; Acosta and Morales, 1996). Salt-free miscible PEO/PBE 60/40 blend exhibits an exothermic crystallization peak at –31 °C, but is progressively shifted to lower temperature with a broader peak when 1–2.5 wt% of $LiClO_4$ is added into the blend. This behavior is the result of the sequential heterogeneous and homogeneous nucleation with reduced rate of crystallization and smaller number of heterogeneous nuclei induced by low salt concentration (Rocco et al., 2004). The addition of salt drastically reduces the ΔH_m and the crystallinity of the miscible PEO/PMVE-MAc 60/40 blend and the melting peak completely disappears at high salt concentration indicative of a strong ion-dipole interaction between the salt and the polymer (Rocco et al., 2002). Immiscible blend systems do experience a reduction of the rate of crystallization, the melting temperature and the crystallinity upon addition of an inorganic salt. Table 7 shows a reduction in the crystallization rate represented by reciprocal half time of crystallization ($t_{0.5}^{-1}$), suppression of T_m° and crystallinity of the PEO/PAc blends by the addition of 15 wt% $LiClO_4$.

TABLE 7 Values of T_m°, X_c, and $t_{0.5}^{-1}$ for salt free-PEO/PAc blends and PEO/PAc blends added with 15 wt% $LiClO_4$. The salt-added blend is not able to crystallize at a reasonable rate at 43 °C , therefore, it is crystallized at a lower T_c = 35 °C (Sim et al. 2011)

PEO/PAc blend	Salt-free blends			Blends added with 15 wt% $LiClO_4$		
	T_m°	X_c	$t_{0.5}^{-1}$ at T_c = 43 °C	T_m°	X_c	$t_{0.5}^{-1}$ at T_c = 35 °C
100/0	87	58	1.5	67	39	1.9
90/10	80	59	1.0	66	40	1.1
80/20	77	58	0.5	72	31	0.16
70/30	71	59	0.2	60	15	0.24
60/40	73	59	0.02	66	12	0.66
50/50	79	58	0.01	-	-	-

18.8.4 MORPHOLOGY

The microstructure of PEO spherulites growing under the influence of a salt encounters dramatic changes especially at high salt concentrations. Rocco et al., (2002, 2004) in their work on the miscible blends of PEO/PMVE-MAc/$LiClO_4$ with 60 PEO and PEO/PBE/$LiClO_4$ have reported an increase in the size and perfection of the crystallites at salt content < 1 and 7.5 wt%, respectively, which correlates with the increase in X_c likely due to an increase in the orderly arrangement of the lamellae induced by low concentration of salt. At salt content ≥ 1 and 7.5 wt%, morphology of the blends drastically changes to smaller and sparsely spaced crystals which tend to disappear at salt concentration of approximately 2.5% and 10 wt%, respectively. Figure 26 shows the difference in the morphology of the PEO/PBE 60/40 blend at salt content 5 and 7.5 wt%. Transformation in the morphology with increasing salt content is the result of reduction in the nucleation and growth rate of the crystallites. Smaller, nonspherical and sparsely spaced spherulites with dark regions are also observed by Reddy et al., (2006) when potassium iodide (KI) is incorporated into the blend of PEO/PEG. This result shows the transformation of the internal crystallite structure from straight or weakly bent lamellae to randomly oriented lamellae as a result of disrupted crystallization caused by the salt which has penetrated into and entrapped within the spherulite. The amorphous regions continue to grow with increasing salt content and the spherulites getting smaller with distorted shape. The diminishing birefringence reflects the development of shorter and randomly oriented lamellae. The spherullites continue to decrease in size until at 20 wt% KI content, only the amorphous region (observed as the dark region) prevails. The SEM micrographs obtained by Mahendran and Rajendran, (2003) and Rajendran et al., (2002d) for the PMMA/PVdF/$LiClO_4$ and PMMA/PVdF/$LiClO_4$/dimethyl phthalate (DMP) films, respectively show a well dispersed phase indicating the presence of a high amorphous region in the salt-added blend which leads to higher conductivity.

FIGURE 26 *(Continued)*

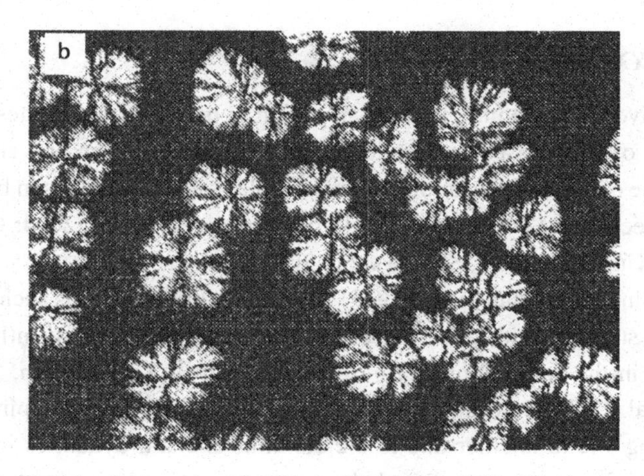

FIGURE 26 POM micrographs of PEO in PEO/PBE 60/40 blends doped with (a) 5 wt% and (b) 7.5 wt% $LiClO_4$. Reproduced from Rocco et al., 2004, copyright (2004), by permission Elsevier Ltd.

The addition of $LiClO_4$ to PEO/ENR and PEO/PAc blends causes the PEO spherulites to grow anisotropically, thus, forming nonspherical spherulites (Chan and Kammer, 2008). Similar amorphous regions which appear as dark spots within the spherulites are also observed in both the electrolyte systems at salt content ≤ 12 wt% (see Figure 27):

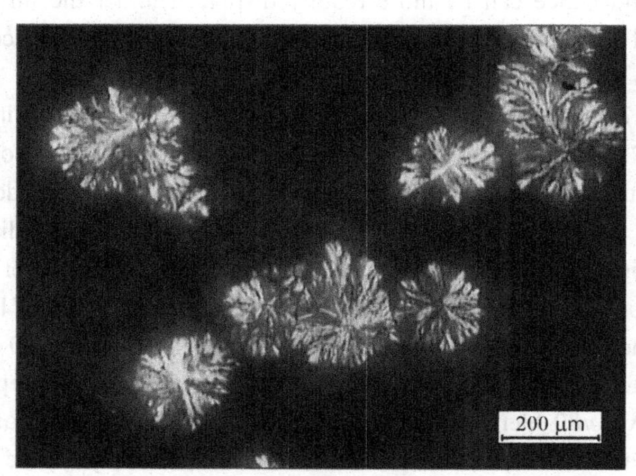

FIGURE 27 POM micrographs of PEO spherulite in immiscible PEO/PAc 60/40 blends doped with 15 wt% $LiClO_4$ after isothermal crystallized at 45 °C (Sim et al., 2011). Magnification: 5 X. The bar corresponds to 200 μm.

18.8.5 CONDUCTIVITY

The intensive studies on the ion transportation in polymer electrolytes have been carried out over the last two decades. To date, it has not been proven otherwise that both the cation and the anion are responsible as charge carriers in blend-based polymer electrolytes (Armand et al., 1979; Florjanczyk et al., 2004; Cameron et al., 1989; Kim et al., 1996; Stoeva et al., 2003).

According to Armand et al., (1979) and other authors, cations such as Li^+ ions coordinate strongly with the ether oxygen of PEO, and consequently retard its movement in the amorphous phase of the electrolyte (Kim and Oh, 2000; Florjanczyk et al., 2004). Ion conduction is realized with the cations hopping from one coordinating site to another due to the continuous chain dynamics in the amorphous phase formed by the PEO chain coiled into a regular helix. A linear log σ versus $1/T$ plot confirms the ion hopping mode of transportation for the cations (Armand et al., 1979; Kumar et al., 1993; Armand, 1986). On the other hand, mobility or rather diffusion of anions through the amorphous phase facilitated by the segmental motion of the polymer chain exhibits a curvature of the log σ versus $1/T$ plot following the Vogel-Tamman-Fulcher (VTF) relation (Fulcher, 1925), which describes transport property in a viscous matrix (Reddy et al., 2006; Mahendran and Rajendran, 2003; Stoeva, 2003). The higher transport number of anions and its mode of mobility reveal that the anions are loosely associated to the polymer segments, and hence can be more readily displaced under the influence of an electric field (Kim et al., 1996). This explains its higher transference number as compared to the cations of a polymer electrolyte.

The ion conductivity of the three PEO/PMMA blend compositions, 25/75, 50/50, and 75/25 studied by us increases with ascending $LiClO_4$ content due to increasing number of free mobile ions as shown in Figure 28. In addition, blend composition with 75 wt% PEO doped with 0 to 12 wt% $LiClO_4$ displays comparatively higher σ values in the order of 10^{-7} S cm^{-1} than the other two blends. Nevertheless, Tan and Johan, (2011) studied the ion conductivity of PEO/PMMA blend polymer electrolyte found that the composition 20 wt% PEO and 80 wt% PMMA is the most miscible proportion for the blend. Figure 28 depicts that the PEO/PMMA 20/80 blend achieves a maximum ion conductivity of 7×10^{-4} S cm^{-1} at 10 wt% $LiClO_4$, further enhancement in conductivity can be achieved by the addition of a low molecular weight, low viscosity, and high dielectric constant plasticizer, EC. The incorporation of EC facilitates the dissociation of the salt leading to higher number of free mobile ions. Besides, it provides more free volume to the polymer which lowers the glass transition temperature. The highest σ value obtained for the PEO/PMMA/$LiClO_4$/EC 20/80/10/20 blend is 2×10^{-3} S cm^{-1} which is higher than 4×10^{-4} and 6×10^{-5} S cm^{-1} at ambient temperature for the blend systems of PEO/PMMA/$LiClO_4$/DMP 79/10/10/1 mol% and PEO/PMMA/$LiBF_4$/DMP 17.5/7.5/8/67 mol% reported by Rajendran et al., (2002c, 2001), respectively.

Similar ion conductivity in the order of 10^{-4} S cm^{-1} has also been reported by Morita et al., (1996) and Peled et al., (1995) for a PEO/PMMA/$LiClO_4$/EC, and PEO/PMMA/LiI/EC complexes, respectively. Higher σ values of the plasticized PEO/PMMA blend-based polymer electrolyte are ascribed to the penetration of the small plasticizer molecules into the polymer matrix, interacting with the chain segments, thus, reducing the cohesive forces between the inter- and intra-polymer chains, and consequently, lowering the T_g and increases segmental mobility (Rajendran et al., 2001; Rajendran et al., 2002c; Tan and Johan, 2011).The ion conductivity of both the PEO/PMMA/$LiClO_4$/DMP and PEO/PMMA/$LiBF_4$/DMP electrolyte systems as a function of temperature agrees with the Arrhenius plot, and thus, can be explained by the VTF equation (Equation 12):

$$\sigma = AT^{-\frac{1}{2}} \exp[-\frac{B}{T} - T_g] \tag{12}$$

where T_g is the reference temperature, constant A is related to the number of charge carriers in the electrolyte system while constant B is related to the activation energy of ion transport associated with the configuration entropy of the macromolecular chain. The ion conductivity of the plasticized blends increasing with ascending temperature suggests that ion transport occurs in the plasticizer-rich phase.

The effect of adding 15 wt% and 12 wt% $LiClO_4$ on the ion conductivity of the immiscible blends of PEO/PAc (Sim et al., 2011) and PEO/ENR (Chan and Kammer, 2008), respectively, is shown in Figure 29. It is interesting to note that the σ values at room temperature of neat PAc and neat ENR are approximately 10^{-11} S cm^{-1}, however, with an addition of 15 wt% $LiClO_4$, the σ value of PAc is 10^{-10}, one order lower than that of ENR doped with 12 wt% $LiClO_4$. Further, both the salt-free and the salt-added PEO/PAc blends obtain lower σ values than those of the PEO/ENR blends.

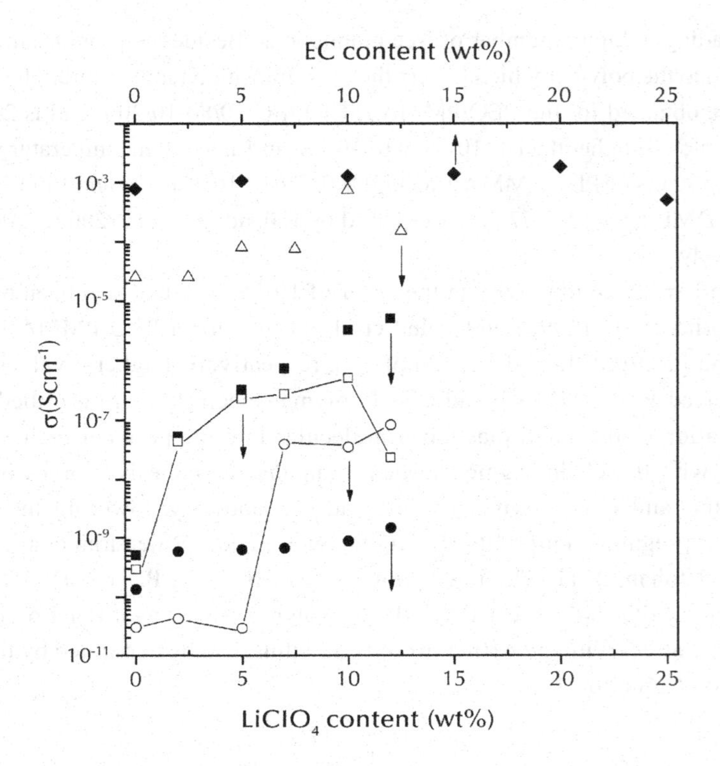

FIGURE 28 Plots of ion conductivity of PEO/PMMA blends as functions of $LiClO_4$ and EC composition. (■) neat PEO and (●) neat PMMA. Blends of PEO/PMMA (□) 75/25, (○) 25/75 (unpublished results), (Δ) 20/80 (Tan and Johan, 2011), and (◆) PEO/PMMA 20/80 blend doped with 10 wt% $LiClO_4$. Modified from Tan and Johan, (2011).

Additionally, the σ value of PEO declines when blended with PAc, contrary to an increase in σ value when ENR is added to PEO in the absence or presence of $LiClO_4$. The T_g values, as discussed earlier, show that the salt is more soluble in PAc as compared to ENR. Therefore, with a fixed salt content, the amount of salt dissociated in the PEO amorphous phase is definitely higher for the PEO/ENR blend compared to the PEO/PAc blend. Besides, the T_g values of PAc in the presence of salt is raised to a range of 29–37 °C which means the PAc is in its glassy state when ion conductivity of the blend is measured leading to restricted ion mobility in the PEO amorphous phase which forms the predominant percolating pathway of the blend electrolyte. It can be concluded that the ion conductivity of miscible or immiscible PEO-based blend electrolyte is governed by the charge carrier density, segmental, and ion mobility in the PEO amorphous phase (Chan and Kammer, 2008; Sim et al., 2011).

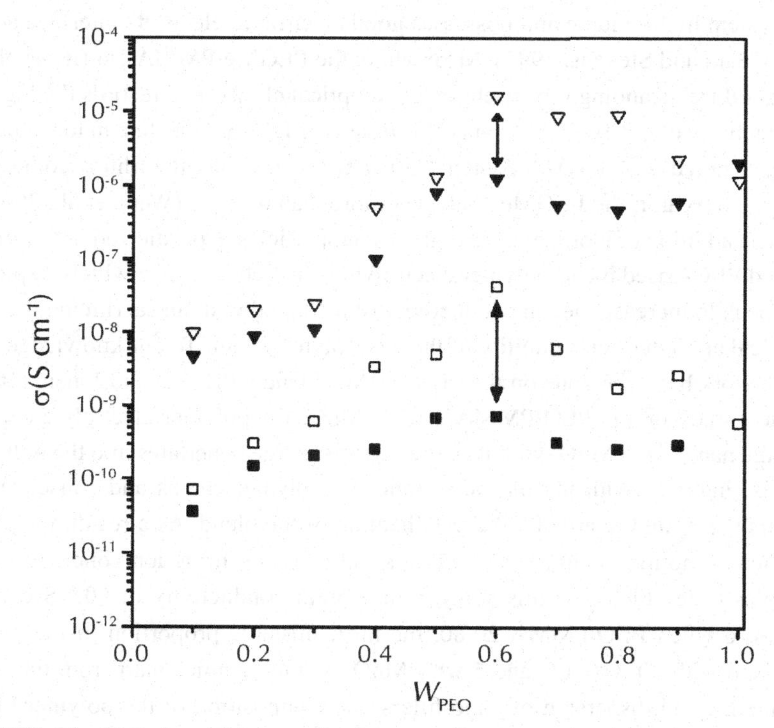

FIGURE 29 Variations of ion conductivity as a function of weight fractions of PEO. (□,▽) denote blends of PEO/ENR and PEO/ENR with 12 wt% $LiClO_4$. (■, ▼) correspond to blends of PEO/PAc and PEO/PAc added with 15 wt% LiClO, respectively. The two headed arrows indicate the difference in σ values between the two blends at blend composition PEO/ENR and PEO/PAc 60/40.

The blends of low molecular weight poly(propylene glycol) (PPG) (M_w = 2000 g mol^{-1}) (Mani and Stevens, 1992) and poly(ethylene glycol) methyl ether (PEGMe) (Wang et al., 2005) with PMMA doped with lithium salt have been applied for smart windows and all-solid electrochromic devices due to their transparent, adhesive, and viscoelastic characteristics coupled with high conductivity. Owing to the hazy appearance arising from the phase separation of the crystalline micro-domain from the PEO matrix, blends of PEO/PMMA are not suitable for electrochromic applications. The composition of PMMA, salt content, and the low molecular weight polymer, each plays an important role in the preparation of a free standing electrolyte with good optical clarity and high conductivity. For the PPG/PMMA/$LiCF_3SO_3$ blend system, incorporation of 8 wt% of PMMA makes the system hydrophobic and possesses good electrolyte-electrode interface adhesion (Mani and Stevens, 1992). Meanwhile, the PEGMe/PMMA 1:4 (w/w) blend has good free-standing film strength, good optical clarity, and records the highest conductivity of 6×10^{-5} S cm^{-1} at [Li]/[O] ratio of 1:13.6. A decline in ion conductivity is noted with PMMA content >30 wt%, meanwhile, the film becomes too tacky and weak if the PEGMe content is more than 69 wt% (Wang et al., 2005).

The addition of inorganic and organic nanoparticles to plasticized or non-plasticized PEO-based blend polymer electrolyte is in recent years a widely explored approach to increase the ion conductivity of the blend without sacrificing the mechanical and interfacial stability of the electrolyte system. To our knowledge, not much work has been done on the PEO/PMMA blend CPE. Tan and Johan, (2011) in their study on the PEO/PMMA/$LiClO_4$/MnO_2/EC polymer electrolyte depicts that the nano-sized MnO_2 with its large surface charge penetrates into the polymer matrix, interacts with the plasticizer and the polymer chains and consequently lowers the T_g and retards PEO crystallization of the blend. As a result, the chain segmental motion is enhanced leading to an increase in the ion conductivity of the plasticized blend. In this study, a maximum conductivity of 10^{-3} S cm^{-1} is achieved when PEO/PMMA 20/80, the most miscible proportion of the blend, is mixed with 20 wt% EC and 5 wt% MnO_2. On the whole, apart from the type, grain size, and dispersion of nano-fillers, the composition of the polymer blend components, salt, plasticizer, and filler do play an important role in improving the ion conductivity of a composite polymer electrolyte. For PEO/PMMA-salt complexes to be of practical application in solid-state batteries and electrochemical devices, the relatively high T_g of the blend has to be lowered without sacrificing the good mechanical and interfacial stability of the blend. As such, an optimum composition of an appropriate plasticizer and nano-fillers used will be crucial in determining the physical properties and the quality of the nano-CPE.

18.9 CONCLUSION

All miscible blends of PEO/polyacrylate exhibit a single T_g which show negative deviation from the theoretical Fox relation especially for blends with intermediate compositions (blend compositions with PMMA content in the range of ≈ 30–70 wt%). The T_g values of these blend compositions which deviate from Fox equation are usually found closer to that of PEO. In view of the weak dipole-dipole interactions between the blending components and a small negative χ_{12} close to zero, miscibility in the PEO/polyacrylate system is considered to be marginally miscible. Besides, the miscible blends also exhibit progressive decrease in the rate of crystallization, X_c, and T_m^o with ascending polyacrylate content. On the contrary, the immiscibility of PEO/PAc blend is marked by the presence of two T_gs and non-systematic variations of the values of crystallization rate and T_m^o as functions of ascending PAc. Addition of an optimum amount of nano-particles enhances tremendously the thermal, physical, and mechanical properties of the PEO/PMMA blend through the formation of an intercalation network between the nano-particles and the blending components.

Incorporation of low concentration of inorganic salt to the PEO/polyacrylate blends strengthen the linkages between the two blending components as the cations form complexes of the type PEO-Li^+-polyacrylate *via* ion-dipole attractions. However, phase separation occurs at higher salt concentration as the stronger ion-dipole interactions between Li^+ ion and the individual blending component replace the weak dipole-dipole interactions between the two blending components especially for blends with intermediate compositions. Therefore, miscible blends with composition 30 wt% > PMMA content >70 wt% are usually selected as the matrix for polymer electrolytes. Owing to the high T_g of the blend which leads to poor segmental mobility, ceramic fillers or a combination of both fillers and a small amount of plasticizer are added to the PEO/PMMA/salt system to lower the T_g, hence, enhances the ion conductivity and meanwhile, improves the mechanical, thermal, and interfacial properties.

In view of the many parameters involved *viz.* blend composition, types of plasticizer, and nano-fillers, grain size, and dispersion and so on, an in-depth study of the conductivity mechanism of the electrolyte system and an understanding of the characteristic and behavior of each additive are important in the pursuit of an applicable PEO/PMMA composite polymer electrolyte system.

ACKNOWLEDGEMENT

The authors would like to express our gratitude towards Ministry of Higher Education (MOHE) for the “Dana Pembudayaan Penyelidikan” grant (RAGs) (600-RMI/RAGS 5/3 (14/2012)) and Research Management Institute, Universti Teknologi MARA for the “Dana Kecemerlangan grant” (600-RMI/ST/DANA/ 5/3/Dst (427/2011)) supporting the research work.

KEYWORDS

- **Poly(ethylene glycol)**
- **Poly(methyl acrylate)**
- **Poly(methyl methacrylate)**
- **Poly-2-ethylhexyl acrylate**
- **Small angle X-ray scattering**

REFERENCES

1. Abraham, K. M. and Alamgir, M. *J. Electrochem. Soc.*, **137**, 1657–1658 (1990).
2. Acosta, J. L. and Morales, E. *Solid State Ionics*, **85**, 85–90 (1996).
3. Agnihotry, S. A., Ahmad, S., Gupta, D., and Ahmad, S. *Electrochim. Acta*, **49**, 2343–2349 (2004).
4. Agrawal, R. C., Mahipal, Y. K., and Ashrafi, R. *Solid State Ionics*, **192**, 6–8 (2011).
5. Alfonso, G. and Russel, T. P. *Macromolecules*, **19**, 1143–1152 (1986).
6. Ali, A. M. M., Yahya, M. Z. A., Bahron, H., Subban, R. H. Y., Harun, M. K., and Atan, I. *Material Letters*, **61**, 2026–2029 (2007).
7. Appetecchi, G. B., Croce, F., Persi, L., Ronci, F., and Scrosati, B. *Electrochim. Acta*, **45**, 1481 (2000).
8. Appetecchi, G. B., Croce, F., and Scosati, B. *Electrochimica Acta*, **40**, 991–997 (1995).
9. Armand, M. B. *Ann. Rev. Mater. Sci.*, **16**, 245–261 (1986).
10. Armand, M. B., Chabagno, J. M., and Duclot, M. In: P. Vashisha, J. N. Mundy, and G. K. Shenoy (eds.), *Fast Ion Transport in Solids*. New York: North Holland, pp. 131–136 (1979).
11. Avrami M. *J. Chem. Phys.*, **7**, 1103–1112 (1939).
12. Bailey, Jr. F. E. and Koleske, J. V. *Poly(ethylene oxide)*. New York, San Francisco, London, Academic Press, Inc., pp. 37–38, 115 (1976).
13. Bamford, D., Reiche, A., Dlubek, G., Alloin, F., Sanchez, J. Y. and Alam, M. A. *J. of Chem. Phy.*, **118**, 9420–9432 (2003).
14. Bandara, L. R. A. K., Dissanayake, M. A. K. L., and Mellander, B. E. *Elctrochim. Acta*, **43**, 1447 (1998).
15. Baskaran, R., Selvasekarapandian, S., Kuwata, N., Kawamura, J., and Hattori, T. *Materials Chemistry and Physics*, **98**, 55–61 (2006).
16. Berthier, C., Gorecki, W., Minier, M., Armand, M. B., Chabagno, J. M., and Rigaud, P. *Solid State Ionics*, **11**, 91–95 (1983).
17. Bohnke, O., Frand, G., Rezrazi, M., Rousselot, C., and Truche C. *Solid State Ionics*, **66**, 105–112 (1993b).
18. Bohnke, O., Frand, G., Rezrazi, M., Rousselot, C., and Truche, C. *Solid State Ionics*, **66**, 97–104 (1993a).
19. Borkowska, R., Laskowski, J., Płocharski, J., Przyłuski, J., and Wieczorek, W. *J. Appl. Electrochem.*, **23**, 991–995 (1993).

20. Brandrup, J., Immergut, E. H., and Grulke, E. A. *Polymer Handbook*, Taylor and Francis Group, LLC (2009).
21. Burgaz, E. *Polymer*, **52**, 5118–5126 (2011).
22. Calahorra, E., Cortazar, M., and Guzmán, G. M. *Polymer*, **23**, 1322–1324 (1982).
23. Cameron, G. G., Ingram, M. D., and Harvie, J. L. *Faraday Discuss. Chem. Soc.*, **88**, 55–63 (1989).
24. Capiglia, C., Mustarelli, P., Quartarone, E., Tomasi, C., and Magistris, A. *Solid State Ionics*, **118**, 73–79 (1999).
25. Caruso, T., Capoleoni, S., Cazzanelli, E., Agostino, R. G., Villano, P., and Passerini, S. *Ionics*, **8**, 36–43 (2002).
26. Chan, C. H. and Kammer, H. W. *J. Appl. Polym. Sci.*, **110**, 424–432 (2008).
27. Chan, C. H., Kummrlowe, C., and Kammer, H-W. *Macromol. Chem. Phys.*, **205**, 664–675 (2004).
28. Chandra, A., Srivastava, P. C., and Chandra, S. *J. Mater. Sci*, **30**, 3633–3638 (1995).
29. Chandra, S. and Chandra, A. Proceedings of National Academy of Sciences, India, Section A, Part II. Solid state ionics: materials aspect, pp. 141–181 (1994).
30. Chaurasia, S. K., Singh, R. K., and Chandra, S. *Solid State Ionics*, **183**, 32–39 (2011).
31. Chen, X., Yin J., Alfonso, G. C., Pedemonte, E., Turturro, A., and Gattiglia, E. *Polymer*, **39**, 4929–4935 (1998).
32. Chen, Y. J., Mays, J. W., and Hadjichristidis, N. *J. Polym. Sci. Polym. Phys. Ed.*, **32**, 715–719 (1994).
33. Chen-Yang, Y. W., Wang, Y. L., Chen, Y. T., Li, Y. K., Chen, H. C., and Chiu, H. Y. *J. Power Sources*, **182**, 340–348 (2008).
34. Choe H. S., Giaccai, J., Alamgir, M., and Abraham, K. M. *Electrochim. Acta*, **40**, 2289–2293 (1995).
35. Choi, B. K. and Kim, Y. W. *Electrochim. Acta*, **49**, 2307–2313 (2004).
36. Choi, B. K., Kim, Y. W., and Shin, K. H. *J. of Power Sources*, **68**, 357–360 (1997).
37. Choi, B. K. *Solid State Ionics*, **168**, 123–129 (2004).
38. Cimmino, S., Di Pace, E., Martuscelli, E., and Silvestre, C. *Makromol. Chem., Rapid Commun.*, **9**, 261–265 (1988).
39. Cimmino, S., Iodice, P., Martuscelli, E., and Silvestre, C. *Thermochimica Acta*, **32**, 89–98 (1998).
40. Cimmino, S., Pace, E. D., Martuscelli, E., and Silvestre, C. *Makromol. Chem.*, **191**, 2447–2454 (1990).
41. Colby, R. H. *Polymer*, **30**, 1275–1278 (1989).
42. Cortazar, M. M., Calahorra, M. E., and Guzman, G. M. *Eur. Polym. J.*, **18**, 165–166 (1982).
43. Croce, F., Brown, S. D., Greenbaum, S. G., Slane, S. M., and Salomon, M. *Chem. Mater.*, **5**, 1268–1272 (1993).
44. Croce, F., Curini, R., Martinelli, A., Persi, L., Ronci, F., Scrosati, B., and Caminiti, R. *J. Phys. Chem. B*, **103**, 10632–10638 (1999).
45. Croce, F., Gerace, F., Dautzemberg, G., Passerini, S., Appetecchi, G. B., and Scrosati, B. *Electrochimica Acta*, **39**, 2187–2194 (1994).
46. Croce, F., Persi, L., Scrosati, B., Serraino-Fiory, F., Plichta E., and Hendrickson, M. S. *Electrochimica Acta*, **46**, 2457–2461 (2001).
47. Datasheet on Elvacite, ICI, (Technical Data Sheet) (1996).

48. Deepa, M., Agnihotry, S. A., Gupta, D., and Chandra, R. *Electrochim. Acta*, **49**, 373–383 (2004).
49. Deepa, M., Sharma, N., Agnihotry, S. A., Singh, S., Lal, T., and Chandra, R. *Solid State Ionics*, 152–153, 253–258 (2002b).
50. Deepa, M., Sharma, N., Agnihotry, S. A., and Chandra, R. *J. Mater. Sci.*, **37**, 1759–1765 (2002a).
51. Dey, A., Karan, S., and De, S. K. *Solid State Communications*, **149**, 1282–1287 (2009).
52. Dey, A., Karan, S., and De. S. K. *Solid State Ionics*, **178**, 1963–1968 (2008).
53. Dionisio, M., Fernandes, A. C., Mano, J. F., Correia, N. T., and Sousa, R. C. *Macromolecules*, **33**, 1002–1011 (2000).
54. Dissanayake, M. A. K. L., Jayathilaka, P. A. R. D., Bokalawala, R. S. P., Albinsson, I., and Mellander, B. E. *Journal of Power Sources*, **409**, 119–121 (2003).
55. Dreezen, G., Fang, Z., and Groeninckx, G. *Polymer*, **40**, 5907–5917 (1999).
56. Drzewinski, M. A. *Blends of polycarbonates containing fluorinated-bisphenol A and poly(methyl acrylate)*. US Patent. 5,292,809, (1994).
57. Drzewinski, M. A. *Blends of polyetherimides, aromatic alkyl methacrylates, and polycarbonates*. US Patent. 5,248,732, (1993).
58. Dumoulin, M. M. Polymer blends forming. In: L. A. Utracki (ed.), *Polymer blends handbook vol. 2. Dordrecht*, The Netherlands: Kluwer Academic, pp. 653–755 (2002).
59. Fang, D. Master Thesis, KTH Institute of Technology, (2009).
60. Fischer, B., Ziadeh, M., Pfaff, A., Breu, J., and Altstädt, *J. Polymer*, **53**, 3230–3237 (2012).
61. Fischer, E. W., Wendorff, J. H., Dettenmaier, M., Lieser, G., and Voigt-Martin, I. *J. Macromol. Sci. Phys.*, **12**, 41–59 (1976).
62. Floriańczyk, Z., Such, K., Wieczorek, W., and Wasiucionek, M. *Polymer*, **32**, 3422–3425 (1991).
63. Floriańczyk, Z., Marcinek, M., Wieczorek, W., and Langwald, N. *Polish J. Chem.*, **78**, 1279–1304 (2004).
64. Flory, P. J. *J. Chem. Phys.*, **10**, 51–61 (1942).
65. Forger, G. R. *Mater. Eng.*, **85**, 44- (1977).
66. Fox, T. G. *Bull. Am. Phys. Soc.*, **1**, 123–135 (1956).
67. Fulcher, G. S. *J. Am. Ceram. Soc.*, **8**, 339–355 (1925).
68. Gaborieau, M., Graf, R., Kahle, S., Pakula, T., and Spiess, H. W. *Macromolecules*, **40**, 6249–6256 (2007).
69. Gan, S. N. *Water-reducible acrylic copolymer dipping and rubber products coated with same*. Malaysia Pat, P120,055,440 (2005).
70. Gao, Z., Molnar, A., Morin, F. G., and Eisenberg, A. *Macromolecules*, **25**, 6460–6465 (1992).
71. Gargallo, L., Martínez-Piña, F., Leiva, A., and Radić, D. *Eur. Polym. J.*, **32**, 1303 (1996).
72. Gargallo, L. and Russo, M. *Makromol. Chem.*, **176**, 2735 (1975).
73. Gedde, U. W. *Polymer Physics*, 1st edition. UK: Chapman and Hall, (1995).
74. Glasse, M. D., Idris, R., Latham, R. J., Linford, R. D., and Schlindwein, W. S. *Solid State Ionics*, **147**, 289–294 (2002).

75. Groeninckx, G., Vanneste, M., and Everaert, V. Crystallization, morphological structure, and melting of polymer blends. In: L. A. Utracki, (ed.), *Polymer Blends Handbook*, vol. 1, pp. 203–294 (2002).
76. Grundmeier, M., Binsack, R., and Vernaleken, H. *Polycarbonate molding compositions*. U.S. Patent 4,056,504 (1977).
77. Guilherme, L. A., Borges, R. S., Mara, E., Moraes, S., Goulart Silva, G., Pimenta, M. A., Marletta, A., and Silva, R. A. *Electrochim. Acta*, **53**, 1503–1511 (2007).
78. Guo, Q., Harrats, C., Groeninckx, G., and Koch, M. H. Z. *Polymer*, **42**, 4127–4140 (2001).
79. Gupta, R. K. and Rhee, H. W. *Electrochimica Acta*, **76**, 159–164 (2012).
80. Hamon, L., Grohens, Y., Soldera, A., and Holl, Y. *Polymer*, **42**, 9697-9703 (2001).
81. Han, H. S., Kang, H. R., Kim, S. W., and Kim, H. T. *J. Power Sources*, **112**, 461–468 (2002).
82. Hanake, E. *Semin. Cutan. Med. Surg.*, **23**, 227 (2004).
83. Hatakeyama, T. and Quinn, F. X. *Thermal Analysis: Fundamentals and Applications to Polymer Science* (2nd Edition), John Wiley & Sons Ltd. pp. 45–46 (1999).
84. Hoffman, D. M. Ph.D Thesis. University of Massachusetts, Amherst, (1979).
85. Hoffman, J. D. and Weeks, J. J. *Journal of Research of the National Bureau of Standards - A. Physics and Chemistry*, **66**, 13–28 (1962).
86. Hsieh, K. H., Ho, K. S., Wang, Y. Z., Ko, S. D., and Fu, S. C. *Synthetic Metals*, **123**, 217–224 (2001).
87. Huang, G. S., Jiang, L. X., and Li, Q. *J. Appl. Polym. Sci*, **85**, 746–751 (2002).
88. Huggins, M. L. *J. Phys. Chem.*, **46**, 151–158 (1942).
89. Ito, H., Russell, T. P., and Wignall, G. D. *Macromolecules*, **20**, 2213–2220 (1987).
90. Jacob, M. M. E., Prabaharan, S. R. S., and Radhakrishna, S. *Solid State Ionics*, **104**, 267–276 (1997).
91. Jeon, J. D., Kim, M. J., and Kwak, S. Y. *J. Power Sources*, **162**, 1304–1311 (2006).
92. Johan, M. R., Oon, H. S., Ibrahim, S., Mohd Yassin,S. M., and Tay, Y. H. *Solid State Ionics*, **196**, 41–47 (2011).
93. Joykumar Singh. Th. and Bhat, S. V. *J. Power Sources*, **129**, 280–287 (2004).
94. Karey, F. E. and MacKnight, W. J. *Macromolecules*, **1**, 537 (1968).
95. Katime, I. and Cadenato, A. *Materials Letters*, **22**, 303–308 (1995).
96. Kato, S. and Kawaguchi, M. *J. of Colloid and Interface Science*, **384**, 87–93 (2012).
97. Kennedy, C. *ICI- The company that changed our lives*, Paul chapman Publishing, London, (1993).
98. Kern, R. J. *Homogeneous polymer blends*. US Patent. 2,806,015, (1957).
99. Ketabi, S. and Lian, K. *Solid State Ionics*, **227**, 86–90 (2012).
100. Kim, C. S. and Oh, S. M. *Electrochimica Acta*, **45**, 2101–2109 (2000).
101. Kim, D. W., Park, J. K., and Rhee, H. W. *Solid State Ionics*, **83**, 49–56 (1996).
102. Kim, H. B., Choi, J. S., Lee, C. H., Lim, S. T., Jhon, M. S., and Choi, H. J. *European Polymer Journal*, **41**, 679–685 (2005).
103. Kim, H. S., Kum, K. S., Cho, W. I., Cho B. W., and Rhee, H. W. *J. Power Sources*, **124**, 221–224 (2003).
104. Kim, I. T., Lee, J. H., Shofner, M. L., Jacob, K., and Tannenbaum, R. *Polymer*, **53**, 2402–2411 (2012).

105. Kim, J. W., Noh, M. H., Choi, H. J., Lee, D. C., and Jhon, M. S. *Polymer*, **41**, 1229 (2000).
106. Ko, S. J., Kim, S. J., Kong, S. H., and Bae, Y. C. *Electrochimica Acta*, **49**, 461–468 (2004).
107. Koningsveld, R. Ph.D. Thesis, University of Leiden, (1967).
108. Krause, S., Gormley, J. J., Roman, N., Shetter, J.A., and Watanabe, W. H. *J. Polym. Sci. Part A: Polym. Chem*, **3**, 3573–3586 (1965).
109. Krivorotova, T., Vareikis, A., Gromadzki, D., Netopilík, M., and Makuška, R. *European Polymer Journal*, **46**, 546–556 (2010).
110. Krückel, J., Starý, Z., Triebel, C., Schubert, D. W., and Münstedt, H. *Polymer*, **53**, 395–342 (2012).
111. Kruse, W. A., Kirste, R. G., Haas, J., Schmitt, B. J., and Stein, D. *J. Makromol. Chem.*, **177**, 1145–1160 (1976).
112. Kryszewski, M., Galeski, A., Pakula, T., and Grebowicz, J. *J. Colloid Interface Sci.*, **44**, 85–94 (1973).
113. Kumar, B., Schaffer, J. D., Munichandraiah N., and Scanlon, L. G. *J. Power Sources*, **47**, 63–78 (1994).
114. Kumar, B., Weissman, P. T., and Marsh, R. A. *J. Electrochem. Soc.*, **140**, 320–323 (1993).
115. Kumar, M. and Sekhon, S. S. *Eur. Polym. J.*, **38**, 1297–1304 (2002).
116. Kumar, R. and Sekhon, S. S. *Ionics*, **14**, 509–514 (2008).
117. Kumar, Y., Hashmi, S. A., and Pandey, G. P. *Electrochimica Acta*, **56**, 3864–3873 (2011).
118. Labreche, C., Levesque, I., and Prud'homme, J. *Macromolecules*, **29**, 7795–7801 (1996).
119. Lartigue, C., Guillermo, A., and Cohen-Addad, J. P. *J. Polym Sci B: Polym Phys*, **35**, 1095–1105 (1997).
120. Lewis, O. G. *Physical constant of Linear Hydrocarbons*, Springer-Verlag, (1968).
121. Li, J. and Khan, I. M. *Macromolecules*, **26**, 4544–4550 (1993).
122. Li, W., Yuan, M., and Yang, M. *Eur. Polym. J.*, **42**, 1396–1402 (2006).
123. Li, X. and Hsu, S. L. *J. Polym. Sci.:Polym. Phys. Ed.*, **22**, 1331–1342 (1984).
124. Li, Z., Li, P., and Huang, J., *Polymer*, **47**, 5791–5798 (2009).
125. Lim, S. K., Kim, J. W., Chin, I., Kwon, Y. K., and Choi, H. J., *Chem. Mater.*, **14**, 1989–1994 (2002).
126. Longworth, R. *Compatible blends of 1-olefins*. U.S. Patent 3,299,176, (1967).
127. Lü, H., Zheng, S., and Tian, G. *Polymer*, **45**, 2897–2909 (2004).
128. Ma, Z. and Dezmazes, P. L., *Polymer*, **45**, 6789–6797 (2004).
129. Mahendran, O., Chen, S. Y., Chen-Yang, Y. W., Lee, J. Y., and Rajendran, S. *Ionics*, **11**, 251–258 (2005).
130. Mahendran, O. and Rajendran, S. *Ionics*, **9**, 282–288 (2003).
131. Mandal, T. K., Kuo, J. F., and Woo, E. M. *J. Polym. Sci. Polym. Phys. Ed.*, **38**, 562–572 (2000).
132. Mani, T. and Stevens, J. R. *Polymer*, **33**, 834–837 (1992).
133. Marcinek, M., Ciosek, M., Żukowska, G., Wieczorek, W., Jeffrey, K. R., and Stevens, J. R. *Solid State Ionics*, **176**, 367–376 (2005).

134. Marco, C., Fatou, J. G., Gomez, M. A., Tanaka, H., and Tonelli, A. E. *Macromolecules*, **23**, 2183–2188 (1990).
135. Marco, G., Di Lanza, M., and Pieruccini, M. *Solid State Ionics*, **89**, 117–125 (1996).
136. Martuscelli, E., Demma, G., Rossi, E., and Segre, A. L. *Polymer Communications*, **24**, 266–267 (1983a).
137. Martuscelli, E. and Demma, G. B. In: E. Martuscelli, R. Palumbo, M. Kryszewski, and A. Galeski. (eds.), *Polymer Blends: Processing Morphology and Properties*. New York: Plenum Press., pp. 101–122 (1980).
138. Martuscelli, E., Marchena, C., and Nicolais, L. *Future Trends in Polymer Science and Technology–Polymers, Commodities or Specialties*. Basel, Switzerland, Technomic Pub. Co., p. 247 (1987).
139. Martuscelli, E., Pracella, M., and Wang, P. Y. *Polymer*, **25**, 1097–1105 (1984).
140. Martuscelli, E., Silvestre, C., Addonizio, M. L., and Amelino, L. *Makromol. Chem.*, **187**, 1557–1571 (1986).
141. Martuscelli, E., Silvestre, C., and Bianchi, L. *Polymer*, **24**, 1458–1468 (1983b).
142. Marzantowicz, M, Dygas, J. R., Krok, F., Łasińska, A., Florjańczyk, F., Zygadło-Monikowska, E., and Affek, A. *Electrochim Acta*, **50**, 3969–3977 (2005).
143. McMaster, L. P. *Macromolecules*, **6**, 760–773 (1973).
144. Menard, K. P. *Dynamic Mechanical Analysis, A Practical Introduction (2nd Edition)*, Taylor and Francis Group LLC, pp. 1–4 (2008).
145. Mishra, R. and Rao, K. J. *Solid State Ionics*, **106**, 113–127 (1998).
146. Mohan, V. M., Bhargav, P. B., Raja, V., Sharma, A. K., and Narasimha Rao, V. V. R. *Soft Materials*, **5**, 33–46 (2007).
147. Molnar, A. and Eisenberg, A. *Macromolecules*, **25**, 5774–5782 (1992).
148. Morita, M., Tanaka, H., Ishikawa,EH M., and Matsuda, Y. *Solid States Ionics*, **86**, 401 (1996).
149. Mu, D., Huang, X. R., Lu, Z. Y., and Sun, C. C. *Macromolecules*, **45**, 2035–2049 (2012).
150. Munichandratah, N., Scanlon, L. G., Marsh, R. A., Kumar B., and Sirear, A. K. *J. Appl. Electochem.*, **25**, 857–863 (1995).
151. Munich-Elmer, A. and Jannasch, P. *Solid State Ionics*, **177**, 573–9 (2006).
152. Münstedt, H., Köppl, T., and Triebel, C., *Polymer*, **51**, 185–191 (2010).
153. Murata, K., Izuchi, S., and Yoshihisa, Y. *Electrochimica Acta*, **45**, 1501–1508 (2000).
154. Nicotera, I., Ranieri, G. A., Terenzi, M., Chadwick, A. V., and Webster, M. I. *Solid State Ionics*, **146**, 143–150 (2002).
155. Nishi, T. and Wang, T. T. *Macromolecules*, **8**, 909 (1975).
156. Okerberg, B. C. and Marand, H., *J. Mater. Sci.*, **42**, 4521–4529 (2007).
157. Olabisi, O., Robeson, L. M., and Shaw, T. *Polymer-polymer Miscibility*. New York, Academic Press, (1979).
158. Olabisi, O. and Simba, R., *Macromolecules*, **8**, 206 (1975).
159. Osman, Z., Ansor, N. M., Chew, K. W., and Kamarulzaman, N. *Ionics*, **11**, 431–435 (2005).
160. Othman, L., Chew, K. W., and Osman, Z. *Ionics*, **13**, 337–342 (2007).
161. Pandey, G. P., Hashmi, S. A., and Agrawal, R. C., *Solid State ionics*, **179**, 543–549 (2008).

162. Panero, S., Scrosati, B., and Greenbaum, S. G. *electrochim. Acta.*, **37**, 1533–1538 (1992).
163. Parizel, N., Laupretre, F., and Monnerie, L. *Polymer*, **38**, 3719–3725 (1997).
164. Patnaik, P. *Dean's Analytical Chemistry Handbook (2nd Edition)*, The McGraw-Hill Companies, Inc. pp. 21–25 (2004).
165. Paul, D. R. and Newman, G. *Polymer Blends*. New York, Academic Press, (1979).
166. Pedrosa, P., Pomposo, J. A., Calahorra, E., and Cortazar, M. *Polymer*, **36**, 3889–3897 (1995).
167. Peled, E., Goloditsky, .D., Ardel, G., and Eshkenazy, E., *Electrochim. Acta*, **40**, 2197 (1995).
168. Pereira, R. P. and Rocco, A. M. *Polymer*, **46**, 12493–12502 (2005).
169. Pfefferkorn, D., Kyeremateng, S. O., Busse, K., Kammer, H. W., Thurn-Albrecht, T., and Kressler, J. *Macromolecules*, **44**, 2953–2963 (2011).
170. Pitawala, H. M. J. C., Dissanayake, M. A. K. L., and Seneviratne, V. A. *Solid State Ionics*, **178**, 885–888 (2007).
171. Porter, D. *Group Interaction Modelling of Polymer Properties*, Dekker, 283 (1995).
172. Quartarone, E., Mustarelli, P., and Magistris, A. *Solid State Ionics*, **110**, 1 (1998).
173. Rahman, M. Y. A., Ahmad, A., and Wahab, S. A. *Ionics*, **15**, 221–225 (2009).
174. Rajendran, S., Kannan, R., and Mahendran, O. J. *Power Sources*, **96**, 406–410 (2001).
175. Rajendran, S., Mahendran, O., and Kannan, R. *J. Phys. Chem. Solids*, **63**, 303–307 (2002a).
176. Rajendran, S., Mahendran, O., and Kannan, R. *Fuel*, **81**, 1077–1081 (2002b).
177. Rajendran, S., Mahendran, O., and Mahalingam, T. *Eur. Polym. J.*, **38**, 49–55 (2002d).
178. Rajendran, S. and Uma, T. *Materials Letters*, **45**, 191–196 (2000).
179. Rajendran, S., Mahendran, O., and Kannan, R. *Solid State Electrochem*, **6**, 560–564 (2002c).
180. Ramana Rao, G., Castiglioni, C., Gussoni, M., Zerbi, G., and Martuscelli, E. *Polymer*, **26**, 811–820 (1985).
181. Ramesh, S., Koa,y H. L., Kumutha, K., and Arof A. K. *Spectrochimica Acta Part A*, **66**, 1237–1242 (2007).
182. Rao, G. R., Castiglioni, C., Gussoni, M., Martuscelli, E., and Zerbi, G. *Polymer*, **26**, 811 (1985).
183. Reddy, M. J., Kumar, J. S., Subba Rao, U. V., and Chu, P. P. *Solid State Ionics*, **177**, 253–256 (2006).
184. Robenson, P. and Lloyd M. *Polymer Blend, A Comprehensive Review*, Hanser Gardner Publications, Inc., pp. 109–111 (2007).
185. Robeson, L. M., Noshay, A., and Merriam, C. N. Matzner, *Angew. Makromol. Chem.*, **29**, 47–62 (1973).
186. Rocco A. M. and Pereira, R. P. *Felisberti, M. I. Polymer*, **42**, 5199–5205 (2001).
187. Rocco, A. M., Fonseca C. P., Loureiro, F. A. M., and Pereira, R. P. *Solid State Ionics*, **166**, 115–126 (2004).
188. Rocco, A. M., Fonseca, C. P., and Pereira, R. P. *Polymer*, **43**, 3601–3609 (2002).
189. Sawyer, L. C. and Grubb, D. T. *Polymer Microscopy (2nd Edition)*, Chapman & Hall London, pp. 17–33 (1996).
190. Schantz, S. *Macromolecules*, **30**, 1419–1425 (1997).
191. Schultz, J. M. *Polymeric Materials Science*. New York: Prentice Hall, (1981).

192. Schwahn, D., Pipich, V., and Richter, D. *Macromolecules*, **45**, 2035–2049 (2012).
193. Shafee, E. E. and Ueda, W. *Eur. Polym. J.*, **38**, 1327–1335 (2002).
194. Shen, C., Wang, J., Tang, Z., Wang, H., Lian, H., Zhang, J., and Cao, C. N. *Electrochimica Acta*, **54**, 3490–3494 (2009).
195. Shetter, J. A. J. Polym. Sci. Part B. Polym. Letter, **1**, 209 (1962).
196. Shultz, A. R. and Young, A. L. *Macromolecules*, **13**, 663–668 (1980).
197. Silvestre, C., Cimmino, S., and Di Pace, E. *Crystallization polymer blends*. In: J. C., Salamone (ed.), Polymeric Materials Encyclopedia, Vol. 2. New York, CRC Press, (1996).
198. Silvestre, C., Cimmino, S., Martuscelli, E., Karasz, F. E., and Macknight, W. J. *Polymer*, **28**, 1190–1199 (1987).
199. Sim, L. H., Chan, C. H., and Kammer, H. W. *Materials Research Innovations*, **15**, 71–74 (2011).
200. Sim, L. H. PhD Thesis, University of Malaya, Kuala Lumpur (2009).
201. Sim, L. H., Gan, S. N., Chan, C. H., Kammer, H. W., and Yahya, R. J. *Material Research Innovations*, **13**, 278–281 (2009).
202. Smith, R. W. and Andries, J. C. *Rubber Chem. Technol.*, **47**, 64–78 (1974).
203. Smith, W. F. and Hashemi, J. *Foundation of Materials Science and Engineering, 5th Edition*, p. 163 (2011).
204. Song, J. Y., Wang, Y. Y., and Wan, C. C. *J. Power Sources*, **77**, 183–197 (1999).
205. Sotele, J. J., Soldi, V., and Pires, A. T. N. *Polymer*, **38**, 1179–1185 (1997).
206. Stallworth, P. E., Greenbaum, S. G., Croce, F., Slane, S., and Salomon, M. *Electrochimica Acta*, **40**, 2137–2141 (1995).
207. Stefanescu, E. A., Tan, X., Lin, Z., Bowler, N., and Kessler, M. R. *Polymer*, **51**, 5823–5832 (2010).
208. Stoeva, Z., Martin-Litas, I., Staunton, E., Andreev, Y. G., and Bruce, P. G. *JACS* articles published on Web 22/03/2003 (2003).
209. Straka, J., Schmidt, P., Dybal, J., Schneider, B., and Spěváček, J. *Polymer*, **36**, 1147–1155 (1995).
210. Suthanthiraraj, S. A. and Sheeba, D. J. *Ionics*, **13**, 447–450 (2007).
211. Takahashi, Y. and Tadokoro, H. *Macromolecules*, **6**, 672–675 (1973).
212. Talibuddin, S., Wu, L., Runt, J., and Lin, J. S. *Macromolecules*, **29**, 7527–7535 (1996).
213. Tan, S. M., Ismail, J., Kummerlowe, C., and Kammer, H. W. *J. Appl. Polym. Sci.*, **101**, 2776–2783 (2006).
214. Tan, S. M. and Johan, M. R. *Ionics*, **17**, 485–490 (2011).
215. Thompson, E. V. *J. Polym. Sci. Part. A-2*, **4**, 199 (1966).
216. Tobishima, S. and Yamaji, A. *Electrochimica Acta*, **29**, 267–271 (1984).
217. Tsuchida, E., Ohno, H., and Tsunemi, K. *Electrochimica Acta*, **28**, 591–595 (1983).
218. Ulrich, H. *Introduction to Industrial Polymers*. Hanser, Munich, (1982).
219. Uma, T., Mahalingam, T., and Stimming, U. *Materials Chemistry and Physics*, **90** 245–249 (2005).
220. Utracki, L. A. *Introduction to polymer blends*. In, Utracki, L. A. (ed.), Polymer Blends Handbook, Vol. 1. Dordrecht, The Netherlands: Kluwer Academic, pp. 123–201 (2002).
221. Utracki, L. A. *J. Appl. Polym. Sci.*, **6**, 399–403 (1962).

222. Utracki, L. A. *Polymer Alloys and Blends, Thermodynamics and Rheology*. Munich, Hanser GmbH and Company, (1989).
223. Utracki, L. A. *Two Phase Polymer Systems*. Munich: Hanser Gardner, (1991).
224. Vondrak, J., Reiter, J., Velická, J., and Sedlaříková, M. *Solid State Ionics*, **170**, 79–82 (2004).
225. Walters, M. H. and Keyte, D. N. *Rubber Chem. Technol.*, **38**, 62–75.
226. Wang, H., Keum, J. K., Hiltner, A., and Baer, E. *Macromolecular Rapid Communications*, **31**, 356–361 (2010a).
227. Wang, M., Braun, H. G., and Meyer, E. *Polymer*, **44**, 5015–5021 (2003).
228. Wang, W., Liu, W., Tudryn, G. J., Colby, R. H., and Winey, K. I. *Macromolecules*, **43**, 4223–4229 (2010b).
229. Wang, Y. J., Pan, Y., and Chen, L. *Materials Chemistry & Physics*, **92**, 354–360 (2005).
230. Wieczorek, W., Lipka, P., Zukowska, G., and Wycislik, H. *J. Phys. Chem. B*, **102**, 6968–6974 (1998).
231. Wieczorek, W., Stevens, J. R. *J. Phys. Chem. B*, **101**, 1529–1534 (1997).
232. Wieczorek, W. *Mater. Sci. Eng.*, **B15**, 108–114 (1992).
233. Wieczorek, W., Such, K., Wycislik, H., and Pocharski, J. *Solid State Ionics*, **36**, 255–257 (1989).
234. Wiley, Processing and Finishing of Polymeric Materials Volume I, John Wiley and Sons, Inc., pp. 94–95 (2011).
235. Wiley, R. H. and Brauet, G. M. *J. Polym. Sci.*, **3**, 647 (1948).
236. Woo, E. M., Mandal, T. K., Chang, L. L., and Lee, S. C. *Polymer*, **41**, 6663–6670 (2000).
237. Wu, C., Xu, T., and Yang, W. *European Polymer Journal*, **41**, 1901–1908 (2005).
238. Xi, J., Qiu, X., Zheng, S., and Tang, X. *Polymer*, **46**, 5702–5706 (2005).
239. Yuan, A. and Ahao, J. *Electrochim. Acta*, **51**, 2454 (2006).
240. Yue, R., Niu, Y., Wang, Z., Douglas, J. F., Zhu, X., and Chen, E. *Polymer*, **50**, 1288–1296 (2009).
241. Zawada, J. A., Ylitalo, C. M., Fuller, G. G., Colby, R. H., and Long, T. E. *Macromolecules*, **25**, 2896–2902 (1992).
242. Zhang, S., Cao, X. Y., Ma, Y. M., Ke, Y. C., Zhang J. K., and Wang, F. S. *eXPRESS polymer Letters*, **5**, 581–590 (2011).
243. Zheng, H., Zheng, S., and Guo, Q. *J. Polym Sci.: Polym. Chem.*, **35**, 3169–3179 (1989).
244. Zhong Z. and Guo, Q. *Polymer*, **41**, 1711–1718 (2000).
245. Zhong, Z. and Guo, Q. *Polymer*, **39**, 517–523 (1998).
246. Zhou, J., Gan, S. N., and Leong, Y. C. (2004) *Recycling newspaper into fiberboard using a water-reducible acrylic resin*. 3rd USM-JIRCAS Joint International Symposium on Lignocellulose: Material for the Future from the Tropics.
247. Zhou, X., Yin, Y., Wang, Z., Zhou, J., Huang, H., Mansour, A. N., Zaykoski, J. A., Fedderly, J. J., and Balizer, E. *Solid State Ionics*, **196**, 18–24 (2011).
248. Zhu, L., Cheng, S. Z. D., Calhoun, B. H., Ge, Q., Quirk, R. P., Thomas, E. L., Hsiao, B. S., Yeh, F., and Lotz, B. *J. of American Chemical Society*, **122**, 5957–5967 (2000).

CHAPTER 19

POLY(TRIMETHYLENE TEREPHTHALATE)—THE NEW GENERATION OF ENGINEERING THERMOPLASTIC POLYESTER

C. SARATHCHANDRAN, CHIN HAN CHAN,
SITI ROZANA BT. ABD. KARIM, and SABU THOMAS

CONTENTS

19.1 POLY(TRIMETHYLENE TEREPHTHLATE) (PTT)

Polyester is the category of polymers with ester functional group on the main chain, although there are many types of polyester, the term polyester in industries specifically refers to poly(ethylene terephthalate) (PET) and poly(butylene terephthalate) (PBT). Polyesters can be classified as thermoplastic or thermosetting depending on the chemical structures. Table 1 shows the industrial production of polyesters, and it is estimated that the production will exceed 50 million tonnes by the year of 2015. Polyesters are made from chemical substances found mainly in petroleum and are mainly manufactured into fibers, films, and plastics. These polyesters are abbreviated as *m*GT, where *m* denotes the number of methylene groups; for example, PET, PTT, and PBT are abbreviated as 2GT, 3GT, and 4GT, respectively.

TABLE 1 The world production of polyester (Adapted from Sirichayaporn (2012))

	Market size per year	
Product type	**2002 [Million tonnes/year]**	**2008 [Million tonnes/year]**
Textile-PET	20	39
Resin, bottle/A-PET (solution casted PET)	9	16
Film-PET	1.2	1.5
Special polyester	1	2.5
Total	**31.2**	**59**

The credit of finding that alcohols and carboxylic acids can be mixed successfully in fabrication of fibers goes to W. H. Carothers, who was working for DuPont at the time and unfortunately when he discovered Nylon, polyester took a back seat. Carothers's incomplete research had not advanced to investigating the polyester formed from mixing ethylene glycol (EG) and terephthalic acid (TPA). It was the two British scientists—Whinfield and Dickson who patented PET in 1941. Later that year, the first polyester fiber – Terylene – was synthesized by Whinfield and Dickson along with Birtwhistle and Ritchiethey. Terylene was first manufactured by Imperial Chemical Industries or ICI. PET forms the basis of synthetic fibers like Dacron, Terylene and polyesters. DuPont's polyester research lead to a whole range of trademarked products, one example is Mylar (1952), an extraordinarily strong PET fiber that grew out of the development of Dacron in the early 1950s.

The industrial production of polyesters involves three steps:

1. Condensation Polymerization: When acid and alcohol are reacted in vacuum, at high temperatures condensation polymerization takes place. After the polymerization, the material is extruded onto a casting trough in the form of ribbon. Upon cooling, the ribbon hardens and is cut into chips.
2. Melt-spun Fiber: The chips are dried completely. Hopper reservoirs are then used to melt the chips. Afterwards, the molten polymer is extruded through spinnerets and cooled down by air blowing. It is then loosely wound around cylinders.
3. Drawing: The fibers consequently formed are hot stretched to about five times of their original length (to reduce the fiber width). This is then converted into products.

The PTT is aromatic polyester prepared by the melt polycondensation of 1,3-propanediol (1,3-PDO) with either TPA or dimethyl terephthalate (DMT). The PTT is synthesized by the transesterification of propanediol with dimethylene terephthalate or by the esterification of propane diol with TPA. The reaction is carried out in the presence of hot catalyst like titanium butoxide ($Ti(OBu)_4$) and dibutyl tin oxide (DBTO) at a temperature of 260°C. The important by-products of this reaction include acrolien and allyl alcohol (Chuah, 2001). Direct esterification of propane diol and TPA is considered as the least economic and industrial method. The reaction is carried out in the presence of a "heel" under a pressure of 70–150 kPa at a temperature of 260°C. The "heel" is usually referred to the added PTT oligomers which act as a reaction medium and increase the solubility of TPA (Chuah, 2001). Recent studies by different groups show that the selection of the catalyst plays a major role on the reaction rate and PTT properties. Commonly used catalysts like titanium (Doerr et al., 1994), tin (Kurian and Liang, 2001; Fritz et al., 1969) and antimony (Karayannidis et al., 2003; Fitz et al., 2000) compounds have their own limitations. Titanium-based catalysts are active but the PTT is discolored, antimony-based catalysts are toxic and only active in polycondensation while tin-based compounds have lower catalytic activity. Karayannidis and co-workers (2003) reported the use of stannous octoate ($[CH_3(CH_2)_3CH(C_2H_5)COO]_2Sn$) as the catalyst for PTT synthesis but its catalytic activity is poor, resulting in a low molecular weight PTT which was confirmed by measuring the content of terminal carboxyl groups (CTCGs). In this study, the catalytic activity was followed by measuring the amount of water generated and characterized by the degree of esterification (DE). Intrinsic viscosity measurement was carried out by using an Ubbelohde viscometer at 25°C with 0.005 g mL^{-1} PTT solution of 1,1,2,2-tetrachloroethane/phenol (50/50, w/w) mixture. The CTCG was determined by titrating 25 mL of chloroform and a few drops of phenolphthalein indicator to a solution of 1.000 g of PTT and 25 mL of benzyl alcohol against a titer of 0.561 g of KOH in 1 L of benzyl alcohol.

Works by different groups show that stannous oxalate (SnC_2O_4) is one of the best for PET (Fitz et al., 2000) and PBT (Paul, 1989) syntheses and also a potential additive for improving the properties of the polymers (Corbiere and Mosse, 1953). Studies by Yong et al. (2008) who used SnC_2O_4 as the catalyst to synthesis PTT, the results show that SnC_2O_4 displays higher polymerization activity than the other catalysts which is clear by the fact that SnC_2O_4 shows the highest intrinsic viscosity ($[\eta]$) and lowest CTCGs (ref. Table 2). Decrease in reaction time and enhancement of $[\eta]$ of PTT are observed as shown in Table 2 and Figure 1 when SnC_2O_4 is used as catalyst. Higher catalytic activity of SnC_2O_4 is attributed to its chelate molecular structure and suggests SnC_2O_4 as a more promising catalyst for PTT synthesis (Yong et al., 2008). The chemical structures of different catalysts used are shown in Figure 2.

TABLE 2 The Effect of various catalysts on the properties of PTT. The amount of catalyst taken was $5 \cdot 10^{-4}$ mol/mol of TPA, and esterification for 1.6 hr; at 230°C (Adapted from Yong and co-workers (2008))

Catalyst	**PTT**		
	Intrinsic viscosity ($[\eta]$) ($dL\ g^{-1}$)	**CTCG ($mol\ t^{-1}$) (content of terminal carboxyl groups)**	**Degree of esterification (DE) after 1.8 hr**
Stannous oxalate (SnC_2O_4)	0.8950	15	97
Stannous octoate ($[CH_3(CH_2)_3CH(C_2H_5)COO]_2Sn$)	0.6155	34	75
Dibutyltin oxide (Bu_2SnO)	0.8192	21	75
Tetrabutyl titanate	0.8491	20	82

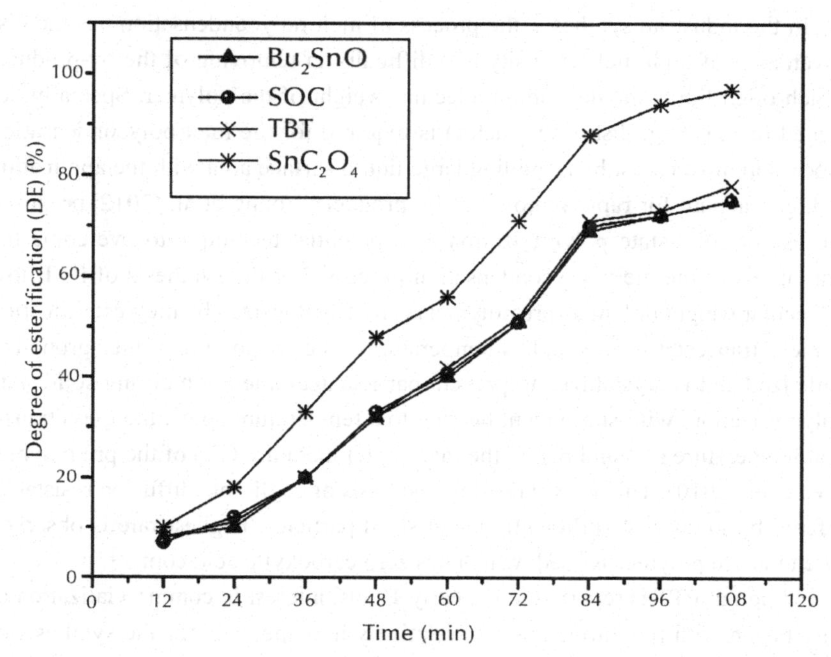

FIGURE 1 The Effect of different catalysts on esterification TPA. The amount of catalyst used was $5 \cdot 10^{-4}$ mol/mol of TPA. Parameter DE denotes degree of esterification (Adapted from Yong and co-workers (2008)).

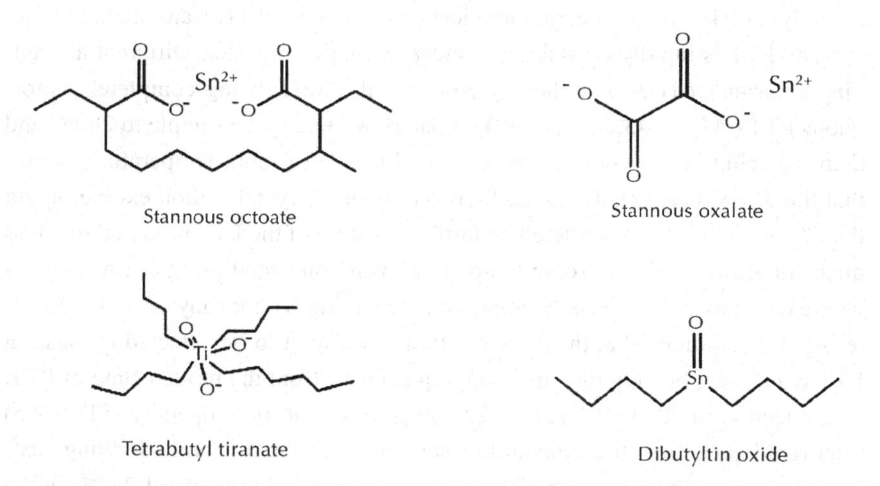

FIGURE 2 Chemical structures of the catalysts used for PTT synthesis.

In the industrial synthesis, the process of melt polycondensation has the disadvantages of high melt viscosity and difficulty in removing of the by-products which often limits the desirable molecular weight of the polymer. Specially designed reactor (e.g. disk-ring reactor) is required for the melt polycondensation process in most cases, by providing large liquid surface area with the application of high vacuum for rapid removal of by-products. Young et al. (2012) proposed the use of solid-state polymerization as a potential technique to overcome the limitations of the melt polycondensation process. For the synthesis of PTT, low molecular weight polymer (pre polymer) is first synthesized by melt esterification or melts transesterification at low temperatures. The pre polymer is then ground or pelletized and is crystallized to prevent particle agglomeration during solid state polymerization, with subsequent heating to a temperature above the glass transition temperature (T_g) and below the melting temperature (T_m) of the pre polymer (Lee et al., 2010). This is explained on the basis of negligible diffusion resistance offered by the use of sufficiently small sized particles. Highest rate is observed when the pre polymer is used, which has zero carboxylic acid content.

Although PTT is reported in the early 1950s, interest in commercialization of PTT began with the introduction of relatively new methods for the synthesis of propane diol by the catalytic hydrogenation of intermediate 3-hydroxypropionaldehyde ($C_3H_6O_2$) and hydroformylation of ethylene oxide (Broun et al., 1998). Recent discovery of fermentive production of 1,3-PDO accelerates the interest of studying PTT for engineering applications (DuPont, 2011; Haas et al., 2005).

The PTT is rapidly crystallizing linear aromatic polyester. Differential scanning calorimeter (DSC) studies by Hong et al. (2002) using completely amorphous PTT ($\overline{M_n}$ = 43,000 g mol^{-1}) prepared by heating the sample to 280°C and then quenching (at a cooling rate of 200°C min^{-1}) to room temperature, shows that the T_g lies between 45–65°C, followed by melt crystallization exotherm and then T_m at 230°C. The completely amorphous nature of the PTT prepared by rapid quenching from 280°C to room temperature was confirmed using X-ray analysis where the sample exhibits only amorphous scattering without any crystalline scattering. It is common that the T_g of the PTT is difficult to be detected by heating PTT with linear heating rate after rapid quenching from the molten state of PTT. The injection molded PTT (M_w = 22,500 g mol^{-1} and polydipersity (PI) = 2.5) after rapid quenching from the molten state was subjected to an underlying heating rate, $<q>$ = 2°C min^{-1}, a period of 60 s, and an amplitude of ± 1.272°C in the temperature range from –50 to 150°C. The T_g observed in the reheating cycle is

50.6°C with change of the heat capacity (ΔC_p) at 0.20 J g^{-1} °C^{-1}. The thermogram for the reheating cycle for this injection molded PTT is shown in Figure 3.

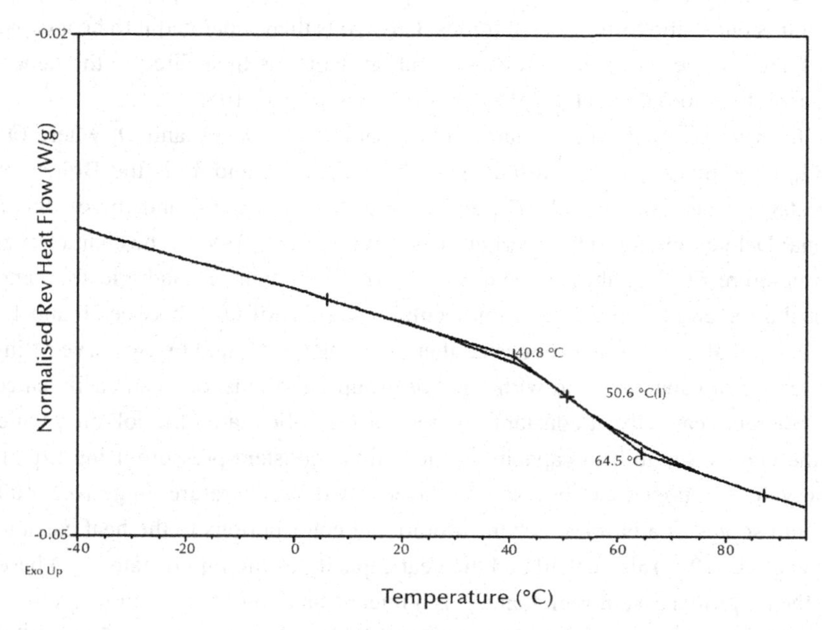

FIGURE 3 The thermogram for the reheating cycle of PTT using TA DSC with an underlying heating rate at 2°C min^{-1}, a period of 60 s, and an amplitude of ± 1.272°C in the temperature range from –50 to 100°C.

Pyda et al. (1998) studied in detail the heat capacity of PTT by adiabatic calorimetry, standard DSC and temperature-modulated differential scanning calorimetry (TMDSC) for this measurement. The computation of the heat capacity of solid PTT is based on an approximate group vibrational spectrum and the general Tarasov approach for the skeletal vibrations, using the well-established Advanced Thermal Analysis System (ATHAS) scheme. The experimental heat capacity at constant pressure is first converted to heat capacity at constant volume using the Nernst–Lindemann approximation

$$C_p(\exp) - C_v(\exp) = \frac{3RA_o C_p^2(\exp)}{C_v(\exp)\dfrac{T}{T^o_m}} \quad (1)$$

where A_o is an approximately universal constant with a value 0.0039 K mol J^{-1}; T is the temperature, T^o_m is the equilibrium melting temperature and R the universal gas constant. Later, based on the assumption that at low temperatures, C_v(exp) contains only vibrational contributions, C_v(exp) is then separated into heat capacity linked to the group and skeletal vibration, which is then fitted to the general Tarasov function (Pyda et al., 1998; Tarasov and Khimi, 1950).

In order to obtain the 3 characteristic parameters Θ_1, Θ_2 and Θ_3 where $\Theta = h\nu/k_B$ (h is the Planck's constant, ν is the frequency, and k_B is the Boltzmann constant). The functions D_1, D_2, and D_3 are the one-, two-, and three- dimensional Debye functions (Wunderlich, 1990; Wunderlich, 1995). The characteristic temperature Θ_3 describes the skeletal contributions with a quadratic frequency distribution and for linear macromolecules, the value of Θ_3 is between 0 and 150 K. Pyda et al. (1998) also suggested that knowing Θ_1, Θ_2 and Θ_3 by curve fitting, skeletal heat capacities and with a list of group vibrations, one can easily calculate the heat capacity at constant volume for the solid state of a polymer, which is then converted to heat capacity of the solid at constant pressure using Eq. (1). The values obtained can be extended over a wide temperature range (0.1–1000 K) and serves as a baseline for the vibrational contributions to the heat capacity. Pyda et al. (1998) also calculated the heat capacity of the liquid state (C_p^L) based on the empirical assumption that C_p^L is a linear function of temperature and using the addition scheme developed for ATHAS. The total heat capacity of liquid PTT is obtained from the contributions of the various structural groups (like CH_2, NH_2, COO, C_6H_4 etc).

$$\frac{C_v(sk)}{NR} = T\left(\frac{\Theta_1}{T}, \frac{\Theta_2}{T}, \frac{\Theta_3}{T}\right) = D_1\left(\frac{\Theta_1}{T}\right) - \left(\frac{\Theta_2}{\Theta_1}\right)\left[D_1\left(\frac{\Theta_2}{T}\right) - D_2\left(\frac{\Theta_2}{T}\right)\right] - \left(\frac{\Theta_3^2}{\Theta_1\Theta_2}\right)\left[D_2\left(\frac{\Theta_3}{T}\right) - D_3\left(\frac{\Theta_3}{T}\right)\right] \quad (2)$$

The mechanical and electrical properties of PTT are close to PET, and even when these polyesters show lower mechanical properties as compared to Nylons, they show better elelctrical properties. The thermal stability of PTT by referring to the onset of decomposition temperature using thermal gravimetry analyzer (TGA), is comparable to polycarbonates (PC) (compare to Tables 3, 4 and Figure 4). Thermogram in Figure 4 shows that PTT is thermally stable up to 373°C with 1.5 wt% of mass loss when it was heated up at 30°C min^{-1} in nitrogen atmosphere. PTT thermally degrades further at around 494°C with 92 wt% of mass loss and the remaining 4 wt% of mass fully decomposes at 600°C.

TABLE 3 Some physical properties of PTT with other polyesters and nylon (Adapted from Hwo (1998))

Physical Property	PET [a]	PTT [b]	PBT [c]	Nylon 6,6 [d]	PC [e]	Nylon 6 [f]
Specific Gravity (measured as per ASTM D792 specifications)	1.40	1.35	1.34	1.14	1.20	1.14
T_m (°C) (Using DSC for injection molded samples at a heating rate of 10°C min^{-1}*)	265	227	228	265	--	230
T_g* (°C)	80	45–60	25	50–90	150	50
Onset of decomposition temperature by using TGA (°C)	350	373	378	375	--	398

*indicates that the same experimental conditions were used in both the studies

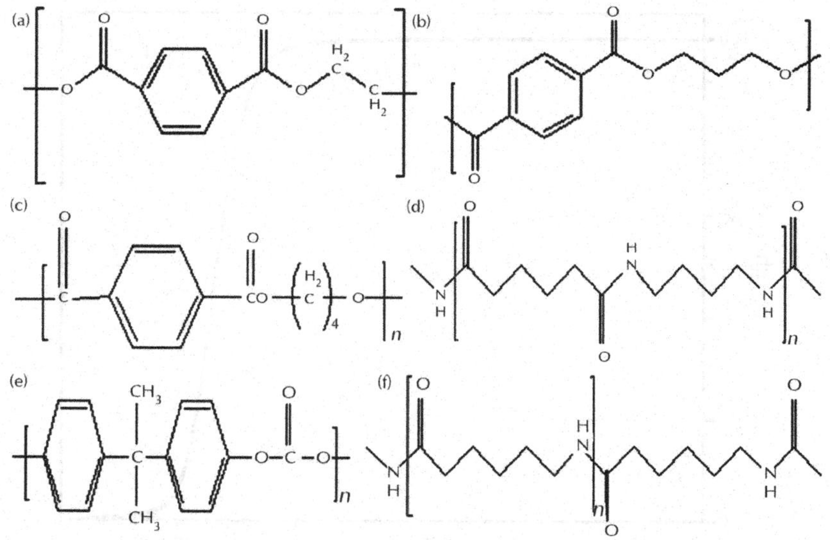

TABLE 4 The other physical properties of PTT (Adapted from Brown and co-workers (1997))

Properties	Values
Melting point (°C)	227
Equilibrium melting point (°C)	248 (Lin, 2007; Dangseeyun et al., 2004), 252 (Srimoaon et al., 2004), 237 (Pyda et al., 1998)
Heat of fusion (ΔH_f) (kJ mol^{-1})	30 ± 2
Fully amorphous heat capacity (J K^{-1} mol^{-1})	94
Crystallization half time at 180°C (min)	2.4
Cold crystallization temperature (°C)	65
Glass transition temperature (°C)	45–60
Thermal diffusivity at 140°C (m^2 s^{-1})	0.99 × 10^{-7}

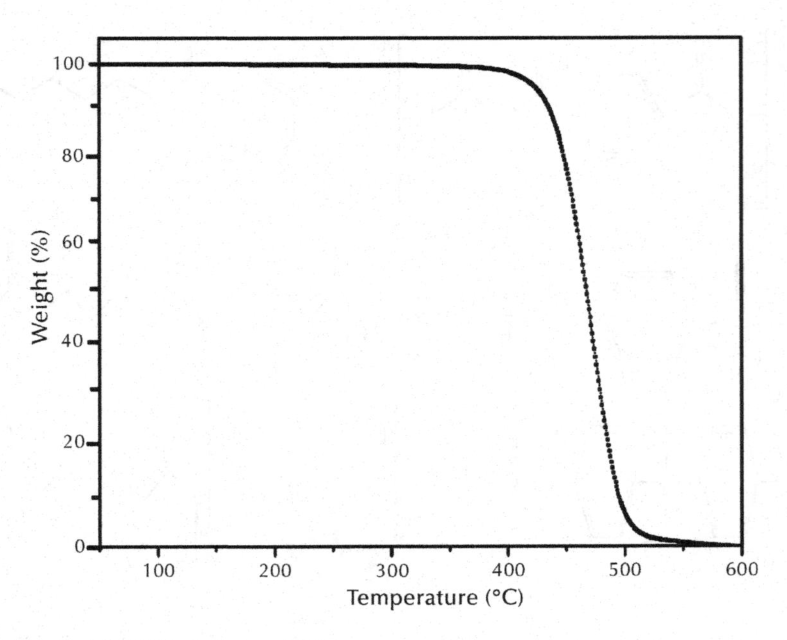

FIGURE 4 The TGA analysis of PTT.

Crystallization of PTT takes place below the T_m and above the T_g of PTT. At temperature below T_m, crystallization of PTT is driven by thermodynamics and at temperature below T_g crystallization ceases due to lack of segmental mobility of PTT chains. The T_m of PTT can be determined by examining the variation in density during cooling or/and heating of PTT at a fixed rate. Lin (2007) studied this relation (variation of density of PTT upon heating or/and cooling) by using atomistic simulations, where the molecular dynamic simulation was used to determine the properties of PTT down to the molecular level. The variation of density with respect to temperature as observed by Lin (2007) is shown in Figure 5. The reported T_m of PTT (with an entanglement M_w between 4900–5000 g mol^{-1}) is around 227–277°C and T_g is around 102°C. Experimentally observed values for T_m is in the range of 226–230°C, and T_g is between 45–65°C. The marked difference in estimated T_m and T_g by using atomistic simulations and DSC, respectively, can be attributed to the extremely fast cooling rate (10^{12}°C s^{-1}) for the former analysis and much slower rate for the later.

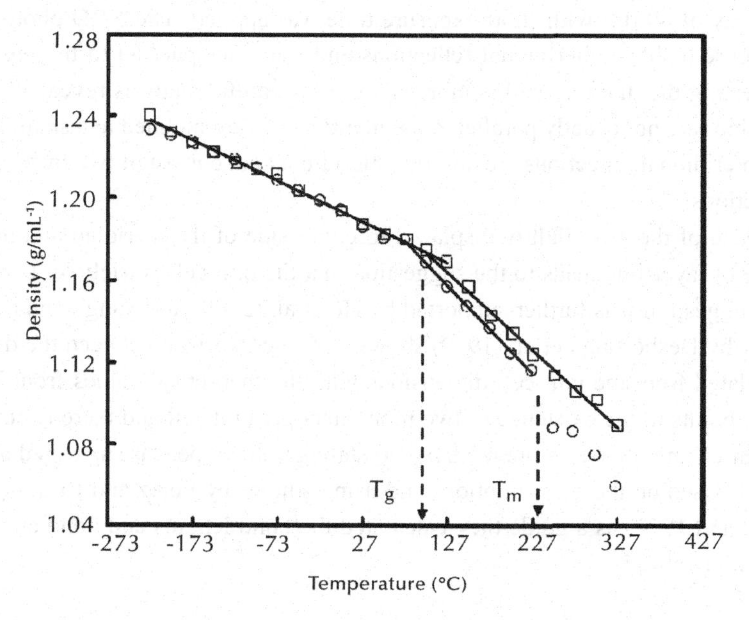

FIGURE 5 The variation of density of PTT fiber with respect to temperature on heating (□) and cooling (○), as per the molecular dynamics simulation studies done by Lin (2007) (Modified from Lin (2007)).

19.2 CRYSTAL STRUCTURE AND STEREOCHEMISTRY OF PTT

The crystal structure and stereochemistry of PTT were studied extensively by different groups (Desborough and Hall, 1977; Hall, 1976; Hall et al., 1978; Mencik, 1975; Yokouchi et al., 1976). As mentioned earlier, PTT is also abbreviated as 3GT where the crystalline PTT has *gauche* and *trans* conformations. The chain conformation of the PTT fiber changes reversibly between two forms when the fiber is strained. This can be followed by using the techniques of X-ray diffraction (XRD). In the unstrained form, the methylene section of the PTT chain has the conformation of *gauche-trans-gauche*; upon straining this conformation changes to *trans-trans-trans* (Hall and Pass, 1976).

The crystal structure and unit cell dimensions using the techniques of XRD were reported by Desborough et al. (1979), PTT was melt spun at 270°C, followed by cold drawn PTT fibers at a draw ratio of 4:1 and then annealed at 185°C for 2 hours. The XRD photographs were taken using Ni filtered CuKα radiation with camera of the type as described by Elliot (1965). A highly monochromatic beam of X-ray of 40 μm with short exposure time was applied. The XRD photograph in Figure 6 shows meridional reflections and low lines parallel to the meridian suggesting that the unit cell is monoclinic but a careful analysis reveals that the row lines are not exactly parallel to the meridian but are inclined to a small angle. The meridional reflections are not truly meridional but consist of two overlapping reflections.

Each of them is slightly displaced to either side of the meridian and the absence of layer line leads to the suggestion that the unit cell is triclinic. Later on, this suggestion was further supported by Ho et al. (2000) and Yang et al. (2001). Study by Desborough et al. (1979) shows that a comparison between the density calculated from the unit cell dimensions with the theoretical values from literatures; points to the existence of two monomers per unit cell and there is one molecular chain with two monomers per crystallographic repeat as suggested in Figure 7. Based on these assumptions and using studies by Perez and Brisse (1975, 1976, and 1977) as a guide they calculated the bond lengths and bond angles as shown in Table 5.

TABLE 5 The values of bond length and bond angle of PTT fiber as calculated by Desborough et al. (1979) (Adapted from Desborough and co-workers (1979))

Bond	Length (Å)		Angle (degree)
C_0-C_1	1.39	α_1	120
C_1-C_3	1.39	α_2	120
C_2-C_3	1.39	α_3	120
H_1-C_1	1.07	α_4	120
H_2-C_2	1.07	α_5	125
C_4-C_3	1.48	α_6	113
O_1-C_4	1.21	α_7	122
O_2-C_4	1.34	α_8	116
C_5-O_2	1.44	α_9	106
H_3-C_5	1.03	α_{10}	113
C_6-C_5	1.54		

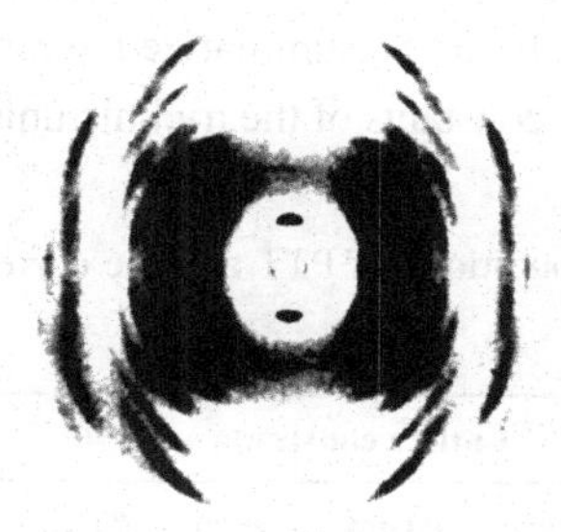

FIGURE 6 Diffraction patterns of PTT fibre drawn so as to reduce the tilt (tilted crystal orientation refers to the position in which the unit cell is tilted away from its usual orientation as compared to the situation where chain and fiber axes coincident).

Source: Reproduced from Desborough and co-workers, copyright (1979), by permission of Elsevier Ltd.

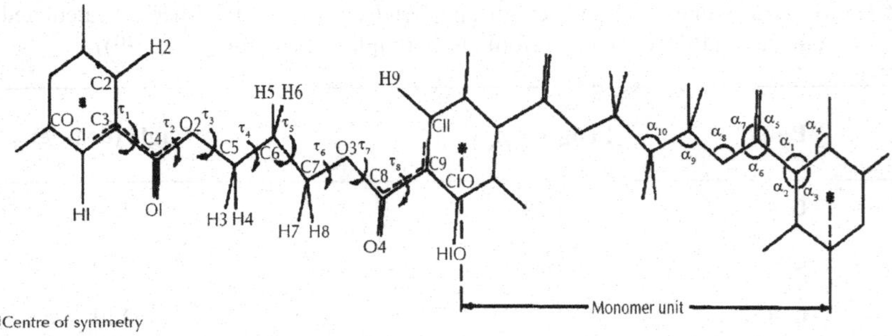

FIGURE 7 Crystallographic repeat of PTT fiber based on the X-ray studies carried out by Desborough et al. (1979).

Source: Reproduced from Desborough and co-workers, copyright (1979), by permission of Elsevier Ltd.

The length ratio between *c*-axis of the unit cell and the extended chain length of PTT is found to be about 75% indicating a big zigzag conformation along the *c*-axis, which has been suggested as the high deformability in crystals when drawn, and this explains why slight deviation in unit cell dimensions are usually observed by different groups. This also accounts for the enhanced tendency of PTT to form fibers when compared to other polyesters, which is evidenced by the exceptional use of PTT as fibers for different applications like sportswear. All these studies by different groups lead us to the conclusion that the unit cell of PTT is triclinic with two monomer units per unit cell (see Figure 8), and the unit cell of the PTT crystal varies slightly based on the preparative conditions of PTT. We summarize the preparative conditions of PTT and the corresponding lattice constants of the triclinic unit cell in Table 6.

TABLE 6 The preparative conditions of PTT and the corresponding lattice constants of the triclinic unit cell

Preparative-conditions	Lattice constants						Characterization technique	Ref
	a (nm)	*b* (nm)	*c* (nm)	*α* (°)	*β* (°)	*γ* (°)		
Melt spun at 270°C, followed by cold drawn at a draw ratio of 4:1 and then annealed at 185°C for 2 hr.	4.5	6.2	18	98	90	111	Electron diffraction	(Elliot, 1965)

TABLE 6 *(Continued)*

Melt polymerization	4.6	6.2	18	98	90	111	Electron diffraction	(Hall, 1984)
Confined thin film melt polymerization (CTFMP) at temperatures between 150–220°C	4.5	6.2	19	97	92	111	Electron diffraction CERIUS simulation program.	(Yang et al., 2001)
Bulk polymerization (180°C and 72 hr)	4.6	6.4	19	99	92	112	WAXD	(Yang et al., 2001)
Polycondensation reaction between TPA and 1,3-PDO	0.46	0.61	1.9	98	92	110	WAXD	(Wang et al., 2001)

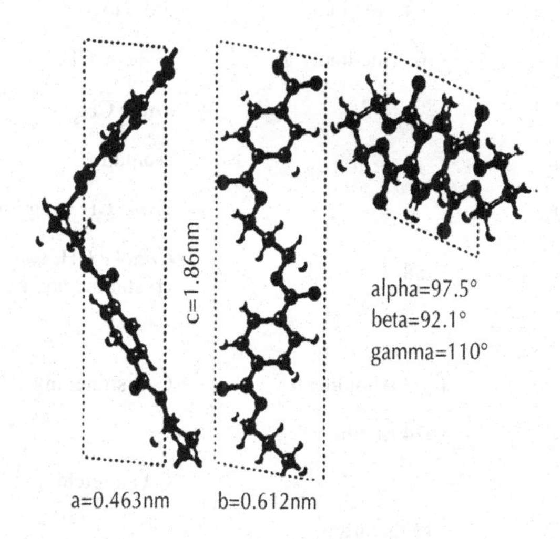

FIGURE 8 Atomic positions of melt-spun PTT chains with triclinic crystal unit cell determined by wide-angle X-ray diffraction (WAXD) by Wang et al. (2001).

Source: Reproduced from Wang and co-workers, copyright (2001), by permission of Elsevier Ltd.

19.3 INFRARED SPECTROSCOPIC (IR) ANALYSIS OF PTT

Fourier transform infrared (FTIR) spectroscopy can be used as a tool to study the crystalline and amorphous fractions (Chuah, 2001; Ouchi et al., 1977; Bulkin et al., 1987; Ward and Wilding, 1977; Yamen et al., 2008) of PTT. The absorption bands of IR between 1750–800 cm^{-1} are helpful to estimate the fraction of the crystalline phase of PTT samples. The assignment of the absorption bands in this region for PTT was proposed by Yamen et al. (2008) (Table 7).

TABLE 7 Wavenumbers and assignments of IR band exhibit by PTT as proposed by Yamen et al. (2008) (Adapted from Yamen and co-workers (2008))

Wavenumber (cm^{-1})		Assignment
Amorphous phase	**Crystalline phase**	
1710 (very strong)	1710 (very strong)	C=O stretch
1610 (strong)	1610 (strong)	aromatic
1577 (weak)	-	-
1504 (medium)	1504 (medium)	aromatic
1467 (medium)	1465 (medium)	*Gauche* CH_2
1456 (medium)	-	*Trans* CH_2
1400 (medium)	-	aromatic
1385 (medium)	-	*Trans* CH_2 wagging
	1358	*Gauche* CH_2 wagging (both crystalline and amorphous)
1173 (weak)	-	
1037 (shoulder)	1043 (shoulder)	C-O stretching
1019 (medium)	1024 (medium)	
976 (weak)	-	C-O stretching
948 (weak)	948 (medium)	

TABLE 7 *(Continued)*

937 (weak)	937 (weak)	
933 (shoulder)	933 (shoulder)	CH_2 rocking (both crystalline and amorphous)

The FTIR spectroscopy studies on PTT (refer Figure 9) subjected to isothermal crystallization for 40 min at 200°C, shows the following result. The area ratio of the absorption band (A_{1358}/A_{1504}), which is assigned to the % of *gauche* conformation, is calculated to be 1.39 while the ratio (A_{976}/A_{1504}), which denotes the *trans* conformation of the methylene groups is found to be 0.28, indicating reasonable amount of crystallinity in the sample (refer Table 8). The crystallinity estimated by DSC analysis after isothermal crystallization at 200°C for 40 min shows a value of 40.8%.

TABLE 8 Peak assignment of PTT crystallized at 200°C for 40 min

Amorphous	Crystalline	Assignment
1708 (very strong)	1708 (very strong)	C=O stretch
1578 (weak)	-	-
1505 (medium)	1504 (medium)	aromatic
1464 (medium)	1465 (medium)	*Gauche* CH_2
1408 (medium)	-	aromatic
1389 (medium)	-	*Trans* CH_2 wagging
	1358	*Gauche* CH_2 wagging (both crystalline and amorphous)
1040 (shoulder)	1043 (shoulder)	C-O stretching
1017 (medium)	1024 (medium)	
976 (weak)	--	C-O stretching
948 (weak)	948 (medium)	
937 (weak)	937 (weak)	
933 (shoulder)	933 (shoulder)	CH_2 rocking (both crystalline and amorphous)
918 (shoulder)	-	amorphous

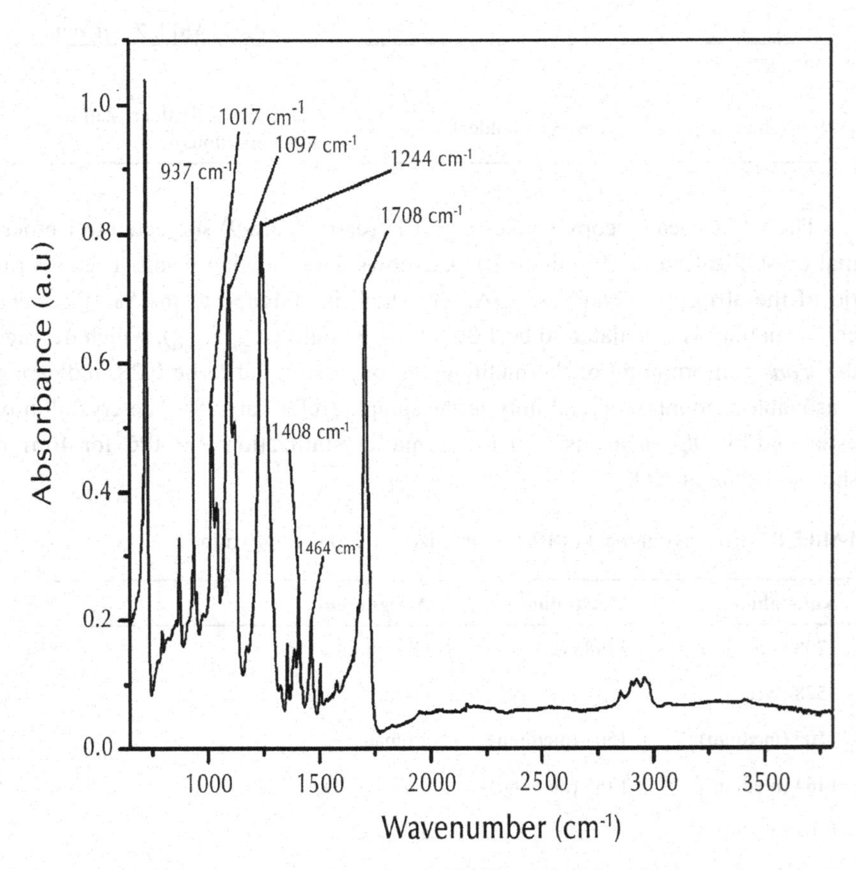

FIGURE 9 The FTIR analysis of PTT crystallized at 200°C for 40 min.

19.4 KINETICS OF ISOTHERMAL CRYSTALLIZATION OF PTT

The overall rate of isothermal crystallization of PTT (semicrystalline polymer) can be monitored by thermal analysis through the evolution of heat of crystallization by DSC as depicted in Figure 10. The sample is isothermally crystallized at preselected crystallization temperature (T_c) until complete crystallization. Half time of crystallization ($t_{0.5}$) for the polymer is estimated from the area of the exotherm at $T_c = const$, where it is the time taken for 50% of the crystallinity of the crystallizable component to develop. The rate of crystallization of PTT can be easily characterized by the experimentally determined reciprocal half time, $(t_{0.5})^{-1}$.

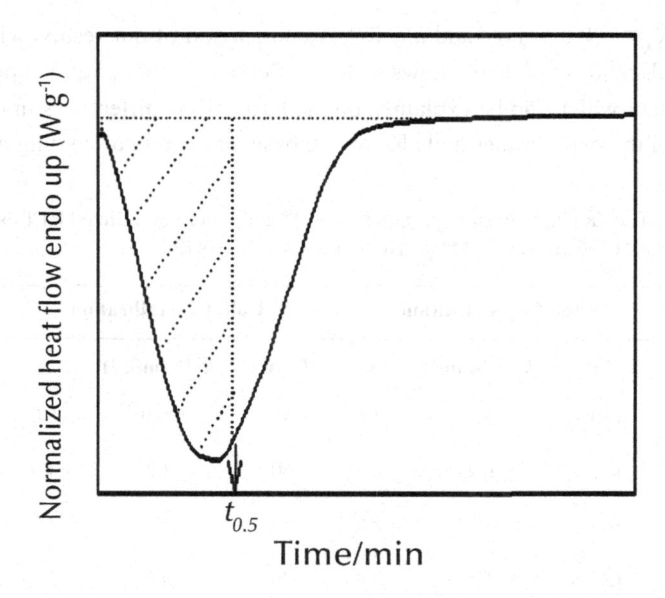

FIGURE 10 Schematic diagram for DSC trace of PTT during isothermal crystallization at preselected T_c.

The crystallization kinetics in polymers under isothermal conditions can be best explained using the equation developed by Avrami and later modified by Tobin. The equation proposed by Tobin is considered as a simplification of the calculations.

$$X(t) = 1 - \exp\left[-K_A^{1/n}\left(t - t_o\right)\right]^n \tag{3}$$

Where $X(t)$ is the normalized crystallinity given as the ratio of degree of crystallinity at time t and the final degree of crystallinity, t_o is the induction period which is determined experimentally and defined as the time where deviations from baseline can be monitored (min), K_A is the overall rate constant of crystallization ($\min^{-n}$), and n the Avrami exponent.

Thus a plot of lg[-ln(1-X)] against lg(t-t_o) gives a linear curve, the slope of which gives the Avrami exponent 'n' and the y intercept gives the rate constant 'K_A'. The values of K_A and n are indicative of the crystallization mechanism. PTT with M_w = 22,500 g mol^{-1} and PI = 2.5 was subjected to isothermal crystallization at 205°C for 65 min. The corresponding Avrami plot is illustrated in Figure

11 with $K_A^{1/n} = 0.07\ min^{-1}$ and $n = 3.9$. A comparison of our results with that of Hong et al. (2002) (Table 9) shows a clear difference in the Avrami exponent and rate constant which can be explained on the basis of the differences in molecular weights of the two samples and also on the basis of the rate of cooling applied.

TABLE 9 The values of Avrami parameters, $K_A^{1/n}$ and n, for crystallized PTT as presented by Hong et al. (2002) (Adapted from Hong and co-workers (2002))

Melt Crystallization			Cold Crystallization		
T_c (°C)	$K_A^{1/n}$ (min^{-1})	n	T_c (°C)	$K_A^{1/n}$ (min^{-1})	n
170	7.9	2.3	55	0.050	5.0
175	6.4	2.7	60	2.2	4.9
180	3.7	2.6	65	13	4.9
185	1.5	2.9	70	45	5.2
190	0.43	3.0	-	-	-
195	0.24	3.0	-	-	-
200	0.010	2.9	-	-	-
205	0.0011	2.9	-	-	-
210	$5.8.10^{-6}$	3.2	-	-	-

Huang and Chang (2000) reported the work of chain folding for PTT at 4.8 kcal mol^{-1}. Hong et al. (2002) studied the isothermal crystallization kinetics of PTT. The DSC analyses were done by melting the samples at 280°C for 5 min and then rapidly cooling with a rate at 200°C min^{-1} to an ambient T_c. For the isothermal cold crystallization, the samples were melted at 280°C and then rapidly cooled to low temperatures using liquid nitrogen so as to get a completely amorphous sample. Avrami model can be adopted to describe primary stage of isothermal crystallization from the melt and glass states adequately. Impingement of the PTT spherulites during the secondary state of the crystallization leads to the deviation from the Avrami model. The values for Avrami parameters as observed by Hong et al. (2002) are given in Table 9.

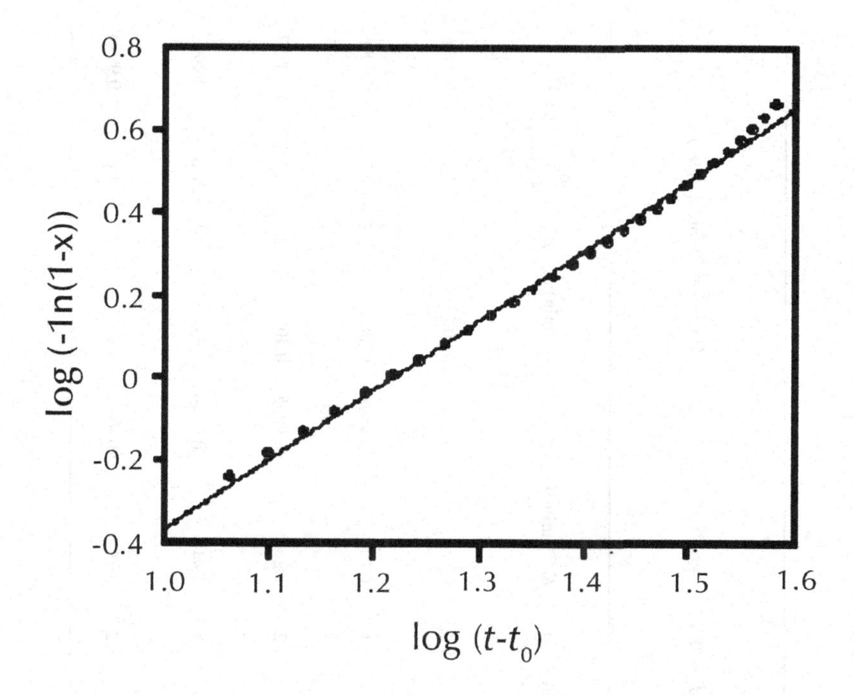

FIGURE 11 Avrami plot for PTT after isothermal crystallization at T_c of 205 °C. Solid curve represents the regression curve after Eq. (3) ($r^2 = 0.9967$).

The Avrami exponent values vary between 2 and 3 corresponding to different T_cs indicating a mixed nucleation and growth mechanism, while the Avrami exponent values of 5 corresponds to a solid sheaf like growth and athermal nucleation for cold crystallization. They reported that the regime I-II and regime II-III transitions occur at temperatures of 215 and 195°C, respectively. The crystallite morphologies of PTT from the melt and cold crystallizations exhibit typical negative spherulite and sheaf-like crystallite, respectively. The regime I-II-III transition is accompanied by morphological change from axialite-like or elliptical-shaped crystallite to banded spherulite and then non-banded spherulite. This is interesting to compare the Avrami exponents and the rate constants after Avrami model for isothermal crystallization kinetics of PET, PBT, and PTT (Dangseeyun et al., 2004) (refer Table 10). At $T_c = const$, the rate constant ($K_A^{1/n}$) of PBT > PTT > PET.

TABLE 10 The Avrami exponents and the rate constants after Avrami model for isothermal crystallization kinetics of PET, PBT, and PTT (Adapted from Dangseeyun and co workers (2004)

	PET (M_w = 84,500 g mol^{-1})				PTT (M_w = 78,100 g mol^{-1})				PBT (M_w = 71,500 g mol^{-1})			
T_c (°C)	$t_{0.5}$ (min)	n	$K_A^{1/n}$ (min^{-1})	r^2	$t_{0.5}$ (min)	n	$K_A^{1/n}$ (min^{-1})	r^2	$t_{0.5}$ (min)	n	$K_A^{1/n}$ (min^{-1})	r^2
184	1.3	1.9	0.63	0.9999	0.58	2.0	1.5	0.9994	0.30	2.1	2.9	0.9996
186	1.4	2.0	0.60	0.9995	0.64	1.8	1.3	0.9998	0.38	2.2	2.2	0.9996
188	1.5	1.7	0.56	0.9993	0.72	2.0	1.2	0.9998	0.40	2.2	2.1	0.9997
190	1.5	2.2	0.59	0.9999	0.90	2.1	0.94	0.9999	0.53	2.1	1.6	0.9996
192	1.5	1.7	0.54	0.9988	1.1	2.0	0.79	0.9999	0.53	1.8	1.6	0.9994

TABLE 10 *(Continued)*

194	1.6	1.7	0.52	0.9994	1.4	2.0	0.63	0.9998	0.78	2.0	1.1	0.9995
196	1.7	1.9	0.49	0.9990	1.6	2.3	0.55	0.9998	0.88	1.7	0.92	0.9993
198	2.0	1.7	0.41	0.9988	2.2	2.0	0.39	0.9999	1.3	1.6	0.64	0.9993
200	2.3	1.6	0.36	0.9990	3.0	2.4	0.29	0.9992	1.5	2.0	0.55	0.9995
202	2.6	2.1	0.33	0.9997	3.7	2.1	0.23	0.9996	2.7	1.8	0.31	0.9992
204	2.8	1.8	0.30	0.9967	5.0	2.3	0.17	0.9996	3.7	1.6	0.22	0.9970
205	3.0	1.8	0.28	0.9997	5.9	2.4	0.15	0.9993				
206	3.0	1.9	0.28	0.9972	6.6	2.2	0.13	0.9998	4.8	1.7	0.17	0.9912
207	3.3	1.8	0.25	0.9998								

TABLE 10 *(Continued)*

208	4.0	1.9	0.21	0.9980	7.6	2.4	0.11	0.9989	7.5	1.9	0.11	0.9971
215	4.7	2.0	0.17	0.9991								
220	10	2.1	0.082	0.9992								

19.5 RADIAL GROWTH RATE OF PTT SPHERULITES

The growth rate of PTT spherulites can be determined by using polarized optical microscopy (POM). During isothermal crystallization, micrographs are captured at suitable time intervals. The increase of spherulite radii is strictly linear with time for all cases. The radial growth rate of the PTT spherulite is shown in Figure 12 (Hong et al., 2002). The radial growth rate of the PTT spherulite decreases exponentially with increasing isothermal T_c from 163 to 221°C.

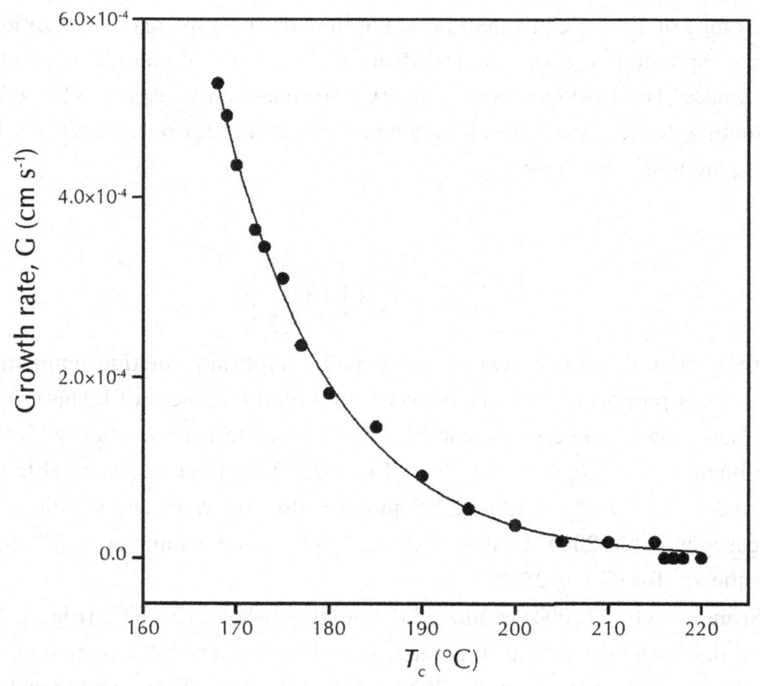

FIGURE 12 Plot of radial growth rate of PTT spherulites as a function of T_c as discussed in Hong et al., (2002) (modified from Hong and co-workers (2002)).

19.6 MELTING TEMPERATURE AND EQUILIBRIUM MELTING TEMPERATURE OF PTT

In contrast to low-molecular substances, melting and crystallization of polymers cannot be observed in equilibrium. This is because the crystallization is extremely low near and below the T^{o}_{m} for the polymer due to the crystal nucleation is greatly inhibited at the proximity of T^o_m. The rate of crystallization for semicrystalline polymer is nucleation rather than diffusion controlled near to T^o_m. Hence, crystallization of a polymer can only proceed in a temperature below T^o_m. Quantity T^o_m of a polymer can be determined experimentally by step-wise annealing procedure after Hoffman–Weeks (Hoffman and Weeks, 1962). Under this procedure, crystallization and melting of polymers proceed under non-equilibrium conditions but near to equilibrium. The sample is isothermally crystallized in a range of (T_c). Half time of crystallization ($t_{0.5}$) for the polymer is determined as described in previous section. Subsequently, the sample is allowed to crystallize again at the same range of T_c's for equivalent period of time until complete crystallization and the corresponding T_ms are obtained from the peak of the endotherms from the DSC traces. The Hoffman–Weeks theory (Hoffman and Weeks, 1962) facilitates calculating the equilibrium melting temperature values for polymers from the T_c. The equation is written as:

$$T_m = \frac{1}{\gamma} T_c + \left(1 + \frac{1}{\gamma}\right) T^o{}_m \tag{4}$$

where T_m and T^o_m are the experimental and equilibrium melting temperatures, while γ is a proportional factor between the initial thickness of a chain fold lamella and final thickness. T^o_m can be obtained by the extrapolation with the $T_m = T_c$ linear curve. Quantity T^o_m for PTT is 229°C, which is comparable to the reported values of T^o_m at 228–232°C in other studies (Ward and Wilding, 1977; Dangseeyun et al., 2004; Lustinger et al., 1989), except Chung et al. (2000) suggests the T^o_m for PTT is 252°C.

Srimoaon et al. (2004) evaluated the melting behavior of PTT (Figures13 (c) and 14 (a) (b)) at different heating rates using DSC and crystallite structures using WAXD, meanwhile Hong et al. (2002) studied the crystallinity and morphology of PTT. Generally, multiple melting peaks are related to various reasons for example:

1. Formation of various crystal structures or dual lamellar stacking during the primary crystallization.
2. Secondary crystallization and recrystallization or reorganization during the heating.
3. Reorganization of the metastable crystals formed during heating resulting in crystal perfection and/or crystal thickening.

4. Multiple melting peaks are observed when the polymer exhibits polymorphism like nylon 6, 6 and isotactic poly(propylene) (*i*-PP).

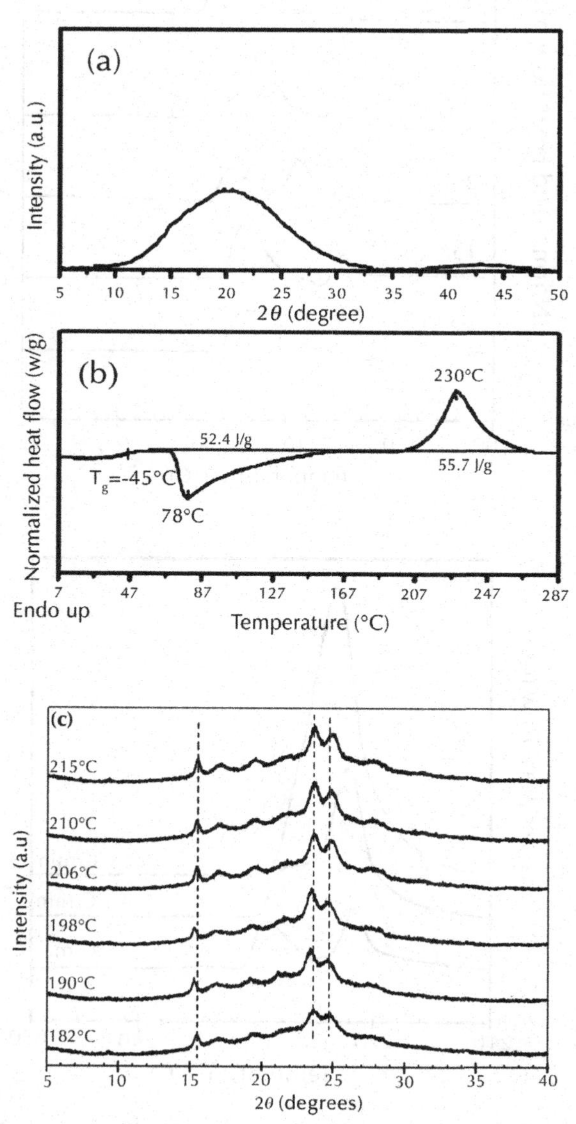

FIGURE 13 (a) The XRD pattern of PTT (b) DSC trace for completely amorphous PTT (Hong et al., 2002) (c) Evaluation of WAXD pattern as a function of T_c (Srimoaon et al., 2004) (adapted from Hong and co-workers (2002) and Srimoaon and co-workers (2004)).

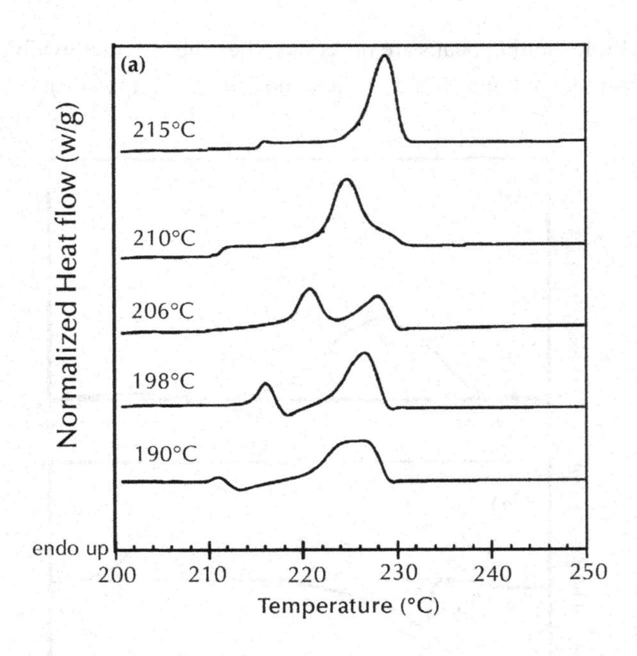

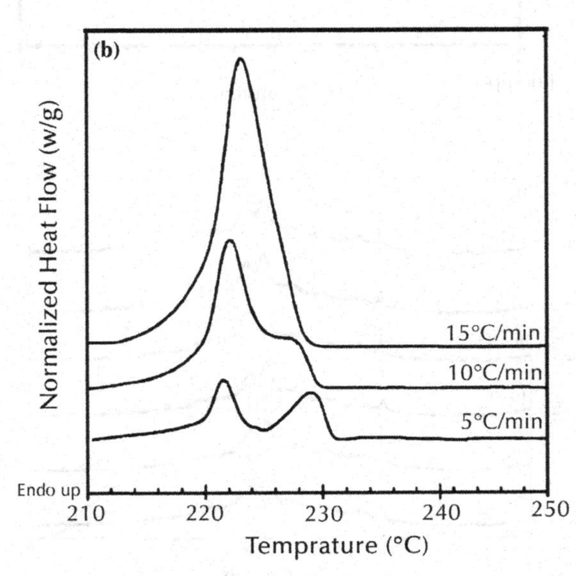

FIGURE 14 (a) The DSC traces of PTT at various T_cs (heating rate 10 °C min^{-1}) (b) DSC heating traces of PTT crystallized at 208 °C at various heating rates (adapted from Srimoaon and co-workers (2004)).

Secondary crystallization can be identified from the deviation of the Avrami plot at the non-linear stage where the spherulites impinge with each other. The WAXD results by Srimoaon et al. (2004) show that there is no shift in the 2 theta (2θ) values indicating that the unit cell of PTT does not change, ruling out the possibility of polymorphism at elevated T_cs and the formation of multiple peaks is explained by the presence of two populations of lamellar stacks, which are formed during the primary crystallization. This can be associated with the lamellar branching effect for the growth of spherulites. This explanation is further supported by Hong et al. (2002) optical microscopy studies.

19.7 MORPHOLOGICAL STRUCTURES OF PTT

The PTT has the unique property of forming banded spherulites and are commonly considered as arising due to chain tilting in the lamellar crystals. The banded structure formation in PTT has been discussed in detail by different groups. A close analysis of the optical images shows that the formation of banded structure indication with arrow as shown in Figure 15 (a) (Hong et al., 2002), which is dependent on the isothermal T_c and as the isothermal T_c increases from 210 to 215°C, the banded structure disappears [c.f. Figure 15 (b) and (c)].

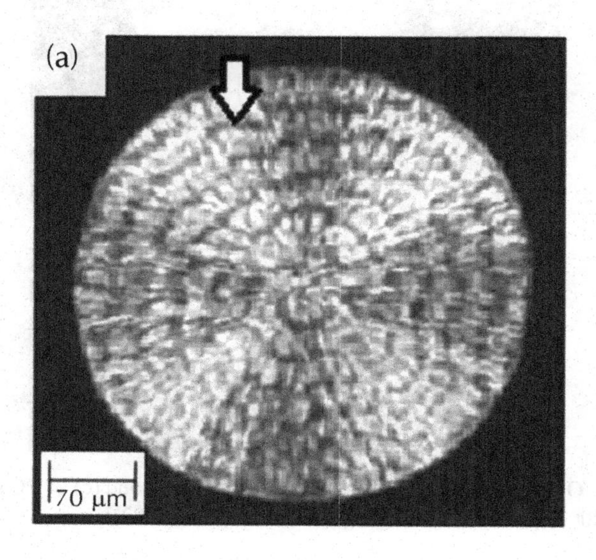

FIGURE 15 *(Continued)*

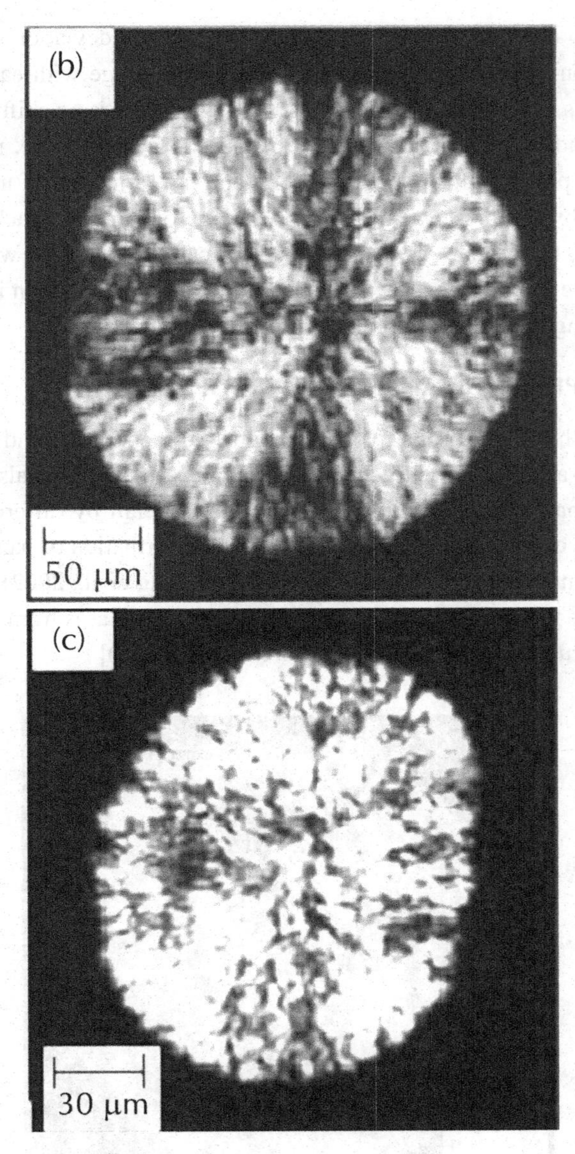

FIGURE 15 Optical images of PTT at (a) 210°C, (b) 215°C and (c) 217°C as observed by Hong et al. (2002) (modified from Hong and co-workers (2002)).

Studies by Ho et al. (2000) point towards lamellar twisting as the reason for the formation of banded spherulites in PTT. The atomic force microscopy (AFM) images for banded spherulite structure of PTT are shown in Figure 16 [as observed by Wang et al. (2001)] for PTT with M_n of 28,000 g mol^{-1} and polydispersity of 2.5 synthesized by polyesterification of TPA and 1, 3-propanediol, and thin films were cast using 0.2–1% (w/w) phenol/1,1,2,2-tetrachloroethane. The films were heated to 30°C above the melting point and then were rapidly cooled to the required T_c and then quenched in liquid nitrogen followed by observation under polarized light. Thin film samples with free surface, where the unrestricted lamellae develop a wave-like morphology. The twisting mechanism is evidenced by the observation of wave-like morphology from polarized optical microscope which confirms the fact that banded spherulite formation is attributed to lamellar twisting along the radius. Schematic representation of the twisting mechanism as proposed by Ho et al. (2000) is shown in Figure 17. Extinction takes place when the direction of rotation axis is parallel to the transmitted light of polarized optical microscope.

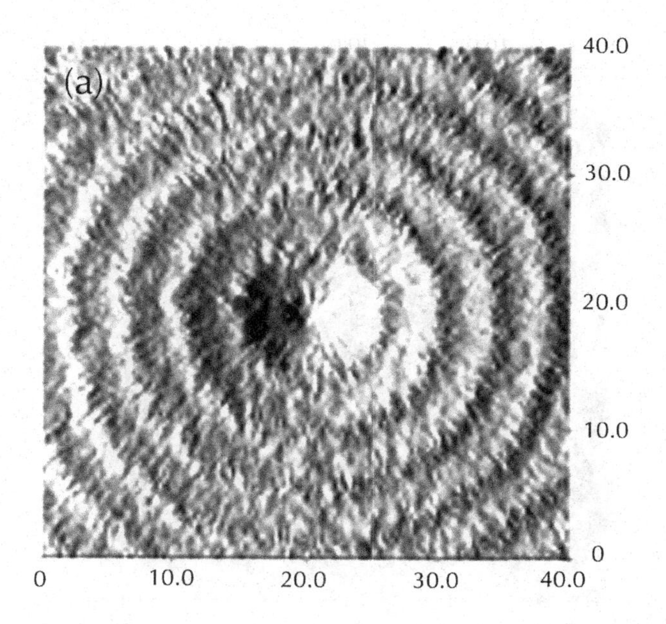

FIGURE 16 *(Continued)*

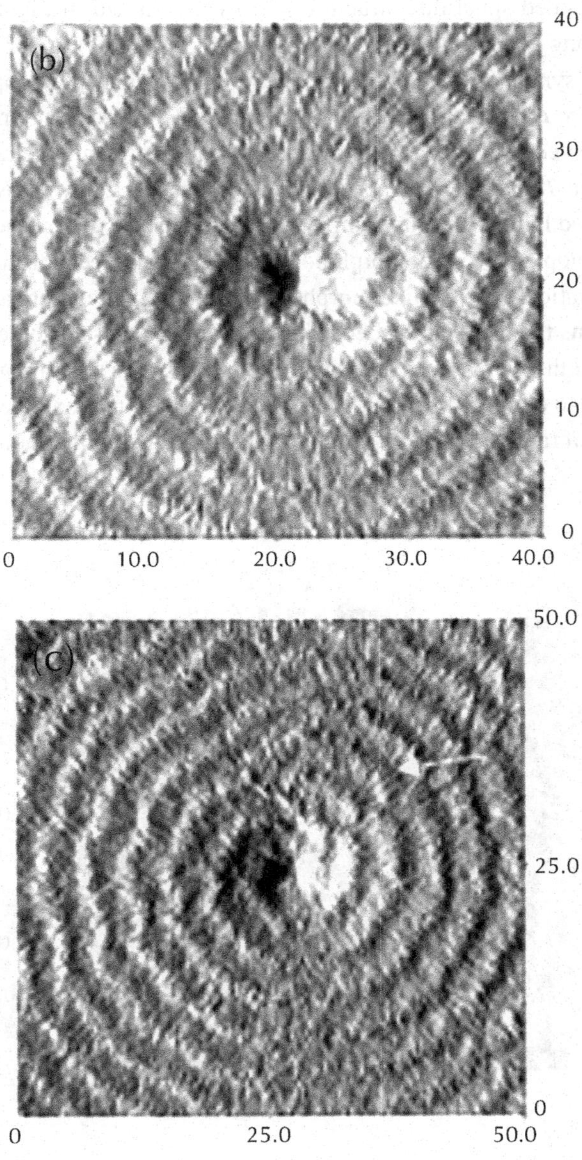

FIGURE 16 The AFM images of the banded spherulites in PTT (a) a regular spherulite (b) a spherulite with a band started at the primary nucleation site and (c) spherulite with band defects along the radial direction (Wang et al., 2001). Reproduced from Wang and co-workers, copyright (2001), by permission of Elsevier Science Ltd.

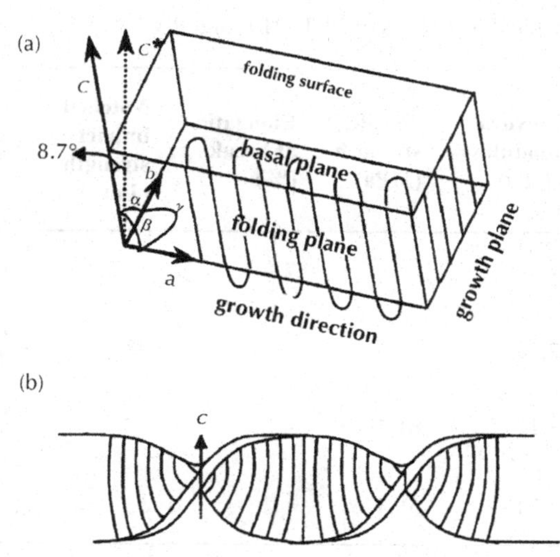

FIGURE 17 Schematic representation of (a) the lamellar geometry of PTT single crystal and (b) the twisting mechanism of the intralamellar model in PTT, as proposed by Ho et al. (2000).

Source: Reproduced from Ho and co-workers, copyright (2000), by permission of American Chemical Society.

The shallow C-shaped and S-shaped textures observed in the crest regions as revealed by the transmission electron microscopy (TEM) images confirm the works of Lustiger et al. (1989) and Ho et al. (2000) speculates that C-shapes and S-shapes are due to thickness limitation of thin film sections so as to form incomplete helical rotations. This helical conformation accounts for the lower modulus of PTT as compared to PET.

19.8 MECHANICAL PROPERTIES

Dynamic mechanical analysis (DMA) of PTT (refer Figure 18) shows high low-temperature (roughly from 30 to 45 °C) modulus of $2.25 \cdot 10^9$ Pa. A drastic decrease in the storage modulus (E') indicates the T_g of PTT is between 50–60 °C which is in agreement with the T_g estimated using DSC at 50.4 °C. A detailed analysis shows that the mechanical properties of PTT is in between those of PET and PBT, with an outstanding elastic recovery which is assumed to be due to its helical structure, as discussed in detail in earlier portions. A comparison of the mechanical properties of PET, PBT, and PTT is given in Table 11.

TABLE 11 Mechanical properties of PET, PTT, and PBT

Polymer	Flexural modulus (GPa)	Tensile strength (MPa)	Elongation at break (%)	Notched impact strength ($J m^{-1}$)	Ref.
Melt-spun PTT	-		7		(Xue et al., 2007)
Hot-press PTT	2.76	59.3	-	48	(Brown et al., 1997)
PET	3.11	61.7	-	37	(Brown et al., 1997)
PBT	2.34	56.5	-	53	(Brown et al., 1997)

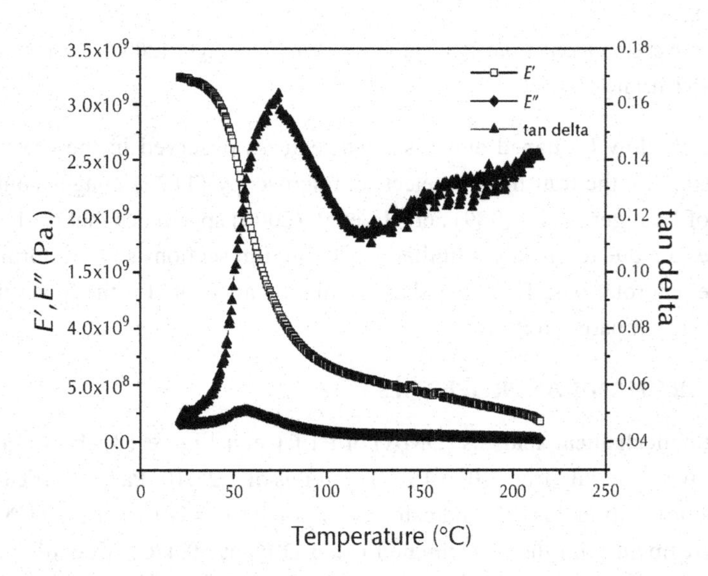

FIGURE 18 The DMA results of PTT.

19.9 PTT-BASED BLENDS

The PTT suffers low heat distortion temperature at 59°C (at 1.8 MPa) (Huang and Chang, 2000), low melt viscosity of 200 Pa·s (at 260°C at a shear rate of 200 s^{-1})

(Huang and Chang, 2000), poor optical properties, and pronounced britilleness at low temperatures. Enhancement of properties for PTT can be achieved by changing the maromolecular architecture and/or be extended by blending with existing polymers. Polymer blends allow combining the useful properties of different parent polymers to be done through physical rather than chemical means. It is a quick and economical alternative as well as a popular industrial practice as compared to direct synthesis in producing specialized polymer systems.

Table 12 summarizes selected PTT/elastomer and PTT/thermoplastic blends followed by the reason(s) for the blending. The purposes for the blending in these cases point toward two directions: i) toughening the matrix of second component with dispersed phase of PTT and ii) increase the strength of PTT matrix with dispersed phase of the second component.

TABLE 12 The PTT-based blends.

Blends	Reason(s) for blending	Ref.
	PTT/elastomer blends	
PTT/ABS	Acrylonitrile-butadiene-styrene (ABS) is associated with good processability, dimensional stability, and high impact strength at lower temperatures.	(Xue et al., 2007)
PTT/EPDM	Improve the toughness of the thermoplastic	(Ravikumar et al., 2005)
	PTT/thermoplastic blends	
PTT/PC	Improve the heat distortion temperature and modify the brittle nature of PTT.	(Xue et al., 2004; Aravind et al., 2010)
PTT/PEI	Improve the optical properties and mechanical properties	(Ramiro et al., 2003)
PTT/PBT	Improve the miscibility of the blends	(Krupthun and Pitt, 2008)
PTT/PEO	Improve the thermal stability	(Szymczyk, 2009)

Xue et al. (2007) studied the PTT/ABS blend system in detail. Blends were prepared in a 35 mm twin screw extruder at the barrel temperature between 245–255°C at a screw speed of 144 rpm. Two separate T_gs in the DSC thermogram indicate that the blends are phase separated in the molten state. First glass transitions is observed at lower temperatures between 40–46°C, which is attributed to the T_g of the PTT amorphous phase, while the second glass transitions at higher temperatures between 100–103°C is attributed to the SAN phase. Increasing the ABS content causes an increase in T_g of the PTT phase, whereas the T_g of the ABS phase decreases with the addition of PTT indicating that PTT is partially miscible with ABS and miscibility can be improved with the addition of ABS content. Decrease in T_m of the PTT phase (226 to 224°C) indicates that the solubility of ABS in PTT phase slightly increases with ascending ABS content. Epoxy resin and styrene-butadiene-maleic (SBM) anhydride copolymer were used as compatibilizer. As the epoxy content is increased from 1 to 3 wt% the cold crystallization temperature (T_{cc}) of PTT shifts to higher temperatures while for 5 wt% of epoxy content, a decrease in T_{cc} of the PTT is observed. PTT/ABS blends with 3 wt% of SBM shows a similar effect to that of 1 wt% epoxy system, indicating the compatibilization of SBM to PTT/ABS blends.

Studies by Ravikumar et al. (2005) show that PTT/ethylene propylene diene monomer (EPDM) blends are immiscible, which is supported by an increase in the free volume and constancy in crystallinity of PTT with increasing EPDM content and the use of ethylene propylene monomer grafted maleic anhydride as compatibilizer is found to produce significant improvement in properties by modifying the interface of the blends.

Xue et al. (2003) studied the PTT/PC blend systems, which form a compatible pair, has a negative effect on the mechanical properties. Thereby they used epoxy containing polymer as the compatibilizer of the blends. The possibility of cross-linking reactions strengthens the interface of the blends and results in the improvement of properties. Miscibility studies using DSC on PTT/PC blends with 2.7 wt% of epoxy shows that the T_g of the PTT rich phase increases from ~ 50 to ~ 60°C with increasing PC content and further addition of epoxy to the blends causes the decrease in the T_g of the PTT rich phase. The DMA shows that the addition of epoxy to the blends causes a significant increase in the T_g of the PTT rich phase from around 70 to 90°C while the T_g of the PC rich phase decreases from around 130 to around 110°C. Morphological studies using SEM and TEM show that the addition of epoxy modifies the interface dramatically.

Huang et al. (2002) studied the miscibility and melting characteristics of PTT/PEI blend systems. The DSC studies show that the miscible blends show single and compositional-dependent T_g over the entire composition range. The

Young's modulus decreases continuously from around 3,200 MPa for pure PEI to around 2,200 MPa for pure PTT. The addition of PEI affects the crystallinity of PTT (decreases from around 27% for neat PTT to around 3% for 25 wt% blend), but the mechanism of crystal growth is seen to be unaffected. The blends shows a synergistic behavior in modulus of elasticity (which is attributed to a decrease in specific volume upon blending). Additionally synergism is observed in the yield stress of PEI rich blends, and ductile nature.

Krutphun et al. (2008) studied the miscibility, crystallization and optical properties of PTT/PBT blends. The presence of a single and compositional dependant T_g by using DSC indicates miscibility of the blends in the molten state. Fitting the experimental T_g results with Gordon–Taylor equation shows a fitting parameter of 1.37 indicating the miscibility. The crystallinity of PTT decreases with the addition of PBT and the banded spherulite structure of PTT becomes more open as the amount of PBT in the blends is increased.

19.10 PTT COMPOSITES AND NANOCOMPOSITES

Table 13 summarizes selected PTT-based micro and nanocomposites and the reason(s) behind the preparation of the composites.

TABLE 13 The PTT-based composites and nanocomposites.

PTT composites	The preparation of composites	Ref.
PTT composites		
PTT/chopped glass fiber (CGF)	Improvement of the thermo-mechanical properties Improvement in tensile strength, impact strength and flexural strength.	(Mohanty et al., 2003)
PTT/short glass fiber (SFG)	1. Improvement of the crystallinity of PTT	(Run et al., 2010)
PTT nanocomposites		
PTT/clay nanocomposites	Improvement of thermal and mechanical properties by addition of small amount of filler.	(Liu et al., 2003)
PTT/multi-wall carbon nanotube (MWCNT)	Improvement of mechanical properties	(Wu, 2009)

Recently, Mohanty et al. (2003) studied the properties of bio-based PTT/chopped glass fiber composites (CGF). Glass fiber modified with PP-*g*-MA (polypropylene-grafted maleic anhydride) was used for the study. PTT/CGF composites with varying amounts of CGF (0 wt%, 15 wt%, 30 wt%, and 40 wt%) were prepared by using twin screw extruder at temperature of 230–245°C and at the screw speed of 100 rpm. The composite pellets obtained were subjected to injection moulding at the barrel temperature of 235°C and mould temperature at 35°C. With the addition of CGF, the tensile strength of the bio-based PTT increases from around 50 MPa to around 110 MPa (for composites with 40 wt% of CGF). The flexural strength also increases from around 80 MPa (PTT) to around 150 MPa (PTT/40 wt% CGF). Composites with 40 wt% CGF shows very high heat distortion temperature (HDT) at around 220°C. The impact strength shows an increase from 30 J m^{-1} for PTT to around 90 J m^{-1} for the PTT/CGF. Morphological analysis of the tensile fractured samples indicates good dispersion of the CGF in the matrix of PTT. Thus, all these results lead to the conclusion that the PP-*g*-MA acts as a coupling agent improving the interfacial adhesion between the CGF and the PTT. The thermo-mechanical properties shown by the composites indicate that they can be promising materials for future automobiles and building products, and can be used as a replacement for the currently used glass-nylon composites materials.

Studies by Run et al. (2010) on PTT/short glass fibers (SGF) composites show that the SGF acts as nucleating agents, which significantly accelerates the crystallization rate of PTT. The DSC results obtained for the increase in rate of crystallization were further confirmed by the WAXD experiments.

The PTT-based nanocomposites have been studied extensively by different groups. Run et al. (2007) investigated the rheology, meling behavior, and crystallization of PTT/nano $CaCO_3$ composites and shows that the presence of nano $CaCO_3$ increases the crystallization rate of PTT. Further studies by Run et al. (2010) adding short carbon fibres to PTT also lead to the conclusion, where the rate of crystllization of PTT acceleates with addition of SGF.

Study by Liu et al. (2003) shows that nano-size clay layers act as nucleating agents to accelerate the crystallization of PTT, and an increase in T_g and modulus PTT/clay (98/02 parts by weight) nanocomposites were prepared by melt intercalation using a co-rotating twin screw extruder with a screw diameter of 35 mm and L/D of 48 at abarrel temperature of 230–235°C and screw speed of 140 rpm. The clay used in the present study is an organic modified clay. The organo-modifier is methyl tallow bis(2-hydroxyethyl) ammonium, and DK2 (organo-clay) has the cation exchange capacity of 120 meq/100 g. Isothermal crystallization studies using the Avrami equation show that the Avrami exponent (n) increases from 2.52 to 2.58 as the T_c of the nanocomposite increases from 196 to 212°C while the $K_A^{1/n}$ decreases from 3.63 to 0.01 min^{-1}. The XRD analysis of the organo-modified clay shows a strong diffraction peak at 2θ = 4.10° corresponding to the (001) plane. This shows exfoliation of the clay in the PTT matrix and the TEM images also confirms this. DMA studies show that the T_g shifts from ~ 60°C for neat PTT to ~80°C for the PTT/clay nanocomposites. Similarly a ten fold increase in E' values is also observed which is explained on the basis of improvement in crystallization capacity of the PTT matrix.

After the discovery of carbon nanotube (CNT) by Ijima (1991), extensive works have been devoted in extracting the optimum properties of the CNTs. Wu (2009) studied PTT/MWCNT composites. The hydroxyl functionalized (MW-CNT-OH) behaves as anchoring sites for the PTT grafted with acrylic acid (PTT-*g*-AA) (compare to Scheme 1). The functionalization of MWCNT improves the compatibility and dispersibility of the MWCNT in the matrix of PTT. The thermal and mechanical properties (compare to Tables 14 and 15) show a dramatic increase leading to the conclusion that functionalized MWCNT can be used for preparing high performance PTT nanocomposites.

TABLE 14 Thermal properties of PTT/MWCNT and PTT-*g*-AA/MWCNT-OH as proposed by Wu (2009) (Adapted from Wu (2009))

MWCNT or MWCNT-OH (wt%)	PTT/MWCNT			PTT-*g*-AA/MWCNT-OH		
	Initial decomposition temperature (IDT) (°C)	T_g (°C)	T_m (°C)	(IDT) (°C)	T_g (°C)	T_m (°C)
0.0	379	49	219.1	362	45	218.2
0.5	392	53	217.9	420	55	215.9
1.0	410	54	216.5	451	59	213.8
1.5	415	52	217.1	459	55	214.8
2.0	421	51	217.8	466	53	215.6

TABLE 15 Mechanical properties of PTT/MWCNT and PTT-*g*-AA/MWCNT-OH as proposed by Wu (2009) (Adapted from Wu (2009))

MWCNT or MW-CNT-OH (wt%)	PTT/MWCNT			PTT-g-AA/MWCNT-OH		
	Tensile strength (MPa)	Elongation at break (%)	Intermediate Modulus (IM) (GPa)	Tensile strength (MPa)	Elongation at break (%)	(IM) (GPa)
0.0	50.6 ± 1.3	12.5 ± 0.3	2.26 ± 0.03	45.8 ± 1.5	11.9 ± 0.4	2.08 ± 0.06
0.5	56.8 ± 1.5	11.6 ± 0.4	2.46 ± 0.04	70.6 ± 1.8	8.3 ± 0.5	2.86 ± 0.05
1.0	61.6 ± 1.6	10.5 ± 0.5	2.65 ± 0.05	82.6±1.9	4.9 ± 0.6	3.32 ± 0.06
1.5	57.1 ± 1.8	10.8 ± 0.6	2.53 ± 0.07	72.3 ± 2.1	6.7 ± 0.7	2.98 ± 0.08
2.0	53.8 ± 1.9	11.2 ± 0.7	2.43 ± 0.08	65.6 ± 2.3	7.8 ± 0.8	2.78 ± 0.09

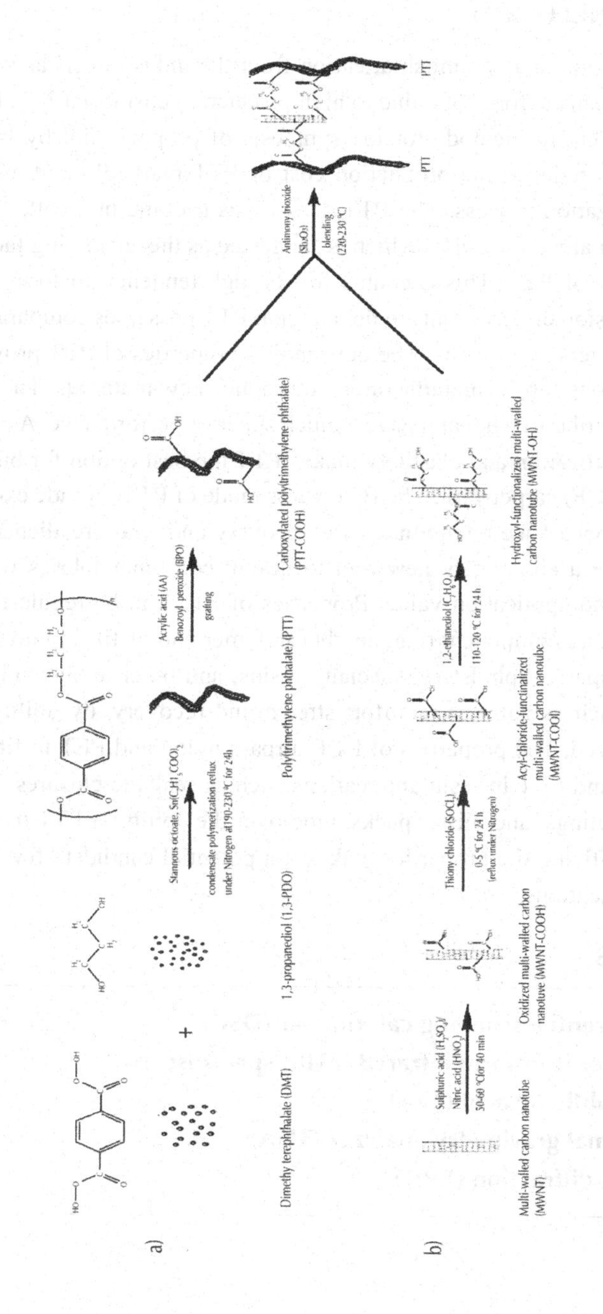

SCHEME 1 The synthesis and modification of PTT and MWCNT and the procedure to prepare the blends as proposed by Wu et al. (2009) (adapted from Wu and co-workers).

19.11 CONCLUSION

The PTT has not attained much attention from the industrialists as well as from the academicians before 2000 due to high production cost of PTT. The discovery of relatively cheap methods for the synthesis of propane diol by bioengineering route has reduced the production cost of PTT markedly and expedites the commercialization process. The PTT crystal has triclinic unit cell, a big zigzag conformation along the *c*-axis which is suggested as the attributing factor of high deformability of PTT. This accounts for its high tendency to form fibers. The above discussion clearly points to the fact that PTT possesses comparable properties of polyesters and nylons. The outstanding properties of PTT provide a wide range of options for the manufacturers to produce new materials. The PTT fibers are almost similar to wool and have a much superior performance. A combination of fast crystallization and elasticity makes PTT the best option for bulk continuous fibers (BCF) carpet yarn. The BCF yarns made of PTT provide excellent bulk resistance, appearance retention, elastic recovery and strain resilience. The PTT yarns provide a completely new tool to design consumer fabrics with optimal processing and application value. Properties of PTT can be regulated easily by adding a second component (e.g. another polymer and/or filler) into it. The PTT is used in apparel, upholstery, specialty resins, and other applications in which properties such as softness, comfort stretch and recovery, dyeability, and easy care are desired. The properties of PTT surpass nylon and PET in fiber applications, PBT, and PET in resin applications such as sealable closures, connectors, extrusion coatings, and blister packs, moreover the ability of PTT to be recycled without sacrificing the properties makes it a potential candidate for future engineering applications.

KEYWORDS

- **Differential scanning calorimeter (DSC)**
- **Fourier transform infrared (FTIR) spectroscopy**
- **Terephthalic acid (TPA)**
- **Thermal gravimetry analyzer (TGA)**
- **X-ray diffraction (XRD)**

ACKNOWLEDGMENT

Some of the experimental works discussed in this chapter were supported by Dana Pembudayaan Penyelidikan (RAGS) 2012 (600-RMI/RAGS 5/3 (14/2012)) from Ministry of Higher Education, Malaysia and Principal Investigator Support Initiative (PSI) 2013 (600-RMI/DANA 5/3/PSI(26/2013)) from RMI, UiTM.

REFERENCES

1. Aravind, I., Grohens, Y., Bourmaud, A. and Thomas, S. *Industrial & Engineering Chem. Research*, **49**, 3873–3882 (2010).
2. Broun, P. J., Blake, M. S., Richard, W. P., Cleve, E. D., and Pedro, A. J. Process for Preparing 1,3-propanediol. US Patent 5,770,776, June 23, 1998.
3. Brown, H. S., Casey, P. K., and Donahue, J. M. Poly(trimethylene terephthalate), Polymer for Fibers. In Meltblowing and Spounding Research, Proceedings of the 1997 TANDEC Annual Nonwovens Conference, Knoxville, Tennesse, November 19–21, 1997, Nonwovens World, (1998).
4. Bulkin, B. J., Lewin, M., and Kim, J. *Macromolecules*, **20**, 830–835 (1987).
5. Chuah, H. H. *Macromolecules*, **34**, 6985–6993 (2001).
6. Chuah, H. H. Poly(trimethylene terephthalate) (2001). Encyclopedia of Polymer Science and Technology, John Wiley & Sons Inc, New York (2003).
7. Chung, W. T., Yeh, W. J., and Hong, P. D. *J. Appl. Polym. Sci.*, **83**, 2426–2433 (2002).
8. Corbiere, P. J. and Mosse, P. Compositions Derived from Polymers and Copolymers of Acrylonitrile. FR Patent 1,043,435, October 4, 1953.
9. Dangseeyun, N., Srimoaon, P., Supaphol, P., and Nithitanakul, M. *Thermo. Acta*, **409**, 63–77 (2004).
10. Desborough, I. J. and Hall, I. H. *Polymer*, **18**, 825–830 (1977).
11. Desborough, I. J., Hall, I. H. and Neisser, J. Z. *Polymer*, **20**, 545–551 (1979).
12. Doerr, M. L., Hammer, J. J., and Dees, J. R. Poly(1,3-propylene terephthalate). US Patent 5,340,909, August 23, 1994.
13. DuPont. http://npe.dupont.com/pr-renew-factsht.html (accessed on March 10, 2011).
14. Elliott, A. J. *Sci. Instrum.*, **42**, 312–316 (1965).
15. Fitz, H. and Kalle, A. G. Process for the Manufacture of a Linear Polyester Using Stannous Oxalate as a Polycondensation Catalyst. US Patent 3,425,994, February 4, 1969.
16. Fritz, W., Eckhard, S., Hans, R., Ulrich, T., Klaus, M., and Ross, D K. Process of Producing Poly(trimethylene terephthalate) (PTT). US Patent 6,800,718, October 5, 2000.
17. Haas, T., Jaeger, B., Weber, R., Mitchell, S. F., and King, C. F. *Appl. Catal. A-Gen.*, **280**, 83–88 (2005).
18. Hall, I. H. *Structure of Crystalline Polymers*; Elsevier Applied Science, London, pp. 39–78 (1984).
19. Hall, I. H. and Pass, M. G. *Polymer*, ***17**, 807*–816 (*1976*).
20. Hall, I. H., Pass, M. G., and Rammo, N. N. *J. Polym. Sci. Polym. Phys. Edn.*, **16**, 1409–1418 (1978).

21. Ho, R. M., Ke. K. Z., and Chen, M. *Macromolecules*, **33**, 7529–7537 (2000).
22. Hoffman, J. D. and Weeks, J. J. *Res. Natl. Bur. Stand.*, **66**, 13–28 (A1962).
23. Hong, P. D., Chung, W. T., and Hsu, C. F. *Polymer*, **43**, 3335–3343 (2002).
24. Huang, J. M. and Chang, C. F. *J. Polym. Sci. Part B: Polym. Phys.*, ***38***, *934*–941 (2000).
25. Huang, J. M. and Chang, F. C. *J. of Appl. Polym. Sci.*, **8**, 850–856 (2002).
26. Hwo, C., Forschner, T., Lowtan, R., Gwyn, D., and Barry, C. Poly(trimethylene phthalates or naphthalate) and Copolymers: New Opportunities in Film and Packaging Applications, In Shell Chemicals, Proceedings of the Future–Pak 98 Conference, Chicago, USA, November pp. 10–12 (1998).
27. Iijima, S. *Nature*, **354**, 56–58 (1991).
28. Karayannidis, G. P., Roupakias, C. P., Bikiaris, D. N., and Achilias, D. S. *Polymer*, **44**, 931–942 (2003).
29. Krutphun, P. and Pitt, S. *Adv. in Sci. and Tech.*, **54**, 243–248 (2008).
30. Kurian, J. V. and Liang, Y. F. *Preparation of Poly(trimethylene terephthalate).* US Patent 6,281,325 August 28, 2001.
31. Lee, H. S., Vermaas, W. F., and Rittmann, B. E. *Trends Biotechnol.*, **28**, 262–271 (2010).
32. Lin, T. S. *Atomistic Molecular Dynamics Simulations for the Morphology and Property Relationship of Poly(trimethylene terephthalate) Fiber*. PhD Thesis, National Taiwan University (2007).
33. Liu, Z. J., Chen, K. Q., and Yan, D. Y. *Europe. Polym. J.*, **39**, 2359–2366 (2003).
34. Lustinger, A., Lotz, B., and Duff, T. S. *J. Polym. Sci. Polym. Phys. Ed.*, **27**, 561–579 (1989).
35. Mencik, Z. *J. Polym. Sci. Polym. Phys.Ed.*, ***13***, *2173*–2181 (*1975*).
36. Mohanty, A. K., Liu, W., Drazal, L. T., Misra, M., Kurian, J. V., Miller, R. W. *Biobased Poly(trimethylene terephthalate): Opportunity in Structural Composite Applications.* Proceeding of the 3rd Annual SPE Composites Conference, Troy, Michigan.
37. Ouchi, I., Hosoi, M., and Shimotsuma, S. *J. Appl. Polym. Sci.*, **21**, 3445–3456 (1977).
38. Paul, E. E. *Method for Esterifying Hindered Carboxylic Acid.* EP Patent 0,331,280 A1, September 6, 1989.
39. Perez, S. and Brisse, F. *Acta Cryst.*, **B32**, 470–474 (1976).
40. Perez, S. and Brisse, F. *Acta Cryst.*, **B33**, 3259–3262 (1977).
41. Perez, S. and Brisse, F. *Canadian Journal of Chem.*, **53**, 3551–3556 (1975).
42. Pyda, M., Bartkowiak, M., and Wunderlich, B. *J. Thermal Anal.*, **52**, 631–656 (1998).
43. Pyda, M., Boller, M., Grebowicz, A. J., Chuah, H., Lebedev, B. V., and Wunderlich, B. *J. Polym. Sci. Part B: Polym. Phys.*, **36**, 2499–2511 (1998).
44. Ramiro, J., Eguiaza´bal, J. L., and Nazábal, J. *J. Polym. Adv. Technol.*, **14**, 129–136 (2003).
45. Ravikumar, H. B., Ranganathaiah, C., Kumaraswamu, G. N., and Thomas, S. *Polymer*, **46**, 2372–2380 (2005).
46. Run, M. T., Yao, C. G., Wang, Y. J., and Gao, J. G. *J. of Appl. Polym. Sci.*, **106**, 1557–1567 (2007).
47. Run, M., Hao, Y., and He, Z. *Polym. Comp.*, **31**, 995–1002 (2010).
48. Sirichayaporn, T. Polyester Fiber Market Demand to Drive Global Paraxylene Growth: Source ICB. http://www.icis.com/Articles/2012/03/05/9537632/polyester-

fiber-market-demand-to-drive-global-paraxylene.html, part of Chemical Industry News & Chemical Market Intelligence (accessed on May 03, 2012).
49. Srimoaon, P., Dangseeyun, N., and Supaphol, P. *Europe. Polym. J.*, **40**, 599–608 (2004).
50. Szymczyk, A. *Europe. Polym. J.*, **45**, 2653–2664 (2009).
51. Tarasov, V. V. and Khimi, F. Z. *Theory of the Heat Capacity of Chain and Layer Struct.*, **24**, 111–128 (1950).
52. Wang, B., Christopher, Y. L., Hanzlicek, J., Stephen, Z. D. C., Geil, P. H., Grebowicz, J., Ho, R. M. *Polymer*, **42**, 7171–7180 (2001).
53. Ward, I. M. and Wilding, M. A. *Polymer*, **18**, 327–335 (1977).
54. Whinfield, J. R and Dickson, J. T. *Improvements Relating to the Manufacture of Highly Polymeric Substances*. Br. Patent 578,079, July 29, 1941.
55. Wu, C. S. *J. of Appl. Polym. Sci.*, **114**, 1633–1642 (2009).
56. Wunderlich, B. *Thermal Analysis*. Academic Press; San Diego, (1990).
57. Wunderlich, B. *Pure and Applied Chem.*, **67**, 1919–1926 (1995).
58. Xue, M. L., Jing, S., Chuah, H. H., and Ya, Z. X. *J. Macromol. Sci. Part B: Phy.*, **43**, 1045–1061 (2004).
59. Xue, M. L., Yu, Y. L., Chuah, H. H., Rhee, J. M., Kim, N. H., and Lee, J. H. *Europe. Polym. J.*, **43**, 3826–3837 (2007).
60. Yamen, M., Ozkaya, S., and Vasanthan, N. *J. Polym. Sci. Part B: Polym. Phys.*, **46**, 1497–1504 (2008).
61. Yang, Y., Sidoti, G., Liu, L., Geil, P. H., Li, C. Y., and Cheng, D. Z. S. *Polymer*, **42**, 7181–7195 (2001).
62. Yokouchi, M., Sakakibara, Y., Chatani, Y., Tdokoro, H., Tanaka, T., and Yoda, K. *Macromolecules*, **9**, 266–273 (1976).
63. Yong, J. S., YuRong, R., Dan, Z., Jing, H., Yi, Z., and Wang, G. Y. *Sci. China Ser. B-Chem.*, **51**, 3257–3262 (2008).
64. Young, J. K., Kim, J., and Seong-Geun, O. *Ind. Eng. Chem. Res.*, **51**, 2904–2912 (2012).

INDEX

D

E

F

G

H

I

K

L

M

N

O

P

Q

R

S

T

U

V

W

X

Y

Z

Printed in the United States
by Baker & Taylor Publisher Services